MRI of the Musculoskeletal System

A Teaching File

MRI of the Musculoskeletal System

A Teaching File

Editors

Jerrold H. Mink, MD

Chief, Division of Musculoskeletal Radiology
Cedars-Sinai Medical Center and
Associate Clinical Professor of Radiology
University of California
Los Angeles, California

Andrew L. Deutsch, MD

Attending Radiologist
Division of Musculoskeletal Radiology
Cedars-Sinai Medical Center
Los Angeles, California and
Associate Clinical Professor of Radiology
University of California
San Diego, California

With a foreword by Donald Resnick, MD

Raven Press  New York

Raven Press, 1185 Avenue of the Americas, New York 10036

Made in the United States of America

The material contained in this volume was submitted as previously unpublished material, except in the instances in which credit has been given to the source from which some of the illustrative material was derived.

Great care has been taken to maintain the accuracy of the information contained in the volume. However, neither Raven Press nor the editors can be held responsible for errors or for any consequences arising from the use of the information contained herein.

9 8 7 6 5 4 3 2 1

Library of Congress Cataloging in Publication Data

MRI of the musculoskeletal system.

Includes bibliographical references.
1. Musculoskeletal system—Magnetic resonance imaging—Case studies.   I. Mink, Jerrold H. II. Deutsch, Andrew L.   [DNLM: 1. Magnetic Resonance Imaging. 2. Musculoskeletal System—anatomy & histology. 3. Musculoskeletal System—pathology.   WE 141 M939] RC925.7.M753 1989   617.4′707548   89-24073 ISBN 0-88167-559-8

We would like to dedicate this book to those men and women who have been both a fabulous help, and, on occasion, an incredible hindrance to production—our wives, Barbara and Jeanne, and our children, Justin, Peter, Phillip, and Samantha (the latter two are both aged 2½ years).

Foreword

The value of magnetic resonance (MR) imaging in the assessment of a variety of musculoskeletal disorders in diverse locations no longer is questioned. Superb contrast resolution, excellent spatial resolution, multiplanar imaging capability, and the absence of ionizing radiation are among the characteristics of MR that ensure its success as a diagnostic technique that can be used as a substitute for or a supplement to other imaging methods. When applied to the evaluation of bone, joint, and soft tissue disease, MR provides both a splendid opportunity and a unique challenge. Its exquisite delineation of anatomic structures heretofore difficult or impossible to visualize with routine radiography, ultrasonography, arthrography, and conventional or computed tomography allow the physician to gain important information regarding the nature and extent of musculoskeletal disease processes, underscoring the splendid opportunity provided by MR. The challenge of the technique relates to correct interpretation of the images which requires of the observer knowledge of anatomic, physiologic, and pathologic data.

Dr. Jerrold Mink, widely regarded as a leader in the field, in collaboration with a number of respected authors and investigators, has edited this book which guides us, in a comfortable and useful fashion, through the intricacies of MR images of the musculoskeletal system. That a critical need exists for such a book is readily apparent to those of us who have struggled with interpretation of such images. All of the important musculoskeletal sites to which MR imaging is applied are addressed, including both spinal and extraspinal locations. In each instance, after a preliminary discussion, a series of cases is presented in which important disease entities are addressed. Each case is introduced with an appropriate history and an MR image or images; the findings then are defined and, subsequently, a concise discussion of the disease process (with additional images of that and related disorders) can be found. The format, which is similar to that employed in the highly successful syllabi of the American College of Radiology, makes for interesting reading with great educational impact. When appropriate, the MR abnormalities are correlated with those provided with other imaging methods, allowing the reader to understand why the MR examination is important as well as its precise placement in a proper imaging protocol. A summary and a list of selected references are additional components of this valuable educational experience.

Drs. Mink and Deutsch and their collaborators are to be commended for the excellence as well as the timeliness of this publication. Their efforts will be especially appreciated by those radiologists, orthopedic surgeons, rheumatologists, and other specialists who are called upon to interpret or review MR examinations of the musculoskeletal system. I believe this book will become an essential component of the personal libraries of these physicians, not one that will rest unused on the shelf, but rather one that will be frequently consulted and read.

Donald Resnick, MD

Preface

The inspiration for this book grew out of our perceived need for a comprehensive text on musculoskeletal MR. There is rather scant authoritative material available on the "how to read 'em" aspects of musculoskeletal MR. The large major references on MR devote surprisingly little space to this topic. There are, of course, subspecialty texts on knee and spine MR, but the other major joints are ignored. This lack of consolidated information is most surprising when one realizes that musculoskeletal imaging is by far the fastest growing application of MR. When we began to perform MR in 1985, virtually all of the 10 examinations per day were of the head (7–8) and spine (1–2). By 1988, our volume of neuroimaging had increased to a total of 12 examinations per day, but we were performing 9 bone and joint (nonspine) studies. At the time of this writing, musculoskeletal MR accounts for 12–14 studies, which is approximately 35% of our daily work load. Our experience is not unique; interaction with our colleagues at conferences suggests that this rapid rate of growth can be expected in radiological practices with available scanner time and a radiologist interested in and conversant with osteoradiology.

We hope to benefit as much from this text as will the reader; it has given us the chance to spend time in the library, consolidate information, organize our thoughts, and learn more about the fascinating topic of musculoskeletal MR.

This book is organized as a case study, but by no means is this a text that asks a question, answers it, and then gets on to the next case. Rather, this book is more closely akin to the ACR syllabi. A question is asked, based on the findings in one or more MR images; there is a detailed description of the findings in the following paragraph. In general, the case centers primarily on one clinical entity. Just below the answer is a comprehensive analysis of the entity under discussion, a differential diagnosis and follow-up imaging studies (CT, MR, arthrograms, bone scans) that are pertinent to the case. The discussions are not limited to the *imaging* aspects of the case; we believe that it is essential for the skeletal radiologist to understand what it is that the clinician needs to know.

This text is designed primarily for the practicing radiologist who desires a reference text that he can keep next to the view box. But the case study format also allows one to digest a small kernel of knowledge in the evening, without having to devote a "chapters worth" of time to learn something of value.

Orthopedic surgeons will also find a great value in this book. MR is now a daily part of orthopedic practice, and it is rare to find an orthopedic surgeon who does not order or review an MR on a daily basis. Surgeons first became familiar with MR in regards to the detection of femoral head osteonecrosis and in assessment of spinal disorders. The rapidly expanding spectrum of musculoskeletal applications makes a working knowledge of MR essential to the orthopedist.

The major articulations—hip, knee, shoulder, and spine—are covered as individual chapters. Smaller joints like the foot and ankle and the elbow and wrist are combined together. Three special chapters are included in the text. The first covers the difficult topic of physics; most authors feel obliged to include a detailed analysis of the physical basis of MR. Being practical, practicing radiologists, we abhorred the idea of repeating the concepts of T1 and T2, gradient imaging, and flow. We, instead, chose to instruct radiologists, like ourselves, on the important factors that govern the quality of the images that one produces and to accomplish this in the most painless manner (for us, physics can be painful) possible. We therefore utilized the case study style which characterizes the book.

We have established the chapter on tumors without regard for the body part in which the lesion occurred so as to better consolidate one's thinking about this complex issue. The other special chapter covers dynamic joint imaging, a topic that requires a new presentation of technical considerations. While most of the cases presented in the dynamic chapter do cover the knee and TMJ, it is our belief that the ability of MR to image physiologic motion will be a primary use of this modality.

Acknowledgments

Authors and editors get all the credit; we would like it to be known that many other people made this book happen. Our technologist staff including Joe Fraire, Mark Tournie, Amy Leithen, Kim Fernandez, Roger Parga, Michael Foto, Earnest Fontalera, Fred Schnakenberg, Gina Slimp, Brenda Tompkins, Ilene Dorfman, Lisa Palomo, and Arthur Aguilar spent many hours filming and refilming and tolerated our excessive demands. Francisco Chanes and the Audio Visual Department of Cedars-Sinai Medical Center devoted untold hours to photographing and illustrating to our standards. Julie Sassoon and Rene Siegel became very familiar with word processing over this year. To all of these individuals, we are deeply in your debt.

Contents

Contributors

Jay L. Amster, MD *Center for Diagnostic Imaging, St. Louis Park, Minnesota*

Javier Beltran, MD *Associate Professor, Musculoskeletal Division, Ohio University Hospital, Columbus, Ohio*

Thomas H. Berquist, MD *Professor of Diagnostic Radiology, Mayo Medical School, and Consultant, Diagnostic Radiology, Mayo Clinic, Rochester, Minnesota*

Phillip Chao, MD *Department of Radiology, Hospital of the University of Pennsylvania, Philadelphia, Pennsylvania*

Andrew L. Deutsch, MD *Attending Radiologist, Division of Musculoskeletal Radiology, Cedars-Sinai Medical Center, Los Angeles, California, and Associate Clinical Professor of Radiology, University of California, San Diego, California*

Kenneth B. Heithoff, MD *Medical Director, Center for Diagnostic Imaging, St. Louis Park, Minnesota*

Robert J. Herfkens, MD *Director, Magnetic Resonance Imaging, Cedars-Sinai Medical Center, Los Angeles, California, and Clinical Professor of Radiology, University of California, Los Angeles, California*

Roger Kerr, MD *Assistant Professor of Radiology, University of Southern California School of Medicine, Los Angeles, California*

J. Bruce Kneeland, MD *Assistant Professor of Radiology, Medical College of Wisconsin, Director of Magnetic Resonance Imaging, Milwaukee, Wisconsin*

Bert R. Mandelbaum, MD *Orthopedic Surgeon, Santa Monica Orthopedic Group, Santa Monica, California, and Assistant Clinical Professor of Orthopedics, Department of Orthopedic Surgery, University of California, Los Angeles, California*

Jerrold H. Mink, MD *Chief, Division of Musculoskeletal Radiology, Cedars-Sinai Medical Center, Los Angeles, California, and Associate Clinical Professor of Radiology, University of California, Los Angeles, California*

Barry D. Pressman, MD, FACR *Chief, Division of Neuroradiology and Head and Neck Radiology, Cedars-Sinai Medical Center, Los Angeles, California, and Assistant Clinical Professor of Radiology, University of California, Los Angeles, California*

Donald Resnick, MD *Professor of Radiology, University of California, San Diego, California, and Chief, Radiology Service, Veterans Administration Medical Center, San Diego, California*

David J. Sartoris, MD *Associate Professor of Radiology, University of California, San Diego, California*

Frank G. Shellock, PhD *Research Scientist–Physiologist, Section of Magnetic Resonance Imaging, Cedars-Sinai Medical Center, Los Angeles, California, and Assistant Professor of Radiological Sciences, University of California, Los Angeles, California*

Michael B. Zlatkin, MD, FRCPC *Attending Radiologist, Memorial Hospital, Hollywood, Florida, and Clinical Associate Professor, University of Miami School of Medicine, Miami, Florida. Formerly, Assistant Professor of Radiology, Hospital of the University of Pennsylvania, Philadelphia, Pennsylvania*

CHAPTER 1

Technical Considerations in Musculoskeletal MRI

Andrew L. Deutsch and Robert J. Herfkens

Optimizing imaging for the musculoskeletal system must balance multiple and often opposing demands between spatial and contrast resolution-oriented techniques. Recent developments in imager hardware and software have now allowed spatial resolution on some imagers to approach that of computed tomography. The various constituents of the musculoskeletal system (e.g., fat, muscle, bone) differ significantly in biophysical properties, providing high inherent contrast capabilities that can be exploited by MR in a manner that exceeds this capacity of any other imaging method. In addition, relative contrast between tissues and lesion conspicuity can be directly manipulated by the appropriate choice of imaging parameters.

In contrast to conventional radiographic imaging methods, image contrast in MR is dependent on the complex interplay of a large number of factors that can be categorized into two principal groups: (1) those under the control of the imaging system, and (2) biophysical properties characteristic of the tissue being imaged. Parameters within the first group include: (1) pulse sequence type (e.g., spin-echo [SE]), (2) timing parameters (e.g., repetition time/echo time [TR/TE], flip angle), (3) slice thickness (ST) and gap, and (4) matrix size and field of view (FOV), among others. The factors reflective of the tissue include: (1) longitudinal or spin-lattice relaxation time, (2) transverse or spin-spin relaxation time (T2), proton density, and physiologic motion. Knowledge of the intrinsic capabilities of the MR pulse sequences and the often conflicting considerations regarding spatial and contrast determining parameters is critical to the successful application of this diagnostic method.

It is not the purpose of this series of cases to serve as a text on the basic physics of magnetic resonance imaging but rather as an introduction to the day-to-day decision making process which is necessary to maximize the diagnostic utility of this flexible imaging modality. Emphasis will be directed toward practical considerations regarding image quality, particularly as they pertain to optimizing imaging of the musculoskeletal system. The complex interplay between resolution and contrast-oriented techniques (e.g., TR/TE, FOV, matrix size, slice thickness) will be considered. The basic principles of the pulse sequences (e.g., SE, gradient-refocused, short tau inversion recovery [STIR] most commonly utilized in musculoskeletal imaging will be reviewed. It is hoped that this section will provide an introduction to the parameters available to maximize the diagnostic capabilities of the equipment, and to the appropriate development of effective imaging protocols.

CASE 1

History: Three sagittal scans of the lumbar spine were accomplished utilizing different spin-echo pulse sequences. Figure 1A is a relatively T1-weighted image of the spine; Fig. 1B is a "spin-density" weighted image; and Fig. 1C is relatively T2 weighted. What is meant by T1 and T2 relaxation? What are the principal determinants of image contrast on spin-echo imaging? How does the choice of the pulse parameters TR and TE affect the resulting image? (Fig. 1A. Sagittal, TR/TE 300/15 msec. Fig. 1B. Sagittal, TR/TE 2,000/20 msec. Fig. 1C. Sagittal, TR/TE 2,000/80 msec.)

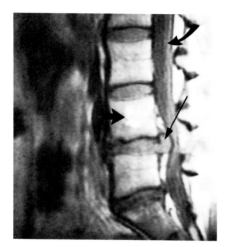 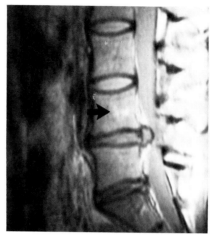 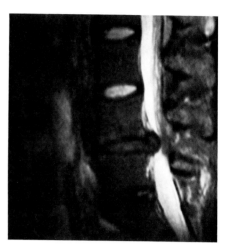

FIG. 1A.   **FIG. 1B.**   **FIG. 1C.**

Findings: In Fig. 1A the marrow of the vertebral bodies is of high signal intensity. Additionally, an irregular focus of higher signal intensity is seen within the L4 vertebral body (arrow). There is a large central herniation of the L4-5 intervertebral disc (long arrow). The cerebrospinal fluid (CSF) within the thecal sac is of relatively low signal intensity. The nerve roots of the cauda equina can be seen in a dependent dorsal position (curved arrow). In Fig. 1B the vertebral bodies remain of relatively increased signal. The focus within the L4 vertebral body is less intense (arrow). The contents of the thecal sac are slightly increased in intensity and the roots can no longer be distinguished. In Fig. 1C the vertebral bodies demonstrate relatively decreased signal intensity. The abnormal focus within the L4 vertebrae has markedly decreased in signal intensity. The upper intervertebral discs and thecal sac now exhibit high signal intensity. The signal within the L4-5 intervertebral disc is decreased in intensity compared with the previous sequences.

Diagnosis: Changes in image contrast related to differences in tissue T1 and T2 relaxation times. The focus within the L4 vertebral body demonstrates imaging characteristics typical of fat.

Discussion: The SE sequence is the most common sequence utilized in current magnetic resonance imaging. This sequence employs a 90-degree radio frequency pulse (excitation pulse) which initially directs the net magnetization into the transverse plane. After a short period of time (TE/2), a 180-degree radio frequency pulse is applied (refocusing pulse) and the remaining magnetization is sampled at a time TE. It is this parameter TE that determines the T2 weighting of an SE sequence. The longer the TE the more heavily T2 weighted the image. The parameter TR is the time between successive repetitions of this sequence. The TR value primarily determines the amount of T1 contrast generated within an image. The advantages of a spin-echo sequence include a great deal of flexibility in the relative amounts of T1 or T2 weighting. In addition, the presence of the 180-degree pulse helps correct for static inhomogeneities in the magnetic field which may be caused by metal implants or direct changes caused by the body itself on the local magnetic field.

The basic signal intensity for a specific point within the images is defined by H(EXP($-$TE/T2))(1 $-$ (EXP $-$ TR/T1), where H represents the spin density; TR, the time between repetitions of the sequence; and TE, the echo delay time. By varying these parameters emphasis can be placed on differences between tissues based on their T1 properties, T2 properties, or spin density. It can be seen from the above equation that as TE increases the signal decreases related to the T2s of the tissues. By increasing the TE, the T2 weighting of an image is increased. Likewise it can be seen that there is a relationship between TR and the T1 signal of an image. By increasing the TR, the contrast in an image related to the T1 components is decreased. By increasing the TR and decreasing the T1 effects, one can de-emphasize the changes resulting from T1 and T2 and therefore leave the image highly dependent upon spin density. It is from these general relationships that the T1-weighted, T2-weighted, and spin-density–weighted images have evolved.

Traditionally, a T2-weighted image is an image with a long TR and long TE. These sequences are generally the most sensitive for detection of disease processes which most commonly are characterized by increased free water. T2-weighted images demonstrate lower signal to noise than either T1 or spin-density images and are more readily degraded by motion. A T1-weighted image has a short TR and short TE. For optimal image contrast, it is important that the TR and TE be sufficiently short (e.g., TR/TE 300/15). If TR and TE are not chosen properly (e.g., short TR with TE greater than 40 msec.), T1 contrast will be diminished. T1 relaxation times increase with field strength and T1 contrast decreases. This point must be remembered when designing scan protocols. A spin-density–weighted image has a long TR and short TE. These images are characterized by relatively high signal-to-noise ratio (SNR) and good anatomic detail. Spin-density images are sensitive to changes in the number of mobile protons, but are relatively insensitive to changes in T1 and T2. It must be emphasized that these are weighted images and that the choice of TR and TE strongly affects the contrast in a weighted image. One user may define his T1-weighted image as a TR of 800 and a TE of 30, while another individual will define his T1-weighted image as a TR of 250 and a TE of 12. These clearly are not the same images and do not contain the same image contrast. Similarly, in T2-weighted images a TR of 1,500/TE 60 does not contain the same image contrast as an image with a TR of 2,500/TE 100. This becomes extremely important when comparing images from the literature and attempting to duplicate results generated by other MR users.

It should be noted that as TR increases the signal increases. As TE increases the signal decreases. There is a particularly dangerous combination of TR and TE when in general most soft tissues will exhibit quite similar signal intensity and therefore virtually no contrast. This combination occurs with intermediate TRs of approximately 1,000 and TEs of approximately 30–40 (depending on the field strength of the magnet). This applies specifically to soft tissues, and certain exceptions to the rule include fat, CSF, and bone, which allow a much greater latitude in choice of TR and TE.

CASE 2

History: Sagittal T1- and T2-weighted images of the knee are presented. What are the imaging characteristics of the principal constituents (e.g., muscle, tendon, bone) of the musculoskeletal system? What pulse sequences are utilized to optimize differences in soft tissue image contrast for the musculoskeletal system? (Fig. 2A. Sagittal, TR/TE 300/15 msec. Fig. 2B. Sagittal, TR/TE 2,000/80 msec.)

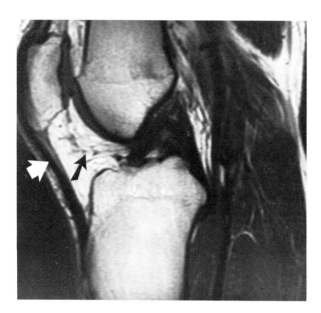

FIG. 2A.

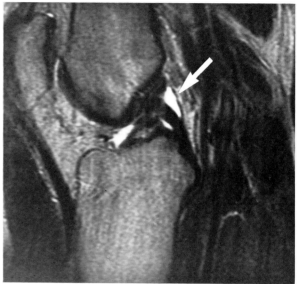

FIG. 2B.

Findings: On the T1-weighted image, the infrapatellar fat pad demonstrates the highest signal intensity (arrow). Muscle is of relative intermediate signal and bone marrow demonstrates a moderately high signal intensity reflecting the predominant aliphatic component in this location. The patella tendon is low in signal (white arrow). On the T2-weighted image (Fig. 2B), the tendon remains low in signal. Muscle, subcutaneous fat, and bone have decreased in intensity. Fluid within the joint now demonstrates high signal intensity reflecting its long T2 relaxation time (arrow).

Diagnosis: The muscle, tendon, and underlying bone are of normal signal intensity. Changes in their appearance between the T1 and T2 images reflect differences in the basic relaxation times of the tissues.

Discussion: The principal constituents of the musculoskeletal system differ significantly in biophysical characteristics. These differences, which can be characterized in terms of T1 and T2 relaxation properties, provide the opportunity to utilize MRI to maximize image soft tissue contrast.

Normal skeletal muscle is characterized by an intermediate to slightly long T1 relaxation time and by a short T2 relaxation time relative to other soft tissues. As a consequence, muscle abnormalities (e.g., injuries, tumors) are most conspicuous on T2-weighted images, which optimize contrast between most pathological processes (e.g., edema) that are characterized by prolonged T2 relaxation times and normal muscle. T1-weighted images are less sensitive for depicting soft tissue abnormalities because most pathologic processes have long T1 relaxation times, similar to normal skeletal muscle. T1-weighted images may be useful in providing specificity in regard to the presence of short T1 processes (e.g., fat, hemorrhage).

Fibrocartilagenous structures (tendons, ligaments, menisci) demonstrate low signal intensity on all pulsing sequences as a consequence of their extremely short T2 relaxation times. Abnormalities of these structures are well characterized on T1-weighted, spin-density, and gradient-echo images. Hyaline cartilage demonstrates intermediate signal intensity on spin-echo sequences reflecting its increased water content. Detection of abnormalities is enhanced on T2-weighted and certain gradient-refocused sequences that provide for a greater degree of contrast with surrounding synovial fluid (arthrogram effect).

The appearance of bone marrow reflects the relative concentration of fat and water, its two principal constituents, and differs according to location within the skeleton and age of the patient. Fat is characterized by both a high proton density and very short T1 relaxation time. The T2 relaxation time of fat is moderately short, and fat demonstrates relatively low signal on T2-weighted images. The water in bone marrow is compartmentalized into different fractions. Bulk water, as a consequence of its long T2, contributes to high signal intensity on T2-weighted images. Most pathological processes are characterized by increases in bulk water concentration, reflected by increased signal intensity on T2-weighted images.

Trabecular and cortical bone are characterized by low signal intensity on all pulse sequences as a consequence of both low proton density and magnetic susceptibility effects. The susceptibility effects of trabecular bone are particularly evident on gradient-refocused sequences. Magnetic susceptibility effects in bone imaging principally relate to the presence and concentration of ferritin and hemosiderin. In sufficient concentration both ferritin and hemosiderin decrease T2 and T2* and may markedly decrease signal intensity on T2 and gradient-echo sequences. In cases of severe iron overload the marrow may even demonstrate decreased signal intensity on T1-weighted images (Fig. 2C).

Areas of calcification within muscle produce low signal intensity and may be extremely difficult to appreciate on MR images. This presents a well described limitation of MR and may significantly decrease specificity of diagnosis in certain situations. Gas within soft tissues presents a similar problem with regard to detectability utilizing MR. Both calcification and gas demonstrate low signal on all pulsing sequences. Conventional radiographs are often required to assist in detection.

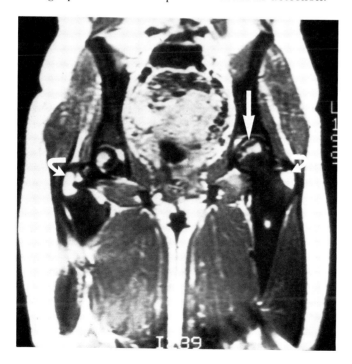

FIG. 2C. Coronal, TR/TE 500/20 msec. A T1-weighted image of the pelvis and proximal femora of a patient with sickle cell anemia demonstrates marked diminution of signal intensity throughout the bone marrow reflecting iron deposition. Note the sparing of the apophyses (*curved arrows*) and evidence of osteonecrosis involving the femoral head (*large arrow*).

CASE 3

History: A 47-year-old man underwent an MR examination for suspected internal derangement of the knee. Figure 3A is from an initial examination performed on the patient. Figure 3B represents a nearly identical section to that depicted in Fig. 3A and was obtained following the initial examination utilizing similar imaging parameters but on a different scanner. What factor accounts for the principal difference in the two scans? What are the principal factors affecting signal to noise in MR imaging? (Fig. 3A. Sagittal, TR/TE 2,000/80 msec. Fig. 3B. Sagittal, TR/TE 2,000/80 msec.)

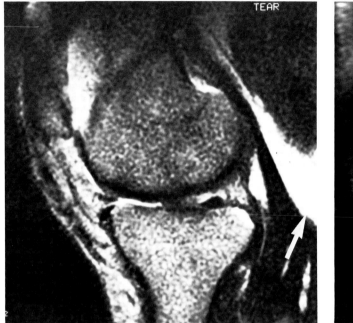

FIG. 3A.

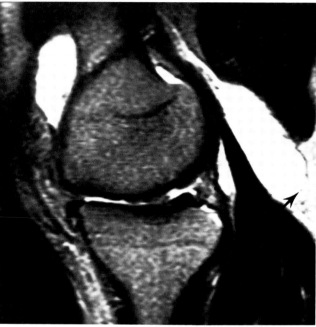

FIG. 3B.

Findings: On the initial examination (Fig. 3A) an eliptical-shaped high signal intensity mass is identified and represents a popliteal cyst (arrow). The image appears "noisy." On the second examination (Fig. 3B) the cyst is again identified. Several small foci of low signal can be seen along the dependent wall of the cyst (arrow) and represent loose osteocartilagenous bodies that could not be identified on the previous study. The overall image quality is improved.

Diagnosis: Popliteal cyst with loose osteocartilagenous bodies. The study in Fig. 3A was accomplished at 0.5T and the study in Fig. 3B at 1.5T. The principal difference in image quality reflects differences in signal to noise.

Discussion: Perhaps the single most important determinant of image quality in MR imaging is the ratio of the signal emanating from the tissue of interest compared to the amount of noise received at the time of imaging. The ratio of the signal received to the background noise is termed the signal-to-noise ratio (SNR). Multiple factors contribute to this ratio including: (1) magnetic field strength, (2) pulse timing parameters (TR/TE), (3) voxel size, (4) signal averages (number of excitations (NEX)), (5) coil efficiency, and (6) system hardware (electronics).

   One of the principal factors affecting the overall signal to noise, and one around which a considerable degree of controversy has occurred, is the effective field strength with which the images are obtained. Although there is a complex relationship between the intricacies of the imaging process and field strength, in general there is a linear increase in signal to noise with an increase in field strength. These two images were obtained with similarly designed systems from the same manufacturer at two different field strengths. The change in signal to noise from 0.5T to 1.5T is clearly visible in these images of the same patient. This increase in signal to noise, related to differences in magnetic field strength, is associated with a number of other potentially offsetting considerations. These include possible increases in chemical shift artifacts for the same band width and changes in image contrast due to differences in T1 relaxation time. There is a general increase in the T1 relaxation time of soft tissue with increases in field strength. For certain soft tissues this roughly averages approximately 400 msec/ T. This factor alone will strongly affect the image contrast obtained at any given repetition time. As discussed in Case 1 of this section, the repetition time in general determines the relative T1 weighting of an individual image. By changing field strength the individual T1s of the tissues increase significantly and the relative contrast for a specific TR will also as a result be changed. Many investigators feel that the potential loss in T1 contrast is overcome by the increase in signal to noise with the resulting image reflecting an overall significant improvement in contrast to noise. It should be noted, however, that these complex changes are interdependent and difficult to analyze in isolation. It is fundamentally important to understand that the image contrast generated at different field strengths for the same given pulse parameters (e.g., TR and TE) may be significantly different.

CASE 4

History: Figure 4A–D represent a series of sagittal sequences obtained of the lumbar spine of a 57-year-old man. Each image is from a different sequence, each of which was accomplished with identical imaging parameters (e.g, FOV, TR/TE) except for one factor that was varied. What parameter was most likely changed? What is the relationship of this factor to the other principal determinants of image contrast? (Fig. 4A. Sagittal, TR/TE 300/15 msec. Fig. 4B. Sagittal, TR/TE 300/15 msec. Fig. 4C. Sagittal, TR/TE 300/15 msec. Fig. 4D. Sagittal, TR/TE 300/15 msec.)

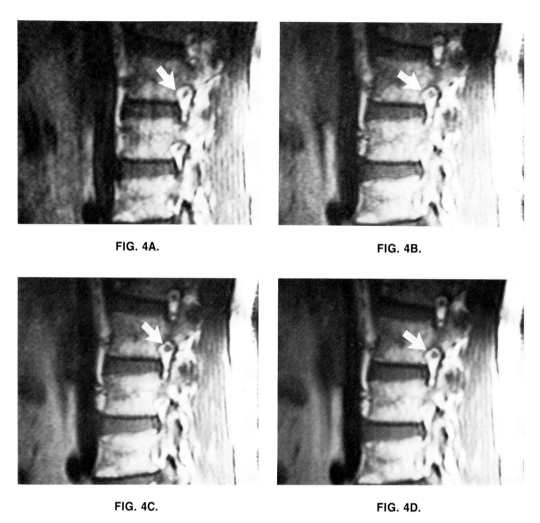

FIG. 4A.

FIG. 4B.

FIG. 4C.

FIG. 4D.

Findings: Each image progressing from Fig. 4A to Fig. 4D demonstrates increasing detail that principally results from an increasing signal-to-noise ratio. Note the improved sharpness with which the nerve root is defined against the perineural fat in the neural foramen (arrow).

Diagnosis: Increased signal-to-noise ratio accomplished by increasing the number of signal averages. (Fig. 4A. 1/2 NEX. Fig. 4B. 1 NEX. Fig. 4C. 2 NEX. Fig. 4D. 4 NEX.)

Discussion: There are a number of basic factors that strongly effect the overall signal to noise in an image. One of the most straightforward of these is the NEX in an image. In order to construct an MRI image, the basic pulse series must be repeated a number of times equal to the Y resolution, that is, 128 times for a 128-matrix image. This process is considered an excitation in imaging. One series of pulses is considered one NEX. For each increase in NEX in an imaging sequence there is an improvement by the square root of two in signal to noise (approximately 1.4). This increase in signal to noise is achieved at the expence of doubling the imaging time. For relatively small increases in NEX (i.e., one NEX to two NEX) there is the square root of two improvement in signal to noise. In order to gain the square root of two again in signal to noise, the imaging time must again be doubled. As a consequence of this relationship, to achieve an increase by a factor of two in signal to noise the scan time must be increased by a factor of four. At some level it becomes impractical to improve image quality simply by increasing the number of excitations because of the eventual problem with patient motion as imaging times become prohibitive.

A number of other factors also contribute significantly to signal-to-noise considerations. Image slice thickness is a linear function of signal to noise. That is, as you decrease the slice thickness by a factor of two there is a reduction in signal to noise by one half. This reduction in slice thickness is quite difficult to regain by increases in signal averaging as is evident from the previous discussion. It is necessary to increase the scan time by a factor of four to account for a reduction in slice thickness of a factor of two.

The efficiency of the receiving antenna can also significantly contribute to signal to noise considerations. The efficiency of surface coils is dependent on multiple factors, one of which is known as the filling factor. The wires of the coil should be placed as close to the body part being examined as possible. Surface coils can result in an SNR as large as four to six times greater than that obtainable with a circumferential body coil. While surface coils assist in achieving improvements in signal to noise, certain shortcomings are noteworthy. Signal strength is greatest close to the coil and drops off rapidly in structures that are more than a few centimeters away. This can result in inhomogeneous images and inadequate resolution of structures at a distance from the coil, factors that place a premium on proper positioning. As noted in previous examples, it is important to keep in mind that in MRI the diagnostic quality of an image is not determined simply by the spatial resolution or the signal to noise alone, but importantly by the complex choice of appropriate imaging parameters that optimizes image contrast in a spatial resolution format that ultimately allows for a clinical diagnosis to be accomplished.

CASE 5

History: Two scans of the knee were accomplished utilizing an identical period of time. All imaging parameters were held constant except for the number of phase encoding steps (doubled) and signal averages (halved) utilized to produce one image as compared to the other. What are the effects of varying matrix size and number of signal averages on image quality and signal-to-noise ratio? What are the principal trade-offs to be considered in producing high resolution studies? (Fig. 5A. Sagittal, TR/TE 300/15 msec, 12 cm FOV, 3-mm slice thickness, 200 window width, 1,100 window level. Fig. 5B. Sagittal, TR/TE 300/15 msec, 12 cm FOV, 3-mm slice thickness, 200 window width, 1,100 window level.)

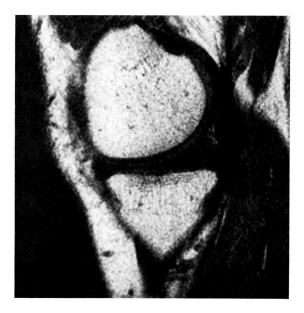

FIG. 5A.

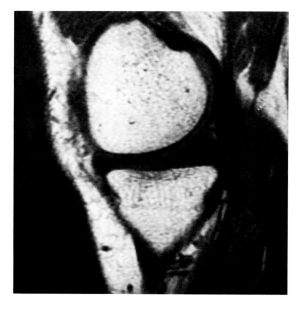

FIG. 5B.

Findings: In Fig. 5A the detail within the subchondral and medullary bone is significantly greater than in Fig. 5B, reflecting higher spatial resolution. The image however appears more "grainy." Image contrast between bone, muscle, and subcutaneous fat is greater in Fig. 5B than in Fig. 5A.

Diagnosis: Figure 5A was accomplished utilizing a 256×256 matrix and 1 signal average. Figure 5B was performed utilizing a 256×128 matrix and 2 signal averages.

Discussion: Spatial resolution is primarily determined by voxel size. A voxel constitutes a three-dimensional sample whose volume is determined by ST, imaging matrix, and FOV. Increasing the number of matrix elements decreases the size of each voxel; increasing either FOV or ST increases voxel size. The smaller each voxel is the better the spatial resolution.

The matrix size of an image is dictated by the number of pictorial elements in each dimension. In general the dimensions in the frequency encoded direction of an image are fixed at approximately 256 resolution elements. The principal choice in most MR imagers is between utilizing 128 or 256 pixels along the phase-encoding axis. In the phase-encoding dimension of an image each additional row of resolution elements requires an additional phase-encoding gradient and therefore adds significantly to the imaging time. In the example above the imaging matrix in the phase-encoding direction has been increased from 128 to 256 elements. When the field of view or size of the object imaged is kept constant this improves the spatial resolution of the image. When the imaging matrix is increased from 128 to 256 pixels, the voxel size decreases in the phase encoding direction by a factor of two. This reduction in voxel size improves the ability to resolve small objects. When imaging time is kept constant, however, there is a reduction of signal within in each voxel by a factor of two. In order to return the overall per voxel signal to noise the imaging time must be increased by increasing the number of excitations. As discussed in Case 4 of this section, doubling the number of excitations gains only the square root of 2 (e.g., 1.4 times increase) in signal to noise. Thus the price for improving spatial resolution can only be countered by significantly increasing the overall scan time.

Changes in voxel size can also be accomplished by changes in the field of view. Reduction of the field of view or the total imaging volume by one half, such as reduction from a 24 cm FOV to a 12 cm FOV, results in a reduction in voxel size by a factor of four. This translates into a reduction in signal to noise by a factor of four. This increase in spatial resolution achieved by reducing the field of view is at a severe penalty of signal to noise. The geometric parameters and the number of acquisitions chosen for each imaging sequence are always a compromise between spatial resolution, signal to noise, and imaging time.

CASE 6

History: Two sagittal gradient-recalled scans through the same level of the cervical spine are presented. Both scans were accomplished utilizing identical pulse sequences except for one parameter that changed in each scan. What is the parameter that was varied in this experiment? In what ways do gradient-recalled scans differ from spin-echo sequences and what are the principal determinants of image contrast utilizing these techniquues? (Fig. 6A. Sagittal, TR/TE 750/15 msec. Fig. 6B. Sagittal, TR/TE 750/15 msec.)

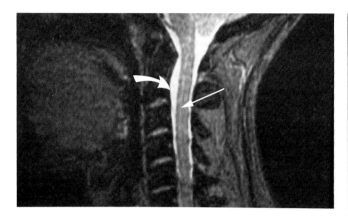

FIG. 6A.

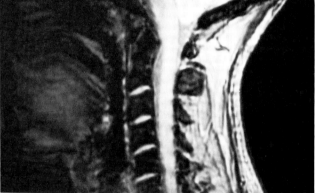

FIG. 6B.

Findings: In Fig. 6A the cerebrospinal fluid (CSF) demonstrates high signal intensity (curved arrow) and the spinal cord, which is of intermediate signal intensity (arrow), is seen in sharp contrast. The vertebral bodies demonstrate markedly low signal intensity and the intervertebral discs are high in signal intensity. In Fig. 6B the cerebrospinal fluid now demonstrates an intermediate signal intensity and the spinal cord and cerebellum are of increased intensity. The subcutaneous fat is of higher signal intensity than in Fig. 6A. The vertebral bodies remain of low signal intensity. A small disc protusion at the C5-6 level is evident.

Diagnosis: The scans differ in the flip angle (theta) with which they were accomplished. (Fig. 6A. Flip angle 10 degrees. Fig. 6B. Flip angle 80 degrees.)

Discussion: The introduction of special imaging sequences that do not utilize the second slice selective 180-degree pulse (refocusing pulse) for generating an echo but rather employ a gradient reversal technique provide a unique opportunity for generating high contrast images in very short periods of time. These techniques are sensitive to local field inhomogeneities but with the introduction of relatively short echos the contrast available in images as short as five seconds can be sufficiently diagnostic. In fact, the sensitivity of these sequences to local field inhomogeneities may add significantly to the image contrast generated. For instance, the increase in local magnetic susceptibility caused by focal iron deposition can significantly alter the image intensity (e.g., marked decreased signal on T2- and T2*-weighted images), making areas of iron deposition much more prominent. Similarly, the effects of paramagnetic contrast agents may, under certain circumstances, be significantly accentuated on these fast imaging techniques.

Fast imaging techniques represent a wide spectrum of gradient-recalled imaging sequences. The relationship of individual gradient pulses within the sequences can significantly alter the contrast in the images themselves. The number of different acronyms including FLASH (fast low angle shot), GRASS (gradient-recalled acquisition in the steady state) and FISP (fast imaging with steady state precession) all have intrinsically different contrast mechanisms. These can, however, be assigned to two basic categories: (1) spoiling of transverse magnetization (FLASH), or (2) sustaining of transverse magnetization (FISP, GRASS). In addition to the previously mentioned variables of TR and TE, these imaging sequences use limited flip angle excitation rather than a 90-degree pulse. The choice of the angle of excitation adds an additional parameter under user control for generating optimal image contrast.

In gradient-echo pulse sequences the radiofrequency (RF) is used to tilt the magnetization at varying angles between 5 and 90 degrees. The gradients are then used to focus the signal and generate an echo. A small flip angle (near 0 degrees) maximizes T2* contrast. A larger flip angle closer to 90 degrees maximizes T1 contrast. When utilizing a repetition time significantly shorter than the T1 of the tissue being examined, the flip angle at which there is maximum signal from an individual tissue is typically at a flip angle significantly less than 90 degrees. The Ernst angle for most soft tissues in relatively short TR sequences is approximately 30 degrees. The choice of flip angle to optimize signal may not however represent the most appropriate flip angle to generate contrast between normal and pathologic structures.

As a consequence of not employing a 180-degree refocusing pulse as in standard spin-echo imaging, static magnetic field inhomogeneity effects are not refocused. As a result, gradient-echo pulse sequences are sensitive to T2* rather than T2. T2* is shorter than T2 and represents the decay of transverse magnetization without refocusing of static susceptibility. Another key difference between gradient-recalled imaging and spin-echo imaging is the effect of flow on these imaging sequences. Because a single slice selective RF pulse utilized in a nonselective RF gradient is used for refocusing, there is significant enhancement of signal from flowing structures wherever they occur. With the combination of flow compensation techniques moving structures become increasingly bright within the images. This makes these sequences extremely sensitive to motion during the acquisition. It also allows for the ability to generate significant contrast with flowing fluids such as blood within vascular structures or pulsatile CSF.

It is possible, with the proper choice of TR, TE, flip angle, and flow compensation, to generate excellent contrast in many structures. It should be cautioned that although an image may have a significantly bright signal from fluid and a relatively suppressed signal from soft tissue (superficially resembling a T2-weighted image) these images are not T2-weighted and the mechanisms of contrast are quite different. T2* relaxation includes not only T2 relaxation due to inhomogeneities on a molecular level from the physical-chemical properties of the sample, but also results from magnetic field inhomogeneities due to large fluctuations in the magnetic field independent of the sample (e.g., primary magnetic field inhomogeneity). Normal bone marrow appears of low signal intensity on gradient-echo images. This principally reflects the high magnetic susceptibility of normal marrow. Most pathological processes demonstrate longer T2* and as a consequence will appear of higher signal intensity.

CASE 7

History: A 42-year-old woman with a history of prior lumbar back surgery and chronic low back pain underwent evaluation with MR. Figures 7A and 7B are representative axial images from her initial examination. Figures 7C and 7D are corresponding sections from a repeat examination performed on the same patient one week later. What single factor can account for the astounding differences between the two examinations? What techniques can be utilized to minimize image degradation secondary to this process? (Fig. 7A. Axial, TR/TE 2,000/80 msec. Fig. 7B. Axial, TR/TE 2,000/80 msec. Fig. 7C. Axial, TR/TE 2,000/80 msec. Fig. 7D. Axial, TR/TE 2,000/80 msec.)

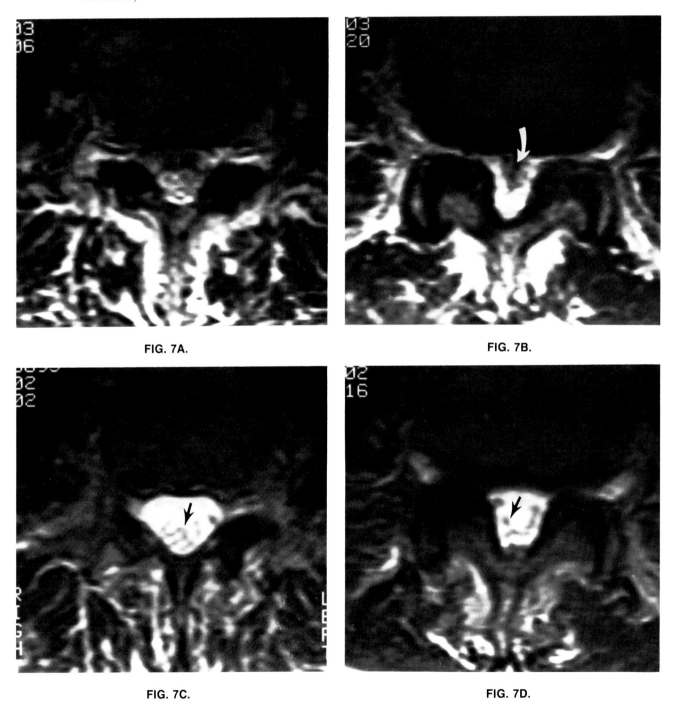

FIG. 7A.   FIG. 7B.

FIG. 7C.   FIG. 7D.

Findings: On the two sections from the initial examination (Figs. 7A and 7B), the intrathecal signal is markedly inhomogeneous. An apparent central irregular focus of low signal intensity within the thecal sac on Fig. 7B (arrow) could suggest the presence of clumped nerve roots or an intrathecal mass. On the repeat examination (Figs. 7C and 7D), the high signal intensity CSF is far more uniform in appearance and individual nerve roots can be demonstrated (arrows).

Diagnosis: Flow induced artifact. On the initial examination, the markedly heterogeneous appearance to the intrathecal contents was thought to be related to flow-related phenomenon. The patient was requested to return and the second examination was accomplished with similar parameters except for the addition of a flow compensation data base.

Discussion: Spin-echo images involve the use of an initial 90-degree slice selective radiofrequency pulse (excitation pulse), followed by a 180-degree slice selective pulse (rephasing pulse). In order for spins to effectively contribute signal to an image these events must occur in a coordinated fashion. Spins must first receive their 90-degree pulse and not have moved outside the imaging plane. If they do move significantly and do not receive their 180-degree pulse they essentially produce a disrupted sequence and do not contribute useful data to the image. In fact, the signal that may be contributed to the image may significantly be displaced and account for flow-motion artifacts. In general, one method for dealing with this flow-motion artifact problem in the body is to apply a pre-saturation pulse outside the imaging volume, so that when spins flowing into the imaging volume encounter radio frequency pulses they already have a transverse magnetization and are dephased. In this case they contribute virtually no signal or artifacts to the imaging volume.

In other instances it may be useful and important to obtain as much signal as possible from flowing spins. In order to avoid artifacts it is critical that this signal be assigned its correct location in the imaging slice. Slow flowing spins moving in a gradient accumulate a small amount of phase which in general displaces them from their normal position in the image. With some prior knowledge of the general velocities encountered a gradient in the opposite direction of the one which has generated this shift in phase can replace these spins into the appropriate position. This flow compensation sequence replaces the potential lost signal intensity from these flowing spins, replacing them to the appropriate position in the image and reducing overall artifact and changing the dark "flow void" to bright flow enhanced signal.

Additional simpler measures to reduce flow artifacts are generally utilized in musculoskeletal imaging. Flow-related artifacts are generated in the phase direction of the image by orienting the phase and frequency directions of the image so that the artifacts are not directed over the area of greatest interest. In general, this technique improves the over-all image quality. For instance, placing the phase direction in the superior-inferior direction of a sagittal image in spine imaging will direct the phase artifacts from vascular pulsations of the aorta and vena cava in a superior-inferior direction. If the phase direction were in the anterior-posterior direction these artifacts would be placed over the spinal canal and in certain cases virtually destroy any useful diagnostic information within the image.

CASE 8

History: Two sagittal sections were obtained utilizing different pulse sequences. Figure 8A is a standard TI-weighted spin-echo sequence. Figure 8B is a STIR image. What are the underlying principles of this sequence? What is the effect on the signal from fat? What is the effect on T1 and T2?

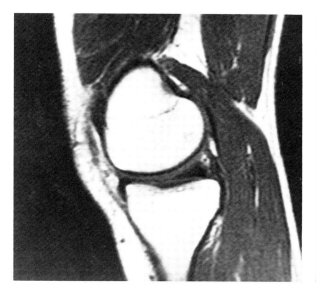

FIG. 8A.

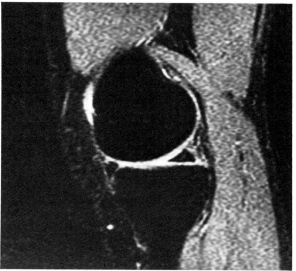

FIG. 8B.

Findings: On the T1-weighted image (Fig. 8A) the bone marrow of the distal femur demonstrates high signal intensity secondary to the predominant fatty component of the marrow in this location in an adult. On the STIR sequence (Fig. 8B) the signal from the fatty marrow has been nulled and the marrow is of marked low signal intensity.

Diagnosis: Differences in image contrast related to pulse sequence utilized.

Discussion: Pulse sequences can be designed to exploit the differences in precessional frequencies and relaxation times of fat and water. One such sequence that has recently received considerable attention with regard to its potential in musculoskeletal imaging is that of STIR. Utilizing this sequence, the signal from the components with short T1 such as the aliphatic component of marrow is nulled and pathological processes characterized by increased bulk water and prolonged T2 relaxation times are accentuated. On STIR sequences the signal from fat is nulled and that from other tissue with longer T1 and T2 is additive. Several recent reports have suggested a high sensitivity for this pulse sequence for bone marrow pathology.

The initial excitation pulse is a 180-degree pulse that is followed by a standard spin-echo pulse sequence at time TI after the initial 180-degree pulse. The inversion time (TI) is set to the crossover time of fat which is approximately two-thirds the T1 of fat. After the initial 180-degree inversion pulse, fat will regain bulk magnetization before water. When a 90-degree pulse is given at the crossover time of fat, the net magnetization of water is directed to the transverse plane where the subsequent 180-degree pulse refocuses signal from the water component only. Longer TR and TI values are required at higher field strength because of longer T1 relaxation times. Longer TE times give better contrast, but shorter TE times give higher signal to noise with STIR. Imaging parameters must be individually adjusted.

RECOMMENDED READING

Bradley WG Jr, Tsuruda JS. MR sequences parameter optimization: an algorithmic approach. *AJR* 1987;149:815.

Bradley WG Jr. Flow phenomena. In Stark DD, Bradley WG, eds. *Magnetic resonance imaging*, St Louis: CV Mosby, 1987;108–137.

Burk DL Jr, Dalinka MK, Schiebler ML, et al. Strategies for musculoskeletal magnetic resonance imaging. *Rad Clin of N Amer* 1988;26(3)653–672.

Crues JV III, Shellock FG. Technical considerations. In Mink JR, et al. eds. *Magnetic resonance imaging of the knee*. New York: Raven Press, 1987;3–27.

Enzmann DR, Rubin JB. Cervical spine: MR imaging with a partial flip angle, gradient-refocused pulse sequence, part 1: general considerations and disc disease. *Radiology* 1988;166:467.

Gomori JM, et al. Variable appearances of subacute intracranial hematomas on high-field spin-echo MR. *AJR* 1988;150:171.

Haacke E. Image behavior: resolution, signal-to-noise, contrast and artifacts. *Magnetic resonance imaging of the spine*. Chicago: Year Book, 1989;1–34.

Haase A, Frahm J, Matthaei D. FLASH imaging: rapid NMR imaging using flow flip-angle pulses. *J Magn Reson* 1986;67:258–266.

Hoult DI, Chen CN, Sank VJ. The field dependence of NMR imaging: II. arguments concerning an optimal field strength. *Magn Reson Med* 1986;3:730–746.

Kanal E, Wehrli FW. Signal-to-noise ratio, resolution and contrast. In Wehrli FS, Shaw D, Kneeland JB, eds. *Biomedical magnetic resonance imaging: principles, methodology and application*. New York: VCH Publishers, 1988.

Mitchell D, et al. The biophysical basis of tissue contrast in extracranial MR imaging. *AJR* 1987;149:831.

Murphy WA, Totty WG, Carroll JE. MRI of normal and pathologic skeletal muscle. *AJR* 1986;146:565–574.

Shields AF, Porter BA, Churchley S, et al. The detection of bone marrow involvement by lyphoma using magnetic resonance imaging. *J Clin Oncol* 1987;5:225–230.

Spritzer CE, Vogler JB, Martinez S, et al. MR imaging of the knee: preliminary results with a 3DFT GRASS pulse sequence. *AJR* 1988;150:597–604.

Sugimura K, Yamasaki K, Kitagaki H, et al. Bone marrow diseases of the spine: differentiation with T1 and T2 relaxation times in MR imaging. *Radiology* 1987;165:541–544.

Unger EC, Cohen MS, Gatenby RA, et al. Single breath-holding scans of the abdomen using FISP and FLASH at 1.5T. *J Comput Assist Tomogr* 1988;12:575–583.

Unger EC, Glazer HS, Lee JKT, et al. MRI of extra-cranial hematomas: preliminary observations. *AJR* 1986;146:403.

Unger EC, Summers TB: Bone marrow. *Top Magn Reson Imag* 1989;1(4):31–52.

Vogler JB, Murphy WA: Bone marrow imaging. *Radiology* 1988;168:679–693.

Wehli FW. Principles of magnetic resonance. In Stark DD, Bradley WG, eds. *Magnetic resonance imaging*, St Louis: CV Mosby, 1987;3–23.

Winkler ML, et al. Characteristics of partial flip angle and gradient reversal MR imaging. *Radiology* 1988;166:17–26.

CHAPTER 2

The Shoulder

Michael B. Zlatkin and Phillip Chao

Shoulder pain is one of the most common patient complaints to orthopedic surgeons. Although there are many different causes for this problem, the presenting complaints are often very similar and thus confusing for the treating physician. It is, however, important to make a correct diagnosis to institute the most appropriate therapy.

Imaging methods utilized in the evaluation of the shoulder have included plain radiographs, bone scintigraphy, ultrasound, and conventional and computed tomography. More invasive techniques include the use of arthrography and conventional and computed arthrotomography.

In the past 2 to 3 years MRI has become more and more the procedure of choice in the evaluation of diseases of the shoulder. This is primarily due to advances in imaging software and improvements in surface coils, as well as work by a number of different investigators describing the efficacy of this technique, particularly in the realm of rotator cuff disease and shoulder instability. Imaging software advances include off-center field of view, which allows good images to be obtained away from the center of frequency of the magnet, which is a less homogeneous region, and the ability to perform oblique images along the axis of the rotator cuff tendons, which do not course along the true orthogonal planes of the body.

With respect to surface coils, we utilize a dual surface coil array employing two 5″ round, commercially available, general purpose receive-only surface coils, on a General Electric 1.5-T system. These are placed one anterior and one posterior to the affected shoulder in a "Helmholtz configuration," in which the signal obtained from the two coils is additive if separated approximately a coil radius apart. The coil cables are connected to a "BNC T connector" to adapt the connection to the surface coil port in the head coil. In testing, this configuration does not result in heating of the cables or appreciable loss in signal-to-noise ratio. The coils themselves are connected via a plastic holder to stabilize them, allow their proper separation, and prevent the anterior coil from directly contacting the anterior chest wall.

We have two standard imaging protocols at our institution designed for the two most common indications, rotator cuff disease and shoulder instability. For patients with suspected rotator cuff disease, a coronal localizing image is performed first in the body coil, with a spin-echo (SE) 600/20 msec sequence and a 40-cm field of view. This determines the location of the off-center field of view and sets up the next pulse sequence, which is an axial surface coil series (SE 600/20 msec). The axial images are helpful in the assessment of the glenoid labrum and anterior capsule and are also utilized as localizers for the subsequent coronal and sagittal oblique images. The coronal oblique images are obtained in a plane perpendicular to the glenoid margin, which is parallel to the supraspinatus muscle and tendon. The coronal obliques are the most important images in the evaluation of the rotator cuff. We utilize two pulse sequences. The first is an interleaved relatively T1-weighted sequence (SE 1,000/20 msec); the second is a variable SE 2,500/70 msec sequence with a 5-mm contiguous (CSMEMP) slice thickness. The shorter TR/TE sequences and proton-density sequences are best for anatomic detail; the long TR/TE sequences highlight any intra- or extra-articular fluid or edema present.

19

The sagittal obliques are obtained in a plane parallel to the glenoid margin, with a SE 800/20 msec sequence. This plane is perpendicular to the supraspinatus. It has proven more useful in assessing the relationship of the acromion and acromioclavicular joint to the cuff muscles and tendons than in the assessment of the integrity of the tendons themselves.

In patients with a history suggesting a glenoid labral tear and/or shoulder instability (recurrent subluxations and/or dislocations), we perform a variable-echo SE 2,500/20, 70 msec sequence in place of the sagittal obliques. The long TR/TE sequences highlight any intra-articular fluid and allow visualization of the normal and abnormal capsular structures, particularly in the presence of a joint effusion. At the present we have replaced the axial SE series used in the rotator cuff protocol with a multiplanar gradient recalled acquisition in a steady state (GRASS) sequence (MPGR), which is a gradient-echo technique. This is obtained with a TR/TE 600/15 msec and a 70 to 80° flip angle, which gives relative T1 weighting and excellent contrast between glenoid labrum, articular cartilage, and marrow.

We utilize a field of view of 14 cm for all the axial imaging sequences and 16 cm for the oblique sequences. The matrix size is 256 × 128 with two signal excitations, except for the coronal body coil localizer, which is performed at one-half signal excitation, which is possible to perform with currently available software.

CASE 1

A 34-year-old man complained of pain in his right shoulder. He also complained of neck stiffness and posterior occipital headaches. His physician ordered plain radiographs, followed by an MRI of his shoulder. To what structure do the straight arrows point? What are the curved arrows demonstrating? (Fig. 1A. Midplane coronal oblique MR, TR/TE 800/20 msec.)

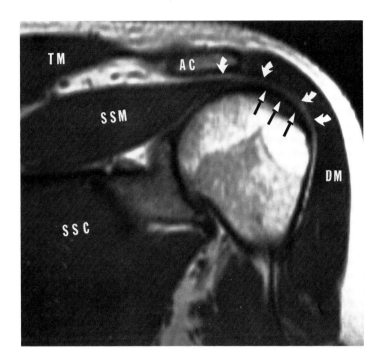

FIG. 1A.

Findings: Figure 1A is a coronal oblique image of a normal shoulder. It reveals the normal bulk and configuration of the supraspinatus muscle. The straight arrows are pointing out the normal homogeneous signal void of the supraspinatus tendon as it inserts into the greater tuberosity. The curved arrows are delineating the fat plane that overlies the tendon, the subacromial-subdeltoid fat plane. (AC, acromion; DM, deltoid muscle; SSM, supraspinatus muscle; SSC, subscapularis muscle; TM, trapezius muscle.)

Diagnosis: Normal shoulder. No evidence of rotator cuff abnormality.

Discussion: Shoulder pain is a very common disorder seen by orthopedic surgeons. The findings are often nonspecific and may be due to a variety of causes. The most common cause is rotator cuff disease, which is usually the result of the impingement syndrome, which this patient was initially suspected of having. The impingement syndrome results from attrition of the rotator cuff tendons as they pass underneath the anterior acromion, coracoacromial ligament, and acromioclavicular joint in overhead motion.

Because many of the important muscles and tendons of the shoulder do not run in the true orthogonal plane of the body, oblique planes of section are acquired. A coronal oblique is obtained by prescribing a plane perpendicular to the glenoid margin, from the axial images. This is parallel to the supraspinatus muscle and tendon (see previous discussion of technique). This is illustrated in Fig. 1B. The coronal oblique plane is the most useful for demonstrating most of the rotator cuff muscles and tendons, including the infraspinatus and teres minor, which insert more posteriorly and inferiorly on the greater tuberosity than the supraspinatus, which inserts anteriorly on this structure. In the normal situation, these muscles are of intermediate signal intensity on T1-weighted images, whereas their respective tendons appear of uniform low signal intensity because of the absence of mobile protons. The subscapularis muscle also helps make up the rotator cuff; however, this muscle and its tendon, which inserts into the lesser tuberosity, are best evaluated on axial images.

An important normal finding in this case is the presence of an intact subacromial-subdeltoid fat plane. This is a constant finding in normal shoulders, and we feel that it most likely represents extrasynovial fat about the subacromial-subdeltoid bursa. Loss of this fat plane has been found to be a helpful secondary sign of incomplete and complete rotator cuff tears. Some of the important muscles not involved in the rotator cuff but well seen on most scans of the shoulder include the deltoid muscle and trapezius muscle. Occasionally, injuries to these muscles result in edema or hematomas and may mimic symptoms of rotator cuff disease.

The sagittal oblique plane is also routinely obtained on most shoulder examinations. This is acquired using a plane of imaging that is parallel to the glenoid margin but perpendicular to the supraspinatus muscle and tendon (Fig. 1C). Although this provides images of the cuff in a different plane, we find it less useful than the coronal obliques in establishing the integrity of the rotator cuff muscles. It is, however, most useful, in our experience, in assessing the orientation and shape of the anterior acromion and determining the relationship of the acromioclavicular joint to the supraspinatus muscle and tendon (Fig. 1D,E). This is particularly helpful in patients with abnormally shaped acromion processes and degenerative changes involving the acromioclavicular joint.

Further evaluation of this patient in fact revealed that this patient's pain was related to a cervical disk disease, which helped to explain his neck stiffness and headaches. His shoulder pain was in fact referred pain from his neck. Other causes for shoulder pain that should be considered in the differential diagnosis of shoulder pain include calcific tendinitis, adhesive capsulitis, and shoulder instability.

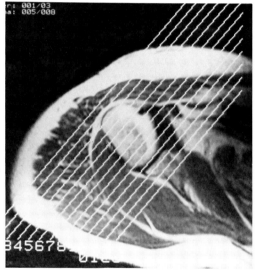

FIG. 1B. Axial image that depicts the plane of section for the coronal oblique sections (TR/TE 600/20 msec). These are obtained perpendicular to the glenoid margin, which is parallel to the supraspinatus muscle and tendon.

FIG. 1C. Axial image depicting the plane of section for obtaining images in a sagittal oblique plane (TR/TE 600/20 msec). These are obtained parallel to the glenoid margin, which is perpendicular to the supraspinatus muscle and tendon.

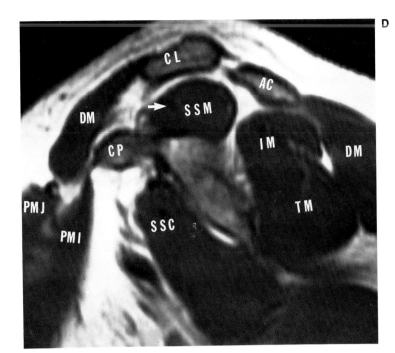

D

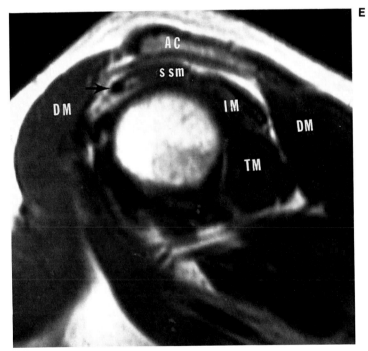

E

FIG. 1D,E. Normal anatomy (TR/TE 800/20 msec). These images illustrate the typical anatomy on medial sagittal oblique images. The sagittal oblique images also demonstrate the rotator cuff muscles and tendons but are usually most helpful in demonstrating the type and appearance of the acromion process and the relationship of the acromioclavicular joint to the supraspinatus muscle and tendon. (AC, Acromion; CL, clavicle; CP, coracoid process; DM, deltoid muscle; IM, infraspinatus muscle; PMI, pectoralis minor muscle; PMJ, pectoralis major muscle; SSC, subscapularis muscle; TM, teres minor muscle; SSM, supraspinatus muscle; arrow, supraspinatus tendon.)

CASE 2

A 34-year-old man, who is an active tennis player, complained of chronic pain in his shoulder, which had been getting worse during the past 3 months. It was exacerbated particularly by serving. What is your diagnosis? (Fig. 2A. Anterior coronal oblique MR, TR/TE 800/20 msec. Fig. 2B. Mid coronal oblique MR, TR/TE 800/20 msec.)

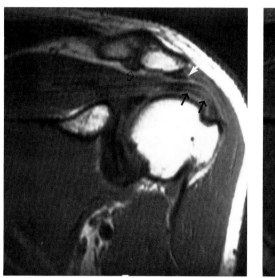 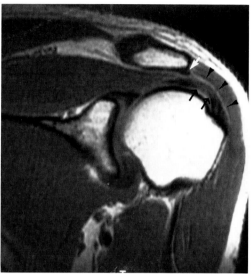

FIG. 2A.   **FIG. 2B.**

Findings: The images demonstrate an intact supraspinatus muscle and tendon; however, there is evidence of diffuse increased signal intensity in the distal tendon (black arrows). This area of increased signal intensity did not increase on T2-weighted images; in fact, it was less well seen. The subacromial-subdeltoid fat plane is intact (arrowheads). A small area of signal void is seen to project from the acromion tip, representing a small subacromial spur (white arrow). The acromioclavicular joint capsule is mildly prominent (open arrow).

Diagnosis: Rotator cuff tendinitis or tendon degeneration, without cuff tear. Findings likely secondary to chronic impingement syndrome.

Discussion: According to Neer, the majority of rotator cuff lesions occur as a result of chronic impingement of the rotator cuff tendons against the structures that make up the coracoacromial arch. These structures consist of the anterior acromion, the coracoacromial ligament (Fig. 2C), and the acromioclavicular joint. Patients who have narrow coracoacromial arches either caused by prominent coracoacromial ligaments, abnormal acromion shapes or slopes, or osteophytes on the acromion tip or the acromioclavicular joint would be predisposed to developing this problem.

Neer believed that rotator cuff disease represents a continuum of disease caused by this impingement. He believed this occurs in three progressive stages, beginning with edema and hemorrhage (stage I), progressing to inflammation and fibrosis (stage II), and culminating in rotator cuff tears (stage III). Stage I usually occurs in young patients who are athletes, usually active in sports utilizing an overhead motion, such as swimming, baseball, or tennis. Often there is a history of recent overuse. With continued impingement the symptoms become more chronic in nature, and progression to a stage II picture occurs.

In the past, imaging techniques were unable to detect the early changes in the tendon in patients with stage I or II disease, as traditional studies such as plain radiographs and arthrograms are normal at this stage. With MR we believe that these changes can now be depicted. They present as increased signal in the distal tendon on T1 and proton-density images, as is seen in this patient. The increased signal is probably related to a number of factors including edema and inflammatory change, as well as mucoid degenerative changes. These changes, we believe, are the MR correlates of stage I and stage II categories described by Neer. Stages I and II, we believe, appear similar on MR but should be separable on clinical grounds.

Bone changes such as spurring on the undersurface of the acromion and the inferior aspect of the acromioclavicular joint may cause bony encroachment on the rotator cuff and may play a major role in the pathogenesis of the impingement syndrome. It is not clear, however, whether the impingement itself leads to the spur formation or whether the degenerative spurring is the cause of the impingement. On MRI, small subacromial spurs are characterized by foci of signal void that project from the acromion tip. A small subacromial spur is identified on the images in this patient.

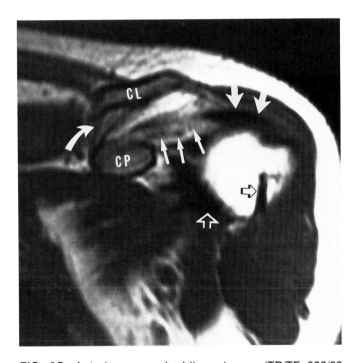

FIG. 2C. Anterior coronal oblique image (TR/TE 800/20 msec) demonstrates the low to intermediate signal intensity coracoacromial ligament coursing obliquely (*white arrows*) and its relationship to the supraspinatus tendon (*large white arrows*). This portion of the supraspinatus tendon at the junction with the biceps tendon (*open black arrow*) and subscapularis tendon (*open white arrow*) is sometimes known as the anterior leading edge, and it is not an uncommon location for rotator cuff tears to begin. Also note the excellent visualization of the coracoclavicular ligaments (*curved arrow*), which are important stabilizers of the acromioclavicular joint. They are torn in third-degree acromioclavicular joint separations. (CL, clavicle; CP, corocoid process.)

CASE 3

A 34-year-old man who works in the stock department of a clothing factory, where he is constantly lifting bales of material and placing them on overhead shelves, complained of chronic shoulder pain. He was suspected of having rotator cuff disease and possibly a small tear. What is your diagnosis based on the MR images? (Fig. 3A. Coronal oblique MR, TR/TE 800/20 msec.)

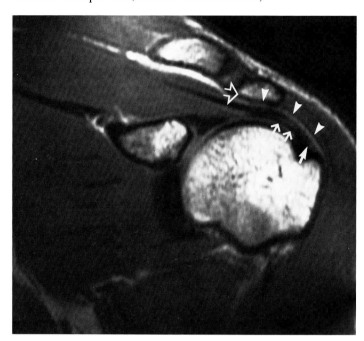

FIG. 3A.

Findings: There is diffuse increased signal intensity in the distal supraspinatus tendon (arrows), but no thinning, irregularity, or discontinuity is seen. The subacromial-subdeltoid bursa is intact (arrowheads). The anterior acromion is very low lying and is seen to impinge on the supraspinatus tendon (open arrow).

Diagnosis: Impingement syndrome.

Discussion: The shoulder impingement syndrome is an important cause of shoulder pain. Patients who have a history of repeated stress to the shoulder particularly in abduction and external rotation are predisposed to develop this disorder. Usually this chronic stress is related to their occupation or to a particular sport in which they are active. Affected patients often complain of pain localized to the region of insertion of the deltoid muscle. Most patients complain that the pain is more severe at night. On physical examination, point tenderness is frequently noted at the insertion of the supraspinatus tendon into the greater tuberosity. A positive impingement sign is elicited by passively lifting the arm and noting pain somewhere between 40 and 120° of abduction (painful arc). The cause of the pain is felt to be impingement of an inflamed and often degenerating rotator cuff tendon, between the greater tuberosity and the coracoacromial arch, which is formed by the coracoacromial ligament, acromioclavicular joint, and anterior acromion. This is also discussed in Case 2.

This patient shows objective evidence of inflammation of the supraspinatus tendon and possibly some degeneration, as manifested by increased signal intensity in the tendon, but the tendon is intact. The differentiating features of these abnormalities from those in rotator cuff tears include the absence of tendon morphologic changes such as tendon thinning or irregularity, and the absence of tendon discontinuity (Fig.

3B–D). No increased signal is seen on T2-weighted images within the abnormal tendon in these patients. In addition the subacromial-subdeltoid fat is intact, and no high signal intensity fluid is identified in the subacromial-subdeltoid bursa on T2-weighted images. Most of these patients are treated conservatively; therefore surgical confirmation of these findings is often lacking. Arthrograms performed in these patients with early changes in the tendon, such as seen in this patient, will be negative (Fig. 3E). However, Kieft and co-workers have indicated that tendon biopsies in patients with similar changes as these reveal mucoid degeneration and inflammatory change.

Note the low-lying anterior acromion, which is at a distinctly lower level than the distal clavicle in this patient. Neer believes that an abnormal size or position of the anterior part of the acromion can predispose a patient to the impingement syndrome. Bigliani has also described abnormal acromion shapes and related this to rotator cuff impingement and cuff tears. In fact a recent study by Seeger and co-workers of MRI in 53 patients with impingement syndrome found the acromion to be at a lower level than the distal clavicle in 14 patients. They concluded that this was corroborative evidence for Neer's theory.

B

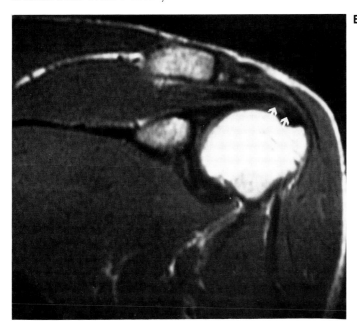

FIG. 3B,C. Tendinitis. Mid and anterior coronal oblique images in another patient (TR/TE 800/20 msec) showing similar findings of increased signal in the distal tendon (**B,C,** *arrows*), consistent with tendinitis, with an intact subdeltoid fat plane (**C,** *arrowheads*). Note again in this patient, who was a very active swimmer, tennis player, and weight lifter, the low lying anterior acromion (**C,** *larger white arrow*), similar to that in the test case.

C

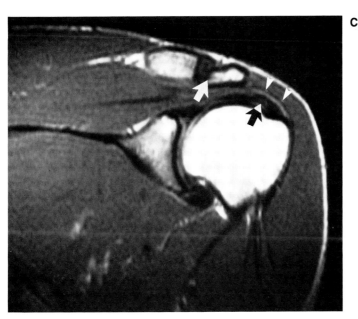

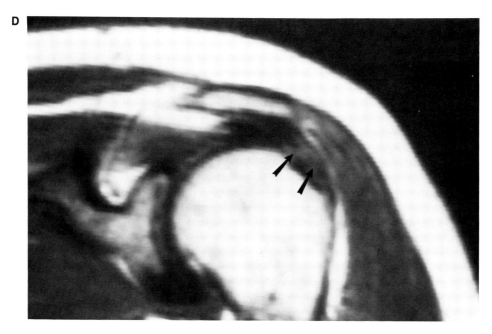

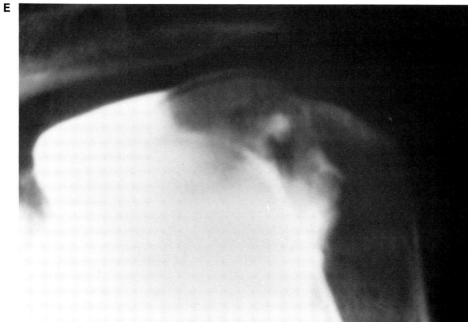

FIG. 3D,E. Tendinitis. Mid coronal oblique image (TR/TE 500/39 msec) demonstrating similar findings of tendinitis in the supraspinatus tendon in this patient (*arrows*). An arthrogram **(E)** performed in this patient 2 days later was normal. (Reprinted with permission from Zlatkin MB, Reicher MA, Kellerhouse LE, et al. The painful shoulder. Magnetic resonance imaging of the glenohumeral joint. *J Comput Assist Tomogr* 1988;12:995–1001.)

CASE 4

A 55-year-old man who had had intermittent shoulder pain for many years slipped on the ice, landing on his shoulder. Subsequently he complained of severe weakness and increased pain in his shoulder, which woke him at night. His range of motion was severely limited, and he could only abduct his shoulder minimally. An MRI scan was performed. What MRI findings are present? What is your diagnosis? (Fig. 4A. Coronal oblique MR, TR/TE 1,000/20 msec. Fig. 4B. Coronal oblique MR, TR/TE 2,500/80 msec.)

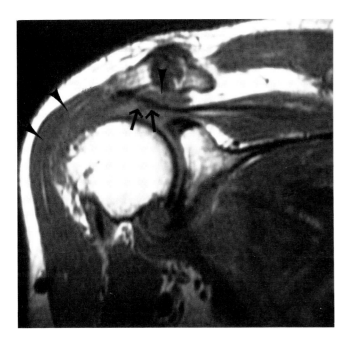

FIG. 4A.

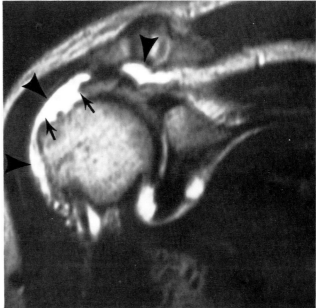

FIG. 4B.

Findings: The relatively T1-weighted coronal oblique image (Fig. 4A) demonstrates complete disruption of the supraspinatus tendon, which has retracted to the level of the acromioclavicular joint. The retracted tendon edges appear frayed and irregular (arrows). The high signal of the subacromial-subdeltoid fat is lost (arrowheads). The T2-weighted image (Fig. 4B) further outlines the large area of disruption of the tendon, as it is outlined by high signal intensity synovial fluid (arrows). Furthermore, high signal intensity fluid is seen in the subacromial-subdeltoid bursa (arrowheads). There are also degenerative changes present in the acromioclavicular joint as well as atrophy of the supraspinatus muscle.

Diagnosis: Large rotator cuff tear with severe retraction, atrophy, and irregular tendon edges (Fig. 4C).

Discussion: Most patients with rotator cuff tears can recall an injury that played a significant role in exacerbating their shoulder symptoms. However, although trauma may play a role in the development of a rotator cuff tear, it will rarely be the initiating event. It is now recognized that the majority of normal tendons do not tear, as 30% of the tendon must be damaged before a substantial reduction in its strength will occur. Therefore it is now believed that most cuff tears are chronic in nature, and the traumatic event usually completes or enlarges a tear in an already diseased tendon. The tendon is usually felt to be degenerated because of chronic impingement.

Magnetic resonance imaging is extremely accurate in the diagnosis of complete cuff tears, and the diagnosis of large tears, such as this case, is usually very obvious. MRI findings associated with complete cuff tears (Fig. 4D–F) include the presence of discontinuity in the rotator cuff tendon, which is usually the supraspinatus. In the majority of cases in which discontinuity is identified, increased signal consistent with fluid is present within the disrupted portion of the tendon on T2-weighted images. Helpful secondary signs that are almost always present in patients with complete tears include loss of the subacromial-subdeltoid fat on T1-weighted images and the presence of fluid in the subacromial-subdeltoid bursa on T2-weighted images.

Magnetic resonance imaging can also demonstrate the presence of retraction of the torn tendon edges and can detect whether the tendon edges are irregular and frayed as in this patient or whether tendon quality is good. It can also accurately assess the size of the tears and their site (4D) and demonstrate the presence of muscle atrophy. Atrophy is manifested by decrease in the muscle bulk and size and the presence of high signal linear bands within the muscle belly, indicative of fatty replacement. Secondary osseous changes of chronic impingement are present in at least 40 to 50% of patients.

The presence of the osseous changes, as well as the muscle atrophy seen in this patient and in other similar cases, supports the notion of an acute event superimposed on a chronically diseased cuff.

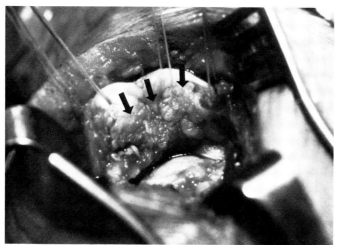

FIG. 4C. An intraoperative photograph of this patient demonstrates the large tear. No supraspinatus tendon is seen over the humeral head. The tendon edges are retracted and appear frayed and irregular (*arrows*), similar to that seen on the MRI scan.

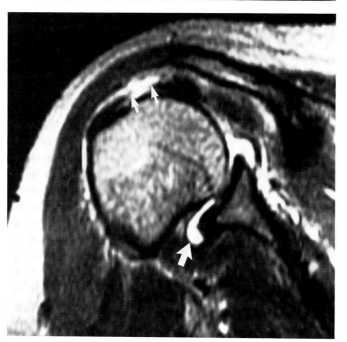

FIG. 4D. Rotator cuff tear. Coronal oblique T2-weighted image demonstrates tear involving the infraspinatus tendon (*small white arrows*). Note the effusion present in the axillary recess (*large white arrow*) and elsewhere in the joint.

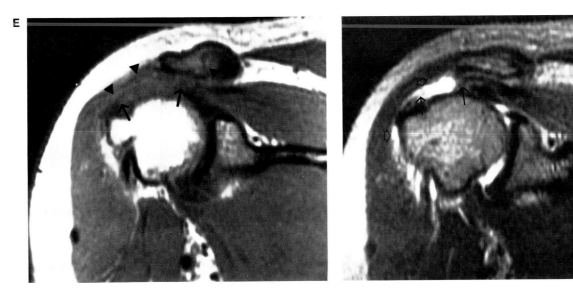

FIG. 4E,F. Rotator cuff tear. T1- (1,000/20 msec) and T2-weighted coronal oblique images (TR/
TE 2,500/80 msec) demonstrate another example of a large cuff tear. There are similar findings
of tendon discontinuity (*arrows*), loss of the subacromial-subdeltoid fat (*arrowheads*), and
fluid in the subacromial-subdeltoid bursa (*open arrows*).

CASE 5

This 46-year-old female patient complained of considerable shoulder pain, which became markedly worse when she reached for objects on the upper shelves in the convenience store where she worked. An MRI exam of her shoulder was performed. A coronal image at the level of the acromioclavicular joint is shown in Fig. 5A (TR/TE 800/20 msec). To what structure are the two arrows to your right pointing? What do you think is the reason this patient's pain is particularly bad when she raises her arm over her head?

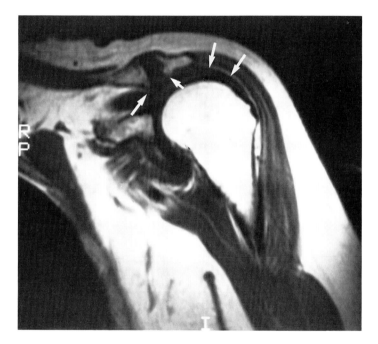

FIG. 5A.

Findings: The two arrows to your right are outlining the supraspinatus tendon. The MR scan in this patient demonstrates prominent osteophytes projecting from the distal clavicle and acromion at the acromioclavicular joint. These bony excrescences are compressing the tendon at the musculotendinous junction (two arrows to your left). Similar findings are evident in another patient (Fig. 5B).

Diagnosis: Secondary bony changes of chronic impingement, with encroachment on the supraspinatus musculotendinous junction.

Discussion: Because patients are scanned in the supine position with the arm at their side, we can see the effects of bony impingement only as they appear in this position. At the present time it is difficult to carry out studies that show the dynamics of the musculotendinous cuff in relation to the bony arch in different degrees of motion, although this may be possible in the future. This image does, however, demonstrate the significant bony impingement evident in this patient. It is not hard to imagine the further compression that would be present with the arm in an overhead position.

Secondary bony changes are often present in patients who have shoulder pain caused by impingement syndrome. These findings include degenerative spurring on the undersurface of the anterior acromion and the inferior aspect of the acromioclavicular joint. Small subacromial spurs on MRI are characterized by foci of signal void that project from the acromion tip. Large spurs frequently contain marrow. They are identified as bone excrescences with high signal intensity similar to marrow. They may be surrounded by a rim of signal void representing cortical bone, as seen in this patient. Hypertrophy of the capsule of the acromioclavicular joint may be present as well. Surgeons who operate on these patients also note thickening of the coracoacromial ligament. Degenerative changes of the humeral head including cystic changes, sclerosis, flattening, and bony proliferation have also been described. Although many of the bony changes can be identified on plain radiographs, the degree of bony encroachment on the supraspinatus musculotendinous junction can be well identified on the MR exam. In addition MRI can better identify a prominent acromioclavicular joint capsule and can often show the relationship of the supraspinatus tendon to the coracoacromial ligament. The sagittal oblique images (Fig. 5C) may be particularly useful in patients with bony impingement. They demonstrate the shape of the anterior acromion, the relationship of the acromioclavicular joint to the supraspinatus muscle and tendon, and the relative volume available to the musculotendinous cuff.

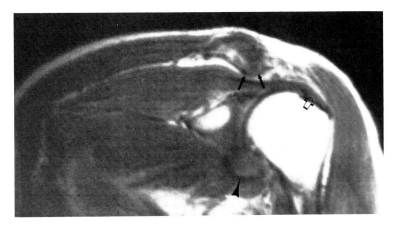

FIG. 5B. Bony impingement. Coronal oblique MR image (TR/TE 850/30 msec) demonstrates a large osteophyte projecting from the acromioclavicular (A-C) joint directly encroaching on the supraspinatus musculotendinous junction (*black arrows*). Intermediate signal intensity is also present in the tendon, consistent with some tendinitis (*open arrow*). A small amount of intermediate signal intensity fluid is also seen in the axillary recess (*arrowhead*). (Reprinted with permission from Zlatkin MB, Reicher MA, Kellerhouse LE, et al. The painful shoulder. MR imaging of the glenohumeral joint. *J Comput Assist Tomogr* 1988;12:995–1001.)

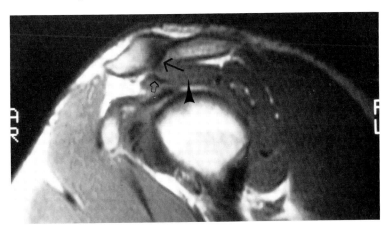

FIG. 5C. Bony impingement. Sagittal oblique image (TR/TE 800/20 msec) demonstrating the relationship of the A-C joint (*black arrow*) to the supraspinatus muscle (*arrowhead*) and tendon (*open arrow*). Note the somewhat narrow volume available for the supraspinatus in this patient as compared to the image of the normal patient in Case 1.

CASE 6

A 34-year-old baseball player has complained of pain in his shoulder for 3 years; however, recently the pain has become worse and he has been unable to throw with any velocity from the outfield. The surgeon requested an arthrogram, but you indicated to him that MRI was superior in the detection of shoulder abnormalities. Figure 6A and B are coronal oblique images of the shoulder (TR/TE 1,000/20 msec). Do you think the supraspinatus tendon is intact? The surgeon indicates he will operate if a partial or small rotator cuff tear is present. What should you tell him?

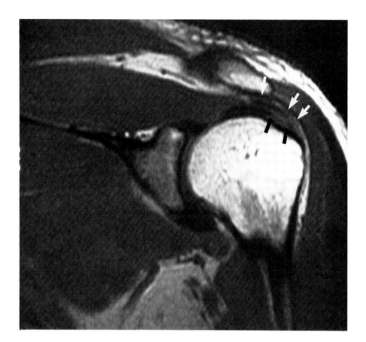

FIG. 6A.

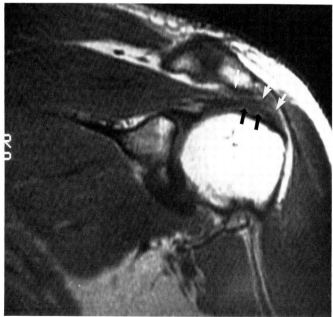

FIG. 6B.

Findings: The images reveal a supraspinatus tendon that is increased in signal intensity, but the tendon also appears thinned and irregular (black arrows). In addition there is loss of the high signal intensity of the subacromial-subdeltoid fat plane overlying the tendon (white arrows). T2-weighted images did not reveal any appreciable increased signal intensity in the cuff or any fluid in the subacromial-subdeltoid bursa. Mild A-C joint disease is present. At arthroscopy an anterior acromioplasty was performed and a thinned cuff with partial tearing of the undersurface was identified.

Diagnosis: Partial tear of the supraspinatus tendon.

Discussion: This case demonstrates findings that correlate with a more advanced stage of the impingement syndrome, sometimes known as a stage 3A lesion, wherein a partial tear of the tendon is present. The early studies of MRI of the rotator cuff indicated that partial cuff tears could not be differentiated from small complete tears. Because disease of the rotator cuff likely represents a continuum, this differentiation as well as the differentiation from tendinitis or tendon degeneration can be difficult. As more data have accumulated, however, this differentiation has become clearer. As distinct from complete tears, which will be discussed more fully in subsequent sections, partial tears of the undersurface present with tendon irregularity, thinning, and increased signal. Tendon discontinuity with increased signal intensity within the area of discontinuity on T2-weighted images, as evident on full thickness tears, should not be present. An associated secondary sign consists of loss of the signal from the subacromial-subdeltoid fat. In distinction from complete tears we have not observed the presence of fluid in the subacromial-subdeltoid bursa in any surgically confirmed cases. At the present time, however, this differentiation of partial tears from small complete tears is based on small numbers of surgically confirmed cases and will need to be confirmed with larger numbers. Whether this differentiation is clinically relevant depends on the treatment philosophy of the referring orthopedic surgeon. Although this patient was treated with surgery because of persistent symptoms, others with similar lesions have returned to normal function with conservative therapy. Follow-up MRI scans in a large number of patients with such abnormalities who do not go on to have surgery may determine some predictive parameters and aid in developing a more rational approach to therapy.

CASE 7

A 38-year-old woman was involved in a motor vehicle accident. She sustained a number of injuries. Radiographs of her shoulder were obtained. As she began recovering from her injuries, she complained of persistent shoulder pain and weakness. This had persisted longer than what was expected based on the injury evident on her plain radiographs. What are the plain radiographic findings in Fig. 7A? A shoulder MRI was ordered. Why was this obtained? What are the MRI findings that explain the reason for this patient's persistent pain and weakness? (Fig. 7B,C. Coronal oblique images of the shoulder, TR/TE 1,000/20 msec [B] and 2,500/70 msec [C]).

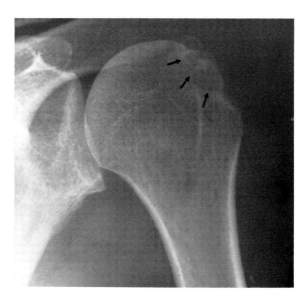

FIG. 7A.

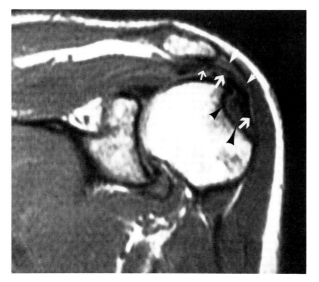

FIG. 7B.

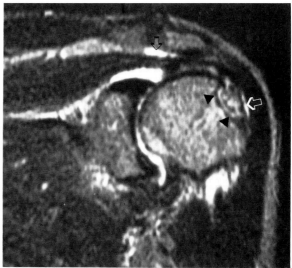

FIG. 7C.

Findings: The plain radiograph (Fig. 7A) reveals a fracture of the greater tuberosity without significant displacement (arrows). The relatively T1-weighted image (Fig. 7B) reveals an irregular thinned tendon, with multiple foci of increased signal intensity within (arrows). The subdeltoid fat is lost (white arrows). There is disruption of the marrow signal in the region of the fracture with depression of the cortex and increased signal in this region on the T2-weighted image (Fig. 7C) (arrowheads). The T2-weighted image highlights fluid in the subacromial-subdeltoid bursa (open arrows). Note also the presence of fluid in the joint and in the biceps tendon sheath.

Diagnosis: Undisplaced fracture of the greater tuberosity associated with a rotator cuff tear.

Discussion: Although rotator cuff tears are generally considered to result from long-term attrition of the tendon caused by chronic impingement, they may occasionally occur acutely, in which case they may be associated with a number of other injuries (Fig. 7D–G). One of these is a shoulder dislocation, with or without a displaced fracture of the greater tuberosity. Another association is the cuff tear that occurs with the seemingly innocuous undisplaced fracture of the greater tuberosity, which occurs without anterior dislocation of the shoulder, as seen in this patient. It is not widely understood that the tendons of the rotator cuff can be torn with this fracture. The fracture itself may appear to be a minor one, and the tendency is to treat it alone. Often the patient may present with persistent pain after the expected period of fracture discomfort has passed. A cuff tear may not be suspected because the patient may have satisfactory abduction of the shoulder. Thus it is important for the surgeon to consider performing an MRI in any patient who has continued pain following what appears to be an insignificant fracture of the greater tuberosity. The same problem has also been seen in patients with fractures of the acromion.

The MRI examination in this patient illustrates the greater tuberosity fracture well. In fact it is more obvious on MRI than on the plain radiograph, where it is rather subtle. The other noteworthy feature of this case is the appearance of the tendon. Although tears may present with more discrete areas of discontinuity, others may appear similar to this case, in which the normal tendon signal is replaced by areas of increased signal intensity, with associated marked tendon thinning and irregularity. The presence of the associated secondary signs of loss of the subacromial-subdeltoid fat, and fluid in the subacromial-subdeltoid bursa aid in arriving at the correct diagnosis of a rotator cuff tear.

We have divided tendon abnormalities into four different grades based on their signal intensity and morphology. This grading system is analogous in concept to that described for the meniscus of the knee by Reicher and co-workers and Crues, Mink, and Stoller. The normal tendon is considered grade 0. Grade 1 is a tendon with increased signal but no change in morphology; grade 2 is a tendon with increased signal and abnormal morphology (i.e., tendon thinning or irregularity). Grade 3 is a tendon with discontinuity, usually with increased signal within the discontinuous region on T2-weighted images. Grade 0, 1, and 2 tendons with a normal subacromial-subdeltoid fat plane are correlated with intact tendons that may have inflammation or degeneration within (grades 1 and 2). Grade 2 tendons with associated loss of the subacromial-subdeltoid fat are correlated with partial cuff tears (see Case 6). When there is associated fluid in the subacromial-subdeltoid bursa, a complete cuff tear should be found. A grade 3 tendon implies a complete cuff tear, which is nearly always further corroborated by seeing loss of the fat plane and/or fluid in the bursa.

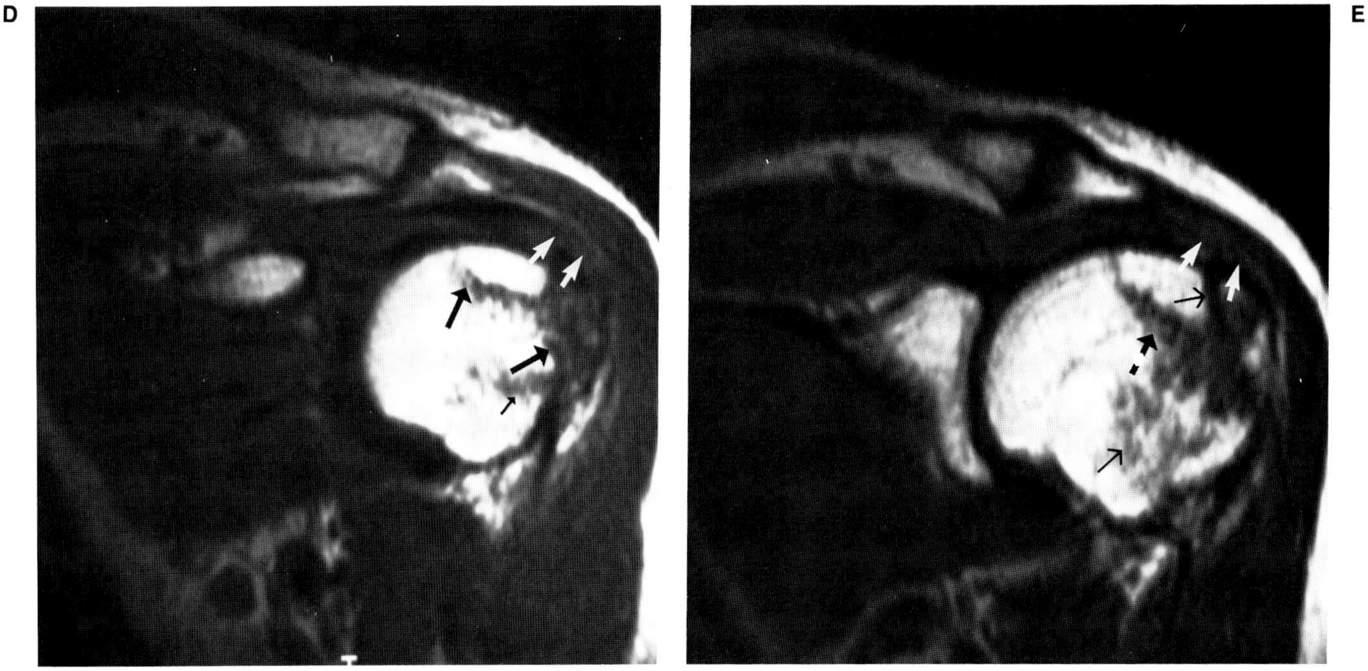

D

E

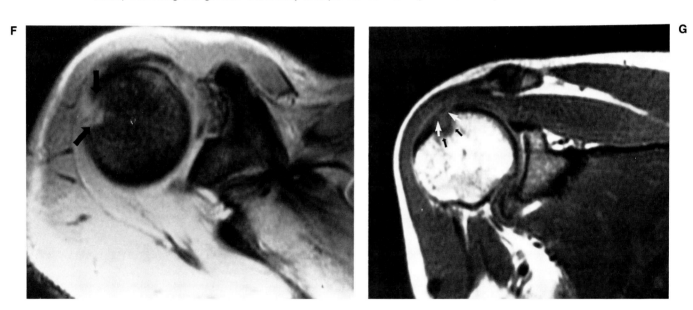

FIG. 7D,E. Humeral fracture and partial cuff tear. Anterior **(D)** and mid **(E)** coronal oblique images (TR/TE 800/20 msec) in another patient who sustained a humeral fracture (*black arrows*) involving the greater tuberosity and partial cuff tear (*white arrows*).

F

G

FIG. 7F,G. Dislocation and incomplete cuff tear. Axial gradient-echo (TR/TE 600/15, 70 msec) image **(F)** and coronal oblique spin-echo image **(G)** (TR/TE 800/20 msec) in another patient who sustained a small incomplete tear (*white arrows*) during an acute dislocation. Note the Hill-Sachs defect (*black arrows*).

CASE 8

A 39-year-old tennis player complained of increasing shoulder pain. An arthrogram was requested by the surgeon (Fig. 8A). Are any other studies indicated? The radiologist suggested an MRI in spite of the findings on the arthrogram. Why? What additional information is provided by the MR scans in Fig. 8B and C. (Coronal oblique MR, TR/TE 2,500/20, 80 msec.)

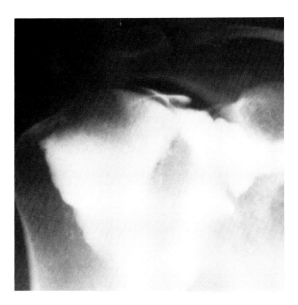

FIG. 8A.

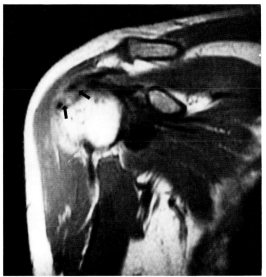

FIG. 8B.

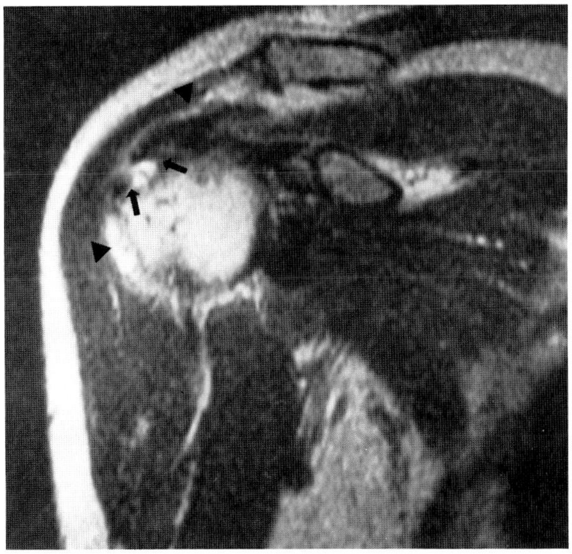

FIG. 8C.

Findings: Figure 8A is a representative film from a single contrast arthrogram in this patient. On this and multiple other views from this study, including those done postexercise, no extravasation of contrast is seen. The MRI scans (Fig. 8B,C), however, depict a small cuff tear at the anterior leading edge of the supraspinatus tendon, at its junction with the biceps and subscapularis tendons. Note the small defect in the supraspinatus tendon (arrows), which shows marked increased signal intensity on T2-weighted images as well as a small amount of fluid in the subdeltoid bursa (arrowheads). At surgery a small tear was found in this location, which was repaired.

Diagnosis: Small, anteriorly located rotator cuff tear. False-negative arthrogram.

Discussion: This case serves to illustrate the fact that in patients with shoulder pain caused by rotator cuff tears an MRI scan can be positive in the presence of a negative arthrogram. A recent study by Zlatkin and co-workers has shown that MRI is more sensitive than arthrography in the diagnosis of rotator cuff tears (91% versus 71%). Magnetic resonance imaging also demonstrated a high degree of specificity (88%). This was found to be particularly true for small, anteriorly located tears, less than 1.5 cm in size, as seen in this patient.

Both single contrast and double contrast arthrography have been found to be equally accurate studies in the diagnosis of complete cuff tears. Single contrast arthrography is said to be less helpful in assessing the size of the cuff tears and the status of the torn tendon edges. The fact that arthrography does have some associated morbidity and is proving less sensitive than MRI in the diagnosis of cuff tears is leading to the conclusion that MRI should be chosen over arthrography in the evaluation of cuff tears.

The exact clinical significance of the small tears identified only on MRI is as yet unclear. In fact many surgeons believe that patients with small cuff tears will do well with conservative management, which includes injections of steroids and physiotherapy. If these patients fail with this type of therapy, surgery can be considered at that time. More recent evidence, however, suggests that with the demonstration of a cuff tear, only 50% of these patients will respond to conservative management, and therefore surgery would be indicated.

The fact that many of these small cuff tears are seen anteriorly in the tendon correlates with other studies (Fig. 8D–F). The excellent contrast that can be obtained with MRI allows us to visualize them in this early stage, at a time when they are perhaps too small to be evident on arthrography.

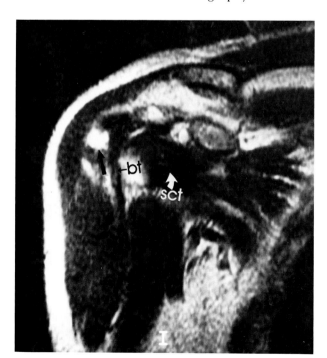

FIG. 8D. Rotator cuff tears. Coronal oblique MR in the same patient slightly more anterior, further illustrating the presence of the anterior cuff tear in the supraspinatus tendon (*black arrows*). (bt, Biceps tendon; sct, subscapularis tendon [*white arrow*].)

E

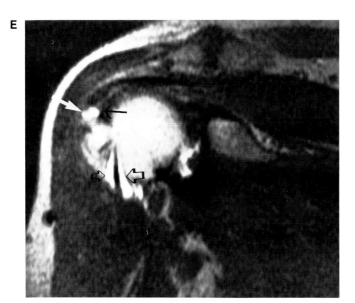

F

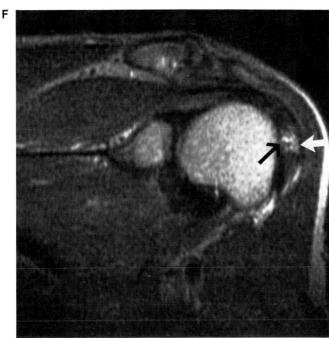

FIG. 8E,F. Rotator cuff tears. Two other examples of small anteriorly located cuff tears (*arrows*) similar to the test case (coronal oblique images, TR/TE 2,500/70 msec). Note the fluid in the biceps tendon sheath (*open arrows*) in **E**, which is a nonspecific finding.

CASE 9

A 62-year-old housewife complained of long-standing shoulder pain and weakness, which now woke her at night. An arthrogram was suggested, but she refused because she was frightened of needles. The surgeon referred her to the MRI center for a scan. The MR images revealed a cuff tear, as he suspected. The surgeon now wants to know how large the tear is, which tendons are involved, and the quality of the torn edges in order to plan his surgery. Based on the MR exam, what do you tell him? (Fig. 9A,B. Anterior coronal oblique MR, TR/TE 2,500/20 msec [A] and 2,500/80 msec [B]. Fig. 9C,D. Anterior and posterior coronal oblique MR, TR/TE 2,500/80 msec.)

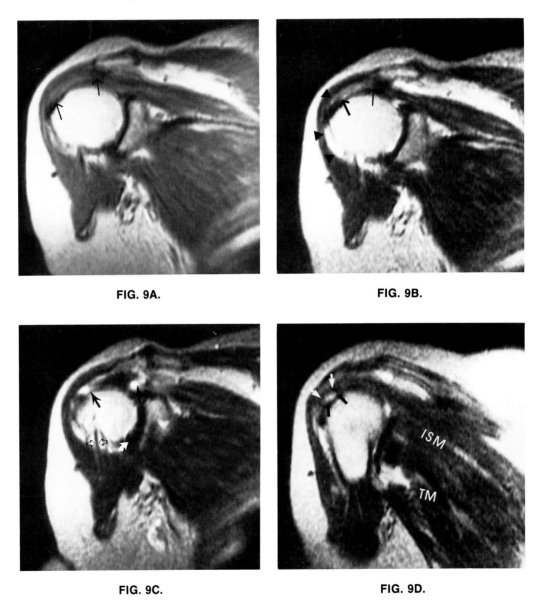

FIG. 9A.

FIG. 9B.

FIG. 9C.

FIG. 9D.

Findings: The first and second echo images in Fig. 9A and B reveal a large area of discontinuity in the supraspinatus tendon, which is highlighted by the higher signal intensity indicating the presence of fluid on the second echo (arrows) (Fig. 9B). Fluid is present in the subacromial-subdeltoid bursa in Fig. 9B (arrowheads). The tendon is retracted to the level of the acromion, but the edges of the tendon appear good (arrows). The size of the tear from medial to lateral at this point is 3.0 cm. The anterior images (Fig. 9C) show the tear to extend to the anterior leading edge of the supraspinatus (black arrows) at the junction with the subscapularis tendon (white arrow) and biceps tendon (open arrows) and more posteriorly to involve the infraspinatus tendon insertion (Fig. 9D, white arrows). (ISM, infraspinatus muscle; TM, teres minor muscle.)

Diagnosis: Large rotator cuff tear involving the supraspinatus tendon and extending to the infraspinatus tendon as well.

Discussion: In addition to being an accurate study in the diagnosis of rotator cuff tears, MRI can also offer (Fig. 9E,F) significant information regarding the size of cuff tears, the specific tendons involved, the degree of atrophy of the muscle, the degree of retraction, and the quality of the torn tendon edges. The procedures the surgeon may perform include a primary repair suturing tendon to tendon or tendon to bone; mobilizing other tendons to fill the defect from the tear, such as the biceps, subscapularis, or other tendons; or inserting an allograft or prosthetic material into the defect. The procedure performed is usually determined by the size and site of the tear; therefore knowledge of this data as determined from the MRI scan can be of value to the surgeon. It may allow him or her to predict preoperatively the difficulty of the repair and the type of repair to perform, to choose the appropriate surgical candidates, and to decide the best type of incision to make.

Mink and others have shown that double contrast arthrography is an accurate way to assess the size of tears as well as the quality of the torn tendon edges. We have recently studied the ability of MRI to assess these factors. We found that MRI performs extremely well in evaluating the size of the tendon defects (R = 0.95) when correlated with the results of surgery. In addition the correlation between MRI and surgery regarding the quality of the torn tendon edges was excellent. Magnetic resonance imaging can also indicate the degree of atrophy as well as determine which rotator cuff tendons are involved in the tear. These factors, as mentioned, are important to the surgeon, as knowledge of them may alter the type of incision and the type of repair or even dissuade him or her from performing surgery. In the presence of very large tears, some surgeons such as Rockwood have recommended only performing an acromioplasty and debriding the torn tendon edges. This has resulted in satisfactory relief of pain in a large percentage of these patients.

E 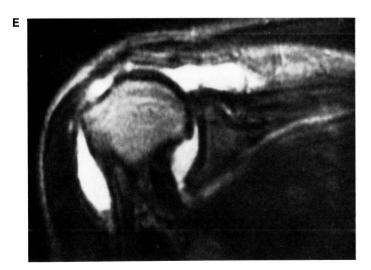 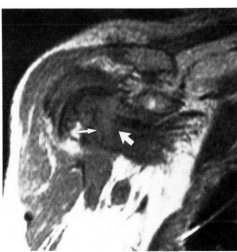 F

FIG. 9E,F. Massive rotator cuff tear (**E.** TR/TE 2,500/80 msec; **F.** 1,000/20 msec). This patient had a massive tear of the rotator cuff, which was seen to even extend to involve the subscapularis tendon on this anterior coronal oblique image (**F**, *arrows*).

CASE 10

A 44-year-old man who operates a forklift complains of shoulder pain and weakness since a work-related accident 3 months ago. What do you think is causing his pain? What MRI findings lead to that conclusion? (Coronal oblique MR, TR/TE 2,500/20 msec [Fig. 10A] and 2,500/80 msec [Fig. 10B].)

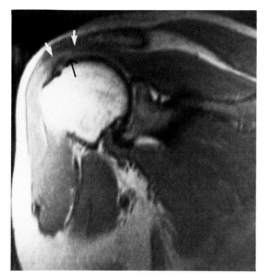
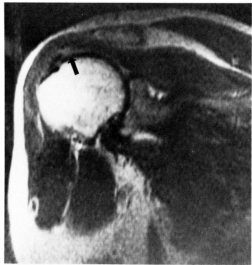

FIG. 10A. **FIG. 10B.**

Findings: The MRI demonstrates a small focus of discontinuity in the distal supraspinatus tendon (black arrow). On the T2-weighted images, this region shows increased signal consistent with some fluid within it. The subacromial-subdeltoid fat plane is lost (white arrows). These findings are consistent with a small rotator cuff tear.

Diagnosis: Small tear of the supraspinatus tendon.

Discussion: Disease of the rotator cuff represents a continuum. At the two ends of the spectrum when patients have either a normal cuff or a moderate- to large-sized tear, there is generally no difficulty in establishing a precise diagnosis, assuming good quality images are obtained. In between are those patients with less severe tendon abnormalities, such as tendinitis or tendon degeneration, partial cuff tears, or small complete tears. These patients may be more difficult to evaluate both clinically as well as with MRI because there can be overlap in their clinical presentations and because their MRI findings may be less dramatic than in patients with large tears. The most important finding in this patient leading to the diagnosis of a cuff tear is the presence of discontinuity in the supraspinatus tendon. In small defects such as this, one needs to distinguish this finding clearly from the increased signal in the tendon present in patients with tendinitis. In patients with tendinitis, the increased signal is generally more diffuse in nature and is contained within the substance of the tendon. The most specific finding, however, is the presence of further increased signal intensity on T2-weighted images, as seen in this case. This is most likely due to fluid present in the cuff defect and is diagnostic of a cuff tear. In small tears in which little retraction of the tendon edges has occurred, this may not always be present or if present may not be significant enough to be appreciated on T2-weighted images because of the considerable loss of signal-to-noise ratio that occurs. In these situations it is helpful to rely on associated secondary signs. In this patient the subacromial-subdeltoid fat plane is lost. (This fat plane is thought to represent extrasynovial fat about the subacromial-subdeltoid bursa.) In a large number of surgical cases that we have studied, loss of this fat plane was nearly always present in association with cuff tears, either partial or complete. It is also almost always intact in patients with intact cuffs or tendinitis. This observation was first made by Dr. Lee Kellerhouse of San Diego, and it can be very useful in establishing a diagnosis in more subtle cases. Another secondary finding seen in patients with cuff tears, which is also very helpful, is fluid seen below the acromion and deltoid but above the cuff. This is most easily identified on T2-weighted images. We believe that this most likely represents fluid that extends from the joint through a cuff defect into the subacromial-subdeltoid bursa. This is somewhat analogous to what occurs on an arthrogram in a patient with a tear, when the injected contrast and air leaks into this bursa. Although this was not present in this patient, it can be seen in the patient in Fig. 10C and D. This patient also had a small rotator cuff tear that was identified at surgery. In this patient a small area of discontinuity is seen on the proton-density weighted images (Fig. 10C), which is difficult to identify on the T2-weighted images. The associated secondary findings in this case include loss of the fat plane as well as fluid seen in the subacromial-subdeltoid bursa on T2-weighted images (arrows). The presence of fluid in the bursa helped us to establish the correct diagnosis of a small cuff tear in this case.

C

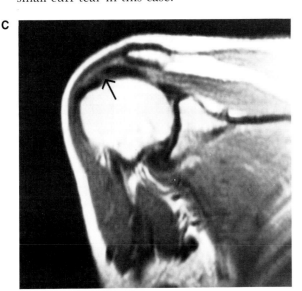

D

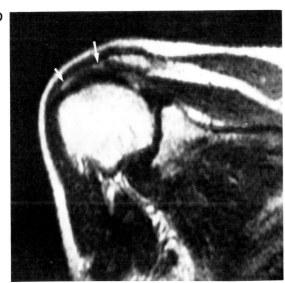

FIG. 10C,D. Small cuff tear. Coronal oblique MR images (**C.** TR/TE 2,500/20 msec; **D.** 2,500/8 msec). Note the small focus of discontinuity in the supraspinatus tendon (*black arrow*) best seen on the proton-density weighted images (**C**) but essentially imperceptible on the T2-weighted images (**D**). High signal intensity fluid is present in the subacromial-subdeltoid bursa in **D** (*white arrows*).

CASE 11

A 34-year-old male construction worker with multiple orthopedic complaints fell on ice outside his home. He sought orthopedic advice for his shoulder pain. An arthrogram was performed. The preexercise views revealed no abnormality. Figure 11A is a postexercise view from the same exam. What are the findings? The patient also had a shoulder MRI performed as part of the protocol for evaluation of shoulder pain, in effect at our hospital at that time. What findings led to the correct diagnosis on MRI? What additional statements about the patient's diagnosis can be made on the basis of the MRI? (Fig. 11B,C. Coronal oblique MR, TR/TE 2,500/20 msec [B] and 2,500/80 msec [C].)

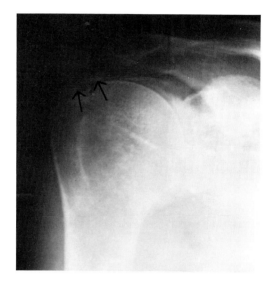

FIG. 11A.

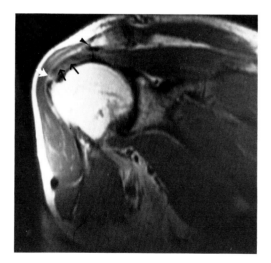

FIG. 11B.

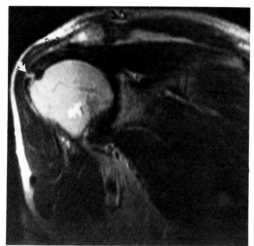

FIG. 11C.

Findings: The postexercise views from the single contrast arthrogram (Fig. 11A) reveal imbibition of contrast into the rotator cuff consistent with a partial cuff tear (arrows). The MRI scans (Fig. 11B,C) reveal irregularity, thinning, and increased signal in the supraspinatus tendon consistent with a partial tear (black arrows) similar to the findings evident on the arthrogram; however, at the distal portion the tendon is seen to be discontinuous (white arrows). There is increased signal extending to both surfaces of the tendon, with further increase in signal on T2-weighted images. This indicates a small complete tear in this portion of the tendon. Note that the fat plane is relatively intact at this level. A small amount of fluid is, however, seen in the subdeltoid bursa, which is secondary evidence for the presence of the complete tear (open arrows). A small subacromial spur is also present (arrowheads). A small humeral head cyst was incidentally noted.

Diagnosis: Partial and small complete tear of the supraspinatus tendon.

Discussion: This case is important because it illustrates a number of points. It again emphasizes that MRI can depict subtle rotator cuff tears such as was present in this case and in many instances in a fashion that is superior to arthrography without the need for injection of contrast material. It also illustrates the fact that there may be morphologic differences that can be identified with MRI that can separate small complete from partial thickness tears. This may not always be clinically relevant but it depends on the philosophy of the treating orthopedic surgeon. It also further illustrates the fact that in cases in which the tears are small, not all the findings that are evident in large tears are present, or if present they may be less apparent. In this case the more obvious findings are related to the tendon itself. There is irregularity, thinning, and increased signal in the portion of the tendon that is partially torn and discontinuity in the portion of the tendon where there is a complete tear. However, the fat plane is relatively well preserved and only a small amount of fluid is seen in the subdeltoid bursa. In other cases, such as that illustrated in Fig. 10C and D, the tendon changes are less obvious, and the secondary signs of a cuff tear, such a loss of the subdeltoid fat plane and fluid in the subdeltoid bursa, are the more obvious findings. This patient went on to have surgery, and both a partial and complete tear was found similar to that identified on the MRI exam (Fig. 11D).

FIG. 11D. Surgical photograph taken from above that demonstrates the portion of the tendon that was partially torn (*arrows*). The probe is in the portion of the distal tendon where the complete tear is present (*arrowheads*).

CASE 12

A 56-year-old man was referred for an MRI because of severe shoulder pain. The plain radiographs revealed vague erosive changes on the humeral head, osteopenia, and a decreased acromiohumeral distance. What are the MRI findings? What is your diagnosis, and what is the cause of this entity? (Coronal oblique MR, TR/TE 800/20 msec [Fig. 12A] and 2,500/80 msec [Fig. 12B].)

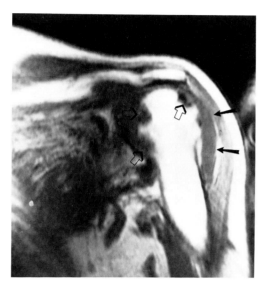

FIG. 12A.

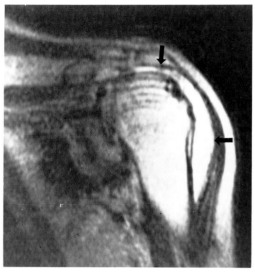

FIG. 12B.

Findings: A chronic long-standing large rotator cuff tear is present in this patient. No definable rotator cuff tendon can be seen, and the acromiohumeral distance is markedly decreased. There is massive atrophy of the surrounding musculature. A large amount of fluid is present in the subacromial-subdeltoid bursa (black arrows). Note the considerable degree of motion artifact on the T2-weighted images (Fig. 12B) as a result of the patients inability to lie still because of pain. Multiple large erosions are present in the humeral head (open arrows), and there is considerable resorption of bone along the glenoid margin.

Diagnosis: Rotator cuff arthropathy secondary to long-standing massive rotator cuff tear.

Discussion: Following a massive tear of the rotator cuff, there is inactivity and disuse of the shoulder, leaking of the synovial fluid, and instability of the humeral head. These events in turn result in both nutritional and mechanical factors that cause atrophy of the glenohumeral articular cartilage and resorption of the subchondral bone of the humeral head. A massive tear allows the humeral head to be displaced upward, causing subacromial impingement that in time erodes the anterior portion of the acromion and the acromioclavicular joint. Eventually, the soft atrophic humeral head collapses. The incongruous head may erode the glenoid so deeply that the coracoid becomes eroded as well. This association of abnormalities is known as rotator cuff tear arthropathy and is the most likely diagnosis in this case. Cuff tear arthropathy is very difficult to treat. Although most tears do not get large enough to cause this problem, it is a factor for the surgeon to consider when planning surgery when a tear is documented. Therefore it is important to diagnose. The best way to treat this problem is a resurfacing total shoulder replacement, with a rotator cuff reconstruction.

The changes seen in this disorder must be differentiated from those caused by rheumatoid arthritis (see Case 22), infection, or other severe erosive, destructive, and degenerative processes, which can also be complicated by atrophy and tear of the rotator cuff. This is probably best done in association with the relevant clinical history, plain radiographs (Fig. 12C), and synovial fluid analysis.

Loss of the acromiohumeral distance is not uncommon in chronic rotator cuff tears. In chronic cuff tears the torn tendon retracts and becomes ineffective in its normal action, as an antagonist to the upward pull of the deltoid muscle. This loss of function, in combination with the decreased soft tissue mass between the humeral head and the acromion caused by the absence of the tendon, results in the decrease in distance between the humeral head and the acromion that is seen in this patient. Loss of the acromiohumeral distance seen on plain radiographs does not always imply a rotator cuff tear, however. Severe degeneration and atrophy of the rotator cuff, without tear, can lead to this finding, which is one reason it is not inappropriate to obtain an MRI even in the presence of this finding. Other plain radiographic findings of rotator cuff tears include reversal of the normal inferior acromion convexity and cystic lesions of the acromion and humeral head.

Other indications for performing MRI in what may appear to be obvious clinical and plain radiographic evidence of this diagnosis include assessing the size of the tear and the quality of the torn tendon edges and the specific tendons involved to aid the surgeon in his or her preoperative planning.

Other complications that may occur in association with rotator cuff tears include lesions of the bicipital tendon, dissecting synovial cysts, interposition of the torn cuff in the glenohumeral joint, and degenerative arthritis.

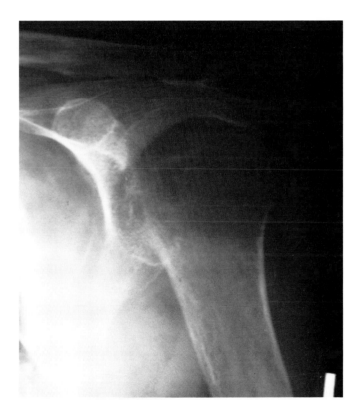

FIG. 12C. Cuff tear arthropathy. Plain radiograph of the patient in the test case, showing osteopenia, vague erosive changes particularly medially, loss of joint space, and decrease in the acromiohumeral distance.

CASE 13

A 67-year-old woman felt a snap in her shoulder while cleaning her house and now complains of severe shoulder pain. She had had an extensive cuff repair in association with a subacromial decompression 2 years earlier. Based on her MRI examination, do you think that her cuff repair is intact? What other findings are present that would make the surgeon reluctant to perform another repair? (Fig. 13A,B. Coronal oblique MR, TR/TE 2,500/20 msec [A], and 2,500/80 msec [B].)

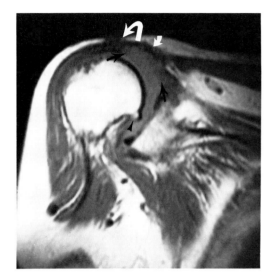
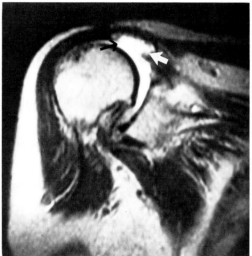

FIG. 13A.   **FIG. 13B.**

Findings: The proton-density and T2-weighted images in this case demonstrate that this unfortunate woman has completely disrupted her repaired rotator cuff (straight arrows). In addition she has evidence of severe atrophy and retraction of the torn edges, which are markedly irregular, indicating they are of poor quality. Note that the acromion and distal clavicle are not present on these cuts as this patient has had an acromioplasty and arthroplasty of the A-C joint. In addition there is distortion of the subcutaneous tissues and surrounding fat planes also related to the previous surgery (curved arrows). A small osteophyte is present at the neck of the humerus (arrowhead).

Diagnosis: Recurrent disruption of the rotator cuff 2 years after cuff repair.

Discussion: This case represents a rather dramatic case of a recurrent tear in a patient who has had prior surgery of the rotator cuff. Although this case is rather obvious, in many cases the assessment of cuff integrity in the patient who has had prior cuff repair can be difficult. Arthrography has been used in these patients; however, arthrography in this setting can be misleading as even patients who have pain relief, return of function, and what at surgery was felt to be an adequate repair often show a leak of contrast. Limited experience with MRI in this clinical setting shows that it may be useful in these patients (Fig. 13C,D). The cuff tendons can be visualized and their integrity directly assessed. As this case illustrates, MRI can demonstrate the disruption of the cuff when present. The presence of severe atrophy and retraction of the tendon edges, which are poor in quality, would likely make the surgeon reluctant to perform any further repair in this case.

A baseline postoperative study is probably useful in many cases to assess the adequacy of the repair, if necessary, in a noninvasive manner. In addition, if the patient later has recurrent symptoms, it would prove useful to separate changes from the prior surgery from those caused by a recurrent tear. This would have been helpful in the case illustrated in Fig. 13C, which is an example of a patient with a cuff that appears essentially intact but shows marked distortion of the repaired tendon, surrounding soft tissues, and fat planes from the prior surgery. Note also the prior acromioplasty. Other complications of cuff repair in addition to recurrent tears that could result in shoulder pain would include postoperative infection, as well as adhesive capsulitis.

Some difficulty may also be encountered in patients who have had prior acromioplasty alone for impingement syndrome without a cuff tear because of the distortion of the soft tissues, fat, and occasionally tendon itself caused by the surgery (Fig. 13E). These scans should be interpreted with more caution and may be a source of inaccuracy in the MRI exam. In such situations a correlative arthrogram may be of value.

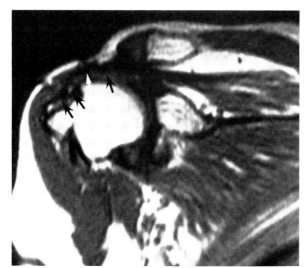

FIG. 13C. Postoperative intact cuff. Coronal oblique MR scan (TR/TE 800/20 msec). Intact tendon post–cuff repair (*arrows*). Note the marked distortion of the tendon and soft tissues present in this case from the repair and acromioplasty.

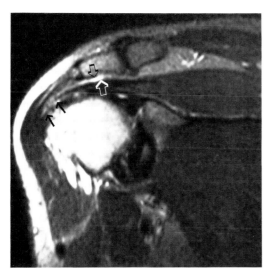

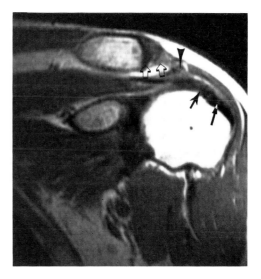

FIG. 13D. Recurrent cuff tear. Coronal oblique MR scan (TR/TE 2,500/70 msec). This patient had a primary repair of tendon to bone and has recurrent symptoms. He now shows a small gap in the distal tendon highlighted on the T2-weighted images (*black arrows*), as well as a small amount of fluid in the subacromial bursa, which is consistent with a small tear (*open arrows*). The patient does not show evidence of an acromioplasty on this scan.

FIG. 13E. Intact cuff, status postacromioplasty. Coronal oblique MR scan (TR/TE 800/20 msec). Note an intact but slightly irregular tendon (*black arrows*) and focal low signal in the overlying fat (*arrowhead*) in this patient who has had a prior acromioplasty (*open arrows*).

CASE 14

A 26-year-old man has a history of recurrent shoulder dislocations. The surgeon examines him and finds no objective evidence of instability, but the plain radiographs suggest a small Hill-Sachs deformity. The surgeon wants to confirm the diagnosis and plan appropriate therapy. He or she orders a CT arthrogram, but the patient is allergic to contrast and has a phobia about needles. You suggest an MRI. What are the MRI findings that confirm the diagnosis? What is their significance? (Fig. 14A,B. Axial MR, TR/TE 2,500/20 msec [A] and 2,500/80 msec [B].)

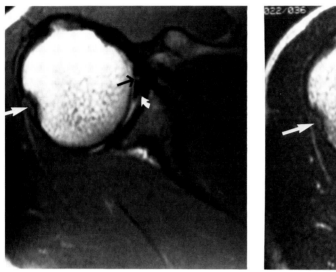

FIG. 14A.

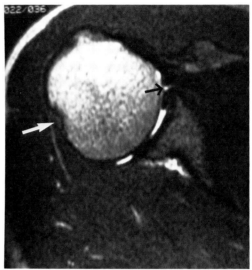

FIG. 14B.

Findings: The proton-density weighted images reveal a linear region of increased signal intensity in the anterior glenoid labrum extending to its surface (black arrows). On the T2-weighted images, there is further increased signal identified, consistent with a glenoid labral tear. A bony defect is also present on the posterolateral surface of the humeral head indicative of a Hill-Sachs deformity (white arrows).

Diagnosis: Anterior instability. Hill-Sachs deformity and anterior labral tear.

Discussion: The shoulder is a very mobile joint; hence it is prone to dislocations. Anterior dislocations of the shoulder account for approximately 50% of all dislocations in the body. When patients have a history of recurrent dislocations, the initial event is usually trauma. If the initial dislocation occurs between the ages of 15 and 35, the dislocations are more likely to become habitual.

There are many lesions that occur in association with the initial and subsequent dislocations that are either markers of or predisposing factors to recurrent dislocations. These include tears and detachments of the glenoid labrum, as well as abnormalities resulting from tearing and detachment of the anterior capsule. Bony defects may also occur, including Hill-Sachs lesions and fractures of the inferior glenoid margin.

This case demonstrates two important signs or sequelae of recurrent dislocation. The first is the glenoid labral tear. The normal anterior labrum is triangular in shape, although it may occasionally appear rounded. It is primarily fibrous and appears as a signal void on MR images. It is usually best evaluated in the axial plane. A normal glenoid labrum is shown in Fig. 14C. Bankart was the first to recognize that labral tears occurred in association with recurrent dislocations. Since then there has been considerable debate as to whether these labral tears are the cause of recurrent dislocations or whether they are just part of the spectrum of injuries that occur in these patients. Nonetheless they are a reliable sign of shoulder instability, as was the case in this patient.

Labral tears can have a number of different appearances on MRI. These include regions of increased signal within the labrum that extend to its surface. These are usually intermediate in signal intensity on T1-weighted or proton-density weighted images, and it often becomes bright on T2-weighted images, as in the test case. Occasionally the increased signal may be more diffuse in nature, as seen in the patient in Fig. 14D. Changes in morphology including blunting, fraying, or attenuation may be seen as well (Fig. 14E).

Smaller foci of increased signal may be seen in the labrum that are likely due to labral degeneration, but they do not extend to the surface of the labrum and do not get bright on T2-weighted images. One pitfall is the normal intermediate signal intensity that occurs at the base of the labrum particularly superiorly, which represents hyaline cartilage (Fig. 14C, curved arrow in Fig. 14A). It is normally continuous with the hyaline cartilage along the glenoid margin.

Hill-Sachs lesions also occur in association with the first and subsequent dislocations. They occur in anterior dislocations when the posterolateral surface of the humeral head impinges on the anterior inferior portion of the glenoid fossa. This actually results in two types of impaction fracture, the more common being the Hill-Sachs defect, which is a vertically oriented wedge-shaped defect, posterior to the greater tuberosity. The incidence of the Hill-Sachs lesion varies from 27 to 100%, depending on the techniques utilized to image it. It tends to be larger in those cases in which the dislocation has been present for a long time and in those in which the dislocation is recurrent. It is best identified on axial images, above the level of the coracoid process. The other impaction fracture is the "bony Bankart" lesion, which is a fracture of the anterior inferior lip of the glenoid fossa. This is demonstrated in Fig. 14F.

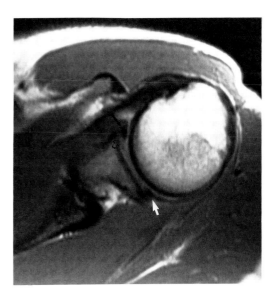

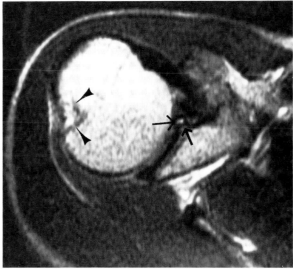

FIG. 14C. Normal labrum (axial MR image, TR/TE 600/20 msec). The normal triangular-shaped low signal intensity anterior glenoid labrum is identified (*black arrow*). Also note the increased signal at the base of the labrum, which represents hyaline cartilage (*open arrow*) and should not be mistaken for a labral tear. The posterior labrum is usually more rounded in appearance (*white arrow*).

FIG. 14D. Torn anterior labrum (axial MR image, TR/TE 2,500/80 msec). Diffuse increased signal is present in the anterior labrum (*arrows*). Also note the Hill-Sachs lesion (*arrowheads*).

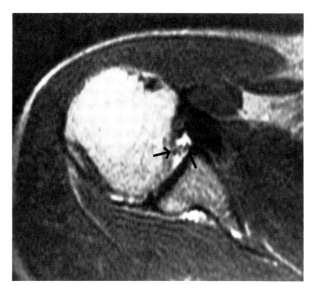

FIG. 14E. Torn anterior labrum (axial MR image, TR/TE 2,500/80 msec). A fragmented and torn anterior labrum is present (*arrows*).

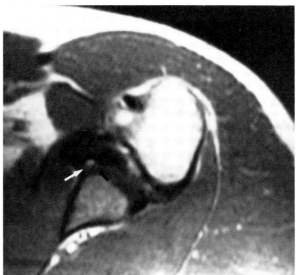

FIG. 14F. Bankart lesion (axial MR image ["Bony Bankart" lesion] TR/TE 800/20 msec). A small bony fragment off the inferior glenoid margin is present (*arrows*).

CASE 15

A 19-year-old football player was seen because of persistent anterior dislocations. He has had previous surgery for this problem. An MRI was done before considering further surgery. What are the findings? (Fig. 15A,B. Axial MR, TR/TE 2,500/80 msec. Fig. 15C. Axial MR, TR/TE 2,500/20 msec.)

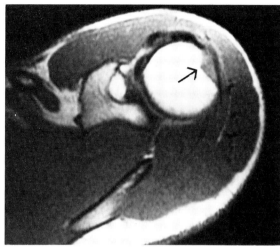

FIG. 15A.

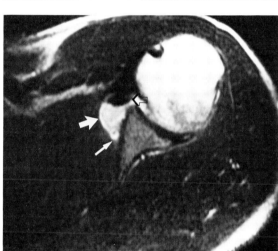

FIG. 15B.

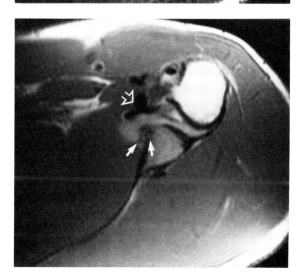

FIG. 15C.

Findings: The axial image at the level of the coracoid reveals a large Hill-Sachs defect (Fig. 15A, arrow). At the mid glenoid level (Fig. 15B), the anterior glenoid labrum is absent, and there is a large "anterior pouch" formed (white arrows) by the detached anterior capsule, highlighted by high signal intensity fluid. The subscapularis tendon is distorted from the prior surgery (Putti Platt). The middle glenohumeral ligament is prominent (open arrow). At the inferior glenoid (Fig. 15C), soft tissue thickening is present at the site of the capsular insertion (white arrows). Some low signal intensity material is also present related to the prior surgery (open arrow).

Diagnosis: Anterior instability.

Discussion: Because of its anatomic structure, the shoulder is one of the most mobile joints in the body. As such it is one of the most unstable and frequently dislocated articulations. Lesions of the anterior capsular mechanism are likely the most important outcome or cause of the first and subsequent anterior dislocations.

The anterior capsular mechanism consists of the synovial membrane, the capsule and glenohumeral ligaments, the glenoid labrum, the subscapularis bursa and related recesses, and the subscapularis muscle and tendon. Although we most commonly look for and think of tears of the glenoid labrum in patients with anterior instability caused by recurrent dislocations, in fact most people now believe that lesions of the entire capsular mechanism are the most important cause of the recurrent dislocations and subluxations that occur in these patients.

Signs specifically related to disruption of the anterior capsule are thus very commonly identified in patients with instability. One of the results of capsular "stripping" or detachment that occurs in these patients after the first and subsequent dislocations is the formation of a large "anterior pouch," usually in association with a large subscapularis bursa. This is a very prominent finding in this patient. This then forms a potential space into which the humeral head can dislocate or sublux. Other findings associated with capsular tears include a distorted appearance of the capsular insertion into the glenoid with either loss or thickening of the intervening soft tissue layer over the scapular margin. We find that T2-weighted images in the axial plane are best for demonstrating soft tissue capsular abnormalities, particularly when joint fluid is present.

One other factor that must be considered is the variation that normally occurs in capsular insertions. In fact there are three types of capsular insertions that can occur. Type 1 inserts near the labrum (Fig. 15D), and types 2 and 3 insert more medially along the scapular margin. Thus, theoretically, some patients who are evaluated may normally have a type 3 insertion without any associated instability. Others, however, believe that even if the type 3 capsule is not the result of traumatic dislocation, people with this type of anatomy are nonetheless more predisposed to developing the initial and subsequent dislocations. This is also because this type of capsular insertion (type 3) may be associated with underdeveloped middle and inferior glenohumeral ligaments, and therefore, in association with a prominent subscapularis bursa, a large recess is formed in front of the glenoid and behind the subscapularis, which offers no passive restraint to anterior dislocation. Thus, theoretically, one would expect a higher incidence of type 3 capsules in patients with instability. Others believe, however, that all type 3 capsules are not developmental but occur as a result of capsular stripping. If a type 3 capsule is found on an MRI scan, the presence of other findings seen in association with shoulder instability, such as Hill-Sachs lesions or labral tears, would help to determine the significance of this anatomy. Figure 15E,F is such a case in a 19-year-old wrestler.

Note the presence of the prominent middle glenohumeral ligament in this patient. The glenohumeral ligaments are actually thickenings of the anterior capsule and act to reinforce it (Fig. 15G). They course from the glenoid to the lesser tuberosity of the humerus and consist of the superior, middle, and inferior ligaments, although there are variations in the size and presence of these ligaments. In the most common variation of this anatomy, the opening of the subscapularis bursa is between the superior and middle ligaments.

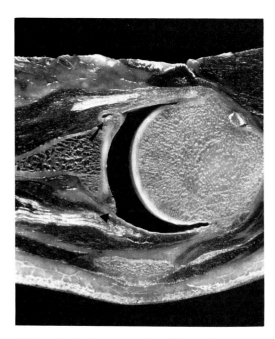

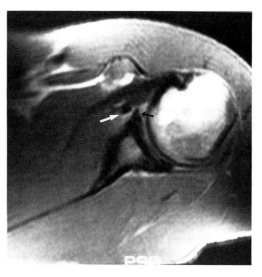

FIG. 15D. Cadaver section illustrating a type I capsular insertion. Note its insertion adjacent to the anterior labrum (*arrow*). Note that the posterior capsule always inserts directly into the labrum (*arrowhead*). (Reprinted with permission from Zlatkin MB, Bjorkengren AG, Gylys Morin V, Resnick D, and Sartoris DJ. Cross sectional imaging of the capsular mechanism of the glenohumeral joint. *AJR* 1988;150:151–158.)

FIG. 15E. Labral tear. Axial MR image (TR/TE 2,500/20 msec) depicts a small anterior labral tear (*black arrow*). A portion of a prominent middle glenohumeral ligament is seen (*white arrow*).

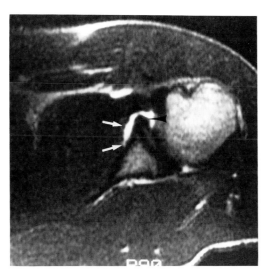

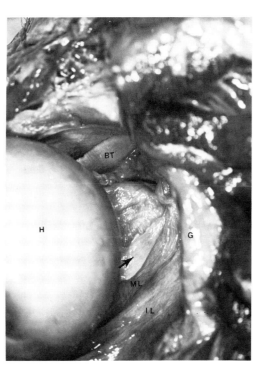

FIG. 15F. Type 3 capsule. Axial MR image slightly inferior in the same patient (TR/TE 2,500/80 msec) demonstrates a very medial capsular insertion (type 3) outlined by fluid (*arrows*) and a blunted, rounded configuration of the anterior labrum (*arrowhead*).

FIG. 15G. Cadaver specimen dissected from posterior, demonstrating some of the important structures and relationships of the anterior capsular mechanism. (BT, biceps tendon, long head; G, glenoid; H, humeral head; IL, inferior glenohumeral ligament; ML, middle glenohumeral ligament; arrow, opening into subscapularis bursa.) (Reprinted with permission from Zlatkin MB, Bjorkengren AG, Gylys Morin V, Resnick D, and Sartoris DJ. Cross sectional imaging of the capsular mechanism of the glenohumeral joint. *AJR* 1988;150:151–158.)

CASE 16

A 32-year-old former college baseball pitcher complains of locking and clicking in his shoulder. Based on his MR exam, what is the cause of his symptoms? (Fig. 16A. Axial gradient-echo MR, TR/TE 600/15 msec, flip angle 70°.)

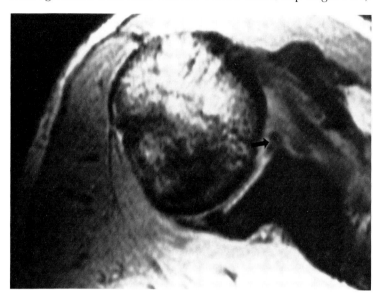

FIG. 16A.

Findings: The MR scan in this case reveals a linear region of increased signal in the anterior superior labrum, extending to the surface (arrow). No other abnormalities were found.

Diagnosis: Anterior labral tear.

Discussion: The shoulder is one of the most commonly injured joints during athletic activities. It may be injured as a result of throwing or of direct trauma. The lesions that occur as a result of recurrent subluxations and dislocations are discussed in Cases 14, 15, and 18. Lesions that have been described with forceful throwing such as in baseball include osteophytes, capsular and tendinous calcifications, and degenerative changes of the articular surface. More recently an additional soft tissue lesion has been described, which is a tear of the anterior superior labrum, as is present in this case. This lesion is thought to be unrelated to shoulder instability that occurs as a result of shoulder dislocations but is related to the throwing mechanism. In fact Andrews has related this injury to traction on the anterior superior labrum by the long head of the biceps tendon, which inserts there. Figure 16B is another example of a detached anterior superior labrum, also in a throwing athlete.

Anatomically the glenoid labrum is a rim of soft tissue that is attached to the bony glenoid, blending with the tendon of the long head of the biceps superiorly and anteriorly. It is fibrous in structure, hence its lack of signal on MR images. It is believed by some to be a fold of the articular capsule. Its function is to deepen the glenoid fossa, hence adding stability to the shoulder. One surface may be contiguous with the capsule, the other with the hyaline cartilage of the glenoid fossa. The anterior labrum is most commonly triangular, and the posterior labrum is more commonly rounded. Occasionally separation of the labrum from the hyaline articular cartilage of the glenoid may occur as a normal anatomic variation. Thus knowledge of the normal and variational anatomy is important in evaluating this structure.

In general labral tears are best evaluated on axial images. We have recently begun using gradient-echo images to evaluate the labrum and hyaline cartilage. Contrast in gradient-echo imaging is determined by a number of factors including TR, TE, and flip angle. The degree of T1 weighting can be controlled via the flip angle. At a given TR, images become progressively more T1-weighted as the flip angle is increased. As the flip angle is decreased the reverse is true. Therefore density-weighted gradient-echo scans can be obtained with TRs of 100 msec or greater and flip angles of approximately 10 or 15°. T2*-weighted sequences can be obtained with similar TR and flip angles as the density-weighted sequences by lengthening the TE to 30 to 60 msec. We have found that utilizing a relatively T1-weighted gradient-echo sequence with a relatively short TE (15 msec), a TR of 600, and a large flip angle (70–80°) gives good signal-to-noise ratio, makes the hyaline cartilage bright, and results in good contrast against the low signal intensity labrum. Labral tears are well appreciated with this technique (Fig. 16C,D), with signal intensity patterns similar to that described for the spin-echo sequences.

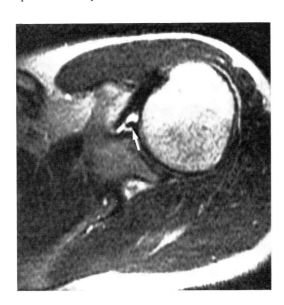

FIG. 16B. Detached labrum (axial MR image, TR/TE 2,500/80 msec). Detached superior labrum in a 44-year-old baseball coach (*arrow*).

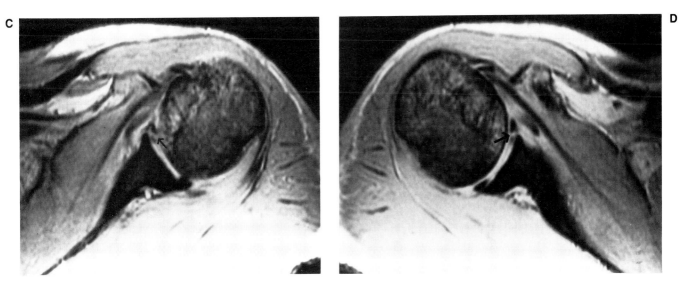

FIG. 16C,D. Labral tears evident with gradient-echo scanning (TR/TE 600/15 msec, flip angle = 70°). **C.** Complex anterior labral tear (*arrow*). **D.** Tear at the base of the labrum (*arrow*).

CASE 17

A 32-year-old man who had a history of prior dislocations now complains of a grating and locking sensation of his joint. An MRI was performed. The arrows indicate the cause of the patient's problem. To what do the arrows point? (Fig. 17A. TR/TE 800/20 msec.)

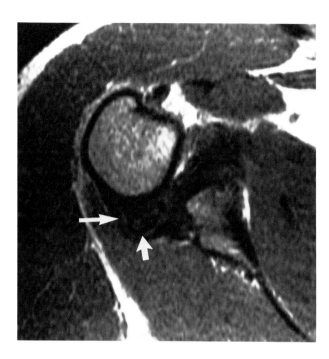

FIG. 17A.

Findings: This is an axial image at the inferior glenoid. The arrows are pointing to an oval-shaped structure that has a signal intensity similar to or slightly lower than the marrow of the glenoid and is surrounded by a low signal intensity rim. It is located just posterior and lateral to the posterior glenoid rim. The posterior glenoid also appears somewhat low in signal intensity.

Diagnosis: Loose body in the posterior inferior glenoid.

Discussion: Loose bodies may occur in the joint in up to 31% of patients with recurrent dislocations. They are usually fragments of bone and cartilage that may be sheared off the glenoid. Other causes of loose bodies would include osteoarthritis and osteochondral fractures from other causes and synovial osteochondromatosis. Loose bodies are nourished by synovial fluid and may grow, calcify, or ossify. Clinically they may cause recurrent effusions, locking, or a grating sensation, as well as a decreased range of motion. They are well identified with the use of plain radiographs, only if they are calcified. CT air arthrography is employed to confirm intra-articular location and to identify noncalcified loose bodies or loose bodies in locations difficult to visualize on plain radiographs.

Magnetic resonance imaging is *not* the procedure of choice to search for loose bodies, although they may be incidentally encountered. Densely calcified loose bodies on MRI may appear as low signal intensity structures, in which case they may at times be difficult to see if small. When ossified they may be of high to intermediate signal with a low signal intensity rim because of the presence of mature marrow elements within. The plain radiographs in this case are shown in Fig. 17B,C. It was thought that the loose body arose from the adjacent glenoid margin because of the low signal intensity seen in this region on the MRI; however, no bony defect was present on the plain radiographs and at surgery, although the loose body was removed, no bony defect was found. It may be that this appearance in the glenoid was as a result of partial volume averaging.

B

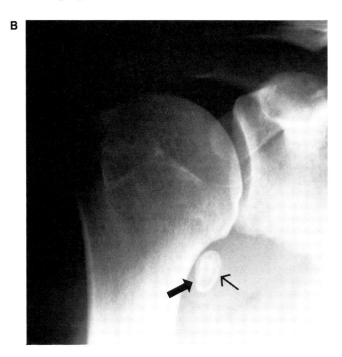

FIG. 17B,C. Loose bodies. Antero-posterior **(B)** and axial **(C)** radiographs demonstrating the loose body in this case (*arrows*).

C

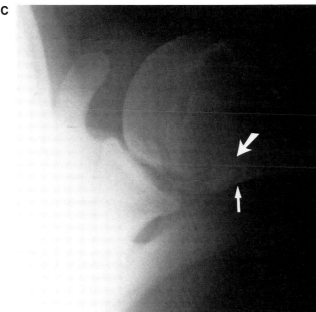

CASE 18

A 43-year-old man with a convulsive disorder is referred for an MRI examination because of a complaint of a subluxing shoulder. The surgeon thinks he may have instability in a posterior direction, but the examination is equivocal. The patient is a poor historian but remembers something about his shoulder "coming out of its socket" years ago. What are the MRI findings? Based on the MRI, can you identify the direction of the instability? (Fig. 18A. Axial MR, TR/TE 800/20 msec. Fig. 18B. Sagittal MR, TR/TE 800/20 msec.)

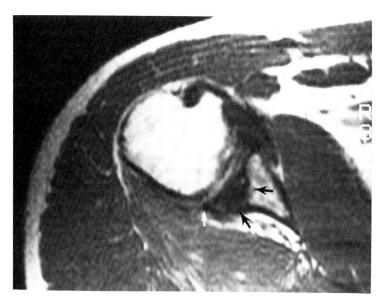

FIG. 18A.

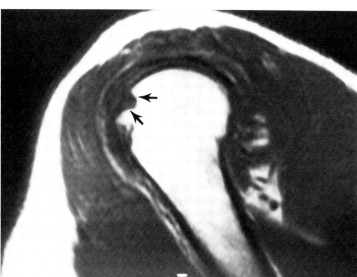

FIG. 18B.

Findings: The axial MRI (Fig. 18A) at the mid glenoid level demonstrates diffuse decreased signal intensity in the mid and posterior aspect of the glenoid (black arrows). The posterior labrum is blunted (white arrow). The sagittal scan (Fig. 18B) reveals an anteriorly and medially located bony defect consistent with a "reverse Hill-Sachs" lesion (arrows).

Diagnosis: Posterior instability, likely posttraumatic, from prior posterior dislocation.

Discussion: Posterior dislocations of the glenohumeral joint are rare. They represent approximately 2 to 4% of all dislocations of the shoulder. They are commonly associated with electric shock or convulsive seizures, and nearly all bilateral posterior dislocations are caused by epileptic convulsions.

The posterior capsular mechanism consists of the posterior capsule, the synovial membrane, the glenoid labrum and periosteum, and the posterior and superior rotator cuff muscles and tendons.

Pathologic findings associated with the first and subsequent posterior dislocations are similar to those for anterior instability and include posterior labral and capsular detachments and tears, as well as laxity of the posterior capsule, in patients with recurrence (Fig. 18C). An impaction type defect may be present on the anteromedial aspect of the humeral head, which is known as a "reverse Hill-Sachs defect" and is present in this patient. Fractures off the posterior glenoid margin may also occur. MRI can show evidence of these lesions and thus can indicate the direction of the instability. This is important, as it is clearly necessary for the surgeon to know the type of recurrent dislocation the patient has before deciding on what type of surgery, if any, to perform. Also knowledge of the lesions that are present may also enable the surgeon to choose the appropriate type of surgical method.

We and others have observed that in addition to the capsular and labral abnormalities that are seen in patients with shoulder instability, MRI may depict changes in marrow signal, as in this patient, in the bony glenoid. This has been particularly common in those patients with posterior instability. The exact etiology of this is not clear but may relate to bony impaction or contusion of the bony glenoid by the humeral head with the episodes of recurrent subluxation or dislocation.

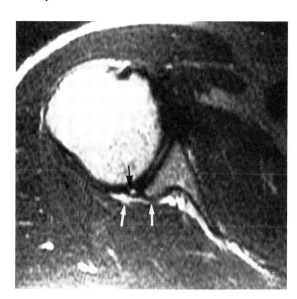

FIG. 18C. Posterior instability. Axial MR image in another patient (TR/TE 2,500/70 msec) demonstrating diffuse increased signal in the posterior labrum consistent with a tear (*black arrow*). The posterior capsule, which normally inserts directly adjacent to the posterior labrum, is inserting more medially in this case (*white arrows*).

CASE 19

A 77-year-old woman was admitted to the hospital because of a history of recent "falls" and a possible stroke. When the admitting intern saw her, she also complained of shoulder pain and immobility but did not recall any recent trauma to that region. A plain radiograph of her shoulder was done. Orthopedics was consulted and an MRI scan was requested. What is your diagnosis? Why did the surgeon request the MRI exam? Why did closed methods of therapy fail in this case? (Fig. 19A. Anteroposterior radiograph of the shoulder. Fig. 19B,C. Axial MR scans, TR/TE 600/20 msec [B] and 2,500/80 msec [C].)

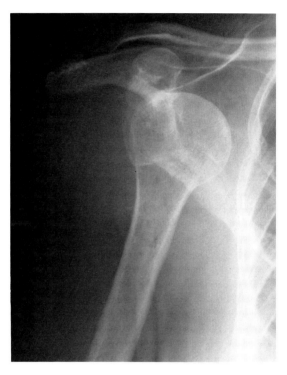

FIG. 19A.

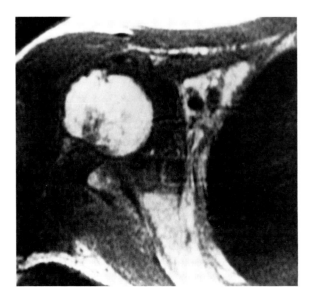

FIG. 19B.

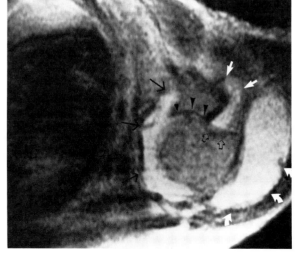

FIG. 19C.

Findings: The plain radiograph (Fig. 19A) reveals that the shoulder is anteriorly dislocated. The MRI scans (Fig. 19B,C) confirm the anterior displacement of the shoulder. In addition a number of other features are present. The anterior capsule is markedly redundant and bowed anteriorly by fluid (black arrows). Similar changes are present in the posterior capsule (white arrows). The glenoid labrum is completely effaced. The anterior glenoid margin is markedly resorbed, and the humeral head is resting in this area of resorption, which has formed a "false glenoid fossa" (arrowheads). A small Hill-Sachs deformity is present (open arrows). There is a massive amount of fluid present in the subdeltoid bursa (curved arrows), caused by large rotator cuff tear.

Diagnosis: Chronic anterior dislocation.

Discussion: Chronic dislocations are primarily found in the elderly. There is no defined time limit when a dislocation is considered chronic; however, Rowe suggested that 3 weeks is an acceptable period to define chronic dislocation. Chronic anterior dislocations are more common than posterior. The patient's complaint in relation to the time of injury and subsequent disability may be confusing and quite inadequate. The usual complaint is loss of motion and pain in the shoulder.

The usual soft tissue and osseous pathology of traumatic dislocation is present in these patients; however, reparative processes and degenerative changes in both the osseous and soft tissue elements of the joint are superimposed, thereby compounding the magnitude of the pathology.

The usual modes of treatment include (a) leaving the shoulder dislocated and attempting to regain as much motion as possible with physical therapy; (b) closed reduction; (c) open reduction; (d) inserting a shoulder prosthesis; and (e) shoulder fusion. The prognosis is often not good with any of these methods.

In this case, because of the patient's lack of history of trauma and her advanced age, the surgeon suspected the dislocation to be chronic. Because of her other medical problems, she was considered a poor surgical candidate, and the MRI was done before any attempted closed reduction to better define the soft tissue and bony anatomy. In spite of the gross pathologic changes evident on MRI, a closed reduction was attempted, which failed. One would have predicted this to happen because of the complete disruption of the capsular mechanism of the shoulder that is present, as well as the evident disruption of the rotator cuff. In addition the formation of the "false glenoid fossa," which is perhaps analogous to the false acetabulum that forms in chronic hip dislocation, likely also played a significant role. The patient is being considered for a shoulder prosthesis or fusion when her condition permits.

CASE 20

A 22-year-old man consulted the orthopedic surgeon because his shoulder has been "popping out" since childhood. He does not recall any significant trauma to the area. He wants to become a professional bowler and wants the surgeon to repair his shoulder. An MRI scan was performed. Can you determine the direction of instability in this case? (Fig. 20A,B. Axial MR, TR/TE 800/20 msec [A] and 2,500/80 msec [B].)

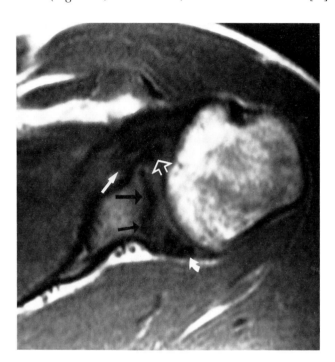

FIG. 20A.

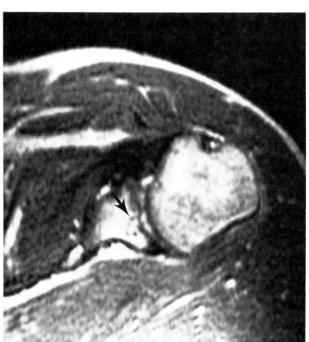

FIG. 20B.

Findings: The anterior labrum is torn (open arrow). The posterior labrum is blunted (curved arrow). There is grossly abnormal signal in the mid and posterior glenoid (black arrows). There is a type 2 anterior capsule insertion (white arrow). No Hill-Sachs lesion was evident on higher cuts. The T2-weighted image (Fig. 20B) demonstrates increased signal intensity in the abnormal marrow of the mid and posterior glenoid (arrow).

Diagnosis: Anterior and posterior (multidirectional) instability.

Discussion: As previously discussed, it is extremely relevant for the surgeon to know the direction of the patient's instability. It is particularly important in patients who are unstable in multiple directions, as they represent a separate subgroup in whom the history and physical exam may be confusing. In addition there is often a psychologic component to the problem in these patients, and surgery may not be helpful.

In patients with anterior and posterior or multidirectional instability, there often is a history of trauma, although in some cases such as in this patient there is not. The basic causative factor is often an abnormality of the soft tissue elements that renders the joint hypermobile. These patients may demonstrate many or all the pathologic lesions recorded for recurrent anterior and posterior dislocations.

In this patient the feature pointing to anterior instability is the fragmented torn anterior labrum. The features pointing to posterior instability are the blunted posterior labrum and the changes in the marrow of the glenoid, which we have observed particularly in patients with posterior instability. The almost necrotic appearance to the glenoid suggests that this is a long-standing process.

Therapy in these patients is usually initially conservative, with an intensive program of physiotherapy, because of the high incidence of failures following operative treatment. If conservative therapy fails and psychologic disturbances and voluntary dislocations are ruled out, some combination of the procedures performed for anterior and posterior dislocations is done.

CASE 21

A 26-year-old man with a renal transplant on high dose steroids complains of increasing pain in his shoulder. Plain radiographs were normal. What is your diagnosis? (Fig. 21A,B. Coronal MR, TR/TE 800/20 msec [A] and 2,500/80 msec [B].)

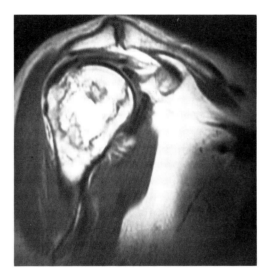 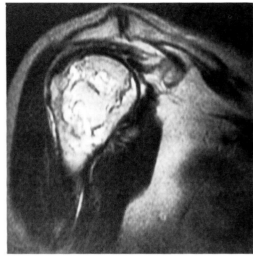

FIG. 21A.   **FIG. 21B.**

Findings: Geographic areas of low signal intensity are seen in the humeral head extending to the cortex. There is increased signal intensity at the margin of these areas on the T2-weighted images.

Diagnosis: Steroid-induced avascular necrosis (AVN) of the humeral head.

Discussion: Avascular necrosis results from a significant decrease or loss of the blood supply to the affected region. AVN is not uncommon in patients receiving high doses of corticosteroids for prophylactic treatment of renal transplant rejection or other processes such as collagen vascular disease. The femoral and humeral head are the two most common sites of disease, although involvement of the humeral head is one-third as common as the femoral head. Ischemic necrosis of the humeral head also often follows trauma, particularly fractures of the anatomic neck. The vessels that supply the humeral head pierce the bony cortex just distal to the anatomic neck, and fractures proximal to this level may result in ischemic necrosis of the articular segment of the humeral head. Some other causes of AVN include sickle cell disease or other hemoglobinopathies, alcoholism, caisson disease, collagen vascular disease, and pancreatitis. Many cases may be idiopathic.

Methods of diagnosing and staging AVN have included plain radiographs, tomography, radionuclide studies, and CT. The early diagnosis of AVN in the past has been difficult, however, in spite of the use of radionuclide studies and CT. Plain radiographic findings of AVN include areas of increased density or lucency in the humeral head or loss of the normal trabecular pattern of bone. A "crescent" sign of bony lucency may be noted in the subcortical bone. Eventually there may be cortical disruption and collapse of the humeral head. These findings, however, usually occur late in the course of the disease. Other modalities such as tomography and CT are useful in assessing the extent and severity of disease but are also not that helpful in the early stages of disease.

It is well accepted that MRI is more sensitive than radionuclide imaging in the detection of osteonecrosis in other joints, most particularly in the hip. Moreover it has the potential to yield a somewhat more specific diagnosis than radionuclide studies in which the presence of increased uptake in the affected joint can be due to a variety of causes, whereas the pattern of AVN on MRI scans is relatively characteristic. Although there are no specific studies assessing the efficacy of MRI in the diagnosis of AVN in the shoulder, it is likely that MRI performs as well in evaluating this joint. The MRI findings in osteonecrosis of the shoulder in our experience are similar to those in other joints and consist of decreased signal intensity in the humeral head at the articular surface caused by disruption of signal intensity from fatty marrow on T1-weighted images, usually with increased signal intensity at the reactive interface at the margin of these low signal regions on T2-weighted images. In general T1- and T2-weighted spin-echo techniques are sufficient to diagnose and characterize AVN. Other helpful pulse sequences include use of chemical shift imaging (echo offset) or the short TI inversion recovery (STIR) technique to accentuate areas of marrow edema. The use of surface coils rather than the body coil (Fig. 21C) and gradient-echo techniques can also help evaluate any degeneration of the overlying articular cartilage or collapse of the humeral head that may be present. The latter application may obviate the need for tomography, which, although generally clinically useful, subjects the patient to a significant degree of radiation.

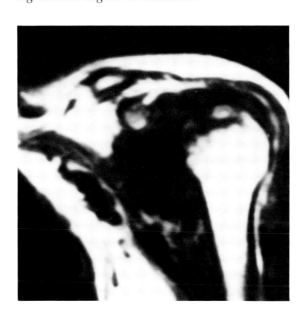

FIG. 21C. Osteonecrosis. Low resolution MR image obtained in the body coil (TR/TE 600/20 msec) that shows characteristic findings of but little detail of the articular cartilage or contour of the humeral head.

CASE 22

A 47-year-old woman with long-standing rheumatoid arthritis in multiple joints now complains of increasing pain in the left shoulder. What are the diagnostic considerations? Does the MRI aid in arriving at a correct diagnosis? (Fig. 22A. Coronal oblique MR, TR/TE 2,500/80 msec. Fig. 22B. Sagittal oblique MR, TR/TE 800/20 msec.)

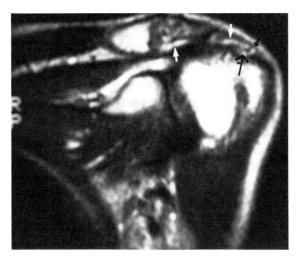

FIG. 22A.

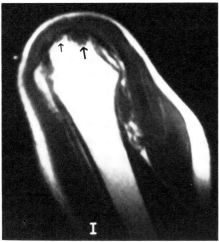

FIG. 22B.

Findings: The coronal oblique MRI scan in this case (Fig. 22A) reveals evidence of a tear in the rotator cuff involving the supraspinatus tendon, identified by a fragmented, discontinuous distal supraspinatus tendon, replaced by high signal intensity fluid (black arrows) and fluid in the subacromial-subdeltoid bursa (white arrows). Fluid is seen in the acromioclavicular joint, and there is erosion of the acromion. The sagittal oblique image (Fig. 22B) demonstrates irregular decreased signal intensity in the marrow of the humeral head caused by erosive changes (arrows).

Diagnosis: Rheumatoid arthritis of the shoulder, associated with a rotator cuff tear.

Discussion: Many forms of inflammatory and degenerative joint processes involve the shoulder. These include rheumatoid arthritis, ankylosing spondylitis and other seronegative spondyloarthropathies, degenerative arthritis, and the crystal deposition diseases.

Clinical symptoms related to the shoulder are not uncommon in rheumatoid arthritis, particularly in long-standing disease. Osseous erosions occur predominantly on the humeral side of the joint, as was the case in this patient (Fig. 22C). Cystic and erosive changes on the superolateral aspect of the humeral head adjacent to the greater tuberosity are characteristic.

Tears of the rotator cuff are also common in long-standing rheumatoid arthritis, secondary to destruction of the adjacent tendons by the inflamed synovium. The acromioclavicular joint is also often involved in rheumatoid arthritis. With advanced disease, extensive erosion of the distal clavicle results in widening of the joint. This case was somewhat atypical in that the acromion was more involved than the clavicle.

Although conventional radiography is still the most convenient and commonly used method of evaluating patients with rheumatoid arthritis and other synovial inflammatory processes, MRI is useful in a number of ways. In the early stages of the disease it may reveal the soft tissue edema and effusions. With progression of disease it can demonstrate rotator cuff tears and synovial cysts (Fig. 22D) that result from the disruption of the tendons and capsule, respectively, in a noninvasive manner. In addition, although loss of articular cartilage and osseous erosions may be visualized with conventional radiography, they may be visualized at an earlier state and their extent may be better assessed with MRI.

Another complication of rheumatoid arthritis of the shoulder is rupture of the long head of the biceps tendon. This can present with acute pain or may be more chronic and present as a mass in the upper arm, when the biceps retracts distally. The entity is discussed in Case 26.

Crystal deposition disease also affects the shoulder and is not an uncommon cause of shoulder pain. In fact the shoulder is the most common site of involvement by calcium hydroxyapatite crystal deposition disease. Crystal deposition in this disease most commonly occurs in the tendinous and bursal structures of the shoulder. It particularly involves the supraspinatus tendon. These crystals incite a synovitis, tendinitis, or bursitis, and periarticular inflammation. The nodular calcific deposits in this disease are usually best evaluated on plain radiographs. On MRI they appear as nodular regions of signal void within the affected tendon, that when small may be difficult to separate from the tendon itself (Fig. 22E). The MRI can, however, demonstrate well the associated inflammatory changes in the tendons and surrounding bursal structures.

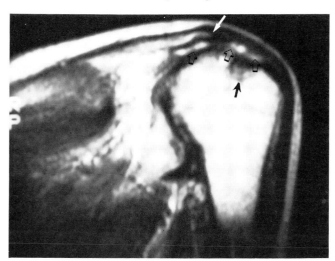

FIG. 22C. Rheumatoid arthritis. Coronal oblique image further posteriorly in the same patient (TR/TE 2,500/80 msec), demonstrating erosion of the humeral head (*black arrow*) and acromion (*white arrow*). The infraspinatus tendon is also seen to be involved in the cuff tear as it is completely replaced by high signal intensity fluid (*open arrows*). There is also a significant amount of muscle atrophy present.

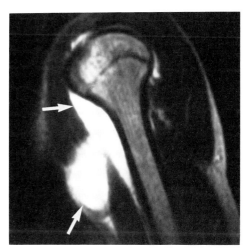

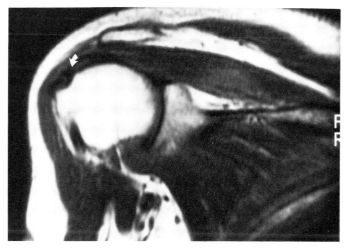

FIG. 22D. Juvenile rheumatoid arthritis. Sagittal MR (TR/TE 2,500/80 msec) illustrating joint fluid dissecting along the humeral shaft and a synovial cyst (*arrows*) in a patient with juvenile rheumatoid arthritis.

FIG. 22E. Hydroxyapatite disease. Coronal oblique MR (TR/TE 800/20 msec) depicting a nodular focus of signal void in the supraspinatus tendon (*arrow*). The plain radiograph (not shown) confirmed a small calcific focus consistent with hydroxyapatite deposition.

CASE 23

A 19-year-old man with an immune disorder was seen because of chronic low grade shoulder pain for the past few months. Plain radiographs showed some osteopenia, minimal loss of joint space, a vague cystic erosion in the humeral head, and some soft tissue prominence. Diagnostic considerations included infection, pigmented villonodular synovitis (PVNS), or possibly a monoarticular arthritis. An MRI scan was performed. What are the findings? What is the value of the MRI in this case? (Fig. 23A. Axial MR, TR/TE 800/20 msec. Fig. 23B. Coronal oblique MR, TR/TE 2,500/80 msec.)

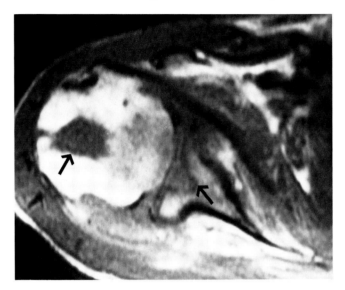

FIG. 23A.

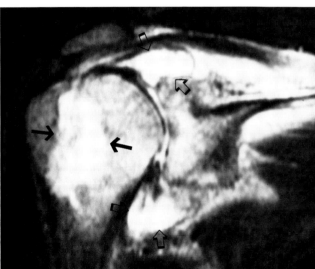

FIG. 23B.

Findings: Large cystic erosions (arrows) of low signal intensity on the short TR/TE images and increased signal intensity on the long TR/TE images are present on both sides of the joint. A huge joint effusion is present (open arrow). There is some loss of joint space. There is considerable muscle atrophy present. No nodular areas of low signal intensity consistent with hemosiderin are seen.

Diagnosis: The patient had a joint aspiration performed, which was unrevealing, and all cultures were negative. He was taken to surgery and open biopsy was performed. Cultures grew out *Mycobacterium tuberculosis*.

Discussion: The history and plain radiographs in this patient suggested some sort of low grade or indolent infection (Fig. 23C). An acute aggressive type of pyogenic infection seemed less likely in view of the protracted course. The MRI was performed to get a better idea of the extent of the disease, to better assess the soft tissue component, and to rule out the presence of hemosiderin, which would point toward a diagnosis of PVNS. Gradient-echo scans would have also been helpful in establishing the presence of hemosiderin, because of increased sensitivity to magnetic susceptibility effects. These could not be obtained in this patient as the patient became ill in the scanner during the study. T2-weighted images with longer echo times are useful in this regard as well. One other advantage of the gradient-echo scans would have been improved evaluation of the articular cartilage.

Hematogenous dissemination of tubercle bacilli is the etiology for most cases of skeletal tuberculosis. In some cases there may be lymphatic spread of tuberculosis. The clinical course is usually insidious and protracted. Tuberculosis most commonly involves the larger joints such as the knee and hip, although any joint including the shoulder may be involved. Delay in diagnosis is frequent, and the joint disease may persist with chronic pain and only minimal signs of inflammation. Correct diagnosis usually requires an awareness of the possibility of mycobacterial infection and the use of synovial fluid and tissue for culture and histologic studies. Tuberculosis is decreasing in prevalence in the Western world, although recently there has been a resurgence of *Mycobacterium tuberculosis* and its variants, particularly in association with immune-related disorders.

A triad of radiographic findings is characteristic of tuberculous arthritis (Phemister triad). This includes juxta-articular osteoporosis, peripherally located osseous erosions, and gradual narrowing of the interosseous space.

A monoarticular process must be considered an infection until proven otherwise. It may be difficult to define the nature of the infective agent; however, the slow progression of the disease as in this case is suggestive of tuberculosis. As discussed previously, other monoarticular processes such as PVNS and synovial osteochondromatosis may simulate tuberculosis, but in PVNS the joint space is usually preserved, there is absence of osteoporosis, and as mentioned previously, nodular masses of hemosiderin should be seen. Hemosiderin itself is also not characteristic of PVNS, because it has also been observed in other disorders such as hemophilia and rheumatoid arthritis. In synovial osteochondromatosis, calcified and ossified intra-articular loose bodies will be present.

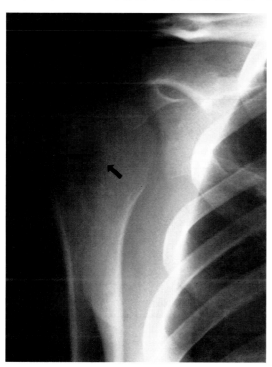

FIG. 23C. Tuberculosis. Plain radiograph of the patient in the test case 2 weeks after initial presentation, demonstrating osteopenia, loss of the cortical margin of the humeral head and glenoid, mild joint space narrowing, and a cystic lesion in the humeral head (*arrow*).

CASE 24

A 46-year-old college professor was hit by a car. After recovering from his other injuries he still complained of persistent shoulder pain. Plain radiographs were unrevealing. He was thought to have a rotator cuff tear, so an MRI was performed. What abnormalities are indicated by the arrows in Fig. 24A and B? Is his rotator cuff intact? (Coronal oblique MR [Fig. 24A], sagittal oblique MR [Fig. 24B], TR/TE 800/20 msec.)

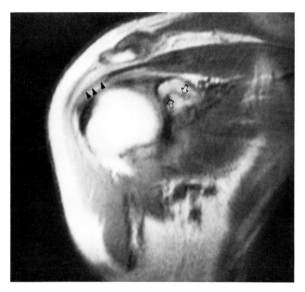

FIG. 24A.

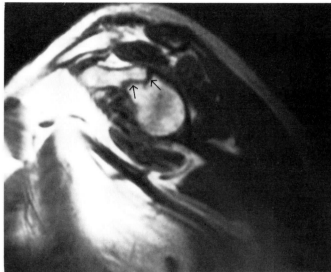

FIG. 24B.

Findings: The open arrows in Fig. 24A illustrate a linear band of low signal intensity in the glenoid extending to the articular surface. The sagittal oblique image (Fig. 24B) reveals a similar band of low signal intensity at the base of the coracoid extending out posteriorly (straight arrows). Diffuse increased signal is seen in the substance of the distal supraspinatus tendon, but no tear is present (arrowheads).

Diagnosis: Occult fracture of the scapula extending into the glenohumeral joint. Rotator cuff tendinitis.

Discussion: Whereas most fractures are well evaluated with plain radiographs and, when necessary, scintigraphy, tomography, and CT, MRI may be useful in certain instances. Because of its multiplanar tomographic capabilities it can assess fractures in locations that are otherwise difficult to visualize, such as in this case involving the scapula. Of particular importance in scapular fractures is assessing whether the glenohumeral joint is involved, which was well demonstrated in this case. The sensitivity of MRI to changes in marrow signal intensity also makes it a valuable technique for bony injuries such as bone contusions, stress fractures, osteochondral fractures, or even complete fractures that are not well seen on plain radiographs. This has been recently well illustrated by Mink and co-workers in the knee. Figure 24C illustrates a fracture of the greater tuberosity, which was not evident on initial plain radiographs.

The appearance of fractures on MRI usually consist of linear regions of low signal that represent the fracture line, which may be outlined by less well-defined regions of intermediate signal on T1-weighted images that represent marrow edema. These intermediate regions will usually get bright on T2-weighted images.

Other posttraumatic causes of shoulder pain and disability can also be evaluated with MRI. For example, we have imaged patients with shoulder pain thought to be caused by rotator cuff disease, who have had recent trauma and in whom the rotator cuff is found to be normal, but other soft tissue causes to explain the patient's pain are found. This includes such entities as edema and hematomas involving the rotator cuff muscles but also the surrounding musculature such as the deltoid and trapezius (Fig. 24D). These hematomas have had high signal intensity on both T1- and T2-weighted images caused by the presence of methemoglobin.

Dislocation of the acromioclavicular joint commonly occurs. In type 1 injuries the acromioclavicular ligaments are strained. In type 2 injuries these ligaments are torn but the coracoclavicular ligaments are intact. In type 3 injuries both ligaments are disrupted. We have imaged a number of patients with dislocations of their acromioclavicular joints, in which the degree and severity of the separation could be seen, including visualization of the torn acromioclavicular ligaments (Fig. 24E). In general this entity is well evaluated with plain radiographs and clinical exam. Patients with this problem are not uncommonly referred, however, to assess the integrity of the rotator cuff or labrum, particularly when surgery is being considered.

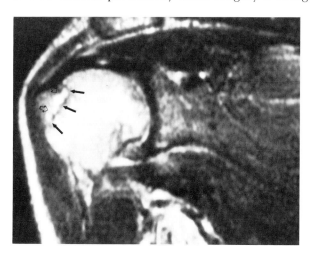

FIG. 24C. Fracture of the greater tuberosity (coronal oblique MR scan, TR/TE 2,500/70 msec). A linear well-defined region of low signal intensity is seen (*black arrows*) with a more diffuse region of increased signal intensity at its margin (*open arrows*).

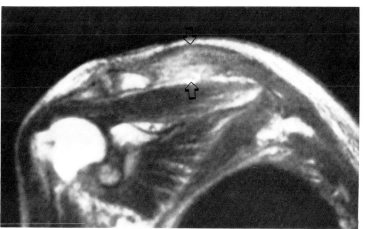

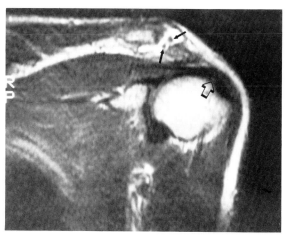

FIG. 24D. Muscle edema and hematoma (coronal oblique MR, TR/TE 850/80 msec). Increased signal intensity and swelling of the trapezius muscle (*arrows*) are seen. The first echo in this case demonstrated less extensive areas of high signal intensity in a similar location. The findings were felt to be due to muscle hemorrhage and edema. (Reprinted with permission from Zlatkin MB, Reicher RA, and Kellerhouse LE. The painful shoulder: MR imaging of the glenohumeral joint. *J Comput Assist Tomogr* 1988;12:995–1001.)

FIG. 24E. Acromioclavicular joint disruption (coronal oblique image, TR/TE 2,500/80 msec). There is a separation of the acromioclavicular joint. The disrupted acromioclavicular ligaments are seen (*black arrows*), and there is fluid in the joint. The supraspinatus tendon is intact (*open arrow*).

CASE 25

A 56-year-old left-handed male pediatrician complained of right shoulder pain of 9 months duration. There was no history of antecedent direct trauma or previous shoulder problems. The patient was an avid handball player who had played singles handball daily for 30 years. He routinely collided with the right sidewall in an attempt to run around his nondominant hand to return volleys. Plain radiographs revealed a vague area of lucency in the posterior glenoid. An MRI was performed. What are the MRI findings? (Fig. 25A,B. Axial MR, TR/TE 600/20 msec [A] and 2,500/80 msec [B].)

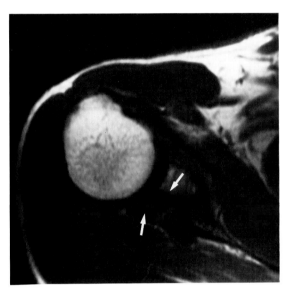 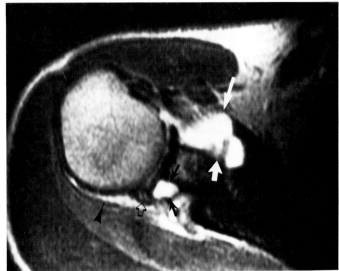

FIG. 25A.   **FIG. 25B.**

Findings: There is a relatively well-defined region of decreased signal intensity on the short TR/TE sequence (white arrows) which increases in signal intensity on the long TR/TE sequences, likely a posttraumatic cyst (black arrows). The posterior labrum is irregular and diffusely increased in signal intensity (open arrow). There is a large amount of fluid in the joint outlining the posterior capsule (arrowhead) and a prominent subscapularis bursa (white arrows).

Diagnosis: Posttraumatic changes in the posterior glenoid and labrum in a handball player.

Discussion: Sports-related injury to the shoulder occurs most commonly in the overhead use sports. The most common injuries result in problems related to the rotator cuff or result in shoulder instability. Acromioclavicular joint abnormalities may also occur.

Throwing subjects the shoulder to tremendous forces. These can result in rotator cuff tears, avulsion injuries of the infraglenoid tubercle, tears of the glenoid labrum, tears of the long head of biceps tendon, and pseudocystic changes in the humeral head, particularly in baseball pitchers. Case 16 discusses anterior superior labral tears in these athletes.

Swimmers are also subject to shoulder problems, usually caused by impingement syndrome. The freestyle and breast strokes are most commonly implicated. Recurrent anterior subluxation may occur in backstrokers doing a reverse flip turn.

In tennis shoulder, the muscles of the dominant shoulder are hypertrophied, the shoulder is drooped, and there is an apparent scoliosis. The shoulder depression may result in impingement on the rotator cuff.

In contact sports such as football, hockey, and soccer, injuries are more likely the result of direct trauma and include fractures of the humerus and clavicle and dislocations of the shoulder, acromioclavicular joint, and sternoclavicular joint. Stress fractures of the coracoid may occur in trapshooters, and resorption of the distal end of the clavicle may occur in weight lifters.

There is little in the literature describing specific injuries that occur in the shoulder related to handball or similar type sports, except for one case report of osteolysis in the distal end of the clavicle in a handball and softball player. In the case illustrated above, the findings are rather unique and likely particular to this individual. The series of findings depicted on the MRI scan in this case were likely related to chronic repetitive trauma. Normally in four-wall handball, volleys are returned with either hand. This patient was left-handed and routinely tried to run around his right hand to return the volleys with his left. When doing this he would repeatedly collide against the wall with the posterior aspect of his shoulder at high speed. This likely resulted in the posttraumatic glenoid cyst and likely also the degenerated torn posterior labrum illustrated on the MRI. The MRI nicely demonstrated these findings and helped to explain the pathophysiology of the injury.

Also note the excellent demonstration of the subscapularis bursa in this case (Fig. 25C). This synovial bursa is the most common bursa of the shoulder capsular mechanism that is visualized on MRI. It extends through an anterior opening in the capsule that exists between the superior and middle glenohumeral ligaments. It extends medially toward the inferior surface of the coracoid process and along the superior tendinous border of the subscapularis muscle. It may also extend over the subscapularis tendon, often in a saddlebag configuration, which is thought to provide a gliding mechanism for this tendon. The size of the bursa tends to vary among different patients. In patients with anterior instability it may be particularly large. There were no clinical findings of anterior instability in this patient.

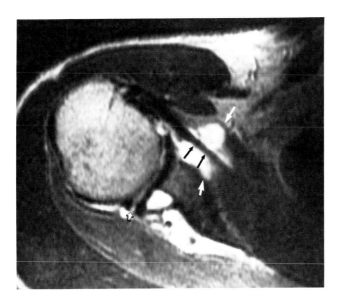

FIG. 25C. Posttraumatic glenoid cyst. Axial MR scan of the same patient, again demonstrating the large glenoid cyst. This image illustrates the extension of the subscapularis bursa (*white arrows*) in front and behind the subscapularis tendon (*black arrows*) in a saddlebag configuration. The posterior labrum is torn at its base on this scan (*open arrow*).

CASE 26

A 82-year-old woman who has a history of arthritis presented with arm weakness and pain. Her granddaughter noticed a "mass" in her mid arm. She was sent for an MRI to rule out tumor. She had plain radiographs at an outside hospital, which she unfortunately did not bring with her. What is the black arrow pointing to? What is your diagnosis? (Fig. 26A. Axial MRI at the superior glenoid level, TR/TE 2,500/80 msec. Fig. 26B. Axial MRI at the mid arm, TR/TE 2,500/80 msec.)

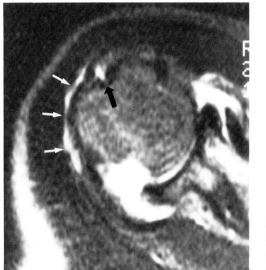

FIG. 26A.

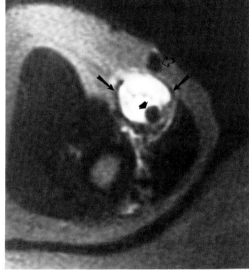

FIG. 26B.

Findings: The black arrow in Fig. 26A is pointing out the bicipital groove, which is filled with fluid. The biceps tendon, which is normally a round area of signal void within the groove, is not seen. Also noted is a large joint effusion outlining the capsule, as well as joint space narrowing. Fluid is seen in the subdeltoid bursa (white arrows) caused by a rotator cuff tear better seen on other images. A defect in the posterior humeral head is also present but of unclear significance. On the axial image of the mid arm (Fig. 26B) a fluid-filled sac is seen in the region of the biceps muscle (long arrows). Within this there is a round area of signal void located posteriorly (short arrow). Note was also made of another nodular focus of signal void located in the subcutaneous fat (open arrow). No tumor mass was seen.

Diagnosis: Biceps tendon rupture.

Discussion: Ruptures of the biceps tendon usually occur in older patients when varying degrees of degeneration are present in the tendon. Although this injury may occur after forceful flexion of the arm, it may also occur spontaneously or from a minor incident (particularly when the tendon is degenerated). Microtrauma to the tendon from years of impingement leads to a weakened degenerated tendon. The tendon actually may become thickened with age, and because the space in the bicipital groove is limited, this may lead to shredding of the tendon. The weakest portion of the tendon is the segment lying just distal to its exit from the joint cavity. Most ruptures occur here. Biceps tendon tears may also occur in association with anterior tears of the rotator cuff, and particularly in middle-aged and older patients who present with such a problem, evaluation of the cuff is indicated. Rupture of the biceps tendon is also associated with rheumatoid arthritis. This is due to synovitis involving the biceps tendon sheath, which eventually weakens the tendon to the point of rupture. After a tear, the intracapsular portion of the tendon lies free in the joint cavity, while the extraarticular portion is pulled distally.

Ruptures of the biceps tendon are easy to recognize clinically on the basis of distal retraction of the biceps muscle and do not require imaging for diagnosis. Arthrography, if performed, may not be that reliable, because even in the normal shoulder arthrogram, visualization of the tendon sheath and tendon of the long head of biceps is not constant. When the diagnosis is in doubt, however, arthrography may help by demonstrating distortion of the synovial sheath with an absent tendon. Evaluation of the biceps tendon is more accurate with ultrasound, as the tendon itself can be directly visualized.

Surgical repair of the biceps tendon is recommended in younger patients, although this is an infrequent occurrence in this age group. Older patients are usually satisfied with the cosmetic deformity of the proximal arm and may accept the decreased force that can be generated with flexion. Open repair usually produces a long scar, and it may be impossible to restore the underlying anatomy because of the degenerated nature of the tendon.

Bicipital tendinitis is also thought to be a common cause of shoulder pain. The long head of biceps is covered by synovium in a separate synovial sheath that communicates with the joint capsule. It may become inflamed as part of other synovial processes such as rheumatoid arthritis, because of trauma or overuse, or as part of the impingement syndrome. Treatment usually involves rest and analgesia and occasionally steroid injections. Surgical therapy is reserved for very recalcitrant cases.

There is not much data available on the efficacy of MRI in the evaluation of biceps tendon ruptures, dislocations, and other abnormalities of the tendon. Because the biceps tendon can be directly visualized, MRI (Fig. 26C) should be as efficacious or better than ultrasound for these problems, although this needs to be studied. Fluid is very commonly observed in the biceps tendon sheath in patients referred for shoulder pain, without any evident abnormalities of the tendon itself. It may be due to tenosynovitis, however, as the tendon sheath communicates with the joint, it may fill with fluid with a joint effusion of any cause; therefore it is often a nonspecific finding. We have not observed any cases of dislocation.

Magnetic resonance imaging did demonstrate the abnormalities associated with rupture quite well in this case (see also Fig. 26D). Unfortunately, in spite of the MRI the surgeon believed the patient had a mass lesion and resected the fluid-filled sac. At pathology a synovium-lined sac containing a tendon-like structure was found. The nodular area of signal void in the subcutaneous tissue was just a focus of calcification of unclear etiology (Fig. 26E).

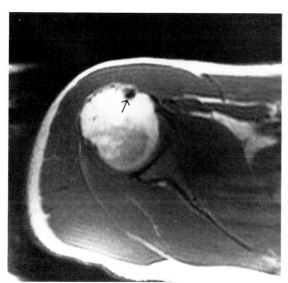

FIG. 26C. Normal biceps tendon. Axial MR scan (TR/TE 600/20 msec) demonstrating the normal biceps tendon in the bicipital groove (*arrow*).

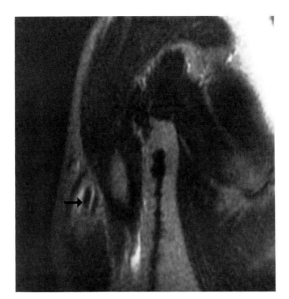

FIG. 26D. Ruptured biceps tendon (coronal MR, TR/TE 2,500/80 msec). Coronal image from the same patient outlining the displaced biceps tendon (*arrow*) surrounded by fluid.

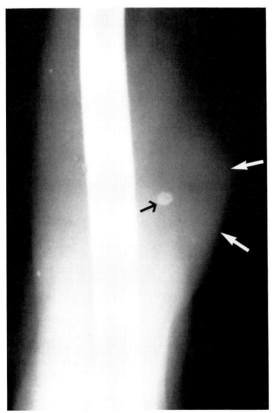

FIG. 26E. Plain radiograph obtained later that day showing the bulging "mass" in the mid arm, which is the abnormally positioned muscle mass of the biceps brachii (*white arrows*). A nodular calcific density is seen (*black arrow*), which corresponds to the small focus of signal void in the subcutaneous tissues on the MRI.

CONCLUSION

High resolution MRI with surface coils is a noninvasive technique that has excellent soft tissue contrast, which makes it ideal for the examination of the shoulder. During the past 2 to 3 years MRI has assumed a very important role in the evaluation of diseases of this joint. The majority of patients with shoulder pain have it as a result of rotator cuff disease, the result of chronic rotator cuff impingement syndrome. This, as we have alluded to, is due to attrition of the rotator cuff tendons underneath the structures of the coracoacromial arch by repeated overhead use of the arms, either in the course of participation in athletics, work, or other daily activities. The result of this is a spectrum of abnormalities ranging from inflammation of the tendon, leading eventually to partial and then complete tears of the rotator cuff. This is usually accompanied by, exacerbated by, or the result of bony changes involving the anterior acromion and the acromioclavicular ligament.

Studies by Kieft and co-workers and Seeger and co-workers demonstrated that MRI could demonstrate the spectrum of changes of patients with impingement syndrome with good clarity. Kneeland and Middleton and Zlatkin, Reicher, and Kellerhouse also demonstrated these findings and developed criteria for diagnosing rotator cuff tears in particular, in distinction from cuff inflammatory changes (tendinitis). Evancho and co-workers recently assessed the diagnostic performance of MRI, using arthrography and arthroscopy as the standard, and found a sensitivity, specificity, and accuracy of 80%, 94%, and 89%, respectively, for complete tears and 69%, 94%, and 84%, respectively, for all tears (partial and complete). In a somewhat more recent study by Zlatkin and co-workers, the performance of MRI was assessed using surgery as the standard. In this study MRI was found to have a sensitivity, specificity, and accuracy of 91%, 88%, and 89%, respectively, for all tears, partial and complete. It was also found to be more sensitive and accurate than arthrography in this regard. In this study MRI was also very accurate in assessing tear site and size, as well as the quality of the torn tendon edges, in comparison with the surgical results.

The other main modalities that are in use in the evaluation of rotator cuff disease are arthrography and ultrasound. However, arthrography is not of value in the early diagnosis of the impingement syndrome. It will not usually show abnormalities until a cuff tear is present. In addition, although arthrography is a good exam in the diagnosis of most complete tears, it now appears that it is less sensitive than MRI, particularly in small anterior tears and partial tears. Arthrography is also not without morbidity, including pain caused by the contrast injection as well as the small but potential risks of contrast reaction and infection. Ultrasound has been shown to be a good technique, particularly as a screening study. It is a relatively sensitive study using arthrography and surgery as the standard; however, in a recent study comparing it to surgery, its sensitivity was 75%. It also requires considerable experience by the examiner. It can only assess the distal part of the cuff not obscured by the acromion process; it cannot assess the bony changes well, nor can it adequately assess the other causes of shoulder pain such as capsular abnormalities, synovial processes, or AVN.

The other main clinical problem for which patients are referred for MRI of the shoulder is shoulder instability. This is usually the result of prior traumatic shoulder dislocations, which result in recurrent shoulder subluxations or dislocations. Another related problem is that of the patient suspected of an isolated labral tear, often the throwing athlete. In the evaluation of these patients MRI has also proven to be very efficacious both experimentally as well as in a moderate-sized series of patients. Magnetic resonance imaging can clearly depict the structures of the capsular mechanism. Magnetic resonance imaging can effectively demonstrate labral tears as well as pathology involving other structures of the capsular mechanism, particularly at the capsular insertion site. Computed tomography arthrography is also an excellent way of evaluating patients with this clinical problem. There is only one published study directly comparing these two modalities. Magnetic resonance imaging diagnosed tears of the glenoid labrum very effectively; however, CT was felt to be superior in evaluating abnormalities of the anterior capsule and its insertion. It is our experience and that of others, however, that MRI can give excellent detail of abnormal findings related to the capsule, particularly with T2-weighted images in the axial plane, especially when an effusion is present, which is quite often the case. Its other advantages over CT arthrography are its noninvasive nature and the fact that other problems that may be the cause of the patients pain, such as rotator cuff disease, can be better evaluated at the same time. The recent application of gradient-echo techniques in the evaluation of the glenoid labrum has also shown considerable promise.

The other main advantage of MRI in the evaluation of the shoulder, as in other joints, is its ability to evaluate the wide spectrum of abnormalities in patients with shoulder pain. This includes other posttraumatic abnormalities such as occult fractures or fractures in locations difficult to image with other techniques. It can also assess patients with infectious or noninfectious synovial inflammatory processes, AVN, and tumors.

In summary, in conjunction with plain radiographs, MRI should be the procedure of choice in evaluating the shoulder. It is noninvasive and accurate, and with advances in surface coils and imaging software it has become much easier to perform. A typical examination at our institution takes 45 min to perform and requires little supervision by the radiologist once technologists are familiar with the imaging protocol. In most cases it should obviate the need for multiple other invasive and noninvasive modalities.

RECOMMENDED READING

Aisen AM, Martel W, Braunstein EM, et al. MRI and CT evaluation of primary bone and soft tissue tumours. *AJR* 1986;146:749–756.

Beltran J, Gray LA, Bools JC, et al. Rotator cuff lesions of the shoulder: evaluation by direct sagittal CT arthrography. *Radiology* 1986;160:161–165.

Beltran J, Simon DC, Katz W, et al. Increased MR signal intensity in skeletal muscles adjacent to malignant tumors: pathologic correlation and clinical relevance. *Radiology* 1987;162:251–255.

Bigliani LU, Morrison DS, April EW. The morphology of the acromion and its relationship to rotator cuff tears (Abstr). *Orthop Trans* 1986;10:216.

Calvert PT, Pacher NP, Stoker DJ, et al. Arthrography of the shoulder after operative repair of the torn rotator cuff. *J Bone Joint Surg [Br]* 1986;68:147–150.

Cofield RH. Current concepts review: rotator cuff disease of the shoulder. *J Bone Joint Surg* 1985;67A:974–979.

Cone RO, Resnick D, Danzig L. Shoulder impingement syndrome: radiographic evaluation. *Radiology* 1984;150:29–33.

Curran JF, Ellman MH, Brown NL. Rheumatologic aspects of painful conditions affecting the shoulder. *Clin Orthop* 1983;173:27–29.

DeJager P, Fleming A. Shoulder joint rupture and pseudothrombosis in rheumatoid arthritis. *Ann Rheum Dis* 1984;43:503.

Depalma AF. *Surgery of the shoulder*. Philadephia: JP Lippincott, 1983;47–64.

Deutsch AL, Resnick D, Mink JH, et al. Computer and conventional arthrotomography of the glenohumeral joint: normal anatomy and clinical experience. *Radiology* 1984;153:603–609.

Edelman RR, Stark DD, Sairi S, et al. Oblique planes of section in MR imaging. *Radiology* 1986;159:807–810.

Evancho AM, Stiles RG, Fajman WA, et al. MR imaging of rotator cuff tears. *AJR* 1988;151:751–754.

Fritts HM, Craig E, Kyle R, et al. MR imaging of the shoulder: clinical experience and surgical correlation. Presented at the 73rd scientific assembly and annual meeting of the Radiological Society of North America. Chicago, IL, Nov 27–Dec 2, 1988.

Fukada H, Mikasa M, Tamanaka K. Incomplete thickness rotator cuff tears diagnosed by subacromial bursography. *Clin Orthop* 1987;223:52–58.

Godsil RD, Linscheid RL. Intratendinous defects of the rotator cuff. *Clin Orthop* 1970;69:181–188.

Goldman AB, Ghelman B. The double contrast shoulder arthrogram. *Radiology* 1978;127:655–663.

Greenway GD, Danzig LA, Resnick D, et al. The painful shoulder. *Med Radiogr Photogr* 1982;58:22–67.

Gschwend N, Ivosevic-Radovanovic D, Patte D. Rotator cuff tear—relationship between clinical and anatomopathological findings. *Arch Orthop Trauma Surg* 1988;107:7–15.

Hall FM, Rosenthal DI, Goldberg RP, et al. Morbidity from shoulder arthrography: etiology, incidence, and prevention. *AJR* 1981;136:59–62.

Harcke A, Grissom LE, Finkelstein MS. Evaluation of the musculoskeletal system with sonography. *AJR* 1988;150:1253–1261.

Hardy DC, Vogler JB III, White RH. The shoulder impingement syndrome: prevalence of radiographic findings and correlation with response to therapy. *AJR* 1986;147:557–561.

Hawkins RJ. The rotator cuff and biceps tendon. In Evarts CM, ed. *Surgery of the musculoskeletal system*. New York: Churchill Livingstone, 1983;5–35.

Hill HA, Sachs MD. The grooved defect of the humeral head. A frequently unrecognized complication of dislocations of the shoulder joint. *Radiology* 1940;35:690–700.

Hodler J, Fretz CJ, Terrier F, et al. Rotator cuff tears: correlation of sonographic and surgical findings. *Radiology* 1988;169:791–794.

Hoult D. The NMR receiver: a description and analysis of design. *Prog Nucl Magnetic Resonance Spectroscopy* 1978;12:41–47.

Huber DJ, Mueller E, Heribes A. Oblique magnetic resonance imaging of normal structures. *AJR* 1985;145:843–846.

Huber DJ, Sauter R, Mueller E, et al. MR imaging of the normal shoulder. *Radiology* 1986;158:405–408.

Imhof H, Hajek PC, Kramer J, et al. Gd-DTPA: help in diagnosis of malignant bone lesions? Presented at the 73rd scientific assembly and annual meeting of the Radiological Society of North America, Chicago, IL, Nov 27–Dec 2, 1988.

Kieft GH, Sartoris DJ, Bloem JI, et al. Magnetic resonance imaging of glenohumeral joint disease. *Skeletal Radiol* 1987;16:285–290.

Kieft GJ, Bloem JL, Obermann WR, et al. Normal shoulder: MR imaging. *Radiology* 1986;159:741–745.

Kieft GJ, Bloem JL, Rozing PM. Rotator cuff impingement syndrome: MR imaging. *Radiology* 1988;166:211–214.

Kieft GJ, Bloem JL, Rozing PM, et al. MR imaging of recurrent anterior dislocation of the shoulder: comparison with CT arthrography. *AJR* 1988;150:1083–1087.

Kneeland BJ, Middleton WD, Carrera GF, et al. MR imaging of the shoulder: diagnosis of rotator cuff tears. *AJR* 1987;149:333–337.

Kottal RA, Vogler JB, Matamoros A, et al. Pigmented villonodular synovitis: a report of MRI in two cases. *Radiology* 1987;163:551–553.

Lie S, Mast WA. Subacromial bursography: techniques and clinical application. *Radiology* 1982;144:626–630.

Mack LA, Matsen FA III, Kilcoyne RF, et al. US evaluation of the rotator cuff. *Radiology* 1985;157:205–209.

MacNab I, Hastings D. Rotator cuff tendinitis. *Can Med Assoc J* 1968;99:91–98.

McNiesh LM, Callaghan JJ. CT arthrography of the shoulder; variations of the glenoid labrum. *AJR* 1987;149:963–966.

Middleton WD, Edelstein G, Reinus WR, et al. Sonographic detection of rotator cuff tears. *AJR* 1985;144:349–353.

Middleton WD, Kneeland JB, Carrera GF, et al. High resolution MR imaging of the normal rotator cuff. *AJR* 1987;148:559–564.

Middleton WD, Reinus WR, Melson LG, et al. Pitfalls of rotator cuff sonography. *AJR* 1986;146:555–560.

Mink JH, Harris E, Rappaport M. Rotator cuff tears: evaluation using double contrast shoulder arthrography. *Radiology* 1985;157:621–623.

Mitchell DG, Kundel JL, Steinberg ME, et al. Avascular necrosis of the hip: comparison of MR, CT and scintigraphy. *AJR* 1986;147:67.

Mitchell MJ, Causey G, Bethoty DP, et al. Peribursal fat plane of the shoulder: anatomic study and clinical experience. *Radiology* 1988;168:699–704.

Moseley HG, Overgaard B. The anterior capsular mechanism in recurrent anterior dislocation of the shoulder: morphological and clinical studies with special reference to the glenoid labrum and the glenohumeral ligaments. *J Bone Joint Surg [Br]* 1962;44:913–927.

Neer CS. Anterior acromioplasty for the chronic impingement syndrome in the shoulder. *J Bone Joint Surg* 1972;54:41–50.

Neer CS. Impingement lesions. *Clin Orthop* 1983;173:70–77.

Nixon JE, Distefano V. Ruptures of the rotator cuff. *Orthop Clin North Am* 1975;6:423–447.

Oveson J, Sojbjerg JO. Lesions in different types of anterior glenohumeral joint dislocations. *Arch Orthop Trauma Surg* 1986;105:216–218.

Ozaki J, Fujimoto S, Nakagawa Y, et al. Tears of the rotator cuff of the shoulder associated with pathological changes in the acromion, a study in changes. *J Bone Joint Surg* 1989;70A:1224–1230.

Pappas AM, Goss TP, Kleiman PK. Symptomatic shoulder instability due to lesions of the glenoid labrum. *Am J Sports Med* 1983;11:279–289.

Penny JN, Welsh RP. Shoulder impingement syndromes in athletes and their surgical management. *Am J Sports Med* 1981;9:11–15.

Preston BJ, Jackson JP. Investigation of shoulder disability by arthrography. *Clin Radiol* 1977;28:259–266.

Rafii M, Firooznia H, Bonamo JJ, et al. Athlete shoulder injuries: CT arthrographic findings. *Radiology* 1987;162:559–564.

Rafii M, Firooznia H, Golimbu C, et al. CT arthrography of capsular structures of the shoulder. *AJR* 1986;146:361–367.

Rose CO, Patel D, Southnayd WW. The Bankart procedure. *J Bone Joint Surg [Am]* 1978;60:1–16.

Rothman RH, Marvel JP, Heppenstall RB. Anatomic considerations in the glenohumeral joint. *Orthop Clin North Am* 1975;6:341–352.

Seeger LL, Gold RH, Bassett LW, Ellman H. Shoulder impingement syndrome; MR findings in 53 shoulders. *AJR* 1988;150:343–347.

Seeger LL, Gold RH, Bassett LW, et al. Shoulder instability: evaluation with magnetic resonance imaging. *Radiology* 1988;168:695–697.

Seeger LL, Ruszkowski JT, Bassett LW, et al. MR imaging of the normal shoulder: anatomic correlation. *AJR* 1987;148:83–91.

Singson R, Feldman F, Rosenberg Z. Recurrent shoulder dislocation after surgical repair: double contrast CT arthrography. *Radiology* 1987;164:425–428.

Spritzer CE, Dalinka MK, Kressel HY, et al. Magnetic resonance imaging of pigmented villonodular synovitis: a report of two cases. *Skeletal Radiol* 1987;16:316–319.

Strizak AM, Danzig L, Jackson DW, et al. Subacromial bursography. *J Bone Joint Surg [Am]* 1982;64A;196–201.

Tang JS, Gold RH, Bassett LW, et al. Musculoskeletal infection of the extremities: evaluation with MR imaging. *Radiology* 1988;166:205–209.

Townley CO. The capsular mechanism in recurrent dislocation of the shoulder. *J Bone Joint Surg [Am]* 1950;32:370–380.

Turkel SJ, Panio MW, Marshall JL, et al. Stabilizing mechanisms preventing anterior dislocation of the glenohumeral joint. *J Bone Joint Surg* 1981;63:1208–1217.

Weiss JJ, Thompson GR, Dauset N. Rotator cuff tears in rheumatoid arthritis. *Arch Intern Med* 1975;135:521–528.

Zlatkin MB, Bjorkengren AG, Gylys-Morin V, et al. Cross-sectional imaging of the capsular mechanism of the glenohumeral joint. *AJR* 1988;150:151–158.

Zlatkin MB, Dalinka MKD, Kressel HYK. MRI of the shoulder. *MR Quarterly* 1989;5(1):3–22.

Zlatkin MB, Dalinka MKD. High resolution magnetic resonance imaging of the gleuohumeral joint. *Top Magn Reson Imaging* 1989;1(3):1–13.

Zlatkin MB, Iannotti JP, Roberts MC, et al. Rotator cuff disease: diagnostic performance of MR imaging. *Radiology* 1989;172:223–229.

Zlatkin MB, Reicher MA, Kellerhouse LE, et al. The painful shoulder: MR imaging of the glenohumeral joint. *J Comput Assist Tomogr* 1988;12:995–1001.

CHAPTER 3

The Elbow, Wrist, and Hand

J. Bruce Kneeland

In comparison with the larger joints, MRI of the elbow and hand has received only a small amount of attention to date. As a result, most of the applications of MRI in these regions remain at a relatively early stage of development.

This lack of attention paid to these structures has probably resulted in large part from the difficulties arising from the attempts to image them. These difficulties result from the small size of the anatomic structures and from problems associated with patient positioning, motion, and discomfort.

The small size requires small fields of view (FOV) from the system and a high signal-to-noise ratio (SNR) from the radio frequency (RF) coils to produce adequate image quality at the small FOV.

The problem with positioning is a complex one. The position in which the patient is prone with his or her arm overhead is ideal from the viewpoint of technical considerations but produces severe shoulder pain when used for more than short time periods. The position in which the patient is supine with the arm at his or her side is much more comfortable. The problems with this position arise from the inhomogeneity of the static field and the RF field transmitted by the whole body coil in this location. Finally, even small degrees of motion can produce bad artifacts at a very small FOV. Good stabilization is required to overcome this problem.

Despite these problems it is possible to obtain high quality images with the commercially available equipment if sufficient attention is paid to technique.

For most of the cases shown in this chapter, we have used small (2″ and 3″) receive-only surface coils. For the remainder, small transmit/receive saddle or solenoid cells were used. For use with the surface coils, the patient is positioned in the prone position with the arm extended overhead. For use with the saddle or solenoid cells, the patient can be placed in the supine or prone position.

We have routinely used T1-weighted (short TR/TE), spin-density (long TR/short TE), and T2-weighted (long TR/TE) spin-echo sequences for patient studies. We have also applied other sequences such as gradient echoes and some form of chemical shift imaging to many of the patients. With such small numbers of patients, however, it is not possible to reach a conclusion in regard to the optimal pulse sequences for the evaluation of a given disorder.

We have only explored the use of motion studies in a small number of cases (none of those shown in this chapter). These studies can be performed in a number of ways using spin-echo or gradient-echo sequences and with varying degrees of spatial and temporal resolution. There must be, in general, some compromise between these two. In addition, such studies are incompatible with the use of tightly fitting whole volume coils and require coils that are larger or do not enclose the area of interest (e.g., flat loops), both of which result in some loss of image quality. The value of such studies can only be determined with further investigation.

In summary, MRI of the hand and elbow is still in the early stages of development. It is probably true, however, that MRI will demonstrate the same value in these locations that it has shown in other parts of the musculoskeletal system.

CASE 1

A 61-year-old woman presented with a history of pain and tenderness involving the proximal portion of the nail on one of her fingers. On physical examination a small focus of hemorrhage beneath the nail was noted. What is the most likely diagnosis? What are some of the classical clinical manifestations of this lesion? What are the most common sites of skeletal involvement? (Fig. 1A. Conventional radiograph. Fig. 1B. Sagittal MR, TR/TE 600/20 msec. Fig. 1C. Sagittal MR, TR/TE 2,000/20 msec. Fig. 1D. Sagittal MR, TR/TE 2,000/70 msec.)

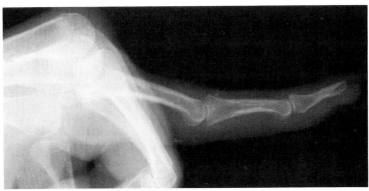

FIG. 1A.

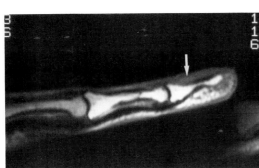

FIG. 1B.

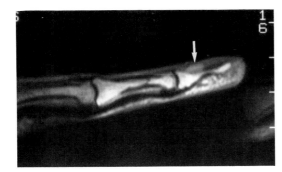

FIG. 1C.

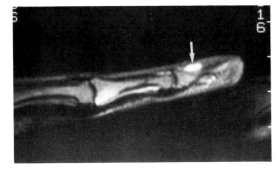

FIG. 1D.

Findings: On the plain radiograph, no soft tissue mass or evidence of bone destruction is identified. On the sagittal MR examinations, a small, well-marginated soft tissue nodule (arrows) is seen in the proximal portion of the nail bed. The mass is best demonstrated on the T2-weighted image as a consequence of its prolonged T2 relaxation time. There is a suggestion of thinning of the cortex. No evidence of marrow invasion is present. The images were obtained with an 8-cm FOV on a 1.5-T system.

Diagnosis: Glomus tumor.

Discussion: The normal glomus functions as a regulator of body temperature and circulation. A myoarterial receptor, it is located within numerous internal organs and within the dermis and superficial subcutaneous tissue of the extremities. When involving the extremities, glomus tumors are most often identified in the region of the fingertips or nail beds. They are most commonly extraosseous lesions with bone involvement representing a secondary manifestation and reported to occur in 15% to 65% of cases.

Glomus tumors are most commonly encountered in the fourth and fifth decades of life. The lesions are typically only a few millimeters in dimension. When located in the nail bed, as in this case, a blue nodule or hemorrhagic area may be visible beneath the nail. Patients typically present with aching pain and point tenderness. Severe paroxysms of pain related to cold exposure is a classical historical feature of the condition. The tumors are benign and are treated by surgical exision with complete relief of symptoms. If incompletely excised, the tumors can recur locally, often within a few months of surgery.

Radiographically, soft tissue glomus tumors may produce shallow, well-defined osteolytic lesions within the adjacent bone that typically demonstrate sclerotic margins. Within a distal phalanx, the radiographic appearance may resemble an inclusion cyst. Glomus tumors may also arise as primary intraosseous lesions, although these are less common than soft tissue tumors. Within bone, the tumor can resemble other well-defined osteolytic lesions including enchondroma, aneurysmal bone cyst, and sarcoidosis. These entities are usually readily differentiated from glomus tumor on the basis of clinical history.

The lesion in this case demonstrates a nonspecific signal intensity pattern. The apparent thinning of the underlying cortex was not evident on the plain radiograph and could represent a partial volume effect. The MR was of value in planning the surgical resection of this lesion as well as in two others that we have studied. The clear demonstration of a normal terminal phalanx with no marrow invasion was very reassuring for the surgeon.

CASE 2

A 29-year-old woman presented with a several month history of aching and numbness over much of the hand, with sparing of the little finger. On physical examination, a positive Tinel sign (the sensation of tingling felt in the distal extremity when percussion is performed over an injured nerve) was obtained. (Fig. 2A,B; Axial MR, TR/TE 2,500/ 25 msec.)

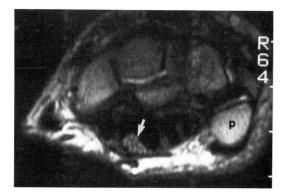 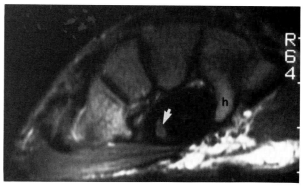

FIG. 2A.   **FIG. 2B.**

Findings: A T2-weighted axial section through the pisiform region (p) of the wrist (Fig. 2A) demonstrates enlargement and a distinct heterogeneous pattern of the median nerve (arrow) that appears to correspond to the neural bundles. On the more distal section through the hook of the hamate (h) (Fig. 2B), the nerve (arrow) is of normal caliber and appearance although interposed among the tendons.

Diagnosis: Carpal tunnel syndrome with segmental swelling of the median nerve.

Discussion: *Carpal tunnel syndrome* is the term given to a symptom complex consisting of pain and paresthesias in the distribution of the median nerve, generally arising from increased pressure on that nerve. The increased pressure has many different causes including tendon sheath thickening, edema, rheumatoid arthritis, and fibrosis. The diagnosis is made on clinical grounds, but electromyography and nerve conduction studies are helpful confirmatory examinations. Treatment of carpal tunnel syndrome consists of release of the flexor retinaculum with surgical debridement.

The role of MRI in the evaluation of carpal tunnel syndrome is unclear at this time. We have imaged a small number of patients with carpal tunnel syndrome, both preoperative cases and those with recurrent symptoms following surgery, and have noted several different findings: (a) segmental swelling of the nerve proximal to the carpal tunnel with a normal diameter nerve within the carpal tunnel as seen in this case; (b) diffuse swelling of the nerve both proximal to and in the carpal tunnel (Fig. 2C); (c) heterogeneous high signal intensity of the nerve (as compared with the muscles) on T2-weighted images with a pattern that appears to correspond to the neural bundles (Fig. 2C) (We have generally found normal nerves to have about the same signal intensity as muscle on all pulse sequences); (d) swelling of the flexor tendon sheaths (Fig. 2D) [compare with normal appearance of contralateral wrist (Fig. 2E)]; and (e) distortion of the normal ovoid shape of the nerve by fibrotic strands in postoperative patients (Fig. 2F). It is possible that MRI may be able to distinguish between those cases in which there is a demonstrable cause of pressure on the median nerve and in which surgery would be of help and those in which conservative management is the preferred treatment.

We have also noted an increase in the signal intensity of the median nerve on T2-weighted images in those cases of Guillain-Barré syndrome (Fig. 2G) with the predominately demyelinating form of the disease (as opposed to the more neuronal degenerative form). The clinical significance of this observation is unknown, but it may be useful in predicting which patients will respond to plasmapheresis.

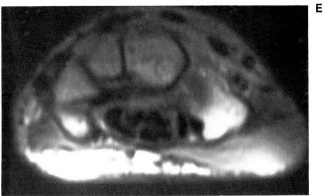

FIG. 2C. Carpal tunnel syndrome. Axial image through the wrist obtained with TR/TE 2,500/80 msec. *Arrow,* median nerve. Although the heterogeneity of the nerve is not optimally depicted, the nerve does have increased signal.

D

E

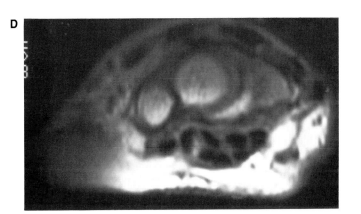

FIG. 2D,E. Carpal tunnel syndrome. Axial sections obtained with TR/TE 2,000/25 msec through the symptomatic (D) and asymptomatic (E) wrists of the same patient. In D, the medium intensity tendon sheaths surrounding the flexor tendons appear thicker than in E.

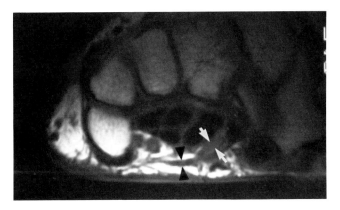

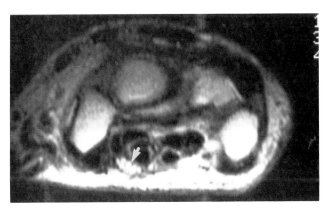

FIG. 2F. Perineural fibrosis. Axial section obtained with TR/TE 500/20 msec through the wrist. The normal ovoid shape of the median nerve (*arrows*) is distorted by adjacent fibrotic tissue (*arrowheads*).

FIG. 2G. Guillain-Barré syndrome. Axial section through the wrist obtained with TR/TE 2,000/80 msec. *Arrow,* median nerve. The nerve demonstrates an increased signal intensity.

CASE 3

A 35-year-old man with a history of a previous suicide attempt by means of wrist slashing presented with pain and paresthesias in the same hand. Physical examination demonstrated motor and sensory deficits as well as muscle atrophy consistent with complete dysfunction of the radial nerve. An ill-defined soft tissue mass was palpated in the volar aspect of the wrist. What is the diagnosis? (Fig. 3A. Axial MR, TR/TE 2,000/20 msec. Fig. 3B. Axial MR distal to Fig. 3A, TR/TE 2,000/20 msec. Fig. 3C. Axial MR, TR/TE 2,000/80 msec.)

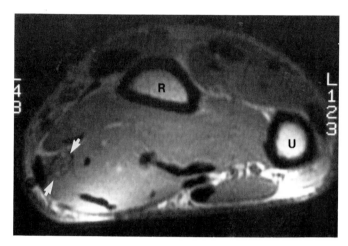

FIG. 3A.

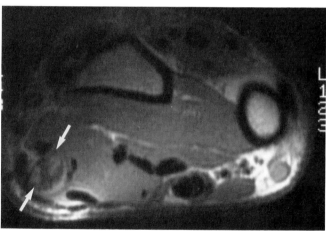

FIG. 3B.

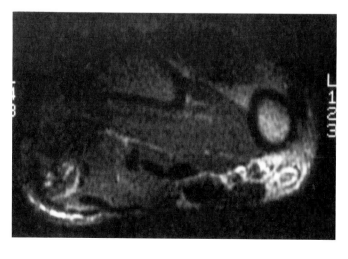

FIG. 3C.

Findings: Axial spin-density weighted section in the distal forearm (Fig. 3A) demonstrates a radial nerve (arrows) with heterogeneous signal intensity. A more distal section obtained with spin-density (Fig. 3B) and T2-weighted (Fig. 3C) sequences demonstrates a heterogeneous medium and low signal intensity mass (arrows) on the radial aspect of the wrist that could be demonstrated on serial sections to be continuous with the radial nerve. (R, radius; U, ulna)

Diagnosis: Posttraumatic pseudoneuroma.

Discussion: There is considerable confusion in regard to naming of masses associated with nerves. A schwannoma, or neurilemmoma, is a true neoplasm arising from the sheath cells of a sensory nerve. The mass is nearly always benign. Because the nerve itself is not involved, except by displacement, a schwannoma can usually be resected without sacrifice of nerve tissue; function is preserved.

A neurofibroma is a mass that arises from neural tissue; it contains an abundance of Schwann cells and perineural cells. The presence of a neurofibroma, especially if multiple, must raise the question of neurofibromatosis.

Two other lesions are usually described as "neuromas" but, in fact, are not neoplasms at all. Morton's neuroma (see Chapter 2, Case 8) represents an irritative response in a nerve, most commonly between the third and fourth metatarsal heads. The lesion may be a result of ischemia, and histologically the lesion is a mass of fibrotic tissue. Resection is curative.

A posttraumatic neuroma is a tangle of regenerating nerve fibers found after the severing of or severe injury to a nerve. These neuromas may result in the symptoms of referred pain and paresthesias. Surgical excision relieves the symptoms although the neuromas may recur following surgery.

We have imaged another case of a posttraumatic pseudoneuroma in the wrist as well as two cases in the supraclavicular region and have found their appearance to have considerable variability. In these cases, MR was of value in confirming the presence of a mass as well as aiding in the presurgical planning by delineating its precise extent. More investigation will be necessary to determine the value of the routine use of MRI in these cases.

CASE 4

A 29-year-old man sustained a scaphoid fracture 6 months ago. He now presents to clinic with complaints of continued pain. A plain radiograph reveals a persistent fracture line in the proximal portion of the scaphoid without evidence of osteonecrosis (Fig. 4C). What is the most likely diagnosis? What are the common complications of scaphoid fracture? What is the role of MR and CT in assessment of these problems? (Fig. 4A. Coronal MR, TR/TE 600/20 msec. Fig. 4B. Coronal MR, TR/TE 2,000/80 msec. Fig. 4C. Plain radiograph of the hand in the anteroposterior projection.)

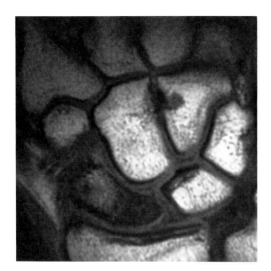

FIG. 4A.

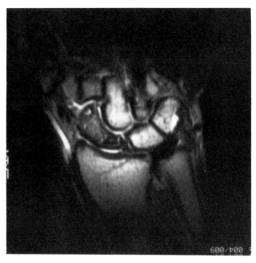

FIG. 4B.

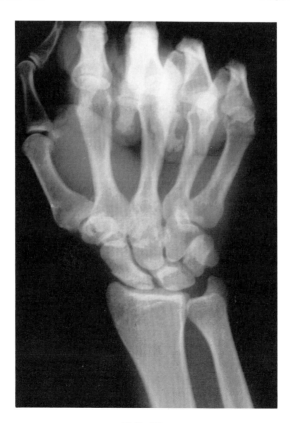

FIG. 4C.

Findings: On the coronal T1-weighted image through the scaphoid, there is a diffuse loss of the normally expected high signal intensity of the medullary marrow. This is most pronounced in the proximal pole. On the T2-weighted image obtained at a slightly different level and with a larger FOV, no significant increased signal intensity is identified.

Diagnosis: Osteonecrosis of the scaphoid.

Discussion: The scaphoid is the most commonly fractured carpal bone, accounting for up to 60% of all carpal fractures. Scaphoid fractures most commonly occur through the middle third (waist), with proximal pole fractures accounting for 20% of cases. Osteonecrosis is reported to occur in approximately 10% to 15% of all scaphoid fractures and is seen in the majority of fractures involving the proximal pole. This relates to the blood supply to the scaphoid, which enters through its distal pole. Radiographically increased density within the proximal pole, indicative of osteonecrosis, is seen in approximately 30% of cases. The diagnosis of osteonecrosis is important for several reasons: first, it is believed to be a predisposing (but not inevitable) factor to nonunion; second, its presence leads to a high failure rate in bone grafting, which is attempted in cases of chronic nonunion.

Magnetic resonance has proven highly accurate for the detection of osteonecrosis of the hip, where it is characterized by several different patterns of abnormal signal including the low signal intensity on both pulse sequences seen here. A preliminary study of the accuracy of MR in the evaluation of carpal bones demonstrated a decreased sensitivity of MR relative to the radionuclide scan. That study, however, was performed at relatively poor resolution by present standards and may well have underestimated the accuracy of MR. More recent investigations have suggested an important role for MR in evaluation of scaphoid fractures and their complications. In our experience, MR has also proven quite sensitive for the diagnosis of osteonecrosis. The most common imaging pattern that we have observed is decreased signal intensity within the proximal pole on both T1- and T2-weighted images (Fig. 4D,E). In cases of more diffuse involvement, low signal is not restricted to the proximal pole.

Nonunion of scaphoid fractures has been considered to be primarily due to delay in diagnosis and/or inadequate immobilization. The tenuous blood supply also is believed to contribute in some cases. Progressive resorption along the fracture site can be demonstrated with the development of cyst-like "cavities" demonstrable by 2 to 3 months. Sclerotic edges along the fracture site are observed later. A number of surgical procedures have been developed for the treatment of scaphoid nonunion. One of the more common is the Matte-Russe graft procedure in which the sclerotic bone and fibrous tissue from the nonunion site is resected and an autogenous (often distal radial donor site) cortico-cancellous graft inserted (Fig. 4F).

Computed tomography has been extensively utilized in the assessment of scaphoid fractures and their complications. Direct sagittal scans along the long axis of the scaphoid are preferred and are particularly valuable for assessment of graft incorporation. The use of MR has also been recently reported in assessment of scaphoid pseudoarthrosis. In a series of patients with established nonunions, MR was able to predict the presence of a true pseudoarthrosis with a high degree of accuracy. In patients with fibrous union, MR demonstrated low signal intensity at the fracture site on both T1- and T2-weighted images. In patients with surgically proven pseudoarthrosis formation, high signal intensity fluid was identified at the fracture site on T2-weighted images. In this regard it should be noted that a fracture line may remain demonstrable on MR long after clinical union has occurred, and the mere persistence of a fine line of low signal at a fracture site should not be equated with nonunion.

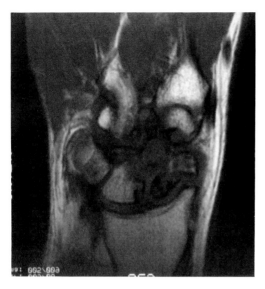

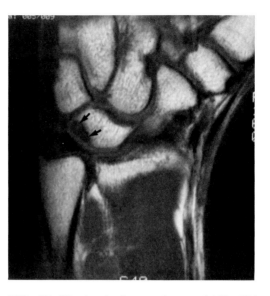

FIG. 4D. Scaphoid osteonecrosis (coronal MR, TR/TE 2,000/20 msec). Coronal section through the wrist of another patient demonstrates evidence of osteonecrosis manifest as focal decreased signal intensity. The apparent loss of signal in the distal carpal row results from partial volume effect.

FIG. 4E. Kienbock disease (coronal MR, TR/TE 500/20 msec). A small region of osteonecrosis (*arrows*) is seen within the proximal portion of the lunate. The finding was radiographically occult.

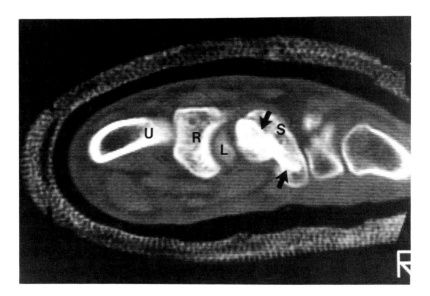

FIG. 4F. Scaphoid graft. A direct sagittal CT section through the long axis of the scaphoid demonstrates incorporation of a scaphoid graft (*arrows*). (R, radius; U, ulna; L, lunate; S, scaphoid.)

CASE 5

A 1-year-old infant with only two fingers on his hand underwent MRI for further evaluation of the nonossified carpal bones. (Fig. 5A. Coronal MR, TR/TE 500/20 msec.) What is the diagnosis?

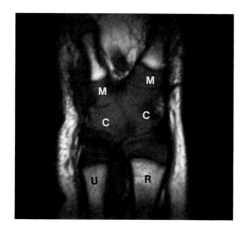

FIG. 5A.

Findings: The coronal T1-weighted image of the wrist demonstrates two major, cartilaginous carpal masses (C) and two small cartilaginous basimetacarpal (Mc) masses. (R, radius; U, ulna.)

Diagnosis: Congenital deformity.

Discussion: In cases of congenital hand deformities in very young children, the structure of the carpal bones will often determine the type of surgical repair that is feasible. The bones are not ossified at this stage of development, however, and are not visualized with plain film radiography. Before MRI, there was no satisfactory method of visualizing the cartilaginous models preoperatively. Magnetic resonance imaging permits the direct visualization of epiphyseal cartilage with all pulse sequences, although T2 weighting is necessary to distinguish between adjacent fluid and cartilage when this determination is necessary.

   A related case is shown in Fig. 5B, in which MRI was used to determine whether a fracture of the metaphysis of the distal humerus extended into the nonossified epiphysis. The fracture line (arrows) is seen to be confined to the metaphysis and physeal line but not to enter the epiphysis (Salter-Harris II). This distinction can have an important role in establishing the prognosis of epiphyseal injuries.

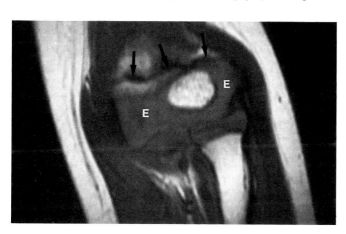

FIG. 5B. Metaphyseal fracture (coronal MR, TR/TE 500/20 msec). A fracture line (*arrows*) is seen in the humeral metaphysis; the lesion was seen on a plain radiograph, but the orthopedist was concerned that a Salter I fracture was also present. The cartilaginous epiphysis (E) is normally related to the metaphysis, excluding a fracture. (Courtesy of Jerrold Mink, MD.)

CASE 6

A 49-year-old man was seen in clinic with chronic wrist pain. Plain radiographs of the hand and wrist were normal. What are the principal diagnostic considerations? What radiographic methods are utilized for further assessment of these conditions? What is their accuracy? How might MR be utilized in assessment of these problems? (Fig. 6A. Coronal MR, TR/TE 2,000/20 msec. Fig. 6B. Coronal MR, TR/TE 2,000/70 msec.)

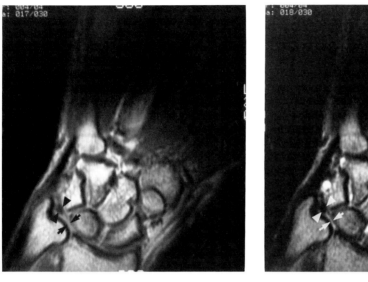 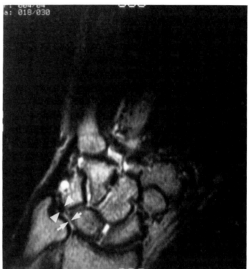

FIG. 6A.   **FIG. 6B.**

Findings: A high signal intensity defect is identified within the triangular fibrocartilage (arrows). Only a small fragment of the normal low signal intensity fibrocartilaginous disk is seen adjacent to the ulna styloid process (arrowhead).

Diagnosis: Tear of the triangular fibrocartilage.

Discussion: The diagnosis of chronic wrist pain can pose a difficult problem for the clinician. One of the more common and identifiable causes is injury to the triangular fibrocartilage complex (TFCC). This complex is composed of the triangular fibrocartilage proper (the articular disk), the meniscus homologue (the ulnar carpal meniscus), the ulnar collateral ligament, the dorsal radial ulnar ligament, the volar radial ulnar ligament, and the sheath of the extensor carpi ulnaris tendon. The TFCC functions as a fundamental stabilizer of the distal radioulnar joint and cushion for the ulnar carpus, preventing ulnar carpal abutment and ulnolunate chondromalacia.

Injuries to the triangular fibrocartilage include tears and detachments. Perforations of the TFCC are most commonly identified in the horizontal portion of the complex in the region of the ulnolunate articulation. Tears within and about the periphery of the radial aspect of the articular disk contiguous with the distal radial attachment have also been commonly encountered in athletes. The diagnosis of tears and detachments of the TFCC has traditionally relied on radiocarpal joint arthrography, in which the diagnosis is established by demonstrating the passage of contrast between the radioulnar and radiocarpal spaces. Several recent investigations have demonstrated that radiocarpal arthrography alone may miss up to 25% of triangular fibrocartilage tears, a finding possibly accounted for by the presence of "one-way" flap tears. Three-compartment arthrography has been suggested for comprehensive wrist evaluation, with a separate injection into the inferior radioulnar joint for optimal assessment of the TFCC. This technique allows for assessment of the proximal surface of the TFCC, and partial tears and full thickness tears not demonstrated on radiocarpal injection have been demonstrated (Fig. 6C).

Magnetic resonance has also been applied to assessment of the TFCC. The anatomy of the wrist has been extensively investigated utilizing MR, and the normal triangular fibrocartilage is characterized by uniform low signal intensity on all pulse sequences. Recent investigations have suggested a high degree of accuracy of MR for detection of TFCC tears. Tears are seen as focal areas of increased signal intensity, replacing the normal low signal intensity fibrocartilage. Coronal T1-weighted images often suffice for diagnosis (Fig. 6D). Fast scan techniques may also be utilized for assessment of TFCC integrity. In addition to evaluation of the TFCC, MR has demonstrated unsuspected osteonecrosis, subluxation of the distal radioulnar joint, and synovitis of the tendon sheaths in patients undergoing examination for wrist pain.

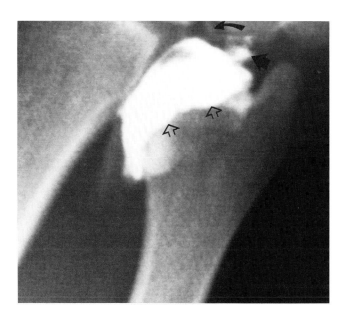

FIG. 6C. Anteroposterior radiograph from a selective distal inferior radioulnar joint arthrogram. In this individual the radiocarpal injection was normal. On the selective distal inferior radioulnar injection (*open arrows*), contrast extends into the TFCC demonstrating a tear (*arrow*). Contrast in radiocarpal joint (*curved arrow*).

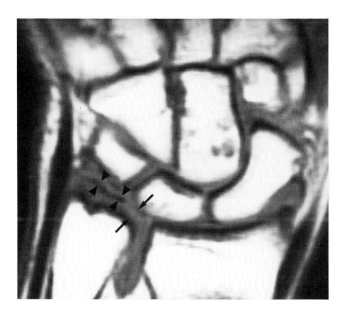

Fig. 6D. Tear of the triangular fibrocartilage (coronal MR, TR/TE 500/20 msec). A small-to-moderate tear is identified (*arrows*), with residual fragment of triangular cartilage (*arrowheads*).

CASE 7

A 31-year-old man was seen in clinic several weeks after significant trauma to his index finger with the complaint of continuing pain in the region of the distal metacarpal. Plain radiographs of the hand were normal. What is the (probable) diagnosis? (Fig. 7A. Plain radiograph. Fig. 7B. Sagittal MR, TR/TE 500/20 msec. Fig. 7C. Sagittal MR, TR/TE 2,000/20 msec. Fig. 7D. Sagittal MR, TR/TE 2,000/80 msec.)

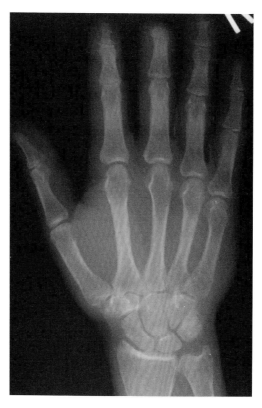

FIG. 7A.

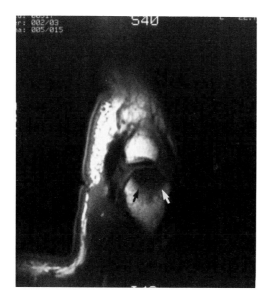

FIG. 7B.

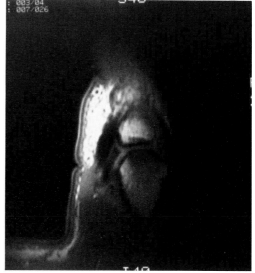

FIG. 7C.

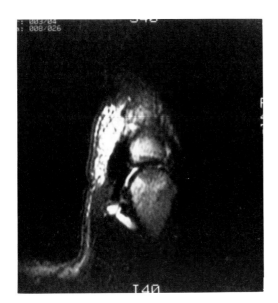

FIG. 7D.

Findings: Sagittal T1, spin-density, and T2-weighted images through the metacarpophalangeal joint of the index finger demonstrate a rounded region of low signal intensity in the metacarpal head abutting the joint space (arrows). A small effusion is also present.

Diagnosis: Osteonecrosis of the metacarpal head.

Discussion: Osteonecrosis of the metacarpal head is an infrequent but well-recognized sequela of trauma. It occurs both in the presence and absence of a fracture. Virtually no cases of osteonecrosis in this region have been described in the radiological literature. As a result, neither its MR appearance nor the accuracy of MRI for its diagnosis is known at this time. The appearance illustrated in this case does correspond to one of the patterns described for osteonecrosis of the hip. In addition, one would anticipate that with adequate resolution, similar accuracy to that obtained with the hip would be achieved in this location as well.

CASE 8

A 23-year-old woman was seen in clinic with dysfunction of the flexor tendons of the index finger. (Fig. 8A,B. Axial MR, TR/TE 500/20 msec.) What is the diagnosis?

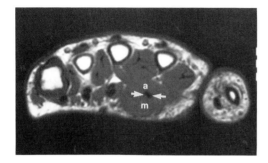

FIG. 8A.

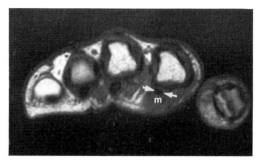

FIG. 8B.

Findings: T1-weighted axial images extend through the midshaft of the metacarpals (Fig. 8A) and the metacarpal heads (Fig. 8B). In Fig. 8A,B a solitary flexor tendon (arrows) to the index finger is present. Deep to the tendon is the normal adductor pollicis muscle (a). Superficial to the tendon is an abnormal muscle (m). In Fig. 8B, the abnormal muscle (m) has decreased in size considerably.

Diagnosis: Absent flexor digitorum profundus (FDP) tendon. Abnormal muscle superficial to flexor tendon felt at surgery to represent an enlarged and anomalously located lumbrical muscle. (The exact nature of the muscular anomaly is somewhat uncertain because of the limited nature of the exploratory surgery.)

Discussion: The diagnosis and treatment of soft tissue malformations of the hand is a challenging task for the surgeon. Magnetic resonance permits the direct visualization of the soft tissue structures at high resolution. In the case shown here, there was a vague history of trauma, and a preoperative diagnosis of ruptured flexor tendon was entertained. The MR images demonstrated complete absence of one of the flexor tendons and the presence of abnormally located muscles, indicating that this represented a developmental anomaly rather than a simple tendinous rupture. Magnetic resonance should prove quite useful in studying complex anomalies in the soft tissues of the hand.

CASE 9

A 25-year-old man who underwent reimplantation of a digit developed signs and symptoms of dysfunction of the single flexor tendon that had been anastomosed at the time of surgery. (Fig. 9. Sagittal MR, TR/TE 500/30 msec.) What is the diagnosis?

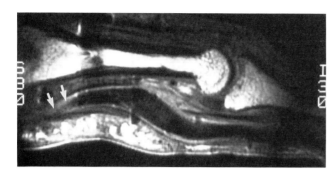

FIG. 9.

Findings: There is a focus of discontinuity in the flexor tendon in the region of the base of the proximal phalanx of the reimplanted digit (arrows).

Diagnosis: Ruptured flexor tendon.

Discussion: Flexor tendon rupture in the hand most commonly results from either trauma or inflammatory arthritis. A variety of pathologic changes occurring in and around the tendon in rheumatoid arthritis (and possibly systemic lupus erythematosis) predispose to rupture. Granulation tissue presents in the paratenon, and lymphocytes and leukocytes characterize tendinitis. Focal areas of necrosis within a tendon may resemble the granulomatous pattern seen in subcutaneous rheumatoid nodules.

Spontaneous tendon ruptures occur in the hand and wrist but may also affect the Achilles tendon, the quadriceps and patellar tendons, and the rotator cuff. Tendon rupture is due to (a) inflammation of the tendon itself; (b) enzymatic proteolysis (enzymes originate from diseased synovium in adjacent joints); (c) injection of local corticosteroids; (d) abnormal stresses secondary to malalignment at inflamed joints; (e) extrinsic erosion caused by adjacent osseous disease; and (f) vasculitis. Although the diagnosis of ruptured flexor tendon of the digit can be suspected on clinical grounds, it can be difficult to establish definitively before surgery.

The flexor tendons and their sheaths can be indirectly examined by means of tenography, a simple radiographic procedure. The flexor tendons, both superficial and deep components, are contained within a sheath extending from distal to the distal interphalangeal joint to just proximal to the metacarpal heads. This arrangement is most commonly seen in the second, third, and fourth fingers; the sheaths of the first and fifth commonly communicate with bursal sacs in the palm. While considerable fibrous tissue separates the metacarpal and proximal interphalangeal joint from the tendon sheath, there is a site near the base of the proximal phalanx at which the joint and tendon sheath are intimately related. This proximity allows infected fluid and proteolytic enzymes to pass from one synovial compartment to another.

Inflammatory arthritis can result in synovial effusions of the tendon sheaths and villous hypertrophy of the lining. The latter leads to the irregularity of the tendon sheath wall found on tenography. Sacculations of the sheath may also occur.

Magnetic resonance imaging permits direct visualization of the flexor tendons and in this case clearly demonstrates the site of rupture. Only a small number of cases have been presented in the literature, but the accuracy of MR for this diagnosis, given the high resolution currently available, should be high.

CASE 10

A 41-year-old man was seen in clinic with a diagnosis of elbow pain. Plain radiographs (Fig. 10A) demonstrated a lytic cortical lesion at the base of the medial epicondyle (arrows). What are the diagnostic possibilities? (Fig. 10A. Plain radiograph. Fig. 10B. Axial MR, TR/TE 800/20 msec. Fig. 10C. Axial MR, TR/TE 2,000/80 msec.)

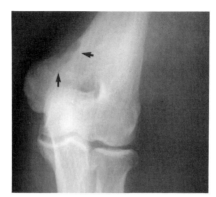

FIG. 10A.

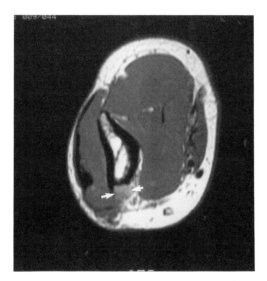

FIG. 10B.

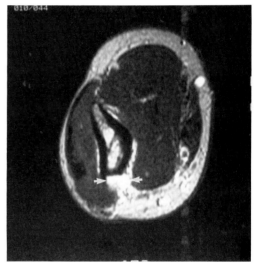

Fig. 10C.

Findings: T1- and T2-weighted axial images demonstrate a small, irregularly marginated soft tissue nodule (arrows) involving the cortex and overlying fat but not invading the marrow. Also noted on axial T1- and T2-weighted sections at a slightly more distal level (Fig. 10D,E) is a small mass with low signal intensity on all sequences in the location corresponding to the loose body seen on plain radiographs and CT (Fig. 10F).

Diagnosis: Carcinoma of unknown primary, metastatic to the cortex. Loose body.

Discussion: Cortical metastases are relatively uncommon. The most frequent source is a primary lung neoplasm. The plain radiographs and CT scan were nonspecific in this case, although in retrospect the tiny soft tissue mass can be delineated on the CT (Fig. 10G). Magnetic resonance imaging clearly and immediately demonstrated the soft tissue mass involving the cortex, indicating the nature of the lesion.

The loose body is an incidental finding but is easily seen with MR.

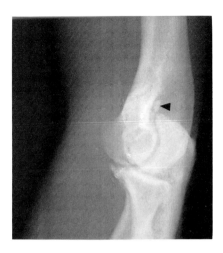

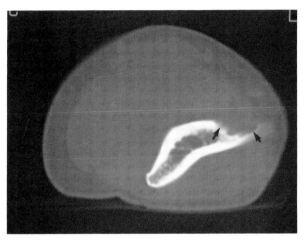

FIG. 10D. Loose body. The plain radiograph easily demonstrates the loose body (*arrowhead*) as well as the metastasis.

FIG. 10E. Metastasis. This CT scan demonstrates the erosion (*arrows*) of the epicondyle but does not reveal the soft tissue mass.

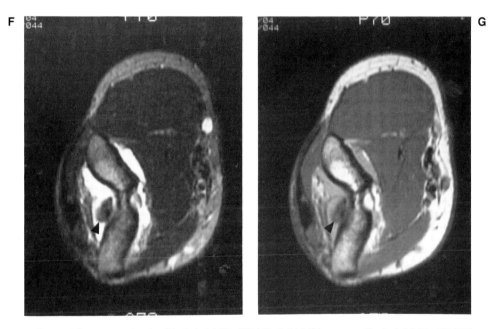

FIG. 10F,G. Loose body (**F.** Axial MR, TR/TE 2,000/20 msec; **G.** Axial MR, TR/TE 2,000/80 msec). The loose body is readily seen on the MR (*arrowhead*).

CASE 11

A 35-year-old man presented with a mass near his right elbow. What is the probable diagnosis? (Fig. 11A. Axial MR, TR/TE 2,000/80 msec. Fig. 11B. Coronal MR, TR/TE 800/20 msec.)

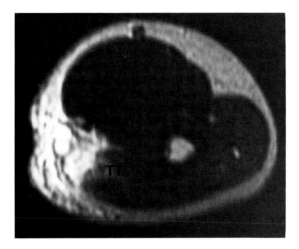

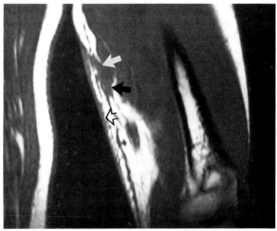

FIG. 11A.   **FIG. 11B.**

Findings: A sharply defined, lobulated soft tissue mass (arrows) is seen in the subcutaneous tissue between the biceps and triceps muscles. The most medial aspect of the triceps (TT) is quite bright, as are the several small loculations in the fat. A small amount of edema is present in the subcutaneous tissue (open arrow). The patient had suffered a puncture wound of the right hand 2 weeks prior. (A marker indents the skin medially.)

Diagnosis: Soft tissue abscess involving the antecubital lymph nodes.

Discussion: Lymphatic dissemination of infections are most commonly manifest by reddish streaks extending cephalad from a wound site; they may extend to the proximal end of an involved limb. Local and systemic antibiotic therapy usually lead to resolution, but occasionally, the local draining nodes undergo suppuration and an abscess results.

The ability of MR to detect bone and soft tissue infections rivals that of nuclear medicine studies; MR depends on the accumulation of excessive water at the site of infections. Magnetic resonance holds an advantage over nuclear studies in its ability to separate bone from soft tissue infections and an abscess from a cellulitis/phlegmon. Magnetic resonance has a greater spatial resolution than nuclear medicine studies; the axial plane imaging and greater anatomic detail make MR the procedure of choice.

There is only one possible differential consideration in this case—a soft tissue hemangioma. Such lesions occur in young individuals, usually in a intramuscular location. There are two MR characteristics that may permit accurate preoperative differential diagnosis. Although hemangiomas, like most other benign and malignant neoplasms, become bright on long TR/TE sequences, the presence of high signal on *short* TR/TE sequences is rather unique. This latter characteristic is thought to be secondary to the variable amount of adipose tissue found in hemangiomas, but slow flow through large vessels and pooling of blood in dilated sinuses may also contribute. In this case, the edema in the surrounding tissues, the lack of high signal on T1-weighted images, and, last but not least, a very important history make abscess the overwhelming diagnostic choice.

CASE 12

A 10-year-old athletic boy had chronic pain in his elbow. He finally sought medical care when he was unable to pitch the "big game." What is the diagnosis? (Fig. 12A,B. Sagittal MR, TR/TE 60/13 msec.)

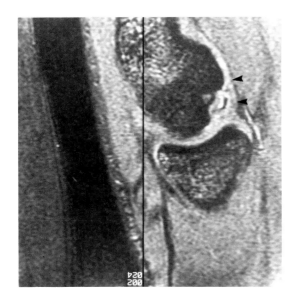

FIG. 12A.

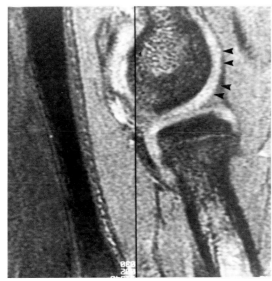

FIG. 12B.

Findings: In Fig. 12A, a focus of osteochondritis dissecans (OCD) is present with a small fragment partially separated from the host; the articular cartilage over the lesion (arrowheads in Fig. 12A) is unlike that over the remainder of the capitellum (arrowheads in Fig. 12B). (The vertical line is a reconstruction artifact.)

Diagnosis: Osteochondritis dissecans of the capitellum.

Discussion: The elbow is the pivot of musculoskeletal motion in baseball, tennis, and other racquet sports. The rapid growth of these sports has led to an increase in attention to posttraumatic abnormalities at the elbow. The type of injury is dependent on the age of the athlete, the magnitude of the forces, and the precipitating stress.

Diffuse, generalized stresses lead to thickening of the cortex of the humerus, radius, and ulna; hypermaturity of open growth centers in the child; and loose bodies. Bone hypertrophy occurs concurrently with the more commonly recognized muscular hypertrophy. Rather than being abnormal, diffuse bony hypertrophy is adaptive and is a manifestation of "being in shape."

Medial tension stresses are the most common localized manifestation of elbow trauma in growing and mature skeletons, but the radiographic findings are different in the two groups. The acceleration phase of pitching produces distractive forces on the ulno-humeral joint and compressive forces on the radio-capitellar articulation. In the adult, a traction spur (osteophyte) arises from the medial aspect of the coronoid tubercle caused by chronic pull from the distal attachment of the ulnar collateral ligament. Occasionally, these spurs may fracture. In the growing skeleton, the lax ligaments and capsule causes the flexor-pronator muscle group to bear the brunt of the excessive distractive forces on the medial side of the elbow. Avulsion of the medial epicondylar apophysis, the site of attachment of the flexor group, is known as "Little League Elbow." Radiographically, the apophysis is pulled away from the humerus. Although the event is usually acute, chronic submaximal forces may produce fragmentation and roughening of the epicondyle.

Although it is distractive forces that produce the changes on the medial side of the elbow, compressive forces laterally result in impaction of the radial head into the capitellum. The result is OCD, a transchondral fracture that results in separation of a segment of the articular surface and often a segment of underlying bone (Fig. 12C–G). The affected fragment may remain *in situ*, or it may separate from its site of origin. If the latter occurs, the fragment may (a) remain loose in the joint, (b) implant on the synovium at a distant site, (c) completely resorb, or (d) grow. Although loose bodies may result from repeated trauma to the synovium with chondral and osseous metaplasia, one must carefully examine both the capitellum and radial head (a less well-known site of OCD) for signs of osteochondral fracture. Treatment of OCD of the elbow depends on the integrity of the cartilage and the presence of loose bodies; MR has the capability to assess both noninvasively.

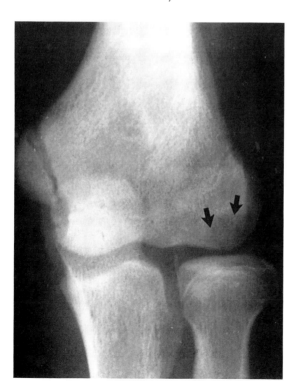

FIG. 12C. Osteochondritis dissecans. This anteroposterior radiograph demonstrates a well-defined lucency in the subchondral bone of the capitellum (*arrows*). No loose bodies were seen.

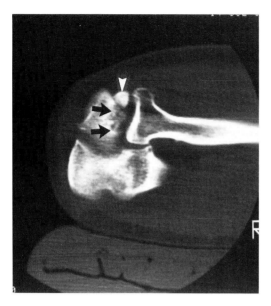

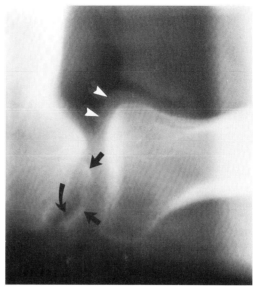

Fig. 12D. Osteochondritis dissecans. This direct sagittal CT image of the elbow demonstrates a large defect in the capitellum (*arrows*) and a large loose body (*arrowhead*).

Fig. 12E. Osteochondritis dissecans. A sagittal image from an elbow arthrotomogram demonstrates several fragments (*arrows*) in close proximity to the capitellar defect. Air delimits the articular cartilage over the radial head (*arrowheads*). Although the articular surfaces are congruous, a small amount of air has insinuated itself deep to the fragment (*curved arrow*), indicating violation of the cartilage.

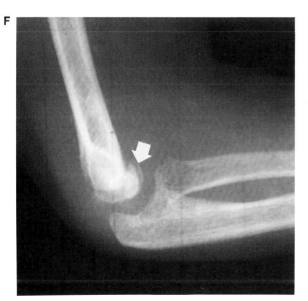

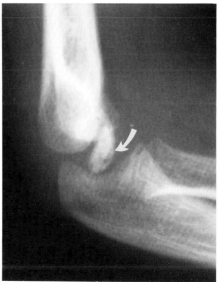

Fig. 12F,G. Panner disease. Osteonecrosis of the capitellar growth center is known as Panner disease. The lateral view **(F)** demonstrates partial collapse of the capitellum. A subchondral lucency (*arrow*) is well seen. Six months later **(G)**, collapse is more evident, as is increased density (*arrow*).

CASE 13

A 20-year-old man had trauma to his hand; after 2 weeks of persistent pain, he sought medical care. Plain radiographs were normal. What is the probable diagnosis? (Fig. 13A. Sagittal MR, TR/TE 700/20 msec. Fig. 13B. Axial MR, TR/TE 700/20 msec.)

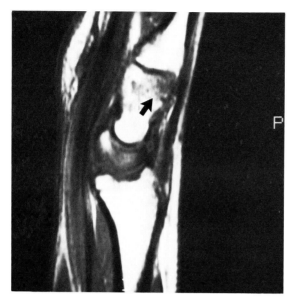

FIG. 13A.

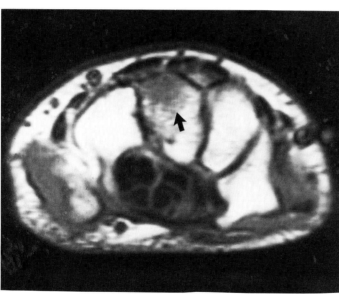

FIG. 13B.

Findings: Both views demonstrate a poorly defined area of decreased signal in the dorsal aspect of the capitate (arrows). The lesion increased in intensity on a long TR/TE sequence (not shown). The patient had had direct trauma to the dorsum of the hand and became asymptomatic 3 weeks after the MR.

Diagnosis: Bone bruise (trabecular microfracture).

Discussion: Although there is no proof in the case, the MR findings and the clinical history are consistent with a bone bruise. A bone bruise is a lesion that was initially described in association with a contralateral collateral ligament tear in the knee. More recently, it has been found laterally in patients with anterior cruciate ligament tears or, less commonly, as a result of direct trauma. Bone bruises are thought to be a result of compressive forces that are transmitted through cartilage and cortical bone to the weaker medullary trabecular bone. The MR findings suggest that a bone bruise is edema, hemorrhage, and trabecular microfracture, although no pathologic material is available. Bone bruises are self-limited nonprogressive lesions that become asymptomatic in several weeks.

The MR findings in this case are nonspecific; the intermediate signal on short and the brighter signal on long TR/TE sequences suggest the presence of fluid, or bone marrow edema. Edema may be the initial manifestation of osteoid osteoma, osteomyelitis, infarction, or malignancy, and the ultimate significance of the MR findings must be assessed in conjunction with the clinical history.

Magnetic resonance has a proven value in detection of radiographically occult fractures at the hip and knee because of the unmatched capability of MR to detect early marrow edema and to provide exquisite anatomic detail. This case demonstrates MR detection of several radiographically occult fractures of the upper limb as well (Fig. 13C).

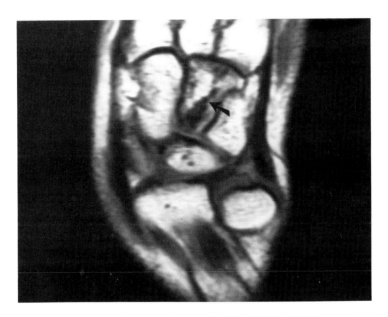

FIG. 13C. Hamate fracture (coronal MR, TR/TE 800/20 msec). A linear focus of decreased signal (*arrow*) is present in the capitate in this MR made 2 weeks after a heavy object fell on this man's hand.

CASE 14

A 24-year-old woman was seen in clinic with signs and symptoms suggestive of ulnar nerve entrapment. (Fig. 14A. Axial MR, TR/TE 500/20 msec. Fig. 14B. Axial MR, TR/TE 2,000/20 msec. Fig. 14C. Axial MR, TR/TE 2,000/80 msec.) Are any abnormalities present?

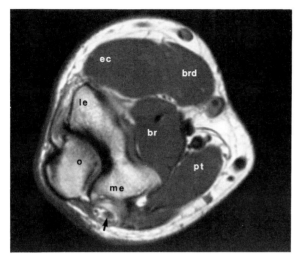

FIG. 14A.

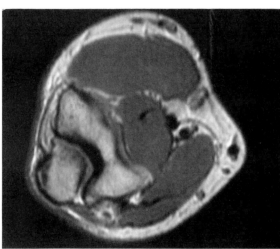

FIG. 14B.

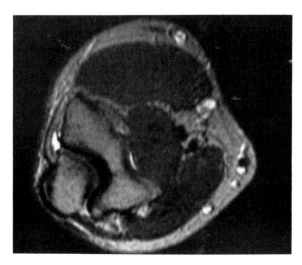

FIG. 14C.

Findings: Axial sections (Fig. 14A–C) through the elbow obtained with T1 (A), spin-density (B), and T2-weighted (C) sequences demonstrate a normal-appearing ulnar nerve (arrow) and surrounding soft tissues (o, Olecranon; me, medial epicondyle; le, lateral epicondyle; br, brachialis muscle; brd, brachioradialis; ec, extensor carpi ulnaris; pt, pronator teres). Additionally, there was no evidence of fibrotic band or other cause of entrapment on any of the MR images.

Diagnosis: Normal-appearing ulnar nerve and surrounding soft tissues. The symptoms resolved over the period of a few weeks without treatment.

Discussion: The diagnosis of ulnar nerve entrapment is, in general, difficult clinically, and frequently it must rely on rather nonspecific nerve conduction studies. Magnetic resonance imaging clearly demonstrated the ulnar nerve and its surrounding tissues in this case. It is unknown whether MRI can demonstrate abnormalities of the nerve itself, as was seen in certain cases of the carpal tunnel syndrome (Case 2), or causes of pressure on the nerve, such as from a fibrous band in the region of the ulnar canal.

CASE 15

A 57-year-old man was assisting a fellow worker lifting a heavy crate. When the crate slipped from the grasp of the fellow worker, the man suddenly bore the full weight of the object and experienced a painful snap in the region of the elbow. What is the most likely diagnosis? What is the current approach to management of this injury? (Fig. 15A. Sagittal MR, TR/TE 2,000/20 msec. Fig. 15B. Sagittal MR, TR/TE 2000/80 msec. Fig. 15C. Axial MR, TR/TE 2,000/20 msec. Case courtesy of H. Fritts, MD, Minneapolis, MN.)

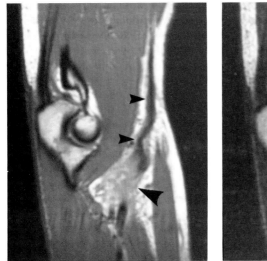 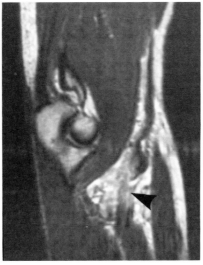 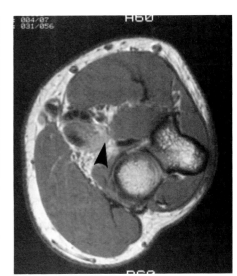

FIG. 15A.   **FIG. 15B.**   **FIG. 15C.**

Findings: On the sagittal intermediate weighted scan (Fig. 15A), there is an irregular area of increased signal intensity in the region of the distal biceps tendon (large arrowhead) and the tendon proximal to this site demonstrates an irregular and serpigenous contour (small arrowheads). On the T2-weighted image (Fig. 15B), the area of irregular increased signal intensity increases, suggesting edema (arrowhead). On the axial scan (Fig. 15C, the biceps tendon is again seen to be disrupted just proximal to its insertion (arrowhead).

Diagnosis: Complete rupture of the distal tendon of the biceps brachii.

Discussion: The biceps tendon may rupture or be avulsed at or near its insertion onto the tuberosity of the radius. The most common mechanism of injury is that of forced flexion of the radius against a strong resistance as in the example described in the case history. The patient commonly experiences a tearing sensation or painful snap at the elbow followed immediately by weakness of active flexion of the elbow and supination of the forearm. The antecubital space is tender, but swelling may be minimal because of the tense overlying fascia. On active flexion, a bulbous swelling forms in the upper arm representing the retracted belly of the the biceps brachii ("popeye sign"). Rupture of the biceps tendon is often preceded by degenerative changes and is associated with advanced age. It has been suggested that the irregular nature of the radial tuberosity may contribute to erosion and fraying of the distal tendon and predispose it to rupture. A partially torn but not yet completely avulsed tendon may produce symptoms of localized pain on attempted flexion and supination particularly if performed against resistance.

The biceps brachii muscle arises from two proximal attachments or heads. The short head arises by a common tendon with the coracobrachialis from the apex of the coracoid process of the scapula. The long head arises by a tendon attached to the supraglenoic tubercle at the apex of the glenoid cavity. This tendon arches over the humeral head and emerges from the joint enclosed in a synovial sheath and descends within the intertubercular sulcus. Each tendon is succeeded by an elongated muscle belly, which while closely applied to each other, can be separated until approximately 7 cm proximal to the elbow joint. At this point they end and contribute to a flattened tendon that attaches to the radial tuberosity. (Fig. 15D,E). At the level of the elbow, the tendon gives rise to a broad aponeurosis that passes obliquely and inferiorly to fuse with the deep fascia covering the origins of the flexor muscles of the forearm.

Following acute rupture of the distal biceps tendon, the marked flexor weakness and partial loss of supinator power is temporary. Following rehabilitation, much of the flexor power is regained. As a result, unless strong flexion and supination is required (e.g., for the athlete or person engaged in a strenuous physical occupation) surgical repair may not be indicated. To regain flexor function, the tendon can be sutured to the brachialis at its insertion. If good rotary power of the forearm is required, the tendon must be sutured to the radial tuberosity.

D
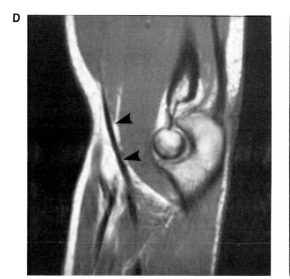

E
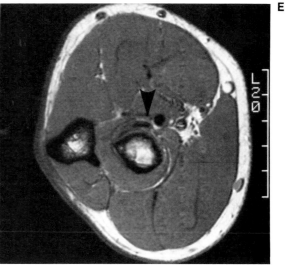

FIG. 15D,E. Normal distal tendon of the biceps brachii. **D:** Sagittal MR, TR/TE 2,000/20 msec. The course of the normal distal tendon of the biceps is seen as it courses to its attachment on the radial tuberosity (*arrowheads*). **E:** Axial MR, TR/TE 2,000/20 msec. The distal biceps tendon is seen just proximal to its attachment on the radial tuberosity (*arrowhead*).

CASE 16

This 35-year-old man was seen in clinic with pain in the region of the proximal interphalangeal (PIP) joint. The diagnosis of ligamentous injury was entertained. (Fig. 16A. Coronal MR, TR/TE 500/20 msec. Fig. 16B. Sagittal MR, TR/TE 500/20 msec.) What is the diagnosis?

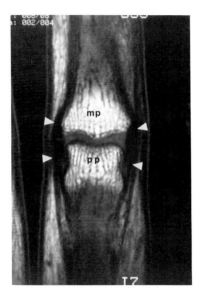 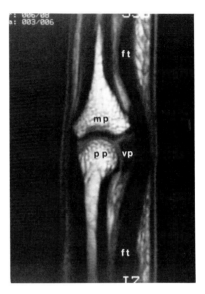

FIG. 16A.   **FIG. 16B.**

Findings: A coronal image of the PIP joint obtained at a 4-cm FOV (Fig. 16A) demonstrates that the collateral ligaments (arrowheads) are intact. A sagittal view through the PIP joint of the same subject (Fig. 16B) demonstrates normal flexor tendons (ft) and volar plate (vp) (pp, proximal phalanx; mp, middle phalanx). (A T2-weighted coronal sequence [not shown here] also demonstrated no abnormalities.)

Diagnosis: Normal-appearing PIP joint and associated soft tissue structures. The patient's symptoms resolved over a several-week course without treatment.

Discussion: The diagnosis of ligamentous, capsular, or tendinous injury to the digit can be difficult to make clinically. This case shows the exquisite detail with which MR performed on a clinical system can demonstrate the soft tissues of the digits. Although the accuracy of MR for the diagnosis of injury to these soft tissues is not known, it is probable that MR will prove very accurate with the level of resolution shown here.

CONCLUSION

In this chapter we have shown a wide variety of cases involving the hand and elbow in which MRI accurately demonstrated the abnormalities. Our experience with any given abnormality is rather limited, with the result that it is difficult to give definitive information in regard to the best techniques for scanning or the appropriate role for MRI in clinical management. Here as elsewhere in the musculoskeletal system, MRI has demonstrated its capability for the imaging of bone marrow and soft tissue structures. It is probably a safe assumption that given sufficient resolution, MRI will prove as accurate for the assessment of disease in the hand and elbow as it has elsewhere in the musculoskeletal system.

Although the capabilities of the currently available commercial systems for imaging these structures are good, we believe that they can be improved with only minor system modifications. To this end we have built and are continuing to build highly efficient RF coils optimized for imaging various parts of the arm. In addition we have written software for our MR system that decreases the minimum FOV from 8 cm to 2 cm and the minimum slice thickness from 3 mm to 1 mm. Finally, we have devoted considerable efforts to developing better methods of patient positioning and stabilization. Using these capabilities we have found it possible to routinely image small parts at FOVs as small as 4 cm. The use of smaller FOVs, however, places considerable strain on the hardware of currently available systems as well as on our patient positioning capabilities. To obtain such images on a routine basis may require the development of high field, small bore systems dedicated to musculoskeletal MRI.

RECOMMENDED READING

Binkowitz LA, Ehman RL, Cahill DR, Berquist TH. Magnetic resonance imaging of the wrist: normal cross section imaging and selected abnormal cases. *Radiographics* 1988;8:1171–1202.

Bowerman J. *Radiology and injury in sports.* New York: Appleton-Century-Crofts, 1977.

Bunnell DH, Fisher DA, Bassett LW, et al. Elbow joint: normal anatomy on MR images. *Radiology* 1987;165:527–531.

Day L, Bovill EG, Trafton PG, et al. Orthopedics. In: Way LW, ed. *Current surgical diagnosis and treatment.* Los Altos: Lange, 1985; 947.

Deutsch A, Mink JH, Waxman A. Occult fractures of the proximal femur: MR imaging. *Radiology* 1988.

Deutsch A, Resnick D. Eccentric cortical metastases to the skeleton from bronchogenic carcinoma. *Radiology* 1980;137:29–52.

Ellman H. Osteochondrosis of the radial head. *J Bone Joint Surg [Am]* 1975;54:1560.

Gore RM, Rogers LF, Bowerman J, et al. Osseous manifestations of elbow stress associated with sports activities. *AJR* 1980; 134:971–977.

King JW, Brefsford HJ, Tullos HS. Analysis of the pitching arm. *Clin Orthop* 1968;67:116–123.

Koenig H, Lucas D, Meissner R. The wrist: a preliminary report on high-resolution MR imaging. *Radiology* 1986;160:463–467.

Middleton WD, Kneeland JB, Kellman GM, et al. MR imaging of the carpal tunnel: normal anatomy and preliminary findings in the carpal tunnel syndrome. *AJR* 1987;147:307–316.

Middleton WD, Macrander S, Kneeland JB, et al. MR imaging of the normal elbow: anatomic correlation. *AJR* 1987;149:543–547.

Mink JH, Deutsch A. Occult cartilage and bone injuries of the knee: detection, classification and assessment with MR imaging. *Radiology* 1989;170:823–831.

Mitchell DG, Rao VM, Dalinka MK, et al. Femoral head necrosis: correlation of MR imaging, radiographic staging, radionuclide imaging and clinical findings. *Radiology* 1987;162:709–715.

Pennes DR, Louis DS, Fechner K. Bone marrow imaging (Letter to the editor). *Radiology* 1989;170:894.

Reinus WR, Conway WF, Totty W, et al. Carpal avascular necrosis: MR imaging. *Radiology* 1986;160:689–693.

Resnick D. Arthrography, tenography, and bursography. In: Resnick D, Niwayama G, eds. *Diagnosis of bone and joint disorders.* Philadephia: WB Saunders Co., 1988;311–324.

Resnick D, Niwayama G. Osteonecrosis: diagnostic techniques, specific techniques, and complications. In: Resnick D, Niwayama G, eds. *Diagnosis of bone and joint disorders.* Philadelphia: WB Saunders Co., 1988;3261–3264.

Vasconez LO, Barton RM, Repogle SL, et al. Plastic and reconstructive surgery. In: Way LW, ed. *Current surgical diagnosis and treatment.* Los Altos: Lange Medical Publications, 1985;1071–1073.

Weiss KL, Beltran J, Lubbers LM. High-field MR surface coil imaging of the hand and wrist. Part II. Pathologic correlations and clinical relevance. *Radiology* 1986;160:147–151.

Weiss KL, Beltran J, Shaman OM, et al. High-field MR surface coil imaging of the hand and wrist. Part I. Normal anatomy. *Radiology* 1986;160:143–146.

CHAPTER 4

The Spine

Kenneth B. Heithoff and Jay L. Amster

Spine Imaging Techniques

Cervical: Patients are positioned head first, supine with their heads resting in a cervical receive-only surface coil. They are centered midsagittal to the bore light, infrasuperiorly at the larynx, and secured with a Velcro strap and tape.

Eleven 3-mm 1.5-mm gap T1-weighted sagittal images are performed through the full width of the cervical spine (time of repetition [TR]/time of echo [TE] 600/20; 256 × 256 matrix; number of excitations [NEX] 2; field of view [FOV] 24-cm).

A presaturation data base is utilized in the superior-inferior direction to minimize blood flow artifact.

Nine multiplanar gradient recalled (MPGR) 3-mm 1-mm-gap peripherally gated partial flip angle sagittal images are obtained through the central canal from facet joint to facet joint (TR/TE 500–800/15; flip angle 20°; 256 × 256 matrix; 2 NEX; 24-cm FOV).

Fourteen to twenty MPGR 3-mm 1-mm-gap peripherally gated partial flip angle axial images are routinely obtained from the C3-4 disk space through the C7-T1 space, to include all affected disks (TR/TE 500–800/15; flip angle 20°; 256 × 256 matrix; 2 NEX; 24-cm FOV).

Thoracic: Patients are positioned feet first, supine with their thoracic spine centered to the 5 × 11-inch receive-only surface coil. They are centered midsagittally to the bore light and infrasuperiorly to midsternum to include C7-T1.

A short T1-weighted 5-mm 5-mm-gap large-FOV sagittal localizer is obtained to provide medial to lateral centering and thoracic vertebrae location (TR/TE 300/20; 256 × 128 matrix; 1 NEX; 48-cm FOV).

Thirteen 3-mm 1.5-mm-gap T1-weighted sagittal images were obtained through the full width of the thoracic spine (TR/TE 800/20; 256 × 256 matrix; 4 NEX; 32-cm FOV).

Nine MPGR 4-mm 1-mm-gap peripherally gated partial flip angle sagittal images are obtained through the central canal of the cord from facet joint to facet joint (TR/TE 500–800/15; flip angle 20°; 256 × 256 matrix; 2 NEX; 32-cm FOV).

Fourteen to twenty MPGR 4-mm 1-mm-gap peripherally gated partial flip angle axial images are obtained through the affected disk levels (TR/TE 500–800/15; flip angle 20°; 256 × 256 matrix; 2 NEX; 24-cm FOV).

Lumbar: Patients are positioned feet first, supine with their lumbar spine centered to the 5 × 11-inch receive-only surface coil. They are centered midsagittally to the bore light and infrasuperiorly at the iliac crests.

A short sagittal T1-weighted sequence is obtained to provide medial to lateral centering.

Eleven to fourteen multiecho 3-mm 1-mm-gap sagittal images are obtained through the full width of the lumbar spine. A flow compensation data base is used to minimize the cerebral spinal fluid pulsation artifact (TR/TE 2,500/30–80; 256 × 256 matrix; 1 NEX; 28-cm FOV).

Seventeen to twenty-five 5-mm 1-mm-gap multiecho axials are routinely obtained from L3-4 to L5-S1, to include all affected disks. A flow compensation data base is used to minimize the cerebral spinal fluid pulsation artifact (TR/TE 3,000/30–80; 256 × 256 matrix; 1 NEX; 24-cm FOV). Additional selected axial images are obtained through more proximal disks showing degeneration on the preceeding sagittal sequence.

CASE 1

This 17-year-old boy was referred with left-sided back and leg pain and neurologic deficit following two traumatic back injuries 6 weeks and 1 week prior to admission (PTA). He had been given a diagnosis of "probable spinal tumor" based on this outside CT. What are the differential considerations and the "Aunt Minnie" diagnosis in this case?

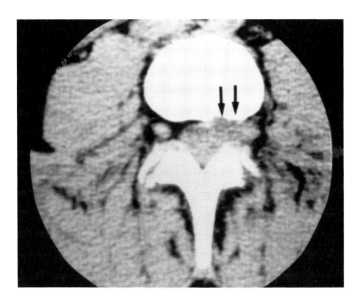

FIG. 1A.

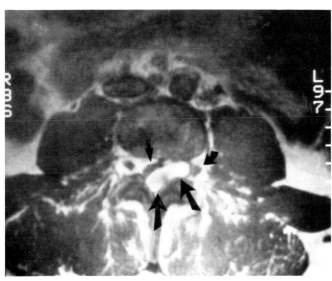

FIG. 1B.

Findings: Axial CT at L4-5 (Fig. 1A) with a large asymmetric soft tissue mass within the central spinal and the left L4-5 intervertebral nerve root canal (arrows), which is isointense with the left L4 nerve root ganglia and the thecal sac. Note that the mass has ill-defined margins and seems to incorporate the exiting nerve root ganglia with infiltration of the fascial planes and complete replacement of the epidural fat. On the MR examination (Fig. 1B), the large arrows point to a large high signal intensity extradural mass lesion at the L4-5 level with a central area of decreased signal intensity. The mass displaces the thecal sac forward and to the right (small arrow) and the left L4 nerve root ganglia laterally (curved arrow).

Diagnosis: Large epidural hematoma.

Discussion: At the time this patient was studied, we had seen approximately 20 cases of lumbar spinal epidural hematoma (Fig. 1C,D), and Levitan had reported three surgically proven cases. All had quite similar and characteristic CT features of a rounded extradural mass lesion, which was largest at the level of the midvertebral body (basivertebral plexus) and had tapering, indistinct margins that extended to the level of the adjacent disk and enveloped the involved nerve root. They were ill-defined and nearly isodense with the thecal sac and indistinguishable from the nerve root and ganglia. Most, but not all, extended superiorly from the disk to the basivertebral plexus of the vertebra above. Review of the first 25 cases (we have now studied more than 70) revealed a notable absence of a history of trauma (this case being the only one). The lesion has been noted in all age groups. Patients present with symptoms indistinguishable from those of a herniated disk. However, the symptoms typically regress rapidly (3 days to weeks) and most commonly (21 of 25 cases) were seen to be associated with an underlying small disk herniation or annular tear on follow-up CT exam after the hematoma had regressed. More recently MR has also shown underlying disk herniation in the vast majority of cases. Typically the disk herniation comprises a small portion of the mass, whereas most of the mass distant to the disk is hematoma. All but four patients in our series regressed spontaneously, in some cases taking up to 4 to 5 months for the mass to resorb entirely. Only one showed an increase in size and symptoms on follow-up. At surgery this lesion had become encysted and had an open basivertebral vein connecting with it, which bled profusely when the hematoma was removed. All three cases reported by Levitan and Wiens had become walled off and chronic and presented with persistent symptoms. Three of the first 20 cases were associated with severe unilateral degenerative facet disease, and the hematoma was based on the medial aspect of the facet joint rather than the disk.

This lesion is remarkably common in our experience, given the paucity of reports of epidural hematoma in the literature. Our hypothesis is that the hematoma results from a torn, fragile epidural vein concomitant with and perhaps secondary to underlying degenerative disk disease (in most cases) or degenerative facet disease. Given the almost universal absence of a history of significant trauma, we have hypothesized that the fragile epidural veins adjacent to the disk or facet are torn by relatively minor torsional injury, as a result of instability (facet disease) or by a minor annular disruption.

We feel that this entity may have been responsible for many of the so-called disappearing disks of myelographic and surgical literature before the availability of CT and MR. The rapid cessation of symptoms in these patients is a probable explanation of why so few of these lesions have been previously noted and reported. In our experience, the patients have severe pain, are seen by their physician, and hospitalized, and were it not for early CT, these lesions would be undetected because myelography was reserved as a preoperative procedure and the vast majority of these lesions and symptoms regress spontaneously and do not require surgery. Therefore, we presume that most are written off as an episode of disk herniation with response to conservative therapy.

Although we became familiar with the CT characteristics of this lesion and made presumptive diagnoses based on (a) the typical initial appearance and (b) spontaneous rapid regression of the symptoms and the lesion on follow-up CT, MR performed early is pathognomonic in most, but not all, cases.

The early MR appearance is typical for blood with a high signal intensity of the lesion on T1 proton-density and T2-weighted images. T1-weighted images allow easy distinction between the epidural hematoma and the adjacent thecal sac and nerve root ganglia (Fig. 1B).

We now perform MRI immediately after the CT when this diagnostic possibility is raised, because the primary differential consideration is a free fragment herniation. The recognition of epidural hematoma and its distinction from a free fragment is crucial given the similarity of symptoms and CT appearance of these two lesions but their disparate treatments (Fig. 1E,F).

When the hematoma is subacute or chronic, however, the signal intensity can be intermediate and similar to that of the disk. They may be difficult to distinguish on the basis of signal intensity alone when degenerative disk disease (i.e., annular tear and dehydration) is present, and one must carefully evaluate the associated disk and determine continuity or lack thereof between the parent disk and the mass and compare its signal intensity with that of the affected disk. If a clear distinction cannot be made and the symptoms are stable or regressing, it is important to follow the patient with MR, as regression of the hematoma virtually always occurs over time. Interestingly, the hematoma may take months to resorb and often persists long after the symptoms have remitted.

In summary, epidural hematoma is much more common than is suggested by the few cases reported in the literature. It presents with symptoms indistinguishable from a herniated disk, which, however, regress more rapidly than would be expected with a free fragment disk. It may take months for the hematoma itself to resorb, persisting long after the patient is asymptomatic. It is an important lesion to recognize to avoid unnecessary surgery because most regress completely and spontaneously. It is probably the lesion responsible for many of the so-called disappearing disks noted in prior surgical literature, in which large extradural masses noted on myelography are not evident at surgery. (We have had several such cases referred to us after the fact in the past 2 years.) Computed tomography has a characteristic appearance. However, only a presumptive diagnosis can be made, as a free fragment disk can have an identical appearance. Magnetic resonance is usually diagnostic in the acute or subacute phase when a typical high signal intensity of T1- and T2-weighted images is noted. With age, the signal intensity decreases and may become isointense with the disk.

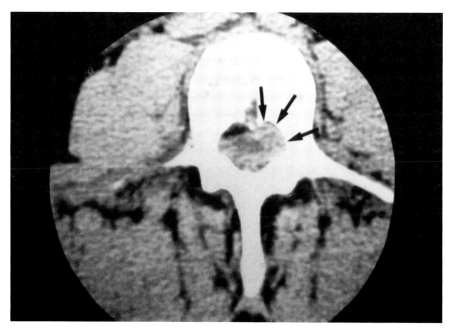

FIG. 1C. Epidural hematoma (axial CT). Note that the epidural hematoma extends up to the midbody of L4 to the level of the basivertebral plexus (*arrows*). It is very typical for these lesions to extend from the margin of the disk up to the basivertebral plexus and not beyond. Note that this lesion again is ill-defined and merges with the margin of the thecal sac. In most cases, the masses are very rounded and may be quite large with marked compression of the thecal sac. The hematoma appears to dissect and be contained within a fascial plane within which lie the epidural veins. This would explain the localization of the hemorrhage, and it causes termination of the hematoma at the level of the basivertebral plexus. This is the most characteristic feature of epidural hematoma. Therefore, one must consider the diagnosis of epidural hematoma when a large and indistinct epidural mass lesion is visualized that seems to originate at or near the basivertebral plexus as a rounded mass and tapers caudally toward the disk with incorporation of the exiting nerve root in a patient with acute sciatica.

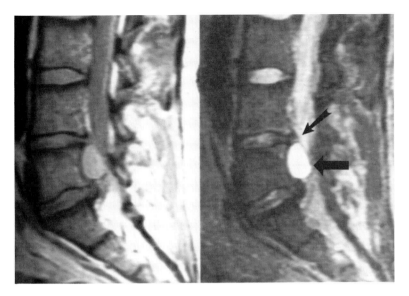

FIG. 1D. Large chronic epidural hematoma (sagittal MR, TR/TE 2,000/30, 80 msec). Large high signal intensity epidural hematoma (*large arrow*). This is the only hematoma in our large series that has failed to resorb. The hematoma increased in size during 8 months with increasing symptomatology. At surgery the lesion openly communicated with a basivertebral vein, which bled profusely when the sac was opened. Note the association of degenerative disk disease with a central annular tear of the L4-5 disk (*small arrow*).

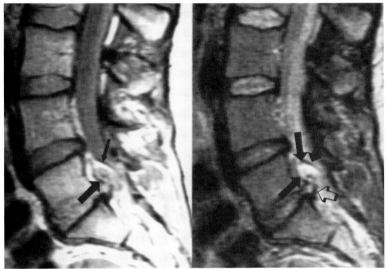

FIG. 1E. Large free fragment disk herniation with associated surrounding epidural hematoma (sagittal MR, TR/TE 2,000/30, 80 msec). The T2-weighted image on the **right** shows the epidural hematoma to be of high signal intensity, whereas it is of intermediate signal intensity on the proton-density image on the **left** (*large arrows*). The superior margin of the hematoma is of high signal intensity on both proton-density and T2 images (*small arrow*). Note the associated extruded disk herniation (*open arrow*). Both the extruded disk herniation and hematoma were surgically proven.

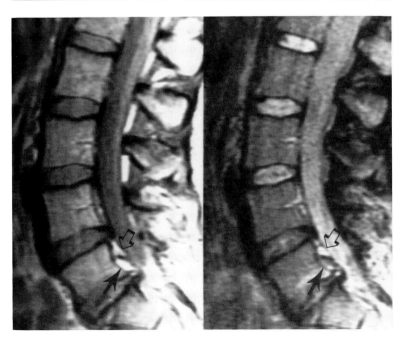

FIG. 1F. Sagittal MR, TR/TE 2,000/30, 80 msec. This sagittal MR was obtained 3 weeks after surgical removal of the disk herniation and epidural hematoma shown in Fig. 1E. The patient had been relieved of symptoms following surgery, then suffered recurrent pain exactly reproducing the preoperative pain 6 days before this MR. This follow-up study demonstrates two components of the epidural hematoma of different ages. More inferiorly, subacute hemorrhage between 3 to 7 days of age is identified, characterized by increased signal intensity on the proton-density image and showing preferential T2 shortening as a result of intracellular paramagnetic substances such as deoxyhemoglobin and/or methemoglobin on the T2-weighted image (*black arrows*). Just craniad to this is a fluid collection that is proteinaceous and consistent with a resolving area of chronic hemorrhage that was demonstrated on the previous examination (Fig. 1E) (*open arrows*).

CASE 2

A 40-year-old man presented with a history of recurrent right leg pain following a right-sided L5-S1 laminectomy 3 months before. A pre- and postgadolinium MR study was performed to rule out a recurrent disk herniation at L5-S1. Is there a recurrent disk herniation present? To what does the small arrow at L3-4 point? (Fig. 2A. Pregadolinium sagittal MR, TR/TE 800/20 msec. Fig. 2B. Sagittal postgadolinium MR, TR/TE 800/20 msec.)

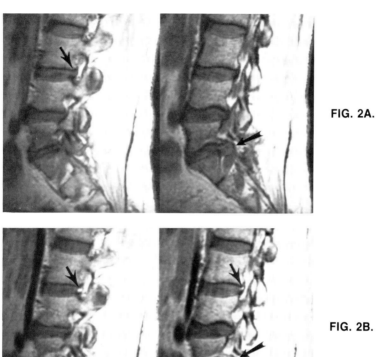

FIG. 2A.

FIG. 2B.

Findings: The large arrows illustrate a large low signal intensity mass lesion lying directly dorsal to the degenerated L5-S1 disk (Fig. 2A). There is an enhancing rim surrounding this mass lesion, which is of uniform thickness and of increased signal intensity with respect to the surrounding fibrosis (Fig. 2B). Note the increased signal intensity in the inferior aspect of the L5 vertebra consistent with inflammation (Fig. 2B). The small arrows delineate an enhancing lesion within the periphery of the L3-4 annulus, which is not visualized on the pregadolinium T1-weighted images (Fig. 2B).

Diagnosis: (1) Recurrent disk herniation with marked enhancement of a fibrous capsule of the disk. (2) Enhancing radial annular tear. The pregadolinium scan shows no evidence of the tear, whereas the postgadolinium scan shows enhancement. This is consistent with uptake by granulation tissue or inflammatory tissue in the margin of the tear. This was visualized on T2-weighted spin-echo images 6 months before as an equally high signal intensity lesion and, in the absence of this new information, was felt to represent inflammatory fluid within an annular tear.

Discussion: The theory of the pathogenesis of degenerative disk and facet disease proposed by Farfan and promulgated by Kirkaldy-Willis indicates that repeated rotational strains produce circumferential fissuring and desiccation of the disk, which leads to radial tears. Because the transmitted weight-loading forces are greatest at the squared-off posterolateral margins of the vertebral bodies (Farfan) and the decussating fibers of the disk are weakest in this location, most herniations occur posterolaterally.

Magnetic resonance has the capability to enhance our understanding of degenerative disk disease by providing *in vivo* visualization of the entire spectrum of degenerative disk disease from early dehydration to fissuring, incomplete and complete radial tears with or without increased signal on T2-weighted images (Fig. 2C–E), contained versus noncontained herniations, and certain differentiation of extruded from free fragment herniations. Magnetic resonance properly performed with two-echo spin-echo sagittal and axial images leaves no excuses for a report that is not *very* specific regarding the type and extent of disk pathology present. Use of words like "suspicious for," "could be," "maybe," or use of the generic term "protrusion" must now be discarded in favor of a very specific descriptive nomenclature made possible by MR.

We have observed a relatively high incidence of incomplete radial tears in our patients with degenerative disk disease. A recent study of cryotome sections of cadaver spines by Haughton (*personal communication*) revealed radial tears in all disk "bulges" greater than 2.5 mm.

A significant number of these radial tears show an increased signal intensity on proton-density and T2-weighted images. Initially, this was presumed to represent very hydrated, contained, herniated nuclear material. However, Bumphrey and co-workers at the Cleveland Clinic showed that the disk material removed from these patients did not have increased water content when compared to patients without this finding and therefore presumed that these collections represented fluid. However, the case presented above indicates that, at least in some patients, the increased signal represents tissue that is either inflamed or undergoing reparative granulation because it enhances with gadolinium. We favor inflammation (Fig. 2F,G) as the most logical explanation because diskography performed in these patients uniformly shows a radial tear that correlates with the MR. In this case, the high signal intensity enhancement persisted for 6 months. Moreover, Dr. Charles April (*personal communication*) has noted that 20 of 21 patients with the finding of increased signal in the annular tear have had positive diskograms with exact reproduction of their symptoms.

Recent experimental work by Dr. Jeff Saal and co-workers (*unpublished data*) has shown very high levels of potent inflammation-inducing substances in herniated disk material, which supports the theory of an inflammatory component of sciatica caused by herniated disks. The MR demonstration of inflammatory reaction in the margins of an annular tear or the presence of inflammatory fluid within the tear raises interesting potential for both specific and highly accurate radiologic diagnosis of symptomatic disk degeneration *and* therapy. The efficacy of intradiskal steroids is being tested in a double-blind prospective study at this time. Finally, we have observed disappearance of the MR visualization of radial tears in several patients on 1-year follow-up examinations, indicating that spontaneous healing does occur in some patients.

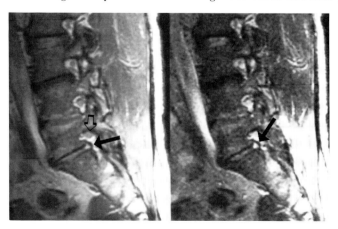

FIG. 2C. Sagittal MR, TR/TE 2,000/30, 80 msec. Sagittal spin-echo images show a small lateral disk herniation with high signal intensity, fluid, and/or granulation tissue contained by the thinned outer annular fibers (*black arrows*). Note the normal location of the L5 nerve root ganglia immediately craniad to the lateral disk herniation within the ventral one-half of the canal as it exits under the L5 pedicle (*open arrow*).

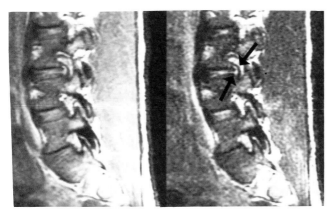

FIG. 2D. Sagittal MR, TR/TE 2,000/30, 80 msec. Note the peripheral circumferential partial annular tear of the L3-4 disk, which shows increased signal intensity on T2-weighted images consistent with inflammation or granulation (*arrows*).

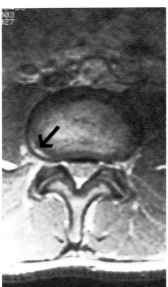

FIG. 2E. Axial MR, TR/TE 1,000/20 msec. Note the high signal intensity peripheral circumferential annular tear of the right side of the L3-4 disk on this proton-density image (*arrow*).

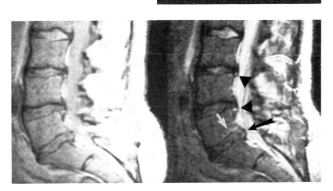

FIG. 2F. Sagittal MR, TR/TE 2,000/30, 80 msec. Note multiple dehydrated and degenerated disks at L3-4, L4-5, and L5-S1 in this 30-year-old woman. Note the presence of a contained central disk herniation at L5-S1 (*large arrow*) with inflammation and/or granulation involving the adjacent marrow of the inferior aspect of the L5 vertebra. This association is often seen in patients with active degenerative disk disease (*white arrow*). There are small fluid collections within the peripheral portions of nearly complete annular tears at L3-4 and L4-5 (*arrowheads*).

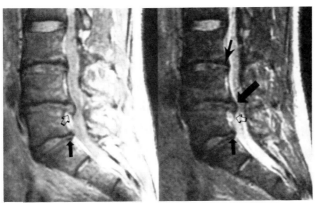

FIG. 2G. Sagittal MR, TR/TE 3,000/30, 80 msec. This case shows the spectrum of degenerative disk disease with marked degenerative dehydration of the L3-4, L4-5, and L5-S1 disks and a nearly complete annular tear of the L3-4 disk (*arrow*), an extruded central disk herniation at L4-5 with a complete tear of the annulus (*small open arrows*), and dorsal displacement of the posterior longitudinal ligament (*large arrow*). There is a small contained central disk herniation at L5-S1 (*small black arrows*).

CASE 3

A 63-year-old man experienced numbness and pain in the right side of the genital and anal region and the right heel for the past 10 days. What lesion is delineated by the arrows?

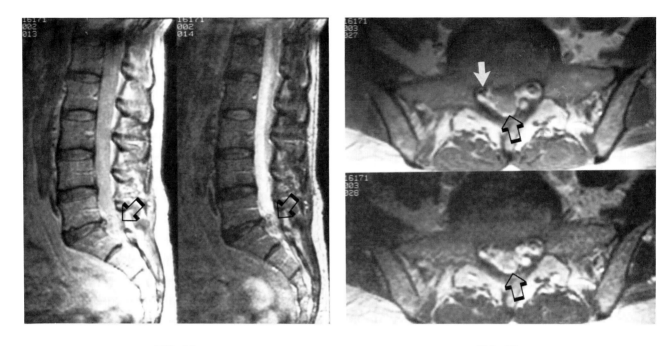

FIG. 3A.   **FIG. 3B.**

Findings: There is a large mass lesion lying within the central spinal canal, dorsal to the S1 vertebra (Fig. 3A,B, open arrows). It is of mixed signal intensity and brighter than cerebrospinal fluid (CSF) on the proton-density images and of lower signal intensity than CSF on T2-weighted images. It is of similar signal intensity to the nucleus of the L5-S1 disk, although no definite continuity between the mass and the disk is identified. Note that the mass simulates the thecal sac on axial images and was misdiagnosed as the thecal sac on a CT scan performed before the MR. There is marked dorsal and left lateral displacement and compression of the thecal sac, and the mass lies between the S1 nerve roots without significant S1 nerve root compression (white arrow). There is marked compression of the remaining sacral nerve roots, explaining the perineal symptoms.

Diagnosis: Very large, free fragment L5-S1 disk herniation, which lies medial to the S1 nerve root and thereby explains the lack of sciatica and neurologic deficit in the S1 distribution but there is involvement of the sacral nerve roots, resulting in genital and perianal symptoms.

Discussion: The differentiation of contained from noncontained disk herniation has assumed greater significance with the advent of percutaneous nucliectomy. Advocates of this procedure feel that best results are obtained in patients with demonstrated contained herniations, and if noncontained, the results drop to approximately 60%.

Because CT cannot resolve annulus from nucleus, this distinction was difficult in many cases before MR. Magnetic resonance performed with double-echo spin-echo sagittal and axial imaging clearly delineates the nonhydrated outer annular fibers from the nucleus and the inner annulus as a sharply defined low signal intensity linear structure forming the posterior margin of the disk. One can also easily differentiate contained from noncontained herniations and annular tears from herniations when the tear is unaccompanied by herniated nuclear material, as discussed in Case 2.

The posterior margin of the disk as visualized on MR is composed of both annulus and posterior longitudinal ligaments, which merge together both anatomically and visually on MR images. The posterior longitudinal ligament is a quite variable structure and tapers centrally from side to side at the level of the disk and often is not an anatomically significant structure. It can be identified as a separate structure in a minority of patients as a very thin, filamentous structure lying dorsal to the vertebral body when it is stripped away by a disk herniation (Fig. 3C–F). I believe it is in these few patients with separate, discrete annulus and posterior longitudinal ligament that the so-called roof disk of surgical literature is found. The MR appearance is that of a complete annular tear, with extrusion of nuclear material beyond the annulus, but an intact, dorsally displaced posterior longitudinal ligament. In all other patients, when a disk herniation is present and associated with complete separation of the annulus–posterior longitudinal ligament complex, one can make an assured diagnosis of a noncontained disk. Conversely, when the outer annulus is visualized to be intact, even though thinned and bulging, that is a contained herniation. In this exercise it goes without saying that *all* sagittal images must be carefully studied for a complete rent because annular tears are often oblique from medial to lateral as they extend dorsally, and often only one of the sagittal images showing the disk herniation will visualize the tear as it reaches the dorsal surface of the disk. When in doubt, consult the T2-weighted axial image of the disk (hopefully, no one is still routinely doing only T1-weighted axials), and the radial tear will be visualized and confirmed by high resolution scanners in most cases.

By using the above scanning techniques and criteria of disk degeneration, we are able to provide a very accurate and complete nomenclature for disk disruption and to clearly define the pathology present for our surgeons. Spine surgeons now try to make early critical decisions in patient management regarding conservative versus surgical treatment. Many of them will attempt conservative management (with excellent success rates) if we diagnose a contained disk herniation (some may offer percutaneous nucliectomy), and they will recommend early surgery for noncontained herniations. Thus, MR can and should play a vital role in this patient management process when utilized to its fullest capabilities by radiologists willing to be definitive with the information presented.

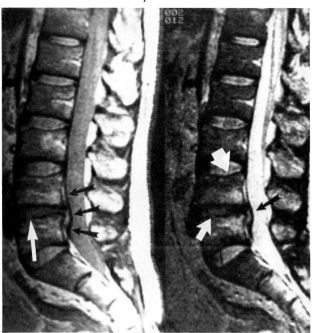

FIG. 3C. Sagittal MR, TR/TE 2,000/30, 80 msec. The MR shows severe degenerative dehydration of the L4-5 disk. There is a moderately large extruded disk herniation, which extends caudally to lie behind the upper one-third of the L5 vertebral body as an intermediate soft tissue mass density, which dorsally displaces the posterior longitudinal ligament (*black arrows*). Thus, although this is an extruded disk herniation, the herniation itself would not be apparent to the surgeon as it lies beneath the posterior longitudinal ligament, which has been stripped away from the posterior margin of the disk by the herniation. Note the erosion of the superior end plate of L5 (*white arrows*) and the active inflammation and/or granulomatous reaction of the marrow of the L4 and L5 vertebra, which presents as low signal intensity on proton-density images and increased signal intensity on T2-weighted images. These inflammatory changes (also known as Modic Type I changes) may be related to the presence of back pain as a prominent component of the patient's symptomatology.

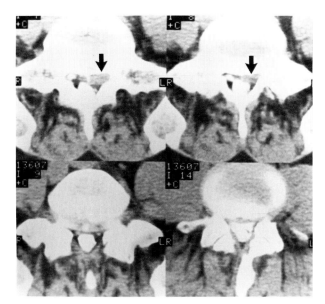

FIG. 3D. Axial CT image showing a left L5-S1 free fragment herniation (*black arrows*). Note the nonfilling of the left S1 nerve root sheath on this contrast-enhanced CT scan and the absence of deformity of the L5-S1 disk annulus caused by the caudal migration of the free fragment herniation. The clear separation of the fragment from the parent L5-S1 disk ensures the diagnosis of free fragment herniation.

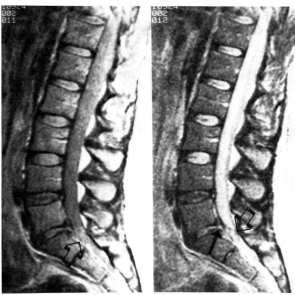

FIG. 3E. Free fragment disk herniation (sagittal MR, TR/TE 2,000/ 30, 80 msec). Both proton density and T2-weighted images show a complete tear in the L5-S1 annulus (*black arrow*). Note that the extruded free fragment L5-S1 disk herniation is nearly isointense with the CSF on the proton-density image but is of decreased signal intensity similar to that of the parent L5-S1 disk on T2-weighted images (*open arrows*).

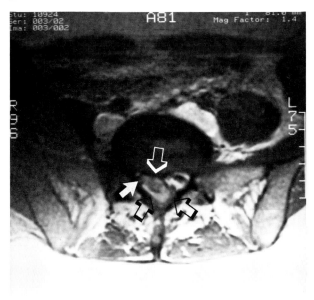

FIG. 3F. Axial MR image (TR/TE 3,000/30 msec) shows marked compression of the right S1 nerve root (*white arrow*) by the large free fragment disk herniation (*open arrows*). The complete rupture of the annulus and posterior longitudinal ligament separates the above extruded disk herniations from contained herniations.

CASE 4

A 56-year-old man presented with a history of severe back pain and 10 prior back surgeries including a fusion from L4 to the sacrum. What process involving the marrow is depicted by the white arrows? What is the diagnosis of the abnormality depicted by the small black arrows? (Fig. 4A,B. Saggittal MR, TR/TE 2,000/30, 80 msec.)

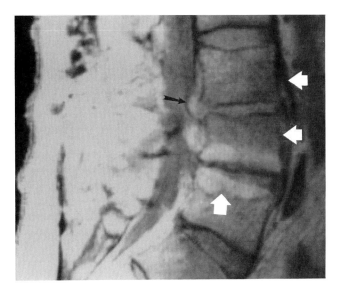

FIG. 4A.

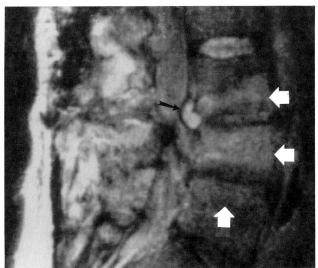

FIG. 4B.

Findings: The white arrows depict diffuse infiltration of the marrow by a process which is of low signal intensity on proton-density images and high signal intensity on T2-weighted images indicating inflammation or granulomatous involvement of the marrow adjacent to the degenerated L3-4 disk. The small black arrows point out a large extruded, free fragment L3-4 disk herniation. This disk herniation and inflammation is occurring at the level above a solid L4 to the sacrum fusion.

At the L4-5 level, the high signal intensity substance noted on proton-density images becomes low signal intensity on T2-weighted images consistent with fatty replacement of the marrow. Plain films and tomograms showed a solid dorsolateral fusion from L4 to the sacrum. This is consistent with fatty replacement of the marrow adjacent to a disk severely narrowed and dehydrated as a result of prior active degenerative disk disease, which is now stabilized by the fusion.

Diagnosis: Transitional syndrome with a central disk herniation and Modic Type I inflammatory changes within the marrow adjacent to the L3-4 disk and fatty infiltration of the L4-5 marrow adjacent to a fused L4-5 motion segment.

Discussion: Modic Type I (Fig. 4C–F) inflammatory changes are very commonly seen in the marrow adjacent to disk spaces affected with active degeneration and/or herniation. It is also seen in patients with segmental instability. Modic has indicated that this process results from infiltration of the marrow by fibrovascular tissue. It is hypothesized that there is permeation of the end plate by inflammatory agents from the disk. The end plates are often abnormal and eroded. Anecdotally, based on experience gained from scanning large numbers of patients but without specific documentation, it is our experience that these inflammatory Modic Type I changes do correlate well with the patient's symptomatology. That is, if the infiltration of the marrow is predominantly right-sided, the patient will have right-sided symptomatic back pain. This is true even in the absence of disk herniation.

Fatty replacement of the marrow is commonly seen in late stage degenerative disk disease. It is thought that fatty replacement of the marrow is a result of marrow degeneration subsequent to preexisting inflammation. In addition to its association with burned-out degenerative disk disease as in the case above, it is also often seen in patients with solid fusions of motion segments previously involved with degenerative disk disease who were successfully fused for incapacitating back pain.

It is now appreciated that the intervertebral disk, one of the two avascular body structures in adult life (the other being the cornea of the eye), receives the majority of its nutrition by the process of diffusion across the end plate. This diffusion is enhanced by a decrease in stress loading on the spine and by exercise. For reasons that are not yet clearly understood, the diffusional resistance can be significantly changed. One means by which this apparently occurs is the production of chemical substances in response to insult or injury to the disk (such as annular tears and/or herniations). Some investigators feel that these substances may, in turn, promote an autoimmune response leading to an increase in diffusional resistance. As this resistance increases, the end plate undergoes sclerosis and the adjacent marrow appears initially to experience an inflammatory response (Modic Type I). Following the inflammatory response, there is marrow degeneration adjacent to these vertebral bodies with fatty replacement.

Research is underway to attempt to verify the observation that inflammatory marrow changes are predictive of and bear a causal relationship to back pain. If there is found to be a positive correlation, MR would then provide a means for determining the efficacy of various conservative treatments and surgical stabilization procedures because one can monitor the response of the inflammatory changes in the marrow. For example, it is known that successful solid fusions are associated with a high incidence of fatty degeneration of the marrow adjacent to fused degenerated disks.

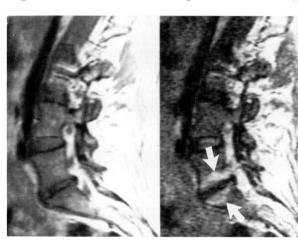

FIG. 4C. Sagittal MR, TR/TE 2,000/30, 80 msec. Note the severe degenerative dehydration of the L5-S1 disk and the Type I inflammatory changes in the adjacent L5 and S1 vertebra with increased signal intensity on T2-weighted images (*arrows*) and decreased signal intensity on proton-density images.

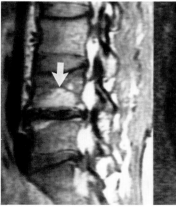
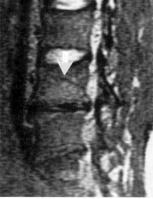

FIG. 4D. Fatty replacement of the marrow (sagittal MR, TR/TE 2,000/30, 80 msec). Note the high signal intensity of the inferior one-half of the L4 vertebral body and the anterior-superior corner of the L5 vertebral body adjacent to a severely degenerated and dehydrated L4-5 disk. There is decay of the fat signal within the marrow on T2-weighted images (*arrows*).

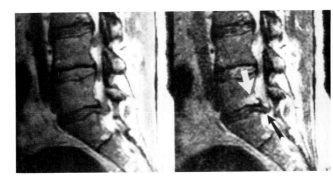

FIG. 4E. Disk herniation and marrow inflammation (sagittal MR, TR/TE 3,000/30, 80 msec). This 51-year-old man had successful chymopapain treatment of a large disk herniation at L5-S1 documented on pre- and postchymopapain CT examinations. The patient has subsequently reherniated this disk, as shown on this MR (*black arrow*), and there are Modic Type I inflammatory changes in the posterior-inferior aspect of the L5 vertebral body (*white arrow*). Such inflammatory changes have been noted in patients as a result of chymopapain injection; however, this study was performed 5½ years after chymopapain. Therefore, the inflammation is felt to be due to the active ongoing degenerative disk disease present in this patient.

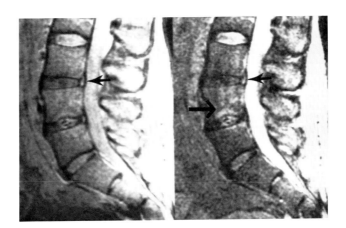

FIG. 4F. Inflammatory changes in the marrow of the L4 vertebral body in a patient with a prior L4-5 diskectomy (sagittal MR, TR/TE 3,000/30, 80 msec). Note the increased signal intensity within the L4-5 disk and the thinning and erosion of the inferior end plate of L4 (*large arrow*). There was no laboratory or clinical evidence of postoperative diskitis nor osteomyelitis. Although this degree of increased signal intensity within the disk with associated erosion of the end plate and inflammation of the vertebral body is very discomforting with respect to possible diskitis and osteomyelitis, it is not uncommon to see such inflammatory changes in the disk and end plate in patients in whom the surgeon has actively scraped the end plate at the time of diskectomy. When these findings are present, therefore, one should very carefully follow the patient both clinically and with MR because one cannot exclude infection on the basis of the initial MR alone. Note the small contained central disk herniation at L3-4 with intact outer annular fibers (*small arrows*).

CASE 5

A 54-year-old woman presented to her physician with a history of severe back pain and systemic symptoms of fatigue and weight loss. What is the etiology of the process involving the L4-5 disk? (Fig. 5A,B. Sagittal MR, TR/TE 2,000/30, 80 msec.)

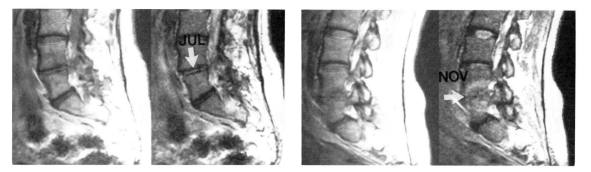

FIG. 5A.   **FIG. 5B.**

Findings: The initial MR study performed in July 1988 (Fig. 5A) shows severe dehydration and degeneration of the L5-S1 disk and marked narrowing of the L4-5 disk (arrow). In November 1988 (Fig. 5B), there was complete resorption of the end plates of the L4 and L5 vertebral bodies adjacent to the L4-5 disk, with complete resorption of the disk and marked inflammatory changes in the adjacent marrow of nearly the entire L4 and L5 vertebra (arrow).

Diagnosis: Isolated lumbar disk resorption syndrome.

Discussion: The isolated lumbar disk resorption syndrome was first described by Harry Crock in 1975 as an entity that was associated with gross narrowing of an isolated disk with sclerosis of the end plates. Patients have a history of multiple bouts of low back pain lasting several days to weeks with resolution. He has since described a possible autoimmune phenomenon. The illustrated case suggests active inflammation as a prominent component of this patient's striking disk resorption. Haughton (*personal communication*) noted that there is a distinct difference between how the marrow responds to an initial exposure to disk material and how it responds to repeated exposure to disk material. If this is validated, this would support an autoimmune mechanism.

In this case, because of the prominent inflammation of the marrow, the possibility of osteomyelitis might be raised; however, the appearance is that of complete and absolute resorption of the disk rather than destruction of the disk by inflammatory fluid, which would be noted as increased signal intensity on T2-weighted images (Fig. 5C). The malaise and weight loss experienced by this patient has been described by Crock as a prominent feature of this disease entity and is also consistent with a systemic disease process such as an autoimmune response.

Juvenile diskogenic disease (Fig. 5D,E) is an interesting entity that is being noted with increasing frequency in our patient population since the advent of MR. The hallmarks of the disease are prominent anterior Schmorl's nodes, marked disk space narrowing at multiple levels throughout the lower thoracic and upper lumbar spine (in some cases involving the entire lumbar spine), degenerative dehydration of the involved disk, and elongation of the vertebral bodies. Patients being seen with this disease are usually individuals in their early 20s to early 30s who typically present with a long history of vague back pain and stiffness. Their MR images show dehydration and degenerative disk disease of multiple vertebra with Schmorl's nodes, limbus vertebrae, molding of the vertebral bodies, and commonly, associated disk herniation. Although Schmorl's nodes occurring in the centrum of the disk end plate appears to relate to persistence of notochordal remnants and although limbus-type vertebrae and other types of Schmorl's nodes may reflect persistence of embyronic vascular channels, the reason for the dehydration and degeneration of the disk at multiple levels is poorly understood at present. If the disease is an acquired apophysitis, as has been thought to be the case, it would appear that the anterior Schmorl's nodes and abnormalities of the end plates may cause poor nutrition of the disk and/or resorption of the disk. This may take place as a manifestation of increased diffusional resistance across the end plate. Alternatively, this may represent a congenital abnormality of development and/or genetic predisposition to degenerative disk disease and weakened disks. These patients tend to present early in life with associated multilevel disk herniations involving the lower lumbar levels, which are not involved with the classical triad of juvenile diskogenic disease. The L4-5 and L5-S1 levels often develop early and severe disk degeneration and herniation. We have seen L4-5 and L5-S1 disk dehydration in an 11-year-old girl who had thoracolumbar abnormalities consistent with juvenile diskogenic disease. We therefore favor genetically induced, biomechanically weakened disk structure in these patients. Because of the frequency and severity of these early degenerative disk changes, we feel vocational counselling is essential when juvenile diskogenic disease is first noted in young individuals.

Chemonucleolysis as a cause of disk resorption has largely been abandoned in this country. This was due to the occurrence of severe and life-threatening complications as well as the complication of severe and unrelenting back pain in some patients. The discovery of marked inflammation of the marrow of the vertebral bodies adjacent to disk spaces injected with chymopapain strongly suggests a casual relationship between back pain and the direct inflammatory response of the marrow to the chymopapain. Dr. Charles Ray at the Institute of Low Back Care first noted and reported direct communication of the disk with large vascular channels during diskography. Marked sclerosis of the lower one-third to one-half of the vertebral bodies adjacent to disks injected with chymopapain occurring years after the injection have been observed. MR now demonstrates marked inflammatory changes in the vertebral bodies in many of these patients (Fig. 5F,G). This provides strong evidence for transgression of the end plates by chymopapain and the production of marked inflammatory changes with subsequent marrow degeneration and sclerosis of the end plates. Because inflammation noted with degenerative disk disease has a strong association with back pain, one should not be surprised to find persistent back pain as a complication of chymopapain in patients with severe inflammatory changes in their marrow.

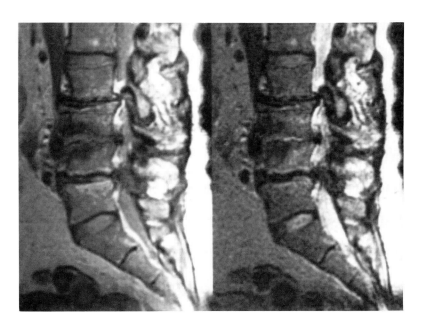

FIG. 5C. Disk resorption syndrome (sagittal MR, TR/TE 2,000/30, 80 msec). There is marked disk resorption at the L3-4 level, and a small amount of fluid within the L3-4 disk. There is marked sclerosis of the end plates and marrow of the adjacent L3 and L4 vertebral bodies as denoted by decreased signal intensity on both proton-density and T2-weighted images. There is also severe degenerative disk disease at L2-3 and L4-5 with marked marginal osteophyte formation adjacent to the involved disks and resultant severe central spinal stenosis at L2-3, L3-4, and L4-5.

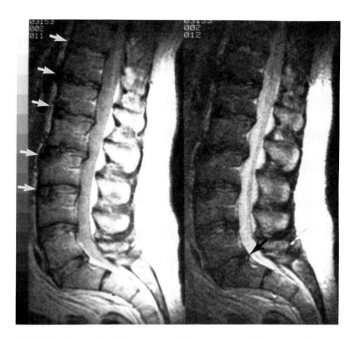

FIG. 5D. Juvenile discogenic disease in a 29-year-old man (sagittal MR, TR/TE 2,000/30, 80 msec). There is marked disk space narrowing of all of the lumbar disks T11-12 caudally through L5-S1. There are large and prominent anterior Schmorl's nodes involving all of the disks from T11-12 through L3-4, and there is anterior osteophytic spurring at the margins of the disks from T12-L1 through L2-3 (*white arrows*). There is also beginning elongation of the vertebral bodies. This is the classic triad of juvenile discogenic disease. Note the associated degenerative disk disease at L4-5 and L5-S1 with a nearly complete central annular tear of the L5-S1 disk (*black arrow*) and moderate dehydration of both L4-5 and L5-S1. In young individuals such as this, advanced disk degeneration at levels not involved with juvenile discogenic disease (L4-5 and L5-S1) is common and suggests an underlying developmental insufficiency of these disks.

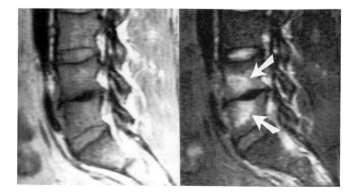

FIG. 5E. Disk resorption following chymopapain (sagittal MR, TR/TE 3,000/30, 80 msec). Note the marked isolated disk resorption of the L4-5 disk in this 26-year-old man with a history of prior chemonucleolysis. There is marked inflammatory response of the adjacent marrow (*arrows*). There is normal hydration of the upper lumbar disks and mild dehydration of the L5-S1 disk without disk resorption. These are typical findings of patients with prior chemonucleolysis. A minority of patients have residual normal hydrated foci of nuclear material within the injected disk.

CASE 6

A 54-year-old man presented with a sudden recurrent right L4 radiculopathy 5½ days following a paralateral removal of a lateral herniated disk. What is the etiology of the recurrent symptoms? (Fig. 6A,B. Axial TR/TE 2,000/30 and sagittal MR TR/TE 2,000/30, 80 msec.)

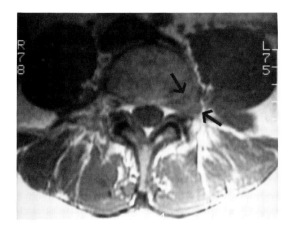

FIG. 6A.

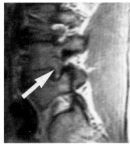

FIG. 6B

Findings: Axial MR image (Fig. 6A) showing a large extruded lateral disk herniation at L4-5 on the left with marked compression and dorsal displacement of the left L4 nerve root ganglia (arrows). The sagittal proton-density image (Fig. 6B) on the left shows the large extruded disk herniation and the compressed L4 ganglia opacifying the entire left L4-5 intervertebral nerve root canal (large arrow). On the T2-weighted image on the right note that there is increased signal intensity of the compressed ganglia by the disk herniation consistent with edema (small arrow).

Diagnosis: Large, recurrent, postoperative lateral annular disk herniation and postoperative collection of fluid and blood lateral to the canal. The herniation was responsible for the recurrent symptoms; the fluid was incidental.

Discussion: Far lateral and lateral disk herniations (synonymous with foraminal herniation) are in fact quite common, although their presence was largely unrecognized by the radiologic or surgical community before CT. We have observed more than 400 cases in our practice. These lesions are very easily detected by CT if one is cognizant of the diagnosis as a cause of asymmetric "enlargement" of the ganglia within the nerve root canal. These patients are often very symptomatic because of the close confines of the bony nerve root canal and the fact that the ganglia is the most sensitive portion of the nerve root to ischemia. Therefore, if there is abnormal opacification of the nerve root canal on axial CT and if sagittal reformatted images confirm that it is not due to (a) up-down (cephalocaudad) bony stenosis with compression of the ganglia or to (b) nerve root anomalies such as conjoined nerve root in a patient with a monoradiculopathy in the L3 or L4 distribution (anterior thigh or groin pain), one can be quite confident of this diagnosis, and additional imaging, i.e., MR, is not necesssary (Fig. 6C,D).

However, when other pathology or nerve root anomalies (conjoined nerve root, nerve root ectasia, etc.) are present, as in our example case, MR is the only procedure that will resolve the issue. Fortunately, most lateral disk herniations are located at L4-5 or L3-4, less often at L2-3, and relatively uncommonly at L5-S1 (unless spondylolisthesis is present), whereas lateral spinal stenosis and nerve root anomalies are less common here than at L5-S1.

When differentiating complex soft tissue anatomy within the intervertebral nerve root canal, i.e., excluding a lateral disk herniation underlying a known conjoined nerve root, or lateral disk herniation from fibrosis or hematoma in postoperative patients with prior foraminotomies (as in the case above), only MR is capable of providing a certain radiologic diagnosis. Plain CT shows only homogeneous opacification of the canal. Neither myelography nor contrast CT is of any benefit because the contrast cannot reach the pathology because the nerve root sheath terminates proximal to the ganglia. This explains the complete inability of myelography to diagnose lateral stenosis and why it cannot suffice as a sole preoperative screening procedure without risking a significant incidence of failed back surgery syndrome (FBSS). Sixty percent of the patients operated for FBSS (in a combined study of 450 patients by Burton and Kirkaldy-Willis) during the era of myelography (pre-CT) had lateral stenosis as the major etiology of their symptoms. In a double-blind study of CT versus myelography which we performed in 1985 in disk herniation and lateral stenosis, 19 of 84 patients had operatively proven lateral spinal stenosis. All 19 were undetected by myelography and all were correctly diagnosed presurgically by CT. A concomitant retrospective study of 100 consecutive laminectomies at the Institute for Low Back Care showed improved salvage rates for patients with multiple previous operations, largely because of detection of previously unrecognized lateral nerve root entrapment (37% had concomitant disk herniation and lateral stenosis and greater than 60% had lateral stenosis).

The soft tissue resolution of MR allows clear distinction between herniated disk material and conjoined nerve root and/or postoperative fibrosis when using double-echo spin-echo technique. This technique provides images with differing signal intensities between recurrent disk and fibrosis rather than the similar low signal intensity of fluid, neural tissue fibrosis, and disk present on T1-weighted images.

Having said this, however, a word of caution regarding the detection of small lateral disk herniations on MR is in order. Because of the relatively thick (5 mm) axial images required to image the entire lower lumbar spine, there is significant partial volume imaging, and the annulus containing the disk is of similar low signal intensity as the bony cortex of the underlying vertebral body and pedicle, making detection difficult. Because CT shows bone as white and disk as gray, axial CT is actually more strikingly abnormal in many cases of lateral disk herniation than axial MR and should be performed if one is unsure of the diagnosis. Also, lateral disk herniations occur at the rounded posterolateral margin of the disk and therefore are not imaged in sharp profile as are classical posterolateral disk herniations. Therefore, partial volume imaging of the normal rounded vertebral body is superimposed on the disk herniation. The information necessary to make the diagnosis is present but may be subtle, and any apparent asymmetries within the nerve root canal or abnormal densities on MR should be carefully scrutinized. This is one place where T1-weighted thin section axial images, as an added sequence, have been very helpful because of improved spatial resolution. (If still in doubt, do axial CT!)

Lateral disk herniation is a surgical enigma. Although radiologic detection is now straightforward, surgery is not. Unfortunately, because of the far lateral location of the disk herniation, the surgeon needs to remove much of the facet to expose the disk herniation, risking instability. Because many of these patients are relatively young and active, this presents a problem of stabilization. Therefore, many spine surgeons now favor a paralateral muscle-splitting approach. This allows preservation of the facets and facet joints if the herniation is focal and far lateral. If this approach is used by your surgeons, you must carefully study the disk to ensure that it is not broad-based and does not extend into the medial one-half of the nerve root canal and/or the spinal (subarticular) recess of the central canal. This type of lateral herniation must be approached from a classical midline approach because the surgeon cannot visualize the medial aspect of the canal from the paralateral approach.

The paralateral approach is not for the uninitiated, as landmarks are scarce, localization can be difficult, and significant vascular structures (arteries as well as veins) surround and overlie the nerve root. Thus, the surgery is often associated with more bleeding than with the midline approach. Finally, the ganglia directly overlies the herniation, being directly in the surgeon's path to the herniation and thus risking nerve root injury for the unwary.

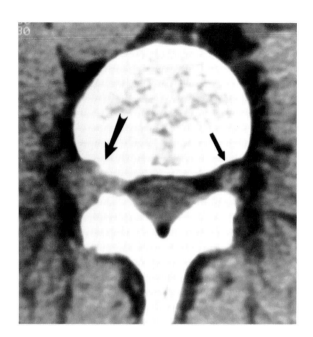

FIG. 6C. Lateral disk herniation. This axial CT image shows a large lateral disk herniation at the L3-4 level on the right with opacification of the intervertebral nerve root canal and marked compression of the exiting right L3 nerve root ganglia (*large arrow*). Note that the disk herniation is isodense with the ganglia. At surgery this was an extruded free fragment lateral (foraminal disk herniation). Note the normal left L3 nerve root ganglia and the surrounding fat (*small arrow*).

Findings: The axial MR image (Fig. 6A) shows a large extruded lateral disk herniation at L4-5 on the left with marked compression and dorsal displacement of the left L4 nerve root ganglia (*arrows*). The sagittal proton-density image (Fig. 6B) on the left shows the large extruded disk herniation and the compressed L4 ganglia opacifying the entire left L4-5 intervertebral nerve root canal (*large arrow*). On the T2-weighted image on the right note that there is increased signal intensity of the compressed ganglia by the disk herniation consistent with edema (*small arrow*).

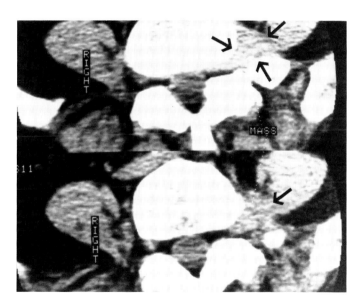

FIG. 6D. Foraminal disk herniation at L5-S1 on the left. The axial CT image shows a very large dense triangular soft tissue mass density within the left L5-S1 intervertebral nerve root canal. At surgery this was a large extruded lateral disk herniation. The iliolumbar ligament forms the lateral border of the L5-S1 intervertebral nerve root canal and contains and delimits any abnormality occurring within that canal, whether it be disk herniation, infection, or neoplasm. Therefore, morphologically all of these structures appear the same. However, in this case, the attenuation values of the mass were very high and consistent with that of disk herniation, allowing the diagnosis of lateral disk herniation (*arrows*).

CASE 7

This is a 35-year-old man with severe right L5 radiculopathy. What is the significance of the spur noted in this case? Is there nerve root compression? What is the normal position of the nerve root within the intervertebral nerve root canal? Is it constant, or is it fixed? (Fig. 7A. Axial CT. Fig. 7B. Sagittal reformatted CT.)

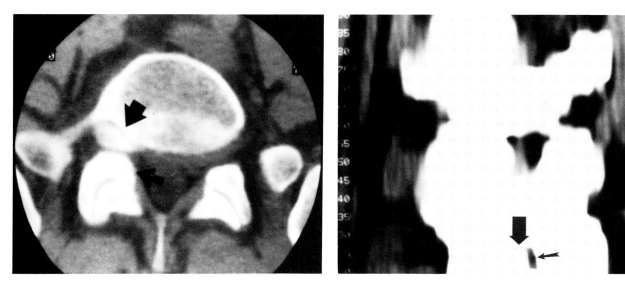

FIG. 7A.   **FIG. 7B.**

Findings: The axial CT (Fig. 7A) shows a large bony osteophyte occluding the right L5-S1 intervertebral nerve root canal (large arrowhead). There is mild hypertrophic overgrowth of the superior articular process (SAP) (small arrow). Note that the osteophytic spur extends laterally to lie immediately adjacent to the medial undersurface of the right L5 pedicle. In Fig. 7B, the sagittal reformatted image shows marked up-down bony lateral spinal stenosis with marked compression of the L5 nerve root ganglia between the bony osteophyte and the undersurface of the pedicle (large arrow). Note that there is fat within the dorsal aspect of the intervertebral nerve root canal, lying ventral to the border forming SAP. The presence of fat within the doral aspect of the intervertebral nerve root canal does not exclude the presence of severe up-down bony lateral spinal stenosis occurring in the ventral one-half of the nerve root canal (small arrow). Note also that a complete facetectomy, which would remove the dorsal margin of the intervertebral nerve root canal, would not affect the severe up-down stenosis occurring in the ventral one-half of the nerve root canal.

Diagnosis: Lateral spinal stenosis.

Discussion: In our experience, lateral spinal stenosis is commonly associated with degenerative lumbar disk disease and herniation. It is the most common cause of the FBSS, being present in more than 60% of FBSS patients in a large combined study of more than 450 patients performed by Burton and Kirkaldy-Willis.

Our initial understanding of lateral stenosis was predicated on the theory of Kirkaldy-Willis, which supposed compression of the nerve root ganglia within the intervertebral nerve root canal by the SAP secondary to the loss of disk height and overgrowth of the SAP. However, in 1982 Dr. Charles Ray of the Institute of Low Back Care in Minneapolis began to note an up-down component of lateral spinal stenosis. On the basis of this information, we began to review our axial and reformatted CT images for the presence of this "up-down" stenosis. We discovered that many patients had large posterolateral osteophytes arising from the dorsolateral aspect of the L5 vertebra, which were causing up-down stenosis of the ventral one-half of the L5-S1 vertebral nerve root canals. Dr. Steve Eisenstein termed these spurs "uncinate spurs" (Fig. 7C–F) at the annual meeting of the International Study of the Lumbar Spine in Cambridge in 1983.

Uncinate spurs as shown in this case arise from the posterolateral aspect of the L5 vertebra. They are constant in appearance and location. They are felt to be traction spurs caused by the attachment of Sharpey's fibers of the posterolateral aspect of the L5-S1 disk to vestigial uncinate processes similar to those in the cervical spine; thus the reason for the likeness of these spurs from individual to individual. Consistent with the theories of Kirkaldy-Willis, these spurs are more prominent on the side of predominant degenerative disk disease, and they are most often bilateral.

These spurs are a very common cause of significant lateral spinal stenosis. Contrary to the theory of impingement by the SAP, we found these ventrally situated spurs to be the primary cause of stenosis in 95% of 300 patients with radiographic lateral spinal stenosis at L5-S1.

Surgical correlation has been obtained in many cases in a number of centers and has confirmed the presence and significance of this lesion. They are difficult to visualize at surgery because they are covered with cartilage and ligamentous attachments of the disk. The nerve roots and ganglia lie directly cranial to these spurs in the superior and ventral aspect of the nerve root canal and are tightly held by extradural attachments (Hoffman ligaments). Therefore, little or no fore and aft movement of the nerve root is possible to avoid compression by these spurs or up-down compression caused by concomitant loss of disk height and spurring at the margin of the L5-S1 disk. Thus, in our experience, by far the majority of nerve root compressions caused by lateral stenosis at L5-S1 is up-down in nature and occurs in the ventral half of the canal, well separated from the SAP. Often the dorsal half of the nerve root canal is patent (within which the nerve root never lies), and a surgeon probing this patent dorsal portion of the canal can incorrectly assume he or she has ruled out lateral stenosis unless forewarned by preoperative imaging and familiarity with these newer concepts of up-down stenosis of the ventral half of the nerve root canal.

The anatomy of the L5-S1 nerve root canal is unique. Although the medial and lateral boundaries of the canal are formed by the L5 pedicle, as is true at other levels, because the L5 pedicle flares laterally, well beyond the confines of the lateral aspect of the S1 SAP at the L5-S1 level, fully one-third of the nerve root canal lies lateral to the SAP. Up-down stenosis occurring between the undersurface of the L5 pedicle and the L5-S1 disk and/or sacral alae is not uncommon. This is the so-called far-out stenosis (Fig. 7G,H) of Wiltse. It is also persistently the most common cause of missed lateral spinal stenosis in the hands of even skilled spine surgeons familiar with lateral stenosis concepts. Less skilled spine surgeons more commonly miss uncinate spurs and hypertrophy of the SAP by assuming that a medial facetectomy will affect relief of lateral spinal stenosis. However, this results only in decompression of the subarticular or spinal recess of the *central* spinal canal in patients with subarticular recess stenosis and does not relieve lateral spinal stenosis. In my experience, neurosurgeons are more likely to make this error, perhaps because Kirkaldy-Willis's precepts have been more widely disseminated and accepted among orthopedists.

Note that the uncinate spur typically lies within the middle one-third of the L5-S1 canal, extending lateral to the SAP of S1, and when large may extend to the far lateral aspect of the nerve root canal.

Once one becomes aware of the presence and significance of the uncinate spur, it becomes clear that lateral spinal stenosis with compression of the exiting L5 nerve root ganglia may occur lateral to the SAP, and therefore even a complete facetectomy does not affect an up-down stenosis occurring between (a) an uncinate spur and the L5 pedicle, (b) the lateral margin of the L5-S1 disk and the L5 pedicle, or (c) laterally situated spurs at the margin of the L5-S1 disk and the L5 pedicle (Fig. 7A–C,E,G,H). A facetectomy relieves front-back stenosis *only* (the minority of our cases), whereas relief of up-down stenosis caused by uncinate spurring requires removal of the uncinate spur, undercutting of the L5 pedicle, or otherwise widening of the canal by distraction, internal fixation and fusion when caused by loss of disk height, lateral disk protrusion, or generalized marginal spurs.

Sagittal reformatted CT images and/or sagittal MR images are of critical importance in the evaluation of up-down lateral stenosis because the presence and significance of the cephalocaudad nerve root compression caused by the spur and the quantification of that compression is poorly appreciated in the axial plane. The failure of the radiologic community to adopt and utilize routine sagittal reformatted images in *all* cases is singularly responsible for our failure to recognize and appreciate the various facets of lateral spinal stenosis. As these can be reasonably obtained only from stacked, nonangled axial images, I would make a strong plea for the abandonment of separated, angled axial images of the disk only. This technique completely ignores imaging of the nerve root canals and belies a failure of understanding of spinal pathology, i.e., the *concomitance* of lateral stenosis and degeneration of the disk. Our routine is stacked 3-mm images from the L5-S1 disk through the pedicle of L5 and 5-mm axial images from the L5 pedicle through the L3 pedicle. Routine sagittal reformatted images are obtained in all cases from the lateral aspect of the pedicles of L5 across the entire spine, including both nerve root canals and the central canal.

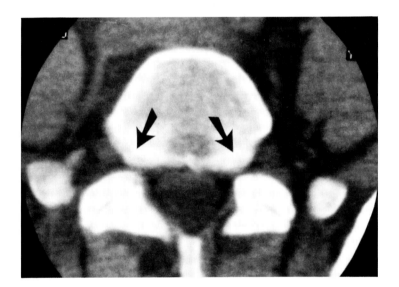

FIG. 7C. Bilateral uncinate spurs. This axial CT at L5-S1 shows bilateral symmetrical uncinate spurs (*arrows*). Note the similarity of position and appearance of these uncinate spurs from patient to patient. As the L5 nerve root ganglia lies immediately adjacent to the posterior margin of the vertebral body as it traverses the L5-S1 intervertebral nerve root canal, it lies directly craniad to these uncinate spurs and becomes compressed between the uncinate spur caudally and the lateral margin of the pedicle cranially.

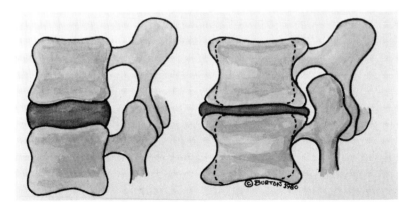

FIG. 7D. Diagrammatic illustration of the mechanism of lateral spinal stenosis secondary to degenerative disk disease with loss of disk height, secondary osteophytic spurring at the margin of the vertebral bodies forming both uncinate and marginal osteophytic spurs, and hypertrophic overgrowth of the superior articular process resulting in up-down and front-back stenosis. (From Burton, with permission.)

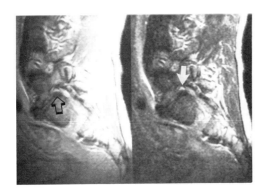

FIG. 7E. Severe up-down lateral spinal stenosis (sagittal MR, TR/TE 2,000/30, 80 msec). There is severe up-down bony lateral spinal stenosis caused by marked loss of disk height of the L5-S1 disk and large uncinate spurring (*open arrow*). Note the marked cephalocaudad compression of the L5 nerve root ganglia, which is flattened in its superior-inferior dimension and of increased signal intensity on the T2-weighted image (*white arrow*). Anatomically compressed ganglia, which show increased signal intensity on T2-weighted sagittal images, correlate with edematous ganglia noted at surgery.

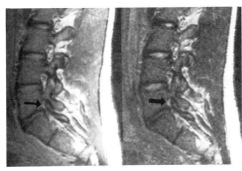

FIG. 7F. Front-back lateral spinal stenosis (sagittal MR, TR/TE 2,000/30, 80 msec). Note the moderately severe front-back narrowing of the intervertebral nerve root canal caused by overgrowth of the superior articular process and thickening of the ligamentum flavum as it attaches to the undersurface of the pedicle. This narrows the anteroposterior dimension of the intervertebral nerve root canal and is producing moderate front-back compression of the exiting left L5 nerve root ganglia (*arrows*).

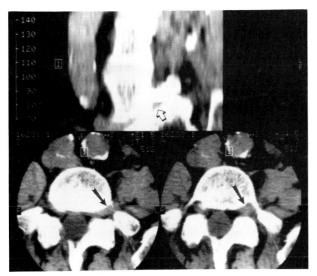

FIG. 7G. Far-out lateral spinal stenosis. The axial and sagittal reformatted CT images show a severe up-down lateral spinal stenosis occurring between the pedicle of L5 and the sacral ala. Note the enlargement of the L5 nerve root ganglia caused by the cephalocaudad compression (*black arrows*). The sagittal reformatted image is obtained in a plane paralleling the sharply marginated undersurface of the L5 pedicle. Note that this far-out lateral spinal stenosis is not related to the L5-S1 facet and lies well lateral and ventral to the superior articular process (*open arrow*). Therefore, a facetectomy would not be expected to result in any improvement of the patient's lateral spinal stenosis.

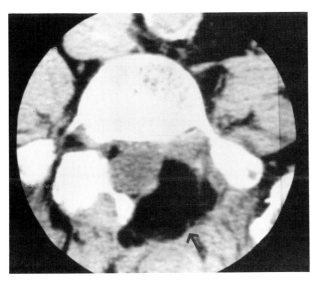

FIG. 7H. Persistent, postoperative, severe far-out lateral spinal stenosis. The postoperative axial CT shows a complete facetectomy at L5-S1 on the left of the patient in Fig. 7G. Note the large fat graft within the extensive operative defect (*arrow*). Note that the surgery, however, did not address the stenotic lesion occurring between the undersurface of the pedicle and the sacrum. The patient had persistent symptomatology. A subsequent operation consisting of removal of the undersurface of the pedicle of L5 was performed, resulting in complete relief of symptomatology.

CASE 8

A 52-year-old man with bilateral L5 radiculopathy. What is the cause of the patient's sciatica? (Fig. 8A. Axial MR, TR/TE 2,000/20 msec. Fig. 8B. Sagittal MR, TR/TE 2,000/ 30, 80 msec.) What is the most common cause of sciatica in patients with lytic spondylolisthesis—or—what is your diagnosis? What demonstrated lesion is responsible for this patient's sciatica?

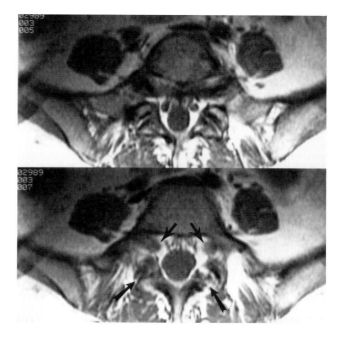

FIG. 8A.

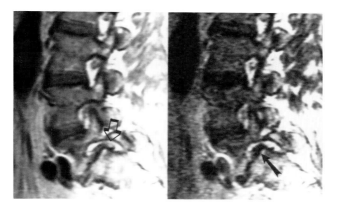

FIG. 8B.

Findings: On the axial image (Fig. 8A) there is bilateral spondylolysis of the pars interarticularis (*arrows*). Note the enlargement of both L5 nerve root ganglia lying medial to the pedicle in the lower image and the lack of visualized L5-S1 foramina on the upper images. The sagittal MR images (Fig. 8B) show marked cephalocaudad narrowing of the L5-S1 intervertebral nerve root canals. There is forward and caudal subluxation of the L5 vertebra on S1, resulting in elongation of the anteroposterior plane of the L5-S1 intervertebral nerve root canal, and there is severe disk space narrowing and dehydration of the L5-S1 disk (black arrow). Note the moderately severe cephalocaudad compression of the exiting left L5 nerve root ganglia (open arrow).

Diagnosis: Lytic spondylolisthesis.

Discussion: In our experience, the most common cause of sciatica in patients with lytic spondylolisthesis is compressive L5 radiculopathy secondary to cephalocaudad lateral spinal stenosis. This results from degenerative disk space narrowing of the L5-S1 disk, which accompanies the forward and caudal subluxation of the L5 vertebral body on S1. Plain films and CT lateral computed radiographs can therefore predict this lesion with a high degree of accuracy.

The L5 nerve root is swept forward with the L5 vertebral body because of its attachment to that structure via extradural Hoffman ligaments. As L5 subluxes on S1, the plane of the posterior disk margin changes from vertical to horizontal and becomes the floor of the intervertebral nerve root canal. Degenerative loss of disk height of the affected L5-S1 disk commonly occurs and results in up-down (cephalocaudad) compression of the exiting L5 nerve root between the caudal margin of the pedicle of L5 and the underlying L5-S1 disk. Degenerative lateral bulging of the L5-S1 disk is common and contributes to the up-down compression of the L5 nerve root ganglia, as the surface of the disk is now projecting in a craniad direction. Lateral disk herniations at L5-S1 are relatively frequent in patients with lytic spondylolisthesis, and when present in conjunction with bony up-down lateral stenosis may cause severe L5 nerve root entrapment within the intervertebral nerve root canal. Also, importantly, because the plane of the disk is fore and aft, classical dorsolateral herniations project in a craniad direction superior to the posteriorly positioned S1 vertebra and not dorsally. Therefore, it is rare for an L5-S1 disk herniation to produce S1 nerve root compression in patients with grade I or greater lytic spondylolisthesis. If nerve root impingement occurs, it is almost always the L5 nerve root that is compressed by lateral and cranial dissection of the herniation into the medial aspect of the L5-S1 intervertebral nerve root canal. Also, when a herniation occurs in the usual position of the dorsolateral aspect of the central canal, the forward subluation of the L5 vertebra produces a large epidural space separating the S1 nerve root from the posterior margin of the disk and often little or no S1 nerve root impingement occurs.

Axial CT images alone can be deceiving with respect to detection of lateral spinal stenosis. The anterior displacement of L5 on S1 produces an elongation of the canal on axial CT images, and the unwary radiologist who foregoes the benefit of sagittal reformatted images may fail to appreciate the presence of significant up-down stenosis with L5 nerve root compression. The only clue from axial images is the apparent enlargement of the L5 nerve root ganglia in the axial plane caused by cephalocaudad compression.

Computed tomography sagittal reformatted images and sagittal MR (Fig. 8C–F) images, on the other hand, clearly depict the presence and severity of up-down stenosis and L5 nerve root compression.

Although disk herniations at L4-5 have been reported to be the most common cause of sciatica, we have found this to be less common than lateral stenosis. Also, because the ventral subluxation of L5 causes enlargement of the epidural space ventral to the thecal sac and L5 nerve roots at the level of the L4-5 disk, unless an L4-5 herniation is large, it ordinarily does not cause L5 nerve root compression.

Dysraphic spondylolisthesis is to be distinguished from lytic spondylolisthesis and is associated with different nerve root compressive lesions. Because the dorsal arch of L5 is intact and the listhesis is secondary to an unstable sagittal or reversed orientation of the L5-S1 facet joints, the entire vertebra, including the dorsal arch, subluxes forward. This may produce a severe central and subarticular stenosis by guillotining the thecal sac and S1 nerve roots. Lateral stenosis is much less common. In this regard, dysraphic spondylolisthesis resembles degenerative spondylolisthesis, i.e., the nerve root compression is secondary to central and subarticular stenosis of the traversing nerve roots.

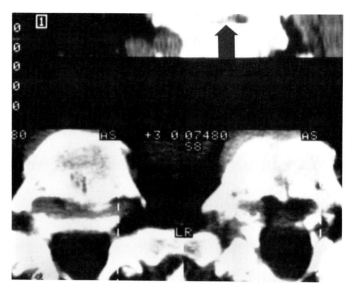

FIG. 8C. Axial and sagittal reformatted CT images showing grade I lytic spondylolisthesis with marked cephalocaudad bony lateral spinal stenosis and compression of the exiting L5 nerve root ganglia. The compression occurs between the pedicle and pars superior of L5 cranially and the L5-S1 disk plate and sacrum caudally. Note the marked narrowing of the cephalocaudad dimension of the intervertebral nerve root canal on the sagittal reformatted image (*large arrow*), despite the apparent wide dimension of the anteroposterior dimension of the nerve root canal on the axial image on the left. Note that the dorsal one-half of the canal contains fat and appears patent; however, the ganglia lies in the very narrowed ventral one-half of the canal (*large arrow*) and is severely compressed.

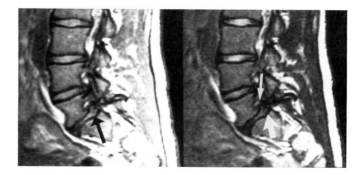

FIG. 8D. Lytic spondylolisthesis (sagittal MR, TR/TE 2,000/30, 80 msec). There is grade I lytic spondylolisthesis and an associated moderately large lateral disk herniation at L5-S1 on the left in this 13-year-old boy. Lateral disk herniations at L5-S1 in patients with lytic spondylolisthesis is not uncommon. Unlike the usual L5-S1 disk herniation, because of the forward subluxation of the L5 vertebra with respect to the sacrum, and lytic spondylolisthesis, the disk herniation as seen in this case (*black arrow* and *large white arrow*) projects in a craniad direction into the L5-S1 intervertebral nerve root canal and impinges on the exiting L5 nerve root ganglia rather than S1. This causes a cephalocaudad compression of the L5 nerve root ganglia (*small white arrow*). Disk herniations occurring in a patient with lytic spondylolisthesis are nearly always lateral disk herniations. In our experience, lateral disk herniations at L5-S1 are more common than the classically described disk herniation at L4-5.

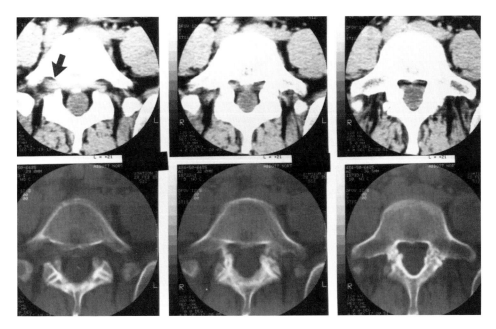

FIG. 8E. Lytic spondylolisthesis. The axial CT images show enlargement of the right L5 nerve root ganglia (*arrow*). This enlargement of the ganglia is an excellent indication of cephalocaudad compression of the ganglia. In patients with lytic spondylolisthesis, this is most commonly due to cephalocaudad bony lateral spinal stenosis, as seen in this case. Less commonly it is due to associated lateral disk herniation. Magnetic resonance is the only means of differentiating between pure bony stenosis and associated lateral disk herniation.

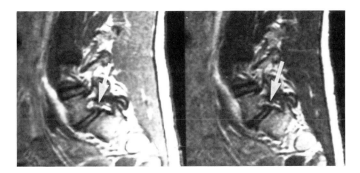

FIG. 8F. Sagittal MR, TR/TE 2,000/30, 80 msec. Sagittal images of the patient in Fig. 8D shows severe cephalocaudad compression of the right L5 nerve root ganglia (*arrows*). The ganglia is compressed between the pars superior and the L5-S1 disk plate. Note that there is no evidence of disk herniation on this side, and the nerve root compression is purely due to up-down bony lateral spinal stenosis.

CASE 9

This 68-year-old woman presented with a history of claudication. What is the diagnosis? (Fig. 9A. Axial MR, TR/TE 1,000/20 msec.)

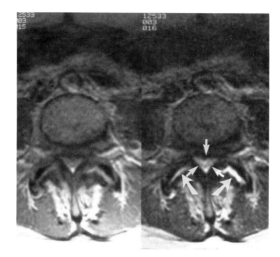

FIG. 9A.

Findings: There is marked degeneration of the facets with high signal intensity fluid within the facet joints (Fig. 9A, large arrows), marked thickening of the ligamentum flavum and posterior bulging of the L4-5 disk annulus, ventral subluxation of L4 on L5 resulting in a double density of the disk, and marked constriction of the thecal sac (small arrows).

Diagnosis: Degenerative spondylolisthesis with marked degenerative facet disease, synovitis of the facets resulting in increased fluid in the facet joints, and marked central and bilateral subarticular recess stenosis.

Discussion: Degenerative spondylolisthesis, a relatively common cause of central and subarticular (spinal recess) stenosis, is characterized by a distinctive CT and MR appearance: marked degenerative hypertrophic overgrowth of the facets, thickening of the ligamentum flavum (which is often marked and associated with ligamentous laxity), and erosion of the medial aspect of the SAP (Fig. 9B). This marked erosion of the supportive SAP of the caudal vertebral body allows ventral subluxation of the intact neural arch of the cephalic vertebral body, with resultant constriction of the central spinal canal and thecal sac. The compression of the thecal sac and nerve roots occurs between the lamina and hypertrophied inferior articular processes dorsally and the posterior aspect of the involved disk and caudal vertebral body ventrally. Bulging or herniation of the involved disk and marked thickening of the ligamentum flavum are responsible for the majority of the neural compression in most cases. Therefore, degenerative spondylolisthesis typically produces central spinal stenosis as well as subarticular recess stenosis with compression of the traversing nerve roots (i.e., L5 nerve roots at the L4-5 level).

Degenerative spondylolisthesis occurs most commonly at the L4-L5 level. The typical patient is an overweight middle-aged to elderly woman with hyperlordosis. The lateral computed radiographs usually show degenerative disk space narrowing at the involved level and, frequently, moderate hyperlordosis. The association of diabetes (reported in up to 30% of these patients), marked degenerative overgrowth of these facets, and the paucity of significant back pain and sciatica before the onset of symptoms of spinal claudication raises the question of a lack of pain sensitivity of the involved joints (i.e., a Charcot-like joint degeneration).

Bony lateral spinal stenosis is relatively unusual in these patients because the pedicles of L4 arise relatively more cephalad on the vertebral body than the pedicles of L5. Therefore, loss of disk height alone at L3-4 and L4-5 does not ordinarily produce a cephalocaudal bony lateral spinal stenosis. Bony lateral spinal stenosis, when it does occur, almost always is secondary to a rotary component of the subluxation. The lateral spinal stenosis then occurs on the side with anterior subluxation of the SAP that rotates into the intervertebral nerve root canal, producing bony lateral impingement of the exiting nerve root ganglion (i.e., L4 at L4-5). The compression occurs between the cephalic margin of the hypertrophied SAP and the undersurface of the pedicle above.

Patients with degenerative spondylolisthesis exhibit significant degenerative soft tissue abnormalities related to their long-standing instability. There is marked thickening of the ligamentum flavum and prominent bulging of the affected disk. In many cases there is also prominent lateral bulging or herniation of the disk. This produces an opacification of the nerve root canals and indistinct visualization of exiting nerve roots on axial CT. On axial CT examinations, this appeared to represent fibrosis around the nerve roots and had been described as fibrocartilaginous overgrowth. However, sagittal MR images show this to be due to up-down compression of the ganglia caused by prominent lateral bulging of the annulus of the disk or marked thickening of the ligamentum flavum at the anterior margin of the SAP as it attaches to the undersurface of the pedicle in most cases and lateral herniation in a few. However, much of the apparent prominence of the disk extending into the *central* spinal canal on the axial images is due to forward subluxation of the cephalic vertebral body with respect to the disk and end plate of the vertebral body below. The diffuse bulging of the disk, however, does extend cephalad because the exposed surface of the involved disk is oriented in an axial plane because of its front-back course between the offset vertebra rather than the normal cephalocaudal orientation. Thus, the thickened bulging annulus projects cephalad into the intervertebral nerve root canals to underlie and flatten the exiting nerve roots (i.e., the L4 ganglia within the L4-5 nerve root canals).

Degenerative spondylolisthesis may be complicated by synovial cyst formation (Fig. 9C,D). In fact, most patients with synovial cysts have underlying degenerative spondylolisthesis. Contrary to previous reports in the literature, synovial cysts and chondromas are often associated with clinical symptoms resulting from nerve root compression within the subarticular recess, i.e., the traversing L5 nerve root at L4-5. Magnetic resonance imaging can clearly visualize the fluid within the synovial cysts on T2-weighted images and, therefore, differentiates synovial cysts from synovial chondromas. This distinction is important in light of recent clinical research, which has shown that injection of steroids into the zygapophyseal joint, which communicates with the synovial cyst, may dramatically decrease or completely relieve the symptoms related to nerve root compression and has become an alternative to surgery. On the other hand, symptomatic synovial chondromas or noncystic focal thickening of the ligamentum flavum (Fig. 9E) that produce nerve root compression do not respond to this treatment. Synovial cysts most commonly arise at the medial margin of the degenerated facet joints. The synovial cyst has a characteristic appearance with its flat base paralleling the ligamentum flavum at the medial margin of the zygapophyseal joint. The rounded medial aspect of the cyst projects in an anteromedial direction rather than dorsally as would be expected with disk herniation. In some patients, the ligamentum flavum and bulging of the posterior margin of the disk are so prominent that they abut each other (Fig. 9F), filling the subarticular recess with abnormal soft tissue density on axial CT images, and the cyst margins are not clearly defined. Very occasionally, they can vary in position and may occur within the lateral aspect of the nerve root canal beneath the lateral attachment of the ligamentum flavum to the base of the superior articular facet. A recent case proven by MRI presented as a mass in the posterolateral aspect of the intervertebral nerve root canal. The cyst was dorsal in position and isodense with respect to the ganglia and ligamentum flavum on CT. Because of its unusual location, the etiology of the mass was indeterminate; MRI showed the mass to be clearly cystic and contained by the ligamentum flavum. The association of these lesions with degenerative spondylolisthesis is so strong that any rounded soft tissue mass projecting into the central spinal canal from the medial margin of the facet joint in patients with degenerative spondylolisthesis at that segment should be considered to be a synovial cyst and/or chondroma until proven otherwise.

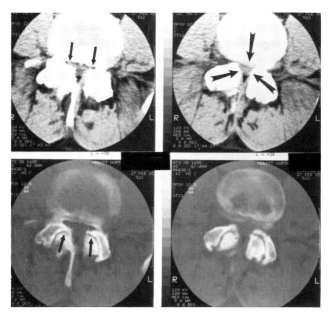

FIG. 9B. Degenerative spondylolisthesis. The axial CT scan of the lumbar spine shows the typical marked degenerative hypertrophic overgrowth of the facets and marked erosive narrowing of the facet joints. Note the marked erosive changes and remodeling of the SAPs, which result in moderate bilateral subarticular recess stenosis with compression of the traversing L5 nerve roots (*small arrows*). There is severe central spinal stenosis secondary to the marked overgrowth of the facets, thickening of the ligamentum flavum, and the central bulging of the disk annulus (*large arrows*). The triangular constriction of the central spinal canal is very typical of degenerative spondylolisthesis.

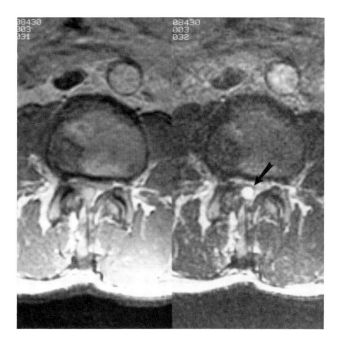

FIG. 9C. Synovial cyst (axial MR, TR/TE 3,000/30, 80 msec). Note the large, rounded, high signal intensity fluid-filled lesion projecting into the central spinal canal at L4-5 on the right. This is arising in a patient with degenerative spondylolisthesis and further compromises an already stenotic central spinal canal. This patient presented with moderately severe right-sided L5 radiculopathy, which was relieved by surgical removal of the cyst (*arrow*). Note the severe degenerative facet disease present at this level, particularly on the right.

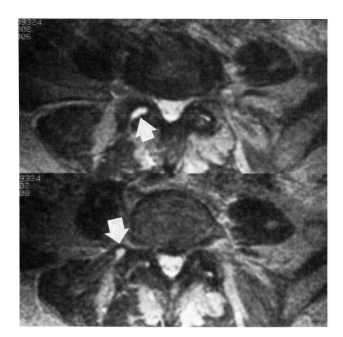

FIG. 9D. Far lateral synovial cyst (axial MR, TR/TE 3,000/30, 80 msec). The axial CT shows increased fluid within the facet joints, which is typical for patients with degenerative spondylolisthesis. Synovial cysts may occur anywhere within the joint capsule and occasionally occur within the intervertebral nerve root canal or even laterally as noted in this case (*arrows*).

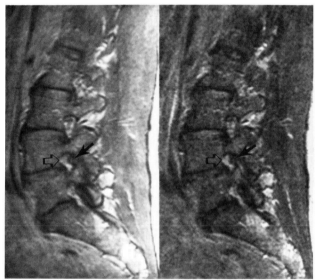

FIG. 9E. Sagittal MR, TR/TE 2,000/30, 80 msec. Focal thickening of the ligamentum flavum at the anterior margin of the joint space as it attaches to the undersurface of the pedicle (*black arrows*). This focal thickening of the ligamentum flavum impinges directly on the dorsal aspect of the right L4 nerve root ganglia (*open arrows*).

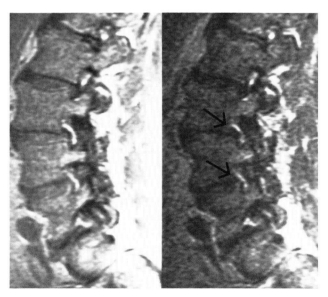

FIG. 9F. Sagittal MR, TR/TE 2,000/30, 80 msec. Sagittal MR images in a patient with degenerative spondylolisthesis at both L3-4 and L4-5 shows the typical lateral nerve root impingements caused by a combination of posterolateral bulging of the disk annuli and marked thickening of the ligamentum flavum at the anterior margin of the joint as it attaches to the pedicle. Note the slit-like character of the intervertebral nerve root canals (*arrows*). The nerve root impingements within the intervertebral nerve root canals noted in most patients with degenerative spondylolisthesis are soft tissue rather than bony.

CASE 10

A 46-year-old man presented with a left L4 and left L5 radiculopathy. There are two abnormalities present on this CT scan; one is obvious, the other is not. What is your diagnosis? (Fig. 10A,B. Axial MR, TR/TE 1,000/20 msec.)

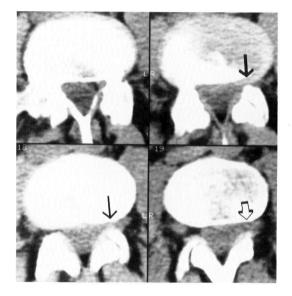

FIG. 10A.

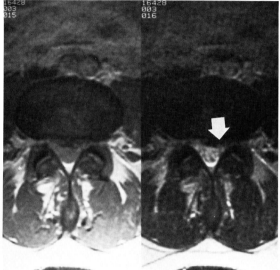

FIG. 10B.

Findings: There is asymmetry of the neural elements at L4-5 on the left (Fig. 10A). The right L4 nerve root ganglia has a normal position within the intervertebral nerve root canal and is surrounded by fat, whereas the left L4 nerve root is ill-defined and appears continuous with the thecal sac, raising the suspicion of a conjoined left L4-5 nerve root (open arrow and thin black arrow). In this patient with acute L4 and L5 radiculopathy, the slight increased prominence of the left side of the L4-5 disk (large arrow) raises the question of an underlying disk herniation. This cannot be resolved by unenhanced CT.

The axial MR (Fig. 10B) shows a moderate-size broad-based low density extradural mass, which is impinging on the thecal sac and the left L5 nerve root (arrow).

Diagnosis: Conjoined left L4-5 nerve root and underlying moderately large central and left-sided disk herniation which extends into the intervertebral nerve root canal; thus impinging on both the left L5 and the left L4 components of the conjoined nerve root.

Discussion: Conjoined nerve roots are common anatomic variants, being present in 8% to 12% of patients. They are most common at L5-S1, less common at S1 and S2, and considerably less common proximal to L5-S1. Conjoined nerve roots have a typical appearance. There is asymmetry of the neural elements where the conjoined nerve root arises from the thecal sac because of a more caudal take-off of the proximal nerve root of the paired conjoined nerve roots. Therefore, as one studies the spine from above down, the normal nerve root will have exited the thecal sac while the conjoined nerve root will still appear part of the thecal sac and cause homogeneous soft tissue density within the ipsilateral central spinal canal and lateral recess of the central canal.

The thecal sac has a "Schmoo-like" appearance, with the head representing the conjoined nerve root and the body the thecal sac. (My apologies to the younger generation who may be unfamiliar with Dogpatch characters.)

Conjoined nerve roots pose difficult diagnostic problems in patients with sciatica in the distribution of the conjoined nerve root. Because of its more caudal take-off, the proximal of the paired nerve roots has a direct lateral course from the thecal sac into the caudal aspect of the intervertebral nerve root canal. This causes it to parallel the posterior margin of the affected disk. This may result in a false-positive diagnosis of disk herniation because of the increased density of the conjoined nerve root and the asymmetric mass effect that it causes being mistaken for a herniated disk. Conversely, because it does parallel the disk and is of higher density than the CSF on axial CT examinations, it is isodense with the underlying disk and may mask an underlying disk herniation. This cannot be resolved on the basis of plain CT. Therefore, when a patient has sciatica in the distribution of the conjoined nerve root, MR is the procedure of choice because it is more accurate than contrast-enhanced CT and is less invasive. On MR studies, the conjoined nerve root is of higher signal intensity than the adjacent annulus of the disk, and they are easily distinguished from each other.

The recognition and exclusion of a disk herniation underlying the conjoined nerve root is of paramount importance (Fig. 10C–F). White and his colleagues reported 70% of 65 patients with conjoined nerve root and underlying disk herniations were unimproved following attempted diskectomy because of nerve root injury. A conjoined L5-S1 nerve root is always stretched over the pedicle of L5, with the L5 component of the conjoined nerve root lying in an abnormally caudal position within the caudal one-half of the intervertebral nerve root canal directly overlying the herniated disk fragment. Unless the nerve root is freed up by removing a portion of the S1 pedicle, removal of the herniated disk is difficult and retraction of the stiff, stretched nerve root may easily lead to nerve root injury.

Magnetic resonance is most useful as an adjunct to CT in complicated spine cases either postoperatively or when multiple concomitant pathological entities are present or suspected. Nerve root impingement by bony stenosis (both front-back and up-down), lateral disk herniation, enlargement of the ganglia caused by conjoined nerve root, and postoperative fibrosis and/or focal thickening of the ligamentum flavum can all be differentiated. This is in contradistinction to CT, in which these pathological lesions, when present singly or in conjunction with each other, show similar attenuation values. Therefore, when CT was previously the only modality available to study the intervertebral nerve root canal, one could not clearly delineate for the clinician and surgeon whether the abnormal soft tissue within the nerve root canal was merely a compressed ganglia in a patient with lateral spinal stenosis or represented a concomitant soft tissue abnormality such as lateral disk herniation, conjoined nerve root, or postoperative fibrosis, which was isodense with the compressed nerve root ganglia. Therefore, we routinely use MR when faced with a complicated postoperative patient with disease within the central spinal canal or intervertebral nerve root canals and in preoperative cases in which these questions arise.

Currently, many MR systems are capable of providing routine images in the orthogonal planes (sagittal, coronal, and axial) as well as nonorthogonal planes (oblique). Primary imaging in the sagittal plane with MR represents a significant improvement in spatial resolution of sagittal images of the spine compared to CT. The spatial resolution of CT in the sagittal reformatted images is equal to the table incrementation (i.e., if the table is moved 3 mm for each image, the spatial resolution of the reformatted images is 3 mm). The spatial resolution of high field strength MR is now less than 1 mm.

Improved axial imaging with flow compensation now allows visualization of individual nerve roots in most cases and, consequently, of clumped nerve roots in patients with adhesive arachnoiditis. In some complicated postoperative cases, MR is the only modality capable of clearly differentiating between the various soft tissue structures within the central spinal canal and intervertebral nerve root canals.

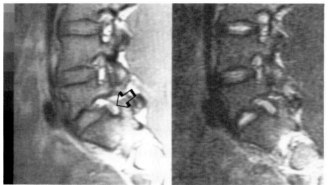

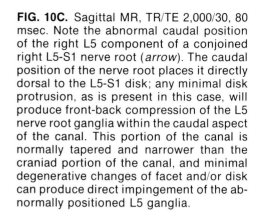

FIG. 10C. Sagittal MR, TR/TE 2,000/30, 80 msec. Note the abnormal caudal position of the right L5 component of a conjoined right L5-S1 nerve root (*arrow*). The caudal position of the nerve root places it directly dorsal to the L5-S1 disk; any minimal disk protrusion, as is present in this case, will produce front-back compression of the L5 nerve root ganglia within the caudal aspect of the canal. This portion of the canal is normally tapered and narrower than the craniad portion of the canal, and minimal degenerative changes of facet and/or disk can produce direct impingement of the abnormally positioned L5 ganglia.

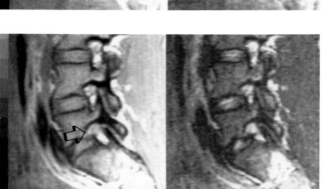

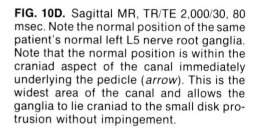

FIG. 10D. Sagittal MR, TR/TE 2,000/30, 80 msec. Note the normal position of the same patient's normal left L5 nerve root ganglia. Note that the normal position is within the craniad aspect of the canal immediately underlying the pedicle (*arrow*). This is the widest area of the canal and allows the ganglia to lie craniad to the small disk protrusion without impingement.

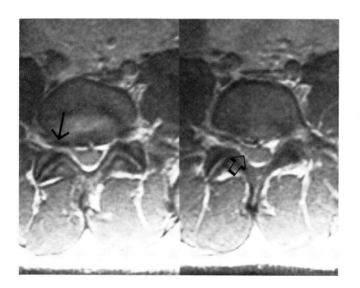

FIG. 10E. Axial MR, TR/TE 2,000/30 msec. Note the asymmetric appearance of the neural elements at L5-S1 on the right caused by a conjoined right L5-S1 nerve root (*open arrow*). Note that the conjoined nerve root produces an abnormal soft tissue density within the lateral recess of the central spinal canal. Note the narrowed anteroposterial dimension of the right L5-S1 nerve root canal (*black arrow*).

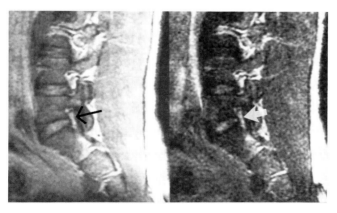

FIG. 10F. Sagittal MR, TR/TE 2,000/30, 80 msec. The sagittal MR in the patient illustrated in Fig. 10E with a conjoined L5-S1 nerve root. Note the marked anteroposterior (front-back) compression of the right L5 nerve root ganglia between the minimal bulging of the disk ventrally and superior articular process dorsally (*black arrow* and *white arrow*).

CASE 11

A 36-year-old patient presented with a history of a prior left L5-S1 hemilaminotomy and diskectomy with recurrent acute S1 radiculopathy. To what structures do the black and the open arrows point? (Fig. 11A,B. Axial MR, TR/TE 3,000/30 msec.)

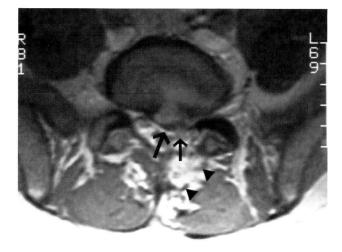

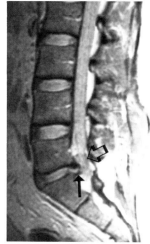

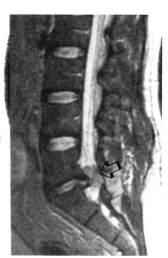

FIG. 11A.   **FIG. 11B.**

Findings: There is a complete tear (small black arrow) of the left side of the L5-S1 annulus (Fig. 11B). There is both nuclear material and annulus projecting dorsal to the vertebral body. There is a large intermediate soft tissue density overlying the disk (open arrows). The axial MR (Fig. 11A) shows high signal intensity of the dorsal paraspinous soft tissues consistent with a prior left-sided hemilaminotomy and associated muscle atrophy and fibrosis (arrowheads). There is a large central and left-sided disk herniation (large arrow), which is producing dorsal displacement and edema of the left S1 nerve root (small arrow).

Diagnosis: Large recurrent central and left-sided disk herniation with dorsal displacement and compression of the left S1 nerve root with moderate edema of the left S1 nerve root. There is moderately dense overlying fibrosis.

Discussion: Magnetic resonance performed with double-echo spin-echo images with long TR, proton-density, and T2-weighted images on both sagittal and axial images is the imaging procedure of choice for evaluating the postoperative patient with recurrent sciatica and suspected recurrent disk herniation. If one uses this technique with a high resolution scanner, one is clearly able to differentiate between recurrent disk herniation and fibrosis in the vast majority of cases. In our experience, less than 3% of our patients require gadolinium for this differentiation. However, if one routinely uses only T1-weighted images in either the sagittal or axial plane in postoperative patients, one is obligated to augment the examination with gadolinium to differentiate between the recurrent disk herniation and fibrosis because fibrosis and recurrent disk herniation, not to mention CSF and neural elements, are all of low signal intensity and difficult to differentiate from one another. In our experience, the routine use of gadolinium in postoperative MRI is unnecessarily costly, invasive, and time consuming unless the scanner being used is incapable of high resolution T2-weighted imaging.

The disk herniation is differentiable from overlying fibrosis on either one or the other of the proton-density or T2-weighted images. Although it may be nearly isointense on one, they are seperable on the other because of the varying characteristics of fibrosis as opposed to disk material.

Magnetic resonance has nearly completely replaced CT and omnipaque-enhanced CT as a procedure of choice in imaging the postoperative spine in our practice because of the almost uniform presence of postoperative fibrosis. In a double-blind study comparing omnipaque-enhanced CT and MR, MR was found to be both more sensitive and accurate than enhanced CT. However, if MR is not available, intrathecal enhanced CT is an excellent substitute, being 85% to 95% accurate in two separate studies we performed.

In many patients with both recurrent herniation and overlying fibrosis there is a fibrotic rim around the extruded disk herniation that lies between it and the fibrosis, which clearly delimits the posterior margin of the disk herniation and makes differentiation between disk herniation and overlying fibrosis quite easy (Fig. 11C–G).

In those patients in which gadolinium is necessary, it is performed with T1-weighted images before and after intravenous injection. The postgadolinium scan is obtained within 5 min of the injection of gadolinium. Gadolinium produces marked enhancement of the epidural and perineural fibrosis but not herniated disk material. The gadolinium is seen to enhance the fibrous capsules surrounding extruded disks (Fig. 2A,B). Although Modic has presented 100% accuracy of gadolinium in the differentiation of disk herniation and fibrosis, I have found it to be confusing in some cases with unusual enhancement patterns, which obscured the differentiation between fibrosis and recurrent disk herniation. In these cases we were able to make the differentiation on subsequent double-echo spin-echo T2-weighted images.

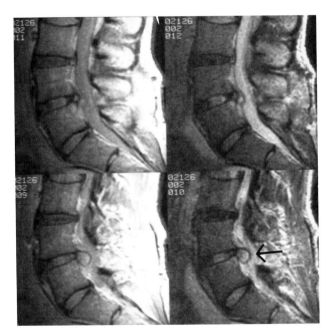

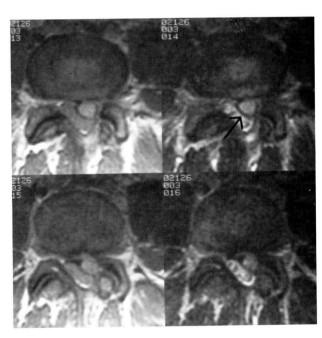

FIG. 11C. Sagittal MR, TR/TE 2,000/30, 80 msec. There is a large recurrent extruded L4-5 disk herniation (*arrow*) that has a low signal intensity fibrotic capsule around the disk. This was surgically proven. This fibrotic capsule simulates the annulus and differentiates the overlying fibrosis from the underlying disk material even though the two are essentially isointense.

FIG. 11D. Axial MR, TR/TE 1,000/20 msec. Axial MR images of the patient in Fig. 11C. Note the thick low signal intensity capsule (*arrow*) surrounding the disk herniation, as well as the marked compression of the thecal sac.

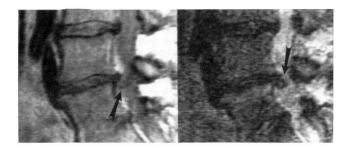

FIG. 11E. Sagittal MR, TR/TE 3,000/30, 80 msec. Moderately large, low density, recurrent disk herniation at L4-5. Note that the recurrent disk herniation is nearly isointense with the CSF on the proton-density image. If one uses T1-weighted sagittal and/or axial images in the postoperative patient, one ensures the need for gadolinium. Note that the disk herniation is apparent and distinguishable from the overlying fibrosis on the T2-weighted image (*arrows*).

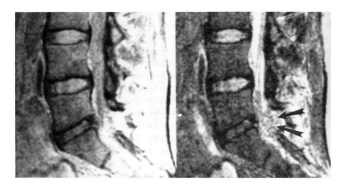

FIG. 11F. Sagittal MR, TR/TE 3,000/30, 80 msec. The patient was suspected to have a recurrent disk herniation at L5-S1 on the left. The MR excludes the presence of a recurrent disk herniation. There is a moderate amount of postoperative fibrosis (*arrows*).

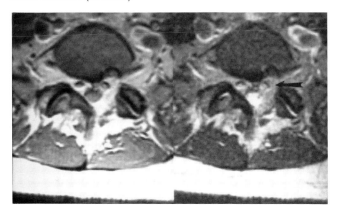

FIG. 11G. Axial MR, TR/TE 1,000/20 msec. The axial MR shows the normal position of the S1 nerve roots. The left S1 nerve root is surrounded by perineural fibrosis. Note the resection of the end plate just ventral to the S1 nerve root. The position of the nerve root is critical in the evaluation of a possible underlying disk herniation. As a general rule, if the nerve root is in the same coronal plane as its companion on the opposite side, no disk herniation is likely present. If it is displaced dorsally, however, one must suspect an underlying disk herniation. In this case it is in a normal position and surrounded by epidural fibrosis (*arrow*).

CASE 12

A 45-year-old physician with progressive weakness and paresthesia. The patient had a history of epidural steroid injections for low back pain and sciatica related to degenerative disk disease. What is the diagnosis? (Fig. 12A. Axial MR, TR/TE 3,000/80 msec. Fig. 12B. Coronal MR, TR/TE 3,000/80 msec.)

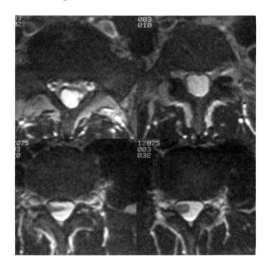
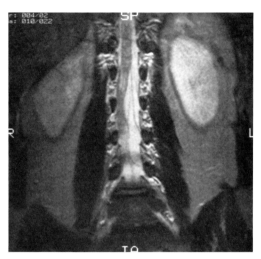

FIG. 12A.   **FIG. 12B.**

Findings: There is a rounded low density mass within the thecal sac (Fig. 12A). The lumbosacral nerve roots are not visualized. There is thickening of the dura. On the coronal section (Fig. 12B), the lumbosacral nerve roots are markedly clumped and adherent to the right side of the dura, with visualization of a few thin strand-like nerve roots extending across the CSF toward the left.

Diagnosis: Severe adhesive arachnoiditis secondary to intrathecal injection of steroids.

Discussion: Adhesive arachnoiditis in our population is most often due to either prior pantopaque myelography and/or surgery. With new software improvements that have gotten rid of CSF pulsation artifacts, visualization of the lumbosacral nerve roots including individual dorsal and ventral rami is now possible with high resolution systems. This is possible in most patients in whom patient motion and/or decreased resolution caused by large body habitus is not present.

Mild adhesive arachnoiditis presents as minimal asymmetric clumping and/or adherence of the nerve roots with either thickening of the individual dorsal and ventral rami or merging of the two as one structure. The rami are often of increased signal intensity where focally thickened. More extensive arachnoiditis presents as generalized clumping of the nerve roots dorsally and/or centrally (Fig. 12C–E). Severe adhesive arachnoiditis as seen in Fig. 12A is associated with either marked mass-like clumping of the nerve roots, or more commonly in the distal thecal sac, an absence of the lumbosacral nerve roots caused by their peripheral adherence to the dura and incorporation within fibrous tissue and scarring adjacent to dura. In these cases the dura is thickened in its appearance with complete absence of the lumbosacral nerve roots.

One must be careful not to misinterpret the paired nerve roots of a conjoined nerve root as abnormal clumping caused by arachnoiditis. In conjoined nerve roots there are four individual but *normal* rami (two dorsal, two ventral) traveling down the thecal sac together. The rami are normal in signal intensity, uniform in size, and exit the thecal sac together. If in doubt, look for the classical signs of the conjoined nerve root.

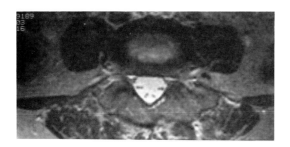

FIG. 12C. Axial MR, TR/TE 3,000/80 msec. Normal T2-weighted MR images of the lumosacral nerve roots. Flow compensation eradicates motion artifact, and the lumbosacral nerve roots are routinely visualized. Both the dorsal and ventral rami are normally seen as paired, small, rounded, low signal intensity structures that transcend the thecal sac to exit anterolaterally. This symmetry and sharp delineation of the nerve roots in this case indicate they are normal.

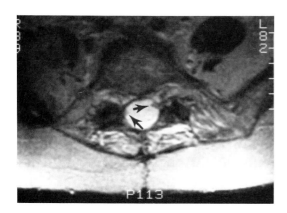

FIG. 12D. Axial MR, TR/TE 3,000/80 msec. Severe adhesive arachnoiditis with peripheral clumping of all of the lumbosacral nerve roots. The nerve roots are assimilated into the dura (arrows).

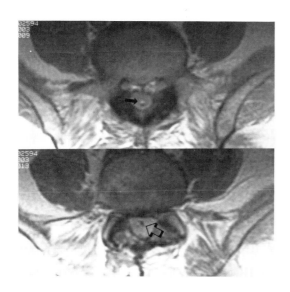

FIG. 12E. Axial MR, TR/TE 3,000/30 msec. Pantopaque-enduced arachnoiditis. Note the presence of the Pantopaque suspended within the thecal sac and adherent to a lumbosacral nerve root (black arrow). Note also the irregular central clumping of the remaining visualized lumbosacral nerve roots (open arrow).

CASE 13

An 82-year-old woman presented to her physician with a 2-month history of low back pain, which began the day following cataract surgery. Plain films showed a compression fracture of T11. An axial CT of T11 was obtained. What is the diagnosis?

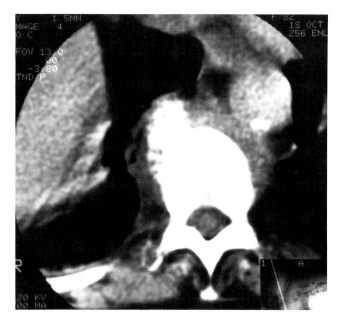

FIG. 13A.

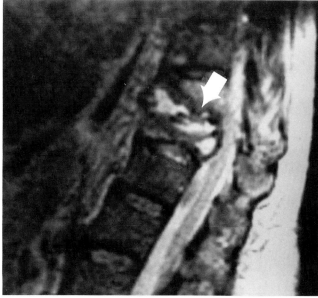

FIG. 13B.

Findings: The axial CT (Fig. 13A) shows a large anterior osteophyte and circumferential prominence of the paravertebral soft tissues. The differential diagnosis included an acute compression fracture with paravertebral hematoma, a large circumferential bulging disk related to an old compression fracture, as noted by the presence of the large osteophyte, and infection (although with the history provided, this was not considered). No bony destruction was noted on this CT scan. An MR was recommended.

The patient refused an MR for 1 month after the CT. She finally agreed because of persistent severe back pain and the onset of neurologic deficit. Sagittal MR (Fig. 13B) shows nearly complete destruction of the T11 vertebral body (arrow) and marked destruction of the caudal and anterior one-half of the T10 vertebral body. There is replacement of the vertebra by high signal intensity fluid collection involving both vertebra and the intervening disk. There is epidural mass effect with dorsal displacement of the cord.

Diagnosis: *Staphylococcus aureus* diskitis and osteomyelitis with progressive destruction of the T10 and T11 vertebra involved by a prior old compression fracture. This represents hematogenous spread of infection to the compressed vertebrae.

Discussion: Magnetic resonance is more sensitive in depicting changes within the vertebra resulting from infection and neoplasm than CT, and MR has become the procedure of choice in the study of diskitis and osteomyelitis, as well as metastatic disease and myeloma of the spine. Inflammation presents as increased signal intensity of the disk and vertebral body on T2-weighted images and decreased signal intensity of the vertebral body on T1-weighted images (Fig. 13C–H). Magnetic resonance detects these abnormalities very early in the disease process and is a very sensitive modality for following the resolution of diskitis and osteomyelitis during antibiotic therapy. With healing, the abnormal increased signal intensity of the disk and vertebra on T2-weighted images reverts to normal. We have followed three patients with severe diskitis and osteomyelitis to complete healing with serial MR scans at 1-month intervals. They all showed an increased signal intensity of the vertebral bodies and increased signal intensity of the disk in the acute inflammatory phase; whereas the involved disk was dehydrated and of low signal intensity on preoperative MR images. There was a gradual decrease in the high signal intensity of the involved vertebral bodies and disks on T2-weighted images during healing. All involved disks showed a degenerated, low signal intensity appearance after complete healing. Persistent abnormal increased signal intensity of the disk or vertebra on T2 MR images indicates residual inflammation/infection.

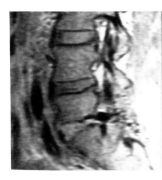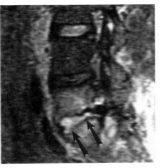

FIG. 13C. Diskitis and osteomyelitis (sagittal MR, TR/TE 2,000/20, 80 msec). There is abnormal increased signal intensity in the previously operated L5-S1 disk. Note also the erosion of the adjacent end plates of both L5 and S1 (*arrows*), and the increased signal intensity of the marrow on T2-weighted images and the decreased signal intensity of the involved marrow on the proton-density image consistent with inflammation. This was a *Staphylococcus aureus* osteomyelitis.

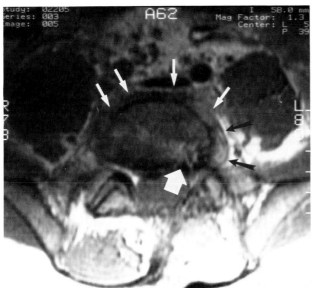

FIG. 13D. Diskitis and osteomyelitis (axial MR, TR/TE 1,800/20 msec). Axial image of the patient in Fig. 5C. Note the paravertebral mass on this axial MR obtained in the patient in Fig. 5C (*arrows*). Thickening of the paraspinous soft tissues caused by granulomatous infiltration is present in nearly all cases with diskitis and osteomyelitis (*large arrow*) and should be looked for as an early sign to verify the suspicion of diskitis and osteomyelitis when the diagnosis is in doubt.

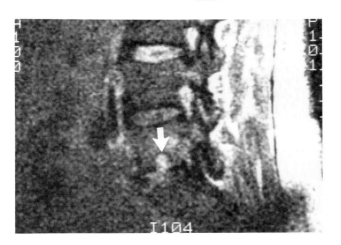

FIG. 13E. Diskitis and osteomyelitis (sagittal MR, TR/TE 2,500/80 msec). The sagittal MR shows a small area of increased signal intensity of the marrow of the L5 vertebral body with erosion of the adjacent end-plate and very slight increased signal intensity of the disk (*arrow*). The patient was a young orthopedic surgeon with excruciating right-sided back pain occurring approximately 6 weeks after a successful left-sided hemilaminotomy for an L5-S1 disk herniation. The pain was severe with jarring and was constant. Review of a preoperative CT did not show any evidence of a Schmorl's node or an end-plate lesion in this area preoperatively. The lesion was consistent with a small focal area of diskitis and osteomyelitis.

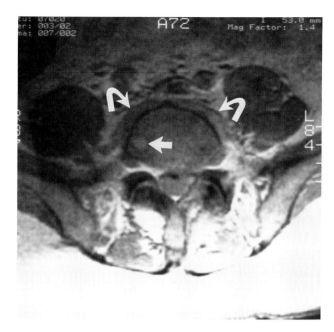

FIG. 13F. Diskitis and osteomyelitis (axial MR, TR/TE 1,000/20 msec). Note the small area of increased signal intensity on the proton-density image (*arrow*) and the thickening of the paraspinous soft tissue (*curved arrows*). Although the focal area of diskitis and osteomyelitis was quite small, the soft tissue mass is prominent. The soft tissue thickening is seen in nearly all patients and is often seen as an *early* accompaniment of diskitis and osteomyelitis. In this case it allowed a definitive diagnosis of diskitis and osteomyelitis with minimal involvement of the disk and vertebra.

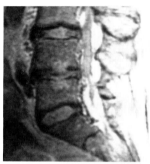 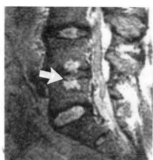

FIG. 13G. Diskitis and osteomyelitis caused by beta hemolytic strep (sagittal MR, TR/TE 2,000/30, 80 msec). Note only slight increased signal intensity of the disk and focal areas of increased signal intensity of the L4 and L5 vertebral bodies immediately adjacent to the disk (*arrow*). Note also the destruction of the end plates adjacent to the disk. Note the relatively localized area of inflammation as opposed to the more generalized inflammation of the vertebra typically seen in *Staphylococcus aureus* osteomyelitis.

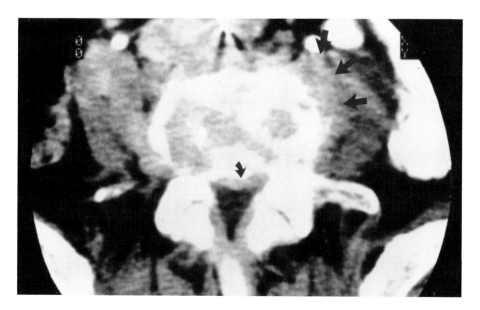

FIG. 13H. Diskitis and osteomyelitis caused by beta hemolytic strep. The axial CT shows irregular destruction of the end plate. This "swiss cheese" appearance of the end plate is very typical appearance of osteomyelitis. Note also the marked paraspinus soft tissue mass thickening, which is very typical of diskitis and osteomyelitis (*arrows* and *curved arrow*). This paraspinus soft tissue is not fluid, as can be seen from its low signal intensity on the MR axial images noted above. This is felt to be reactive granulation. Following successful therapy for osteomyelitis, this paraspinus soft tissue mass disappears.

CASE 14

This 35-year-old man complained of several months of slowly progressive back pain and recent onset of leg weakness and paresthesias. This MRI was obtained for evaluation of possible herniated disk. What is your diagnosis? (Fig. 14A. Sagittal MR, TR/TE 800/20 msec. Fig. 14B. Sagittal MR, TR/TE 2,500/30, 70 msec. Fig. 14C. Axial MR, TR/TE 3,000/30, 80 msec.)

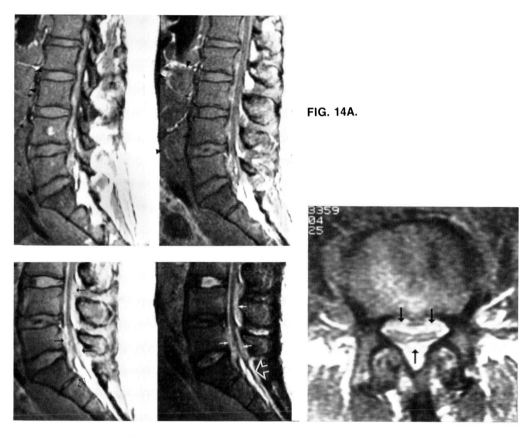

FIG. 14A.

FIG. 14B.

FIG. 14C.

Findings: Circumferential epidural soft tissue (small arrows) compresses the thecal sac over several segments (Fig. 14A–C). The tumor is hypointense relative to fat (open arrows), excluding epidural lipomatosis. Retroperitoneal paraspinal lymphadenopathy is also present (arrowheads).

Diagnosis: Epidural and retroperitoneal lymphoma.

Discussion: The differential diagnosis of an epidural lesion includes HNP, abscess, hematoma, epidural lipomatosis, and tumor. The most common epidural tumor is metastatic disease extending from the adjacent bone. Primary epidural tumor without bone involvement is unusual in metastatic disease but frequent in lymphoma and may be the presenting site of disease in up to 2% of cases. Associated paraspinal mass clearly differentiates lymphoma from other diagnostic entities. Focal destructive lesions and diffuse marrow involvement as seen in myeloma can occur. Occasionally focal epidural lymphoma may mimic a free disk fragment or focal epidural hematoma (Fig. 14D,E). Contrast enhancement following gadolinium administration distinguishes neoplasm from either of these two entities.

The differential diagnosis of spinal epidural tumor also includes hematoma and abscess. Usually, epidural abscess spreads from adjacent disk space infection or osteomyelitis. Primary infection of the epidural space may occur either by hematogenous spread from a distant primary infection or following spine surgery, penetrating injury, or epidural steroid injection (Fig. 14F). The most common etiological organism is *Staphylococcus aureus* although gram-negative bacilli and mycobacterium tuberculosis are not uncommon causes. Magnetic resonance imaging has proven to be extremely effective in early diagnosis as well as evaluation of extent of disease. These cellular and proteinaceous fluids have distinctive signal characteristics, being greater in intensity than CSF on T1-weighted and spin-density sequences, and greater in signal intensity than most epidural soft tissue tumor masses on T2-weighted images. Adjacent paraspinal inflammatory masses containing edema and granulation tissue may have intermediate intensity on T2-weighted images compared to the abscess cavity itself. Morphologic distortion of the disk with reduced volume and height of the disk when associated with marked increased signal intensity relative to adjacent normal disks indicates primary disk space infection, because degenerated disks usually show reduced hydration.

The differential diagnosis of a paraspinal fluid collection in the postoperative patient includes abscess, hematoma, infected hematoma or seroma, and pseudomeningocele (Fig. 14G). Chronic hematoma and seroma are composed of proteinaceous fluid like an abscess and may not be distinguishable on MRI. If, however, residual paramagnetic break-down products of the hematoma are present, it will result in increased signal intensity on T1-weighted images. Very high concentration of protein without paramagnetic moieties may also significantly reduce T1 by enabling more efficient spin-lattice relaxation of the water protons.

A postoperative pseudomeningocele may occur as a result of a tear in the dura with extension of an arachnoidal membrane through the dural tear. Enlargement of the cystic space occurs as a result of a "ball-valve" mechanism, resulting in a type of arachnoid diverticulum. Alternatively, a capsule of fibrogranulation tissue without lining of arachnoid may form. The pseudomeningocele, even when it communicates freely with the subarachnoid space, may be hyperintense relative to CSF elsewhere in the spinal canal. Contributing factors are twofold: dampening of CSF pulsation, which tends to reduce the signal intensity of the subarachnoid space; and slightly greater protein content of the fluid in the pseudomeningocele.

A pitfall in interpretation of the now commonly used protocol of sagittal and axial multiecho T2-weighted sequences needs mentioning. Fatty tissue in the posterior paraspinal musculature may mimic a proteinaceous fluid collection. Because of its close approximation with the surface coil, the signal intensity may be much greater than retroperitoneal or intramedullary fat on heavily T2-weighted sequences (Fig. 14H).

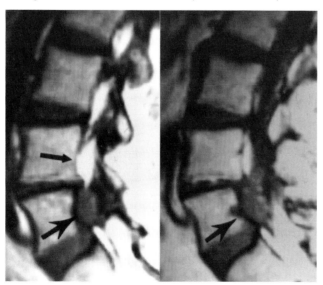

FIG. 14D. Epidural lymphoma (sagittal MR, TR/TE 600/20 msec pre–Gd-DPTA). Epidural soft tissue mass (*large arrows*) replaces epidural fat. Compare with the normal epidural fat of the adjacent level (*small arrow*).

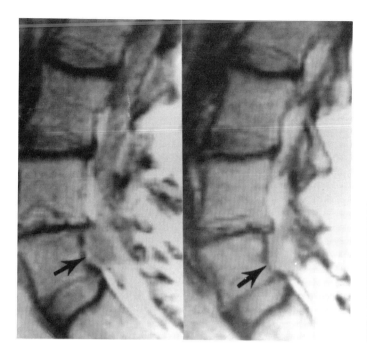

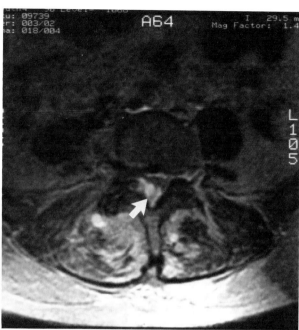

FIG. 14E. Epidural lymphoma (sagittal MR, TR/TE 600/20 post–Gd-DPTA). Enhancement of the soft tissue lesion (*arrows*) indicates neoplasm rather than epidural hematoma or disk herniation. Note, however, postcontrast images alone reduce conspicuity of the lesion relative to epidural fat, indicating the need for precontrast images. (Case courtesy of Gordon Sze, M.D. From *Radiology* 1988;4, with permission.)

FIG. 14F. Epidural abscess (*arrow*; axial MR, TR/TE 3,000/80 msec). Increasing signal intensity on progressively more T2-weighted images, typical of a proteinaceous fluid.

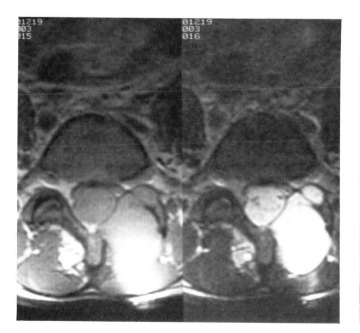

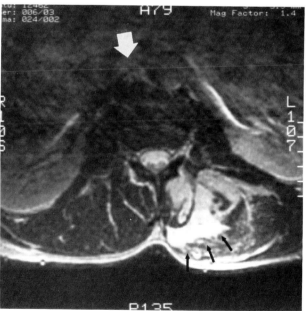

FIG. 14G. Postoperative pseudomeningocele (axial MR, TR/TE 3,000/30, 80 msec). Note the slightly greater signal intensity of the pseudomeningocele relative to subarachnoid space. Slight increase in protein content, reduced CSF pulsations, and/or closer position to the surface coil may account for this.

FIG. 14H. Paraspinal and intraosseous lipoma (axial MR, TR/TE 3,000/80 msec). Lipomatous tissue (*small arrows*) lying close to the surface coil is greater in signal intensity than retroperitoneal fat (*large arrow*), mimicking a fluid collection.

CASE 15

A 72-year-old man had a history of trauma and upper lumbar pain several years before this examination. Currently his symptoms are only low back and left hip pain. What is your diagnosis? What are the differential diagnoses? (Fig. 15A. Sagittal MR, TR/TE 3,000/30, 80 msec. Fig. 15B. Plain radiograph, lumbar spine.)

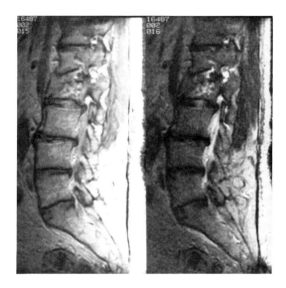

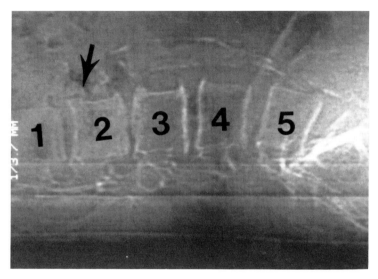

FIG. 15A.　　　　　　　　　　　　　　　　**FIG. 15B.**

Findings: In Fig. 15A, there is abnormal high signal intensity of L1 and L2 extending through the involved disk with disruption of the end plate. The lateral plain film (Fig. 15B) of the lumbar spine at the time of the old injury shows a compression fracture of the superior end plate of L2 and the fracture of the anterior cortex of the L2 vertebra (arrow). The fracture was untreated. Collapse of the L2 vertebra occurred between the time of this examination and the MR.

Diagnosis: Old vertebral end-plate fracture with wedging of the involved vertebra and fracture of the end plates. The high signal intensity represents residual hemorrhage and/or seroma within the vertebrae and disk space.

Discussion: An initial impression when viewing this case might be that of florid diskitis and osteomyelitis because of the involvement of the disk as well as the vertebral bodies by contiguous high signal intensity abnormalities on T2-weighted images, not to mention the disruption of the end plates. It certainly crossed our minds! His referring physician, however, could provide no clinical or laboratory support for this diagnosis, and plain films showed only wedging and collapse of the vertebral bodies and narrowed disk space with end-plate disruption consistent with old end-plate fracture—*not erosive destruction*— as would be expected with diskitis and osteomyelitis. Six-month follow-up revealed no change in his clinical status or plain films.

Review of the MR confirms that the discontinuity of the end plates is sharply marginated and no destruction is seen. The collapse of the vertebral body raises the question of neoplasm. The MR picture, however, is not really compatible with neoplasm because of the irregular distributed pattern of the abnormality rather than marrow infiltration and the fact that it violated the disk. Plain films showed no destruction, and the bone scan was normal.

The lesion showed high signal intensity on T1, proton-density, and T2-weighted images. This is compatible with chronic hematoma or seroma. The patient had lack of any symptoms currently referable to this area. Further questioning of the patient turned up an interesting history of an episode of significant trauma to the upper lumbar spine several years before his current work-up. Plain films and a CT were read as normal. However, retrieval and review of those plain films and CT revealed a missed, mild compression fracture of the superior end plate of the L2 vertebral body and disruption of the end plates. Considerable further collapse had occurred in this individual with an untreated compression fracture (Fig. 15B).

The three-column concept proposed by Denis provides a useful framework for radiographic reporting of vertebral fractures and their significance relating to the question of stability. Computed tomography provides excellent images of bony fractures and is the procedure of choice for evaluation of severe spine injuries including burst fractures with displaced fragments and suspected associated posterior column fractures requiring decompression and stabilization. These patients are often severely injured and require close observation and life support, which precludes the more time-consuming MR examination. In such cases when information about possible cord compression is required, useful information can rapidly be obtained with a contrast-enhanced CT obtained by C1-2 puncture, and dynamic low dose CT can examine several motion segments in as little as 5 to 10 min, with reformatted images obtained later after the patient has left the CT suite. These rapid exams can be obtained with and without traction to demonstrate its efficacy in relieving cord compression.

Magnetic resonance is the modality of choice, however, for stable spine injury patients or follow-up examination of previously injured stabilized spine patients in whom cord compression or injury is suspected (Fig. 15C–F).

In addition to the information provided by CT regarding bony fracture instability and displaced bony fragments, MR provides direct visualization of cord compression in patients treated both conservatively and surgically. Magnetic resonance is also the procedure of choice in the evaluation of post–cord injury patients with increasing pain, sensory, or motor deficit, because many such patients have been found to have syringomyelia or myelomalacia (Case 19). Intramedullary hematomas, as well as epidural hemorrhage, are clearly identified.

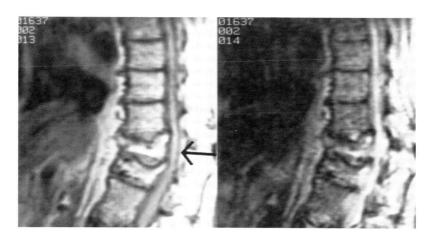

FIG. 15C. Burst fractures with spinal stenosis and cord compression (sagittal MR, TR/TE 2,000/30, 80 msec). T11 and T12 burst fractures with 50% narrowing of anteroposterior dimension of the lower thoracic spine caused by dorsal displacement of the posterior cortical margin of the T11 and T12 vertebra into the central spinal canal with moderate dorsal displacement and compression of the distal thoracic cord and conus (*arrow*).

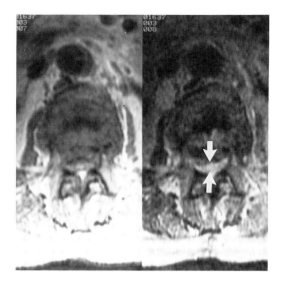

FIG. 15D. Burst fractures with spinal stenosis and cord compression. The axial image (TR/TE 1,000/10 msec) shows moderately severe central spinal stenosis at the superior aspect of T11 with approximately 60% narrowing of the anteroposterior dimension of the central spinal canal caused by the dorsal displacement of the posterior cortical margin of the vertebra into the central spinal canal. Note the mild compression of the distal thoracic cord (*arrows*).

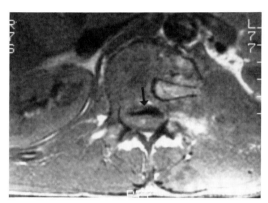

FIG. 15E. Postoperative T12 burst fracture with left-sided anterolateral decompression and rib strut grafting. Note the poor decompression of the central spinal canal with persistent 50% narrowing of the anteroposterior dimension of the canal by the dorsally displaced bone fragment. There is moderate cord compression (*arrow*).

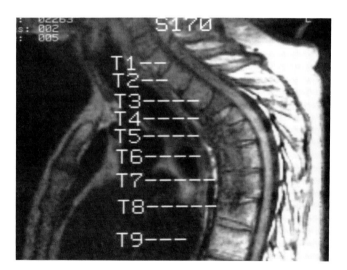

FIG. 15F. The sagittal MR (TR/TE 600/20 msec) of the thoracic spine shows anterior wedge compression fractures of the T4, T5, and T6 vertebra with placement of a rib strut graft. There is moderate kyphotic deformity of the spine; however, no encroachment on the central spinal canal nor compression of the thoracic cord is present. No syrinx is seen. Magnetic resonance provides an excellent means of evaluating the entire thoracic spine and cord in postoperative patients with persistent or recurrent pain.

CASE 16

This 37-year-old patient was referred for a cervical MR to rule out a cervical tumor because of a history of a slowly progressive difficulty with gait disturbance and numbness of the right leg for the past year. What is the etiology of the cord impingement? (Fig. 16A. Sagittal MR, TR/TE 600/20 msec.)

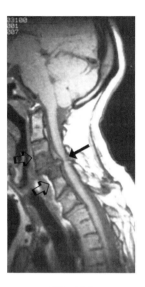

FIG. 16A.

Findings: There are multiple congenital anomalies with congenital blocked vertebra of the C3-4 and C5-6 vertebra in Fig. 16A. The transitional disk is severely degenerated with marked anterior and posterior spurring, as well as marked sclerosis of the adjacent vertebral bodies. The large posterior osteophytic spurs are producing moderately severe compression of the thoracic cord. There are also segmentation abnormalities of all of the upper thoracic vertebra visualized with incomplete disks of the vertebra from T1 through T5.

Diagnosis: Marked central stenosis and cord compression caused by marked posterior osteophytic spurring at the C3-4 level, which is a transitional level between two congenital blocked vertebra.

Discussion: Sagittal MR images provide excellent delineation of the smallest midline disk herniations and definition of cord impingement (Fig. 16B,C). Visualization of the lateral extent of the disk herniation as it extends into the intervertebral nerve root canal and determination of nerve root impingement is, however, difficult because the cervical neural foramina are poorly visualized on sagittal images. Because of the narrowness of cervical disks, disk hydration information is not as easily obtained as in the lumbar region.

The utility of MR in the evaluation of cervical radiculopathy is limited if the MR scanner employed cannot obtain thin section images (less than 2-mm slice thickness) because of poor visualization of structures within pathologically narrowed foramina. Partial volume imaging present on wide (i.e., 5 mm) slice thickness scans coupled with similarity of signal intensity of degenerated herniated disk material and uncinate spurring makes these images all but uninterpretable, other than to state that they are abnormal. Additional quantitative and qualitative data are difficult to extract from these images.

Axial images do provide excellent delineation and visualization of the cervical cord and cord compression and displacement caused by cervical degenerative spurring and disk herniation.

Fortunately, recent software and hardware improvements allow obtaining up to 60 1.5-mm axial images of the cervical spine and infinite choice of slice thicknesses. This has dramatically improved our MR cervical spine images.

Gradient-echo imaging (GEI) is now the MR technique of choice in evaluation of patients with cervical radiculopathy. Gradient refocused partial flip angle techniques generate images with high CSF signal (myelographic effect) and high CSF/cord/disk contrast in relatively short acquisition times compared to conventional spin-echo techniques. These techniques provide increased signal intensity in this sequence. Because TR contributes little to signal intensity, TRs of 20 to 400 msec can achieve this myelographic effect. In addition, shorter TE times can be used with this technique, allowing for increased signal-to-noise ratio and thinner slices compared with spin-echo techniques. This provides better visualization of the intervertebral neural canals and exiting nerve roots.

Single slice GEI techniques require very short TRs, 20 to 80 msec, and flip angles of 6 to 8°. Multislice GEI techniques are now available, allowing the use of relatively longer TRs, 400 to 800 msec. Although lengthening the TR contributes little to signal-to-noise ratio, it does allow the use of longer flip angles without sacrificing CSF cord contrast. Longer flip angles do significantly improve signal-to-noise ratio, a significant advantage of multislice GEI techniques. Flip angles of 11 to 20° can be used.

Unenhanced CT performed with 1.5-mm slice thickness images remains the procedure of choice in our practice when evaluating monoradiculopathy of the upper extremity and impingement of a cervical nerve root is suspected.

Images are routinely obtained from C7-T1 through C2. Slice thicknesses greater than 1.5 to 2.0 mm provide inadequate delineation of cervical disks and also inadequate delineation of the neural foramina. Despite having an MR scanner available in our practice, we continue to utilize CT as a procedure of choice in cervical radiculopathy for two primary reasons: (a) It is considerably less expensive than MR; and (b) the evaluation of narrowed foramina and differentiation of disk herniation from uncinate and/or posterior spurring is easier with CT at the present time (Fig. 16D,E). This certainly was true when we were unable to obtain axial MR images thinner than 3 mm. Continued improvements in MRI including a recent software and hardware update of our scanner, which allows 1.5-mm three-dimensional stacked images of the cervical spine, have recently provided marked improvement in the imaging of the narrowed foramina by eliminating the problem of partial volume imaging. Certainly, MR images of normal foramina provide spectacular images of normal neural anatomy; however, when the nerve root canals become narrowed by either low density disk herniation or osteophyte and both are of similar low signal intensity, MR often cannot differentiate between them. In our experience, this is more easily performed at present with thin section CT imaging as described.

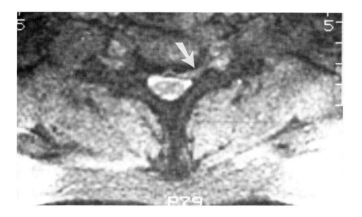

FIG. 16B. The axial MR (TR/TE 150/ 13 msec) shows a moderate-sized disk herniation projecting into the left C5-6 intervertebral nerve root canal (*arrow*). In this case the disk is of higher signal intensity than the adjacent bone and is easily visualized.

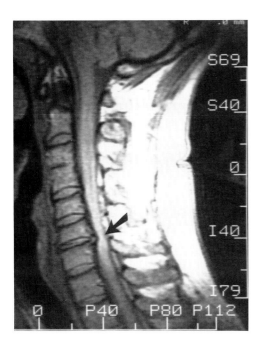

FIG. 16C. Sagittal MR image shows a small central disk herniation at C6-7 with mild cord compression (*arrow*). Note the congenital block vertebra at C3-4. There is degenerative disk space narrowing, as well as anterior and posterior osteophytic spurring.

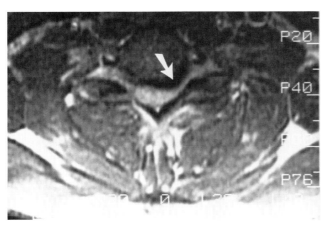

FIG. 16D. Axial MR of the cervical spine in the patient in Fig. 16C. On the axial image, the disk herniation appears low signal intensity and cannot be distinguished from posterior or uncinate spurring (*arrow*). There is moderate lateral spinal stenosis with compression of the exiting left C6 nerve root, and there is mild compression of the left side of the cord. This case illustrates the difficulty in differentiating disk herniation from osteophyte on the basis of axial images.

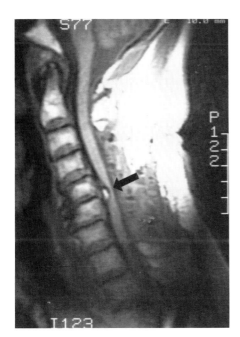

FIG. 16E. A sagittal MR (TR/TE 800/20 msec) shows marked focal impingement on the cord by a high signal intensity mass. This is a large bony osteophyte (*arrow*) with fatty marrow within the osteophyte. The patient was considerably improved following surgery. Note the severe degenerative disk disease at the involved C5-6 level as well as at C6-7.

CASE 17

This 16-year-old girl presented with a 10-day history of increasing ataxia and gait disturbance. What is causing the expansion of the cervical cord from C4 to C6? (Fig. 17A. Sagittal MR, TR/TE 500/20 msec. Fig. 17B. Sagittal MR, TR/TE 500/15 msec, flip angle 15°. Fig. 17C. Axial MR, TR/TE 500/15 msec. Fig. 17D. Axial MR, TR/TE 2,400/80 msec.)

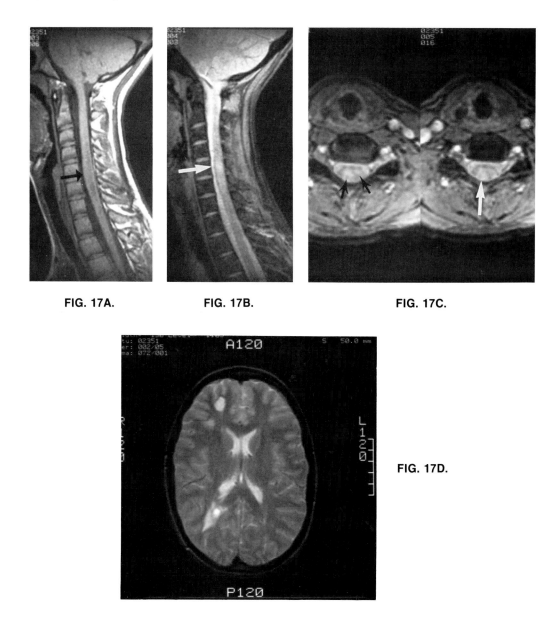

FIG. 17A. **FIG. 17B.** **FIG. 17C.**

FIG. 17D.

Findings: In Fig. 17A, there is local enlargement of the spinal cord (arrow) from C4 to C6 with a long area of decreased signal intensity, which on the gradient-echo sequence (Fig. 17B,C) appears as an ill-defined area (arrows) of increased signal intensity within the spinal cord at the C5 level as a result of demyelination and swelling.

Examination of the brain reveals multiple periventricular and deep white matter demyelinating plaques in this patient with multiple sclerosis.

Diagnosis: Florid multiple sclerosis with brain and cervical cord involvement.

Discussion: When the presenting symptomatology includes myelopathy, MR is the clear imaging procedure of choice, as MR has dramatically improved the visualization of the cervical cord. Images of the cervical spine have been improved by software programs that have alleviated the problem of CSF flow artifacts and decreased motion artifacts from breathing and swallowing. In the evaluation of intramedullary demyelinating processes of the spinal cord such as multiple sclerosis, GEI may not always be adequate in delination of these lesions. Cardiac-gated spin-echo T2-weighted images are usually necessary to delineate demyelinating plaques although they can be demonstrated using GEI (Fig. 17A–D).

The craniovertebral junction is very well imaged by MR. Congenital anomalies such as Arnold-Chiari are easily detected and therefore seen with increasing frequency.

Chiari I malformation is characterized by downward displacement of the cerebellar tonsils through the foramen magnum (Fig. 17E). This malformation is associated with cervical cord hydromyelia. Chiari II malformation is almost always associated with meningomyelocele. This malformation is characterized by downward displacement of the fourth ventricle through the foramen magnum. Frequently a cervicomedullary kink is present. There is a more extensive anomaly of the neural axis, and multiple intracranial anomalies are present and may be associated with agenesis of the corpus callosum. Hydrocephalus usually is a result of aqueductual stenosis. The much more rare Chiari III malformation has an associated occipital encephalocele.

Syringomyelia is easily detected by MR. The size and distribution of the syrinx can be well defined. The diameter and length of involvement of the cord by the syrinx is well delineated, as are any loculations of the syrinx. Magnetic resonance has shown a surprisingly high incidence of syringomyelia in patients with unusual thoracic scoliosis (up to 50% of patients with atypical thoracic curves). In our practice, the first six patients referred to us for the evaluation of atypical thoracic curves, rapidly advancing scoliosis, and/or painful scoliosis had syringomyelia (Fig. 17F). Currently, a research study is underway to determine the incidence of syringomyelia in the patient population with idiopathic scoliosis.

The incidence of posttraumatic syringomyelia and myelomalacia of the cord is surprisingly high. Recent studies of patients with progressive posttraumatic paraplegia or quadriplegia have shown a very high incidence of posttraumatic syringomyelia.

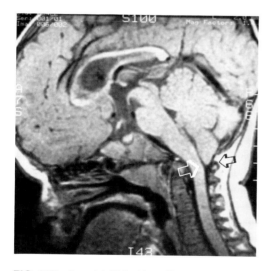 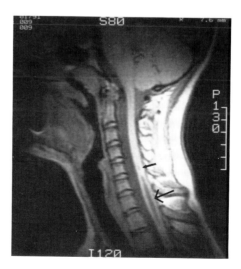

FIG. 17E. Arnold-Chiari I malformation (sagittal MR, TR/TE 800/20 msec). Note the caudal displacement of the cerebellar tonsils. The tip of the tonsils lie at the level of the arch of C1 (*arrows*). There is no associated hydrocephalus or syrinx.

FIG. 17F. Cervical syringomyelia and scoliosis (sagittal MR, TR/TE 1,500/20 msec). An enlongated syrinx expands the lower cervical cord from T2 to C5-6 (*arrows*). Note that the lesion is septated with a small proximal cyst separated from the main syrinx. This syringomyelia was associated with a scoliosis.

CASE 18

This 69-year-old woman with progressive spastic quadriparesis is sent for cervical MR. What is your differential diagnosis? (Fig. 18A. Sagittal MR, TR/TE 600/30 msec.)

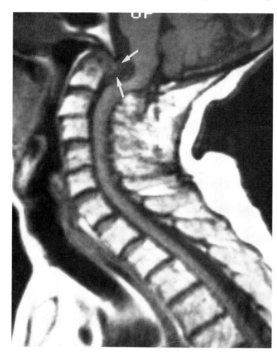

FIG. 18A.

Findings: Erosion of the dens with subchondral replacement of fatty marrow by inflammatory tissue is present (Fig. 18A). Inflammatory soft tissue pannus (arrows) produces ventral compression of the spinal canal. Vertical subluxation of the dens relative to the basiocciput results in acute cervicomedullary angulation and kinking.

Diagnosis: Rheumatoid arthritis of the occipital-atlanto-axial joints with cranial settling.

Discussion: Seropositive rheumatoid arthritis, as well as seronegative spondyloarthropathies such as psoriatic arthritis, is an inflammatory disease of synovial-lined joint capsules and bursae. The occipital-atlanto-axial junction is a common site of disease with potentially serious neurologic sequelae. Subaxial subluxation is also commonly observed in these diseases. Facet joint erosions and instability and secondary degenerative disk disease may lead to multilevel spinal cord compression.

Synovial inflammation of the bursa between the transverse ligament and the posterior aspect of the dens leads to ligamentous laxity and anterior atlanto-axial subluxation. Posterior compression of the spinal canal by the posterior arch of C1 as well as the retrodental soft tissue mass of pannus combine to produce spinal stenosis and cord compression.

Atlanto-axial impaction, cranial settling, and pseudobasilar invagination are terms describing the upward displacement of the dens relative to the basiocciput. The atlanto-occipital and lateral atlanto-axial articulations are synovial joints frequently involved in these inflammatory arthropathies. Subchondral osseous erosions of the occipital condyles, lateral masses of C1, and the articular pillars of C2 lead to progressive loss in vertical height of the supporting structures of the cranio-cervical junction. Subsequent superior migration of the dens and protrusion into the foramen magnum produces brainstem compression. Lateral, rotatory, and posterior subluxation of the atlanto-axial joint are less common features. Magnetic resonance imaging clearly demonstrates these complex mechanical derangements and the degree of compression of the neural axis. Flexion and extension sagittal MRI elegantly depicts dynamic instability. In addition, associated fractures of the dens may be difficult to demonstrate using other imaging modalities. Severe osteopenia may limit the value of conventional tomography. A transverse fracture of the dens parallel to the CT plane of section may go undetected unless sagittal and coronal reformatted images are used.

Replacement of normal fatty marrow by subchondral inflammatory and fibrotic tissue leads to reduced signal intensity on T1-weighted images within the dens. Inflammatory pannus formation leads to osteolysis of the dens and retrodental soft tissue mass compressing the anterior aspect of the spinal cord, findings mimicking tumor. The radiographic differential diagnosis therefore includes metastasis and other tumors such as chordoma.

An os odontoideum (Fig. 18B,C) may represent a congenitally ununited ossicle or the result of previous fracture and pseudoarthrosis. The two may not be radiographically distinguishable. Cortical bone produces no MR signal because of its lack of mobile water protons, making acute cortical fractures difficult to visualize. Nevertheless, interruption of the normal bright marrow signal on T1-weighted images by the opposing hypointense cortical surfaces at the pseudoarthrosis is usually apparent.

Ununited fracture or congenital dysplasia of the dens and secondary degenerative disease may lead to a so-called pseudotumor, which has a typical MRI appearance. Instability of the joint leads to fibroproliferation and bony eburnation, producing a retrodental soft tissue mass that compresses the cervicomedullary junction. The fibrous tissue containing dense collagen fibers has decreased MR signal intensity in keeping with its low content of mobile water protons.

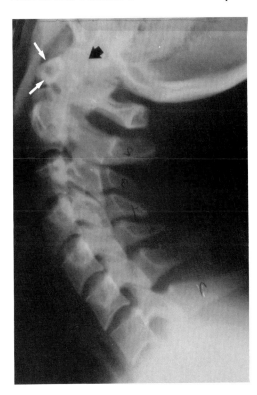

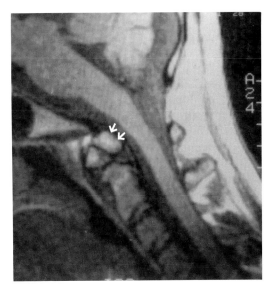

FIG. 18B. Os odontoideum. Lateral cervical spine. Note the ununited os odontoideum (*black arrow*), which is displaced posteriorly and superiorly. The anterior arch of the atlas (*white arrows*) moves as a unit with the separate dens ossicle leading to chronic instability.

FIG. 18C. Os odontoideum (sagittal MR, TR/TE 600/20 msec). Intermediate signal intensity fibrous or loose areolar tissue is interposed between the two opposing hypointense cortical surfaces. Only mild impression of the ventral spinal canal is apparent in this case (*small arrows*).

CASE 19

This 51-year-old woman presented with paresthesia and dysesthesia in her arms bilaterally. What is your diagnosis? (Fig. 19A,B. Sagittal MR, TR/TE 600/15 msec, following intravenous gadolinium administration.)

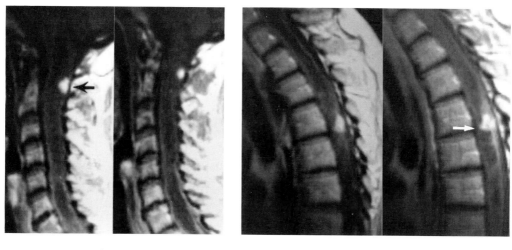

FIG. 19A.   **FIG. 19B.**

Findings: Expansion of the cervical and upper thoracic spinal cord is present by a predominantly cystic lesion (Fig. 19A, B). More inferiorly in the lesion there appears to be multiple septations of the cyst. Two well-defined enhancing nodules are identified in the wall of the cystic lesion (arrows).

Diagnosis: Cystic hemangioblastoma of the spinal cord.

Discussion: Most intramedullary tumors, including ependymoma, astrocytoma, and oligodendroglioma, are of glial origin. Hemangioblastoma is a less common entity, representing less than 5% of the cases, but its MRI features may be characteristic. As in the posterior fossa, the findings of a cystic mass with a well-defined enhancing mural nodule are highly suggestive of this diagnosis. These lesions may also be predominantly solid, diffusely enlarging a long segment of the spinal cord, mimicking a glioma. Large feeding and draining vascular structures of some very vascular lesions may suggest arteriovenous malformation. Spinal hemangioblastoma may occur separately or in association with Hippel-Lindau disease, an entity characterized by retinal capillary angiomatosis, hemangioblastomas of the posterior fossa and spinal cord, multiple renal cysts and/or carcinomas, pancreatic and hepatic cysts, and other benign abdominal tumors. This patient had previous surgery for a posterior fossa hemangioblastoma, presumptive evidence of the syndrome.

As in posterior fossa hemangioblastomas, gadolinium is essential in identifying the mural nodule as this must be resected to avoid recurrence. Gadolinium is also helpful in preoperative assessment of intramedullary tumor in identifying the extent of disease and differentiating tumor cysts from associated syrinx. It also directs the surgeon to an appropriate biopsy site. In this case, without the use of gadolinium, the exact nature and cause of the cystic lesion of the cord may be difficult to evaluate.

Syringomyelia should be differentiated from hydromyelia, which is a dilatation of the central spinal canal, usually extending superiorly to its communication in the fourth ventricle. A hydromyelic cavity is centrally placed within the spinal canal and lined with normal ependyma. It is commonly associated with congenital anomalies of the spine and neural axis such as Chiari I malformation as well as lipomyelodysplasia and diastematomyelia. In Chiari II malformation, with its associated meningomyelocele and tethered cord, the hydromyelic cavity may extend superiorly from the placode.

Syringomyelia, on the other hand, is eccentrically placed and not lined with normal ependyma. It may result from damage to the central nervous system (CNS) by direct trauma with hematoma formation or vascular insult. These cystic spaces may enlarge and extend superiorly, leading to progressive neurologic deficit. Cystic cavitation of the cord may also occur in association with both intramedullary and extramedullary tumors of the spinal canal.

Posttraumatic progressive myelopathy (Fig. 19C,D) is frequently the result of syrinx formation of the cord. T1-weighted images are best to delineate small cysts (arrows) and their extent. T2-weighted images may not clearly differentiate hyperintense areas of myelomalacia from true cystic spaces. In addition, the size and contiguity of the cysts are best depicted using T1-weighted images. Progressive neurologic deficit may require shunting of these posttraumatic cysts, and MRI is invaluable in the preoperative evaluation of these patients.

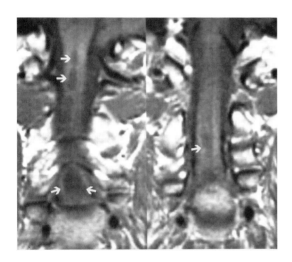

FIG. 19C. Posttraumatic syrinx (*arrows*) (coronal MR, TR/TE 800/20 msec).

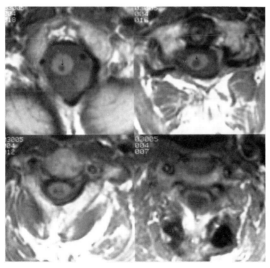

FIG. 19D. Posttraumatic syrinx (axial MR, TR/TE 800/20 msec). Multiple loculated cysts (*arrows*) are present in the spinal cord above the level of the site of injury. This patient presented with progressive ascending neurologic cord dysfunction as a result of expanding cyst formation.

CASE 20

This 28-year-old with known right parietal glioblastoma presented with rapidly progressive weakness in the legs as well as paresthesias. What is your diagnosis? (Fig. 20A. Sagittal MR, TR/TE 500/20 msec, pre–Gd-DPTA. Fig. 20B. Post–Gd-DPTA. Cases courtesy of Gordon Sze, MD.)

 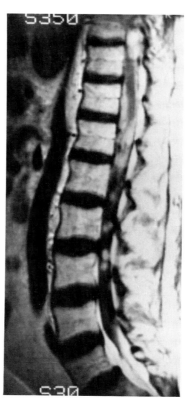

FIG. 20A.   **FIG. 20B.**

Findings: T1-weighted pregadolinium and postgadolinium MR images (Fig. 20A,B) reveal multiple enhancing nodules in the subarachnoid space surrounding the cauda equina as well as completely encasing the conus medullaris.

Diagnosis: Leptomeningeal metastases.

Discussion: Leptomeningeal carcinomatosis or carcinomatous meningitis represents metastatic disease involving the subarachnoid space of the intracranial or intraspinal nervous system arising from either a CNS or distant primary site. Breast and lung carcinoma are the most common sources of spread in adults. In children, primary CNS posterior fossa tumors such as medulloblastoma and ependymoma account for the majority of cases, and other CNS malignancies such as pineal region tumors and gliomas are less common causes.

Clinically, the patient frequently presents with multifocal symptoms involving the intracranial and spinal CNS. Multiple cranial and spinal nerve deficits; headache; changing mental status; neck, back, and leg pain; and gait disturbances are common symptoms. Although CSF cytology and chemical analysis still remains the most sensitive method of detection of leptomeningeal metastases, MRI has proven to be useful in evaluating the extent of disease in these patients.

Myelography has previously been the "gold standard" for the radiological evaluation of intraspinal meningeal spread of malignancy, revealing nodular filling defects; thickened adherent nerve roots; and irregularity, deformity, and constriction of the thecal sac. Unenhanced MRI may show a "ground glass" appearance of the subarachnoid space, as in this case, but is relatively insensitive in detecting subarachnoid seeding, with a true-positive of 27% and a false-negative of 44% by one study. Small, irregular infiltrating tumor nodules of high water content are poorly delineated by T2-weighted images relative to CSF with increased protein content.

Normal noncontrast MRI does not exclude significant intraspinal meningeal disease as high grade or complete myelographic blocks may be present with relatively little changes in signal intensity. Gadolinium-enhanced MRI, however, has been shown to be as sensitive and possibly more sensitive than myelography and intrathecally enhanced contrast CT in the detection and characterization of disease in the intradural extramedullary space. Findings include well-defined enhancing nodules, distortion and clumping of the nerve roots, and diffuse enhancement of otherwise morphologically normal nerve roots (Fig. 20C–F).

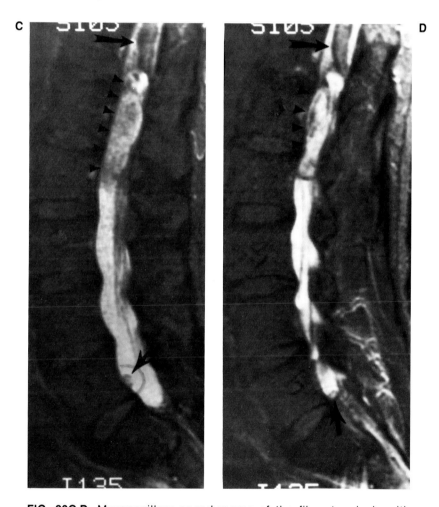

FIG. 20C,D. Myxopapillary ependymoma of the filum terminale with meningeal seeding (sagittal MR, TR/TE 2,500/80 msec). Heterogeneous extramedullary tumor mass (*arrowheads*) displaces the conus (*long arrows*) posteriorly. Small drop mets in the distal thecal sac are identified (*short arrows*). (From G. Sze, *AJNR* 1918;9:153–163, with permission.)

E

F

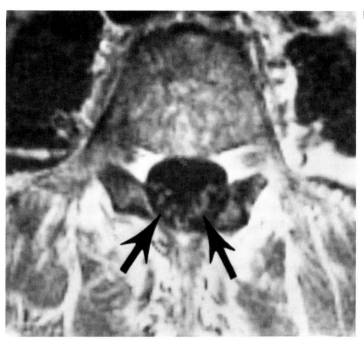

FIG. 20E,F. Sagittal **(E)** and axial **(F)** MR, TR/TE 500/20 msec, post–Gd-DTPA. Homogeneous enhancement of the primary tumor, a small meningeal seed at L5 (**E**, *arrow*) as well as diffuse enhancement of all the nerve roots (**F**, *arrows*). Diffuse meningeal involvement of the cauda equina was found at surgery. (From G. Sze, *AJNR* 1988;9:153–163, with permission.)

CASE 21

This 54-year-old woman complained of back pain and gate disturbance. What does the arrow point to? (Fig. 21A. Sagittal MR, TR/TE 600/15 msec. Fig. 21B. Sagittal MR, TR/TE 300/16 msec, flip angle 10°. Fig. 21C. Axial MR, TR/TE 400/10 msec.)

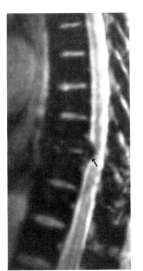

FIG. 21A.

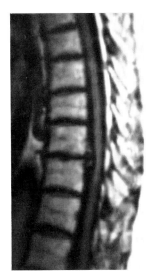

FIG. 21B.

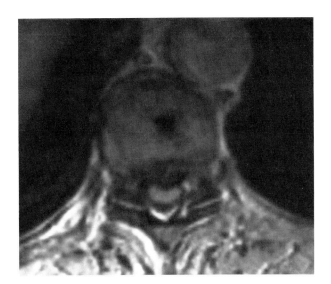

FIG. 21C.

Findings: Marked disk space narrowing, desiccation of the disk, and a focal epidural low signal intensity mass (arrow) centered at the disk space indents the ventral aspect of the spinal cord (Fig. 21A–C).

Diagnosis: Calcified thoracic herniated disk with cord compression.

Discussion: Magnetic resonance imaging has become the imaging procedure of choice in the study of the thoracic spine. Sagittal imaging has enabled evaluation of all of the thoracic disks and the thoracic spinal cord in a relatively short imaging time. As discussed elsewhere in this chapter, gradient refocused partial flip angle techniques provide a myelographic effect, thin slices, and high CSF/cord/disk contrast in shorter imaging time than conventional T2-weighted spin-echo techniques.

Usually, acutely herniated thoracic disks on these imaging sequences are of increased signal intensity and similar in signal intensity to the parent nucleus. In this case, the marked decreased signal intensity may be related either to marked osteophyte formation or calcification of a herniated disk. It should be noted that T1-weighted images in this case are relatively poor in delineating the disk herniation and the degree of compression of the subarachnoid space, which is of similar signal intensity to the calcified disk fragment.

With the widespread use of thoracic MRI, it has become apparent that thoracic disk herniations are much more common than was previously thought. In addition, many of these are asymptomatic and when present may be multiple.

Magnetic resonance imaging of a patient with suspected thoracic disk herniation should include axial and sagittal partial flip angle techniques providing a myelographic effect, as well as sagittal T1-weighted images to evaluate the morphology of the spinal cord. This is essential in identifying syrinx cavities that could occur as a result of hemorrhage or ischemia of the cord adjacent to a thoracic disk herniation (Fig. 21D, E). The GEI in this case also identifies the signal abnormality within the spinal cord itself, but the T1-weighted image identifies this as a true cystic lesion and not an area of myelomalacia or demyelination (arrow, Fig. 21E), which could also be of increased signal intensity on this T2-weighted image. Note also the much better depiction of the disk herniation itself (open arrow, Fig. 21D) on the partial flip angle image compared with the T1-weighted image. A small central annular tear without extrusion of disk material several segments higher is present as well (arrowhead, Fig. 21D). The etiology of the cyst formation in this case was thought to be related to an acute flexion injury. At the time of initial disk herniation, intramedullary hemorrhage or possibly compressive phenomena leading to central ischemia may have occurred.

Other intraspinal thoracic lesions that can be of decreased signal intensity include calcified or ossified meningioma as well as a bone-forming or calcified extradural tumor such as osteochondroma (Fig. 21F,G). These benign lesions typically arise from the posterior elements. The cartilaginous component of these tumors is usually heterogeneous, with areas of increased and intermediate signal intensity on T2-weighted images.

In this case, the signal void of the dense calcified and/or ossified lesion is clearly outlined by displaced high signal of epidural fat.

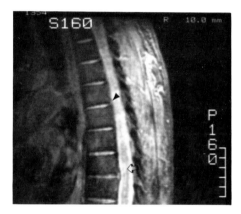

FIG. 21D. Thoracic disk herniation with posttraumatic syrinx (sagittal MR, TR/TE 723/9 msec, flip angle 15°).

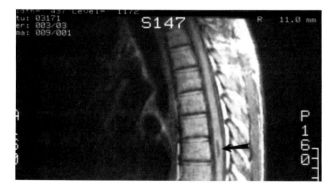

FIG. 21E. Thoracic disk herniation with posttraumatic syrinx (sagittal MR, TR/TE 800/20 msec). Note the improved delineation of the disk herniation on the partial flip angle image compared with the T1-weighted spin-echo image. Although the signal abnormality within the spinal cord is visualized on the partial flip angle image, the T1-weighted image identifies this as a posttraumatic cystic lesion of the cord rather than an area of cord edema, demyelination, or myelomalacia. Surgical drainage of an enlarging cyst may be necessary.

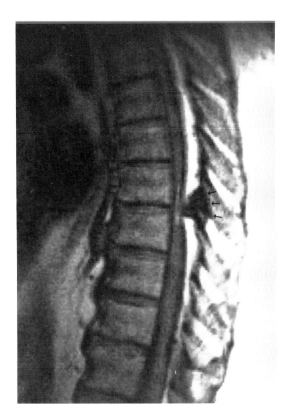

FIG. 21F. Osteochondroma (*arrows*) (sagittal MR, TR/TE 600/30 msec).

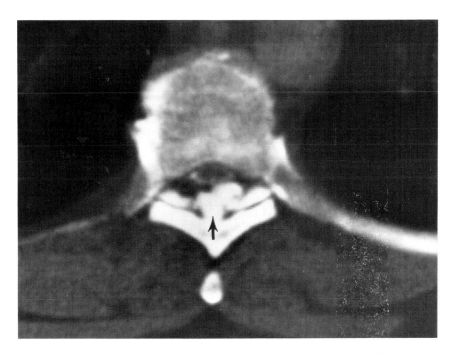

FIG. 21G. Osteochondroma (axial intrathecally enhanced CT). The densely calcified lesion visualized on CT (*large black arrow*) displaces the spinal cord anteriorly, almost completely obliterating the subarachnoid space, which is minimally opacified at this level. This corresponds to the area of signal void (*small arrows*, Fig. 21F) displacing the high signal intensity epidural fat.

CASE 22

Both of these middle-aged women presented with back pain. What is the significance of the areas of signal void within the subarachnoid space (arrows)? (Fig. 22A. Sagittal MR, TR/TE 2,000/30, 70 msec. Fig. 22B. Sagittal MR, TR/TE 750/9 msec, flip angle 15°. Fig. 22C. Sagittal MR, TR/TE 2,150/70 msec. Fig. 22D. Axial MR, TR/TE 2,150/70 msec. Case courtesy of Marsha Guerrein, MD)

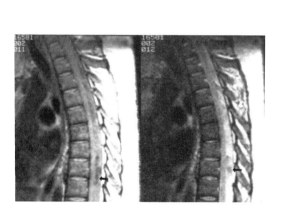

FIG. 22A.

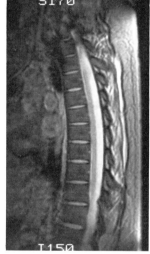

FIG. 22B.

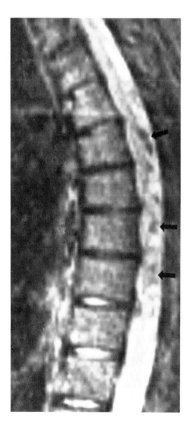

FIG. 22C.

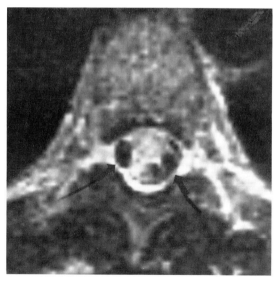

FIG. 22D.

Findings: Multiple round and serpiginous areas of decreased signal intensity in the subarachnoid space surrounding the spinal cord on these noncardiac-gated T2-weighted spin-echo images (Fig. 22A–D) suggest an arteriovenous malformation (arrows). GEI is normal.

Diagnosis: Cerebrospinal fluid pulsation artifacts (arrows, Fig. 22D).

Discussion: Pulsatile motion of CSF in the spinal subarachnoid space results in areas of decreased signal intensity that may mimic arteriovenous malformation. These artifacts can be overcome by the use of CSF gating. This is accomplished by acquiring data at similar phases of the cardiac cycle. This technique has been demonstrated to be of great advantage in evaluating intramedullary spinal cord disease. The disadvantages of this technique, however, are related to relatively long acquisition times for T2-weighted spin-echo imaging. For practical purposes, because most imaging of the cervical and thoracic spine is aimed at depiction of extradural disease, GEI has become the preferred imaging technique in evaluation of cervical and thoracic spine disease. This technique provides adequate delineation of the spinal cord, subarachnoid space, and extradural structures. This technique has the added advantage of reduced imaging time. In this case, nongated T2-weighted spin-echo images were inadequate in the evaluation of the morphology of the spinal cord and subarachnoid space. GEI (Fig. 22B) provided better delineated morphology of the spinal cord. In addition, no CSF motion artifacts are present, thereby excluding the possibility of an arteriovenous malformation.

Despite these advantages of GEI, magnetic susceptibility artifacts may severely degrade these images below diagnostic levels (Fig. 22E,F). Following anterior interbody fusion of the cervical spine, inadvertent contact of the drill bit with metallic operative utensils may cause residual metal shavings at the site of interbody fusion. These small pieces of metal are not detected with plain film or CT but may cause large signal voids, mimicking extradural defects such as recurrent postoperative osteophyte formation. Usually, however, these artifacts are characteristic and easily recognized by a halo of signal void and geometric distortion.

Another important pitfall to be aware of in MRI of the spine is the appearance of subarachnoid Pantopaque (Fig. 22G,H). Pantopaque has a high lipid content and therefore a short T1, giving it a bright signal on T1-weighted images. T2-weighted images demonstrate marked decrease in signal intensity with progressively longer TE, indicating a short T2. Based on its signal characteristics alone, this may mimic subacute hemorrhage between 3 and 7 days or intraspinal lipomatous tissue. Several distinguishing features are characteristic, however. Usually intraspinal lipomatous tissue has signal intensity similar to that of subcutaneous fat. It is usually noted that on heavily T2-weighted images, Pantopaque has much less signal intensity than lipomatous tissue because of its much shorter T2. This is because of a higher water content of naturally occurring lipomatous tissue in the body. In addition, usually a well-defined CSF/Pantopaque fluid level is demonstrated as a result of layering within the subarachnoid space by gravity. Subacute hemorrhage between 3 and 7 days may have similar signal characteristics (see discussion of epidural hematoma).

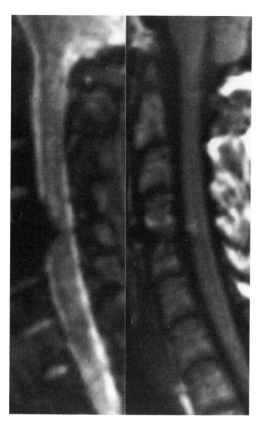

FIG. 22E. Metallic artifact following anterior interbody fusion (sagittal MR, TR/TE 500/20 msec [SE] and 300/15 msec [GE], flip angle 10°). The discrepancy in size between the apparent ventral epidural defect on the gradient-echo (GE) image compared with the spin-echo (SE) image is an important observation in identifying this as a magnetic susceptability artifact. Blooming of the signal void obliterates the ventral aspect of the spinal cord, yet spinal cord morphology is normal on the T1-weighted SE image. (From M. Guerrein, with permission.)

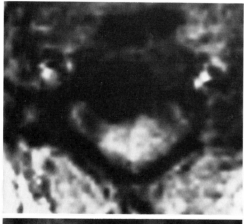

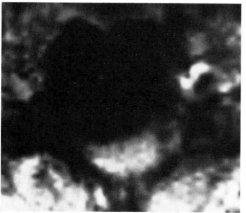

FIG. 22F. Metallic artifact following anterior interbody fusion (axial MR, TR/TE 400/10 msec [GE], flip angle 70°). (From M. Guerrein, with permission.)

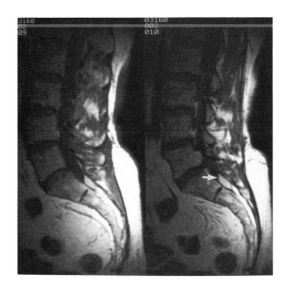

FIG. 22G. Subarachnoid Pantopaque (sagittal MR, TR/TE 500/20 msec). Note the well-defined lipid/fluid level with dependent layering of the Pantopaque posteriorly (*arrow*). (From M. Guerrein, with permission.)

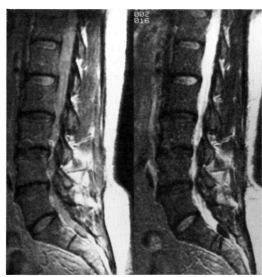

FIG. 22H. Subarachnoid Pantopaque (sagittal MR, TR/TE 2,500/30, 70 msec). Pantopaque has a short T2, reducing its signal intensity greater than subcutaneous fat on heavily T2-weighted images (*arrow*). (From M. Guerrein, with permission.)

CASE 23

This 27-year-old man presented with a right S1 radiculopathy. What is your differential diagnosis? (Fig. 23A. Axial MR, TR/TE 700/20 msec, pre–Gd-DTPA [top] and post–Gd-DTPA [bottom]. Fig. 23B. Axial MR, TR/TE 2,000/20, 80 msec.)

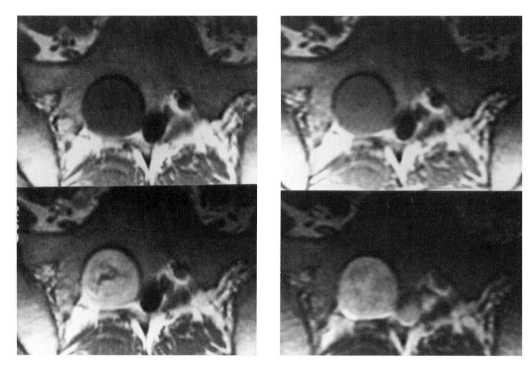

FIG. 23A.   **FIG. 23B.**

Findings: A solid homogeneously enhancing mass expands the right S1 neural canal (Fig. 23A,B).

Diagnosis: Schwannoma of the right S1 nerve sheath.

Discussion: Schwannomas (or neurinomas) may be primarily intradural or may be "dumbbell" lesions, with both intradural and extradural components extending through the intervertebral nerve root canal. Similar to the acoustic schwannoma, the intraspinal lesion usually arises from the sensory nerve, the dorsal root. A true neuroma of a spinal nerve is much less common and the presence even of a single lesion usually indicates von Recklinghausen's disease.

Magnetic resonance signal intensities of these lesions are characteristic. On T1-weighted and spin-density images, the signal intensity is usually similar to that of neural tissue but clearly greater than CSF. The high inherent contrast between an intraforaminal tumor and perineural and intramedullary fat of the sacrum is best depicted on T1-weighted images. Increasing signal intensity on T2-weighted images may reduce conspicuity of an intradural lesion within the near isointense subarachnoid space of the more distal thecal sac. In the case of a dumbbell lesion, however, this sequence may better delineate the extradural component from relatively hypointense adjacent paraspinal muscle. A hypointense central fibrous stroma relative to hyperintense homogeneous tumor tissue is characteristic (Fig. 23A). As with eighth nerve tumors, peripheral lesions frequently show homogeneous vigorous contrast enhancement.

Tarlov or perineural sheath cysts may expand the distal central spinal canal or a sacral neural canal and, on heavily T2-weighted images alone, may be difficult to differentiate from a neurofibroma or schwannoma. Both may become isointense with CSF on these sequences. It is critical to evaluate the spin-density or long TR/intermediate TE images. A perineural sheath cyst should be isointense to CSF on all sequences, whereas schwannomas are usually hyperintense relative to CSF on these spin-density or balanced sequences. One caveat, however, is that pulsatile flow may reduce the signal intensity of CSF relative to fluid in an arachnoid cyst or a Tarlov cyst. For this reason, T1-weighted images are most reliable in distinguishing solid tumor mass from a CSF-containing space. Obviously, gadolinium enhancement of the lesion clearly differentiates the two but is not always necessary.

An epidermoid tumor may mimic a simple Tarlov cyst on CT, as it is of decreased density, similar in density to CSF (Fig. 23C–F), and, as in this case, occasionally contains calcification or ossification. Magnetic resonance imaging, with its improved ability to characterize tissues, can clearly differentiate this from a CSF-containing space or schwannoma. They are clearly of increased signal intensity relative to CSF on spin-density and T1-weighted images. Unlike the smoothly marginated homogeneous schwannoma, they are heterogeneous and lobulated. These lesions can be primary congenital tumors, as in the intracranial subarachnoid space, presumably the result of growth of a congenital rest of epidermal tissue. These have also been described as a complication of previous lumbar puncture, particularly before the routine use of a stylet in lumbar puncture needles.

Although most ependymomas of the spinal canal are intramedullary, a myxopapillary ependymoma is an extramedullary intradural subtype of ependymoma arising in the filum terminale or cauda equina (Fig. 23F). Occasionally they occur as an epidural sacral lesion. Microscopically, they contain cystic, mucin-containing spaces and frequently form papillae around a central vascular network. This feature likely accounts for the heterogeneous signal intensity not seen in other intradural lesions such as schwannomas or meningiomas, which are microscopically more homogeneous.

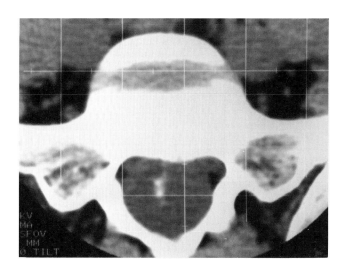

FIG. 23C. Intraspinal epidermoid tumor (axial, noncontrast CT).

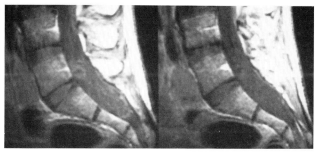

FIG. 23D. Intraspinal epidermoid tumor (sagittal MR, TR/TE 800/20 msec).

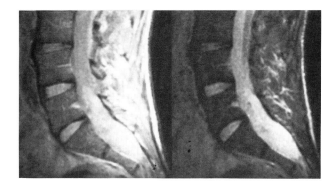

FIG. 23E. Intraspinal epidermoid tumor (sagittal MR, TR/TE 2,500/30, 70 msec). A hypodense mass, nearly isodense with CSF, could be mistaken for a Tarlov cyst if it were not for the presence of the small calcification. Magnetic resonance imaging, however, shows a heterogeneous, lobulated mass expanding the central spinal canal, with clearly different signal intensities than CSF.

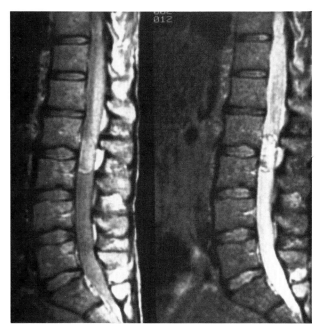

FIG. 23F. Ependymoma of the filum terminale (sagittal MR, TR/TE 2,500/30, 70 msec). Heterogeneous lesion occurring in a typical location at the tip of the conus is isointense to cord on spin-density weighted image. On more heavily T2-weighted images, the lesion gains in signal intensity and becomes isointense with CSF as a result of its mucin-containing spaces.

RECOMMENDED READING

Abdullah AF, Chambers RW, Daut DP. Lumbar nerve root compression by synovial cysts of the ligamentum flavum. Report of four cases. *Radiology* 1984;153:855.

Alarcon GS. The use of magnetic resonance imaging (MRI) in patients with rheumatoid arthritis and subluxations of the cervical spine [Letter]. *Arthritis Rheum* 1988;31(2):304.

Anda S, Nilsen G, Rysland P. Periodontoid changes in rheumatoid arthritis: MRI observations. Report of two cases. *Scand J Rheumatol* 1988;17(1):59–62.

Angtuaco EJC, McConnell JR, Chadduck WM, et al. MR imaging of spinal epidural sepsis. *AJNR* 1987;8:879.

Arnoldi CC, Brodsky AE, Cauchoix J, et al. Lumbar spinal stenosis and nerve root entrapment syndromes: definition and classification. *Clin Orthop* 1976;115:4–5.

Aubin ML, Baleriaux D, Cosnard G, et al. MRI in syringomyelia of congenital, infectious, traumatic or idiopathic origin: a study of 142 cases. *J Neuroradiol* 1987;14:313.

Barkovich AJ, Sherman JL, Citrin CM, et al. MR of postoperative syringomyelia. *AJNR* 1987;8:319.

Beerman R, Batt HD, Green BA. Lumbar vertebral reformation after traumatic compression fracture. *AJNR* 1985;6:455.

Beres J, Pech P, Berns TF, Daniels DL, et al. Spinal epidural lymphomas: CT features in seven patients. *AJNR* 1986;7:327–328.

Berger PE, Atkinson D, et al. High resolution surface coil magnetic resonance imaging of the spine: normal and pathologic anatomy. *Radiographics* 1986;6:573.

Bjorkengren AG, Kurz LT, Resnick D, et al. Symptomatic intraspinal synovial cysts: opacification and treatment by percutaneous injection. *AJR* 1987;149:105.

Bognanno JR, Edwards MK, Lee TA, et al. Cranial MR imaging and neurofibromatosis. *AJNR* 1988;9:461.

Boyd HR, Pear BL. Chronic spontaneous spinal epidural hematoma. Report of two cases. *J Neurosurg* 1972;36:239–242.

Braffman BH, Bilaniuk LT, Zimmerman RA. Central nervous system manifestions of the phakomatoses on MR. *Radiol Clin North Am* 1988;26:773.

Braun IF, Hoffman JC Jr, Davis PC, et al. Contrast enhancement in CT differentiation between recurrent disk herniation and postoperative scar: prospective study. *AJNR* 1985;6:607.

Braun IF, Malko JA, Davis PC, et al. Behavior of Pantopaque on MR: *in vivo* and *in vitro* analyses. *AJNR* 1986;7:997.

Breedveld FC, Algra PR, Vielvoye CJ, Cats A. Magnetic resonance imaging in the evaluation of patients with rheumatoid arthritis and subluxations of the cervical spine. *Arthritis Rheum* 1987;30(6):624–629.

Brodsky AE. Low back pain syndromes due to spinal stenosis and posterior cauda equina compression. *Bull Hosp Jt Dis* 1969;30:66–79.

Bundschuh C, Modic MT, Kearney F, Morris R, Deal C. Rheumatoid arthritis of the cervical spine: surface-coil MR imaging. *AJR* 1988;151(1):181–187.

Burger PC, Vogel FS. *Surgical pathology of the central nervous system and its coverings*, 2nd ed. New York: Churchill Livingstone. 1982;620–633.

Burk DL Jr, Brunberg JA, Kanal E, Latchaw RE, Wolf GL. Spinal and paraspinal neurofibromatosis: surface coil MR imaging at 1.5 T. *Radiology* 1987;162:797.

Burns DA, Blaser S, Ross JS, et al. MR imaging with Gd-DTPA in leptomeningeal spread of lymphoma. *J Comput Assist Tomogr* 1988;12:499.

Burton CV, Heithoff KB, Kirkaldy-Willis W, Ray CD. Computed tomographic scanning and the lumbar spine. Part II: clinical considerations. *Spine* 1979;4:356–379.

Burton CV, Kirkaldy-Willis WH, Yong-Hing K, Heithoff KB. Causes of failure of surgery on the lumbar spine. *Clin Orthop* 1981;157:191–199.

Carrera GF, Haughton VM, Syversten A, et al. Computed tomography of the lumbar facet joints. *Radiology* 1980;134:145–148.

Carrera GF, Williams AL, Haughton VM. Computed tomography in sciatica. *Radiology* 1980;137:433–437.

Casselman ES. Radiologic recognition of symptomatic spinal synovial cysts. *AJNR* 1985;6:971.

Chambers AA. Thoracic disc herniation. *Semin Roentgenol* 1988;23:111.

Chui MC, Bird BL, Rogers J. Extracranial and extraspinal nerve sheath tumors: computed tomographic evaluation. *Neuroradiology* 1988;30:47.

Crock HV. Isolated lumbar disc resorption as a cause of nerve root canal stenosis. *Clin Orthop* 1976;115:109.

Davis PC, Friedman NC, Fry SN. Leptomeningeal metastases: MRI imaging. *Radiology* 1987;163:449–454.

de Roos A, Kressel H, Spritzer C, et al. MR imaging of marrow changes adjacent to end plates in degenerative lumbar disk disease. *AJR* 1987;149:531.

Deeb ZL, Schimel S, Daffner RH, et al. Intervertebral disk-space infection after chymopapain injection. *AJNR* 1985;6:55.

Denis F. The three column spine and its significance in the classification of acute thoracolumbar spinal injuries. *Spine* 1983;8:817–831.

Edelman RR, Shoukimau GM, Stark DD, et al. High resolution surface coil imaging of lumbar disc disease. *AJR* 1986;144:1123.

Eisenstein S. The trefoil configuration of the lumbar vertebral canal. *J Bone Joint Surg (Br)* 1980;62:78–82.

Enzmann DR, Rubin JB. Cervical spine: MR imaging with a partial flip angle, gradient-refocused pulse sequence. Part II: spinal cord disease. *Radiology* 1988;166:473–478.

Epelbaum R, Haim N, Ben-Shahar M, et al. Non-Hodgkins lymphoma presenting with spinal epidural involvement. *Cancer* 1986;58(9):2120–2124.

Epstein JA, Epstein BS, Lavine LS, Carras R, Rosenthal AD, Summar P. Lumbar nerve root compression at the intervertebral foramina caused by arthritis of the posterior facets. *J Neurosurg* 1973;39:362–369.

Epstein JA, Epstein BS, Lavine LS, Carras R, Rosenthal AD. Degenerative lumbar spondylolisthesis with an intact neural arch (pseudo spondylolisthesis). *J Neurosurg* 1976;44:139–147.

Epstein JA, Epstein BS, Rosenthal AD, Carras R, Lavine LS. Sciatica caused by nerve root entrapment in the lateral recess: the superior facet syndrome. *J Neurosurg* 1972;36:584–589.

Farfan HF. The pathological anatomy of degenerative spondylolisthesis: a cadaver study. *Spine* 1980;5:412–418.

Farfan HF, Sullivan JD. The relation of facet orientation to intervertebral disc failure. *Can J Surg* 1967;10:179.

Francavilla TL, Powers A, Dina T, Rizzoli HV. MR imaging of thoracic disc herniations. *J Comput Assist Tomogr* 1987;11:1062–1065.

Franklin EA Jr, Berbaum KS, Dunn V, Smith WL, Ehrhardt JC, Levitz GS, Breckenridge RE. Impact of MR imaging on clinical diagnosis and management: prospective study. *Radiology* 1986;161:377.

Gentry LR, Turski PA, Strother CM, et al. Chymopapain chemonucleolysis: CT changes after treatment. *AJNR* 1985;6:321.

Goy AMC, Pinto RS, Raghavendra BN, Epstein FJ, Kricheff II. Intramedullary spinal cord tumors: MR imaging, with emphasis on associated cysts. *Radiology* 1986;161:381.

Grange EO, Carr JP. Diffuse leptomeningeal carcinomatosis: clinical and pathologic characteristics. *Neurology* 1955;5:706–722.

Haddad P, Thaell JF, Kiely JM, Harrison EG Jr, Miller RH. Lymphoma of the spinal extradural space. *Cancer* 1976;38:1862–1866.

Harris ME. Spontaneous epidural spinal hemorrhage. *AJR* 1969;105:383–385.

Hedberg MC, Drayer BP, Flom RA, et al. Gradient echo (GRASS) MR imaging in cervical radiculopathy. *AJNR* 1988;9:145.

Heindel W, Friedmann G, Bunke J, Thomas B, Firsching R, Ernestus R-I. Artifacts in MR imaging after surgical intervention. *J Comput Assist Tomogr* 1986;10:596–599.

Heithoff KB. Cooperative study of the efficacy of metrizamide myelography and CT in the study of lumbar disc disease and spinal stenosis. 71st Scientific Assembly, Radiological Society of North America, Chicago, November 17–22, 1985.

Heithoff KB. Comparison of MRI and metrizamide enhanced CT scanning in the evaluation of post-operative recurrent disc herniation versus fibrosis. Annual Meeting of the International Society for the Study of the Lumbar Spine, Dallas, Texas, May 1986.

Holtas SL, Kido DK, Simon JH. MR imaging of spinal lymphoma. *J Comput Assist Tomogr* 1986;10(1):111–115.

Holtas S, Nordstrom CH, Larsson EM, et al. MR imaging of intradural disk herniation. *J Comput Assist Tomogr* 1987;11:353.

Johansen JG, Barthelemy CR, Haughton VM, et al. Arachnoiditis from myelography and laminectomy in experimental animals. *AJNR* 1984;5:97.

Kaffenberger DA, Shah CP, Murtagh FR, et al. MR imaging of spinal cord hemangioblastoma associated with syringomyelia. *J Comput Assist Tomogr* 1988;12:495.

Kim KS, Ho Su, Weinberg P, Lee C. Spinal leptomeningeal infiltration by systemic cancer: myelographic features. *AJR* 1982;139:361–365.

Kirkaldy-Willis WH, Heithoff KB, Bowen CVA, Shannon R. Pathological anatomy of lumbar spondylosis and stenosis, correlated with the CT scan, radiographic evaluation of the spine. In: Post MJD, ed. *Current advances with emphasis on computed tomography.* New York: Masson Publishing, 1980;34–55.

Kirkaldy-Willis WH, McIvor GWD. Spinal stenosis. *Clin Orthop* 1976;115:2–144.

Kirkaldy-Willis WH, Paine KWE, Cauchoix J, et al. Lumbar spinal stenosis. *Clin Orthop* 1974;99:30–50.

Kirkaldy-Willis WH, Wedge JH, Yong-Hing K, Reilly J. Pathology and pathogenesis of lumbar spondylosis and stenosis. *Spine* 1978;3:319–328.

Kolawole TM, Teishat WA. Secondary ependymoma presenting as a pre-sacral mass. *Comput Tomogr* 1988;12:108.

Krol G. MR of cranial and spinal meningeal carcinomatosis: comparison with CT and myelography. *AJNR* 1988;9:709–714.

Kulkarni MV, Narayana PA, McArdle CB, et al. Cervical spine MR imaging using multislice gradient echo imaging: comparison with cardiac gated spin echo. *Magn Reson Imaging* 1988;6:517.

Lancourt JE, Glenn WV, Wiltse LL. Multiplanar computerized tomography in the normal spine and in the diagnosis of spinal stenosis: a gross anatomic-computerized tomographic correlation. *Spine* 1979;4:379–390.

Laredo JD, Reizine D, Bard M, Merland JJ. Vertebral hemangiomas: radiologic evaluation. *Radiology* 1986;161:183.

Lau LSW, Petty PG, Australas. *Radiology* 1987;31:83.

Lee BCP, Zimmerman RD, Manning JJ, Deck MDF. MR imaging of syringomyelia and hydromyelia. *AJNR* 1985;6:221–228.

Levine E, Huntrakoon M, Wetzel LH. Malignant nerve-sheath neoplasms in neurofibromatosis: distinction from benign tumors by using imaging techniques. *AJR* 1987;149:1059.

Levitan LH, Wiens CW. Chronic lumbar extradural hematoma: CT findings. *Radiology* 1983;148(3):707–708.

Lewis TT, Kingsley DPE. Magnetic resonance imaging of multiple spinal neurofibromata-neurofibromatosis. *Neuroradiology* 1987;29:562.

Lifson A, Heithoff KB, Burton CV, Ray CD. High-resolution computed tomographic scanning of the lumbosacral spine. *Mod Neurosurg* 1982;1:31–36.

Little J, Dale AJ, Okasaki H. Meningeal carcinomatosis: clinical manifestations. *Arch Neurol* 1974;30:138-143.

Lynch D, McManus F, Ennis JT. Computed tomography in spinal trauma. *Clin Radiol* 1986;37:71.

Mall JC, Kaiser JA. Usual appearance of the postoperative lumbar spine. *Radiographics* 1987;7:245.

Mamourian AC, Briggs RW. Appearance of Pantopaque on MR images. *Radiology* 1986;158:457–460.

Markham JW, Lynge HN, Stahlman GEB. The syndrome of spontaneous spinal epidural hematoma. *J Neurosurg* 1967;26:334–342.

Masaryk TJ, Boumphrey F, Modic MT, et al. Effects of chemonucleolysis demonstrated by MR imaging. *J Comput Assist Tomogr* 1986;10:917.

McArdle CB, Crofford MJ, Mirfakhraee M, et al. Surface coil MR of spinal trauma: preliminary experience. *AJNR* 1986;7:885.

McDonald JV, Klump TE. Intraspinal epidermoid tumors caused by lumbar puncture. *Arch Neuroradiol* 1986;43:936–939.

McNabb I. Negative disc exploration: an analysis of the causes of nerve root involvement in sixty-eight patients. *J Bone Joint Surg* 1971;52:891–903.

Mixter WJ, Barr JS. Rupture of the intervertebral disc with involvement of the spinal canal. *N Engl J Med* 1934;211:210–215.

Modic MT, Feiglin DH, et al. Vertebral osteomyelitis: assessment using MR. *Radiology* 1985;157:157.

Modic MT, Masaryk TJ, Paushter DM. Magnetic resonance imaging of the spine. *Radiol Clin North Am* 1986;24:229.

Modic MT, Pavlieck W, Weinstein MA, Boumphrey F, Ngo F, Hardy R, Duchesneau PM. Magnetic resonance imaging of intervertebral disc disease. Clinical pulse and sequence considerations. *Radiology* 1984;152:103–111.

Modic MT, Weinstein MA, Pavlicek W, et al. Nuclear magnetic resonance imaging of the spine. *Radiology* 1983;148:747–762.

Olson M, Chernik LN, Posner JB. Infiltration of leptomeninges by systemic cancer. *Arch Neurol* 1974;30:122–137.

Park WM, McCall MB, O'Brien JP, Webb JK. Fissuring of the posterior annulus fibrosus in the lumbar spine. *Br J Radiol* 1979;52:382–387.

Paushter DM, Modic MT, Masaryk TJ. Magnetic resonance imaging of the spine: applications and limitations. *Radiol Clin North Am* 1985;23:551.

Paushter DM, Dengel FH, Modic MT, et al. Clinical applications of nuclear magnetic resonance: central nervous system—brainstem and cord. *Radiographics* 1984;4:97.

Pear BL. Spinal epidural hematoma. *AJR* 1972;115:155–164.

Pech P, Haughton VM. Lumbar intervertebral disc: correlative MR and anatomic study. *Radiology* 1985;156:699.

Pettersson H, Larsson EM, Holtas S, Cronqvist S, Egund N, Zygmunt S, Brattstrom H. MR imaging of the cervical spine in rheumatoid arthritis. *AJNR* 1988;9(3):573–577.

Phillips J, Chiu I. Magnetic resonance imaging of intraspinal epidermoid cyst; case report. *Comput Tomogr* 1987;11:181.

Pojunas K, Williams AL, Daniels DL, Haughton VM. Syringomyelia and hydromyelia: magnetic resonance evaluation. *Radiology* 1984;153:679–683.

Quencer RM, El Gammal T, Cohen G. Syringomyelia associated with intradural extramedullary masses of the spinal canal. *AJNR* 1986;7:143–148.

Ray CD. Far lateral decompressions for stenosis: the paralateral approach. In: White AH, Rothman RH, Ray CD, eds. *Lumbar spine surgery: techniques and complications.* St. Louis: CV Mosby Co, 1986.

Ray CD. Methods of resection of bone spurs, osteophytes and bony encroachment. In: White AH, Rothman RH, Ray CD, eds. *Lumbar spine surgery: techniques and complications.* St. Louis: CV Mosby Co, 1986.

Robertson GH, Llewellyn HT, Taveras JM. The narrow lumbar spinal canal syndrome. *Radiology* 1973;107:89–98.

Roosen N, Dietrich U, Nicola N, et al. MR imaging of calcified herniated thoracic disc. *J Comput Assist Tomogr* 1987;11:733.

Ross JS. Thoracic disc herniation: MR imaging: reply. *Radiology* 1988;167:875.

Ross JS, Masaryk TJ, Modic MT. Postoperative cervical spine: MR assessment. *J Comput Assist Tomogr* 1987;11:955–962.

Ross JS, Masaryk TJ, Modic MT, Bohlman H, Delamarter R, Wilber G. Lumbar spine postoperative assessment with surface-coil MR imaging. *Radiology* 1987;164:851.

Ross JS, Perez-Reyes N, Masaryk TJ, Bohman H, Modic MT. Thoracic disc herniation: MR imaging. *Radiology* 1987;165:511.

Rothman SL, Glenn WV Jr. CT multiplanar reconstruction—253 cases of lumbar spondylolysis. *AJNR* 1984;5:81–90.

Rubin JB, Enzmann DR, Wright A. CSF-gated MR imaging of the spine: theory and clinical implementation. *Radiology* 1987;163:784.

Ryan RW, Lally JF, Kozic Z. Asymptomatic calcified herniated thoracic discs: CT recognition. *AJNR* 1988;9:363.

Sato Y, Waziri M, Smith W, Frey EE, Yuh WTC, Hanson J, Franken EA Jr. Hippel-Lindau disease: MR imaging. *Radiology* 1988;166:241.

Schipper J, Kardaun JWPF, Braakman R, van Dongen KJ, Blaauw G. Lumbar disk herniation: diagnosis with CT or myelography? *Radiology* 1987;165:227.

Schornstrom NS, Bolender NF, Spengler DM. The pathomorphology of spinal stenosis as seen on CT scans of the lumbar spine. *Spine* 1985;10:806–811.

Scotti G, Scialfa G, Colombo N, et al. Magnetic resonance diagnosis of intramedullary tumors of the spinal cord. *Neuroradiology* 1987;29:130.

Sherman JL, Citrin CM, Gangarosa RE. MR appearance of CSF pulsations in the spinal canal. *AJNR* 1986;7:879.

Slasky BS, Bydder GM, Niendorf HP, et al. MR imaging with gadolinium-DPTA in the differentiation of tumor, syrinx, and cyst of the spinal cord. *J Comput Assist Tomogr* 1987;11:845.

Svien HJ, Adson AW, Dodge HW Jr. Lumbar extradural hematoma. *J Neurosurg* 1950;7:587–588.

Sze G, Abramson A, Krol G. Gadolinium-DTPA in the evaluation of intradural extramedullary spinal disease. *AJNR* 1988;9:153–163.

Sze G, Brant-Zawadzki M, Wilson CR, Norman D, Newton TH. Pseudotumor of the craniovertebral junction associated with chronic subluxation: MR imaging studies. *Radiology* 1986;161:391.

Sze G, Krol G, Zimmerman RD, Deck MDF. Malignant extradural spinal tumors: MR imaging with Gd-DTPA. *Radiology* 1988;167:217.

Sze G, Krol G, Zimmerman RD, et al. Intramedullary disease of the spine: diagnosis using gadolinium-DTPA-enhanced MR imaging. *AJNR* 1988;9:847.

Tabas JH, Deeb ZL. Diagnosis of sacral perineural cysts by computed tomography. *Comput Tomogr* 1986;10:255.

Tantana S, Pilla TJ, Luisiri A. Computed tomography of acute spinal subdural hematoma. *J Comput Assist Tomogr* 1986;10:891.

Teplick JG, Haskin ME. Spontaneous regression of herniated nucleus pulposus. *AJNR* 1985;6:331.

Van Zanten TEG, Teule GJJ, Golding RP, et al. CT and nuclear medicine imaging in vertebral metastases. *Clin Nucl Med* 1986;11:334.

Verbiest H. A radicular syndrome from developmental narrowing of the lumbar vertebral canal. *J Bone Joint Surg (Br)* 1954;36:230–237.

Verbiest H. Further experiences on the pathological influences of a developmental narrowness of a bony lumbar vertebral canal. *J Bone Joint Surg (Br)* 1955;37:576–583.

Wagle WA, Jaufman B, Mincy JE. Intradural extramedullary ependymoma: MR-pathologic correlation. *J Comput Assist Tomogr* 1988;12:705.

Wang A-M, Lin JCT, Whykal HA, et al. Ependymoma of the filum terminale: Metrizamide-enhanced CT evaluation. *Comput Radiol* 1986;4:239.

Weinstein PR. Diagnosis and management of lumbar spinal stenosis. *Clin Neurosurg* 1983;30:677–697.

Weisz GM, Lee P. Spinal canal stenosis. Concept of spinal reserve capacity: radiologic measurements and clinical applications. *Clin Orthop* 1983;179:134–140.

White JG III, Strait TA, Binkley JR, Hunter SE. Surgical treatment of 63 cases of conjoined nerve root. *J Neurosurg* 1982;56(1):114–117.

Williams AL. CT diagnosis of degenerative disc disease. The bulging annulus. *Radiol Clin North Am* 1983;21:289–300.

Williams AL, Haughton VN, Pojunas KW, et al. Differentiation of intramedullary neoplasms and cysts by MR. *AJNR* 1987;8:527.

Williams AL, Haughton VN, Daniels DL, Grogan JP. Differential CT diagnosis of extruded nucleus pulposis. *Radiology* 1983;149:141–148.

Williams AL, Haughton VN, Syvertsen A. Computed tomographic and the diagnosis of herniated nucleus pulposis. *Radiology* 1980;135:95–99.

Williams CE, Nelson M. Varied computed tomographic appearances of acute spinal epidural hematoma. *Clin Radiol* 1987;38:363.

Williams MP, Cherryman GR. Thoracic disc herniation: MR imaging. *Radiology* 1988;167:874.

Willinsky RA, Grosman H, Cooper PW, et al. Radiology of sacral cysts. *J Can Assoc Radiol* 1988;39:21.

Wiltse LL, Guyer RD, Spencer CW, Glenn WV, Porter IS. Alar transverse process impingement of the L5 spinal nerve: the far out syndrome. *Spine* 1984;9:31–41.

Wiltse LL, Kirkaldy-Willis WH, McIvor GWD. The treatment of spinal stenosis. *Clin Orthop* 1976;115:83.

Wright A. Symposium: the role of spinal fusion. *Spine* 1981;6:292.

Yaghmai I. Spine changes in neurofibromatosis. *Radiographics* 1986;6:261.

Yong-Hing K, Kirkaldy-Willis WH. The pathophysiology of degenerative disease of the lumbar spine. *Orthop Clin North Am* 1983;14:491-504.

Yu S, Haughton VM, Ho PSP, Sether LA, Wagner M, Ho KC. Progressive and regressive changes in the nucleus pulposus. II. The adult. *Radiology* 1988;169:93–97.

Yu S, Sether LA, Ho PS, Wagner M, Haughton VM. Tears of the anulus fibrosus: correlation between MR and pathologic findings in cadavers. *AJNR* 1988;9:367–371.

Yu S, Haughton VM, Lowell SA, Wagner M. Annulus fibrosus in bulging intervertebral disks. *Radiology* 1988;169:761–763.

CHAPTER 5

The Hip

Javier Beltran and Jerrold H. Mink

Routine MRI of the hips can be performed in a relatively short period of time. In most cases an axial scan using spin-echo technique with short time of repetition (TR) (500–800 msec) and short time of echo (TE) (20–40 msec), followed by a coronal spin-echo pulse sequences with long TR (2,000–2,500 msec) and short and long TE (20–40, 80–100 msec), provide adequate T1, proton-density (PD), and T2 information, sufficient for evaluation of most pathologic conditions involving the hip, including avascular necrosis (AVN), septic arthritis, and degenerative osteoarthritis.

A sagittal midline image including the sacrum and lower lumbar spine can be added to the routine protocol to study patients with bone marrow disorders, such as sickle cell anemia (SCA), leukemia, lymphoma, myeloma, and metastases. In these cases, T2-weighted images provide better contrast between the normal and abnormal bone marrow, hence improved sensitivity. Chemical shift imaging with water and fat images can also be used for bone marrow evaluation instead of the spin-echo T2-weighted sequences.

Fast scanning pulse sequences using low flip angle could be used to replace T2-weighted images, which require a longer period of time. In addition, fast scanning techniques provide excellent visualization of the articular cartilage. The *clinical* use of fast imaging of the hip has not yet been investigated.

In very specific clinical situations, such as stress or acute fractures of the femoral neck in an elderly individual, a single coronal plane sequence, using short TR and short TE, with less than 20-min imaging time, has been found highly accurate for the detection of fractures not seen on plain radiographs. Inversion recovery sequences are also used in fracture detection.

Because of the relatively large size of the anatomic structures of the hip, 5-mm slice thickness with 1-mm gaps are sufficient to study the area; however, small lesions may require thinner sections.

The body coil and a relatively large field of view can be used to study both hips and pelvis simultaneously. Recently, surface coil sagittal sections with smaller field of view have been recommended for more accurate evaluation of the articular cartilage and subchondral cortex. This technique provides dramatic depiction of the normal and abnormal anatomy in the sagittal section, and it can be expanded to coronal and/or axial planes if needed. A single or dual circular receive-only surface coil can be applied over the region of the hip with the patient supine. Because of the signal drop away from the surface coil, this technique may not be effective in large or obese patients.

CASE 1

This is a 25-year-old man with a 3-week history of right hip pain. Plain radiographs obtained immediately before MRI were normal. What is the diagnosis? (Fig. 1A. Coronal MR, TR/TE 600/20 msec.)

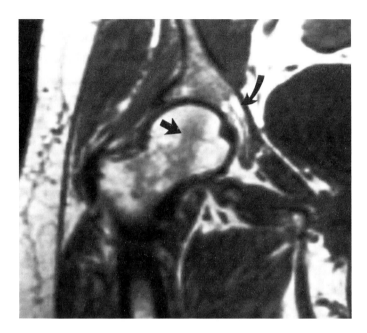

FIG. 1A.

Findings: There is a vertically oriented low signal intensity (SI) band across the head of the femur (straight arrow) contrasting with the hyperintense fatty marrow of the rest of the femoral head. The bone marrow space of the femoral neck and acetabulum show low SI with the exception of the greater trochanter and a small region in the acetabulum (curved arrow in Fig. 1A).

Diagnosis: Normal MRI of the hip in a young adult.

Discussion: Magnetic resonance imaging is a very sensitive test for AVN and other bone and joint disorders of the hip. One must, however, be familiar with the normal MRI anatomy to avoid false-positive diagnoses.

Cellular red marrow is uniformly distributed through all bone marrow spaces at birth. Immediately after birth, fatty infiltration of the red marrow begins in the terminal phalanges of the hands and feet and progresses centrally. This fatty marrow conversion is asymmetric and it is inhomogeneous within a bone. In an adult, red marrow is present in the spine, pelvis, ribs, sternum, areas of the calvarium, and proximal metaphyses of the skeleton.

The MRI appearance of normal bone marrow is related to the percentage of hematopoietic elements and fat. If fat is the main component of the bone marrow, high SI on T1-weighted images and low SI on T2-weighted images will be seen, similar to the subcutaneous fat. If it is composed predominantly of cellular elements, as when hematopoietic red marrow predominates, the SI will be lower than fatty marrow on T1-weighted images and will have low SI on T2-weighted images.

Not infrequently, the distribution of red and yellow marrow in a particular bone is inhomogeneous. This results in a mottled MRI pattern, which reflects an admixture of the different components. Occasionally, differentiation of this normal pattern from bone marrow diseases such as leukemia or aplastic anemia may be difficult based on MRI findings alone.

The primary compression trabeculae of the femoral head are oriented in such a way that the stress forces of the pelvis are transmitted to the femur. This group of trabeculae may be very prominent in young individuals, producing a vertical low SI band across the femoral head as seen in this case. It is not to be confused with AVN or other pathologic conditions. With aging, osteoporosis develops and the trabecular pattern becomes less conspicuous on plain radiographs and MRI.

The distribution of fatty and hematopoietic marrow is closely correlated with temperature and bone marrow blood supply, both of which change with age. As one gets older, the percentage of fatty marrow increases as a consequence of decreasing marrow vascularity. Only 12% of normal patients older than 50 years have significant hematopoietic marrow in the proximal femur, whereas 95% of patients less than 50 years have red marrow in the intertrochanteric region (Fig. 1B,C).

Bone infarction occurs almost exclusively in fatty marrow. Is premature conversion of hematopoietic to fatty marrow a risk factor for AVN? Only 33% of patients under 50 years with AVN had red marrow in the intertrochanteric region as opposed to 95% without AVN. It is not known, however, if this relationship between early fatty marrow conversion and AVN is causal or coincidental.

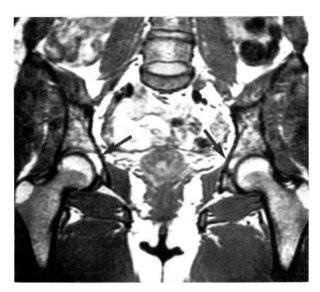

FIG. 1B. Normal marrow distribution in a young adult (coronal MR, TR/TE 600/20 msec). The low signal hematopoetic marrow in this T1-weighted image occupies the neck and intertrochanteric regions of the femora. Fatty marrow is present in both femoral epiphyses, the greater trochanters, and portions of both acetabuli (*arrows*).

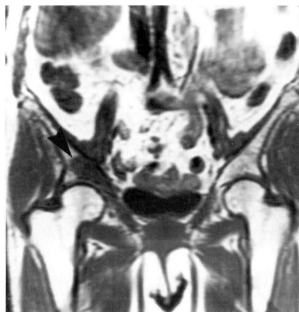

FIG. 1C. Fatty replacement in the proximal femur. (coronal MR, TR/TE 800/20 msec). The hematopoetic marrow has been totally replaced by bright fat in this 65-year-old woman. A metastatic lesion from breast carcinoma is present in the right acetabulum (*arrowhead*).

CASE 2

This 13-year-old boy presented with acute onset of left hip pain and limitation of motion. What is the diagnosis? (Fig. 2A. TR/TE 2,000/20 msec. Fig. 2B. TR/TE 2,000/80 msec.)

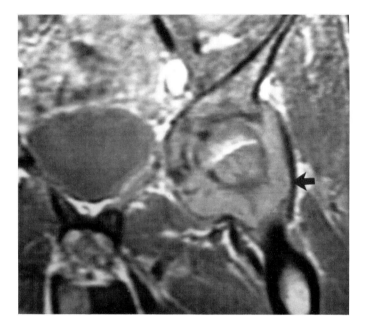

FIG. 2A.

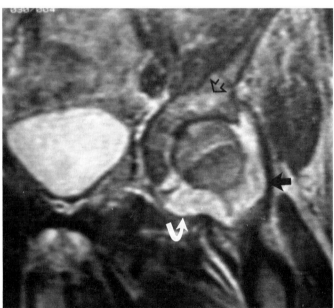

FIG. 2B.

Findings: There is a joint effusion in the left hip, with intermediate SI on a PD image (straight arrow in Fig. 2A) and high SI on T2-weighted images (straight arrow in Fig. 2B), associated with marked capsular distension (curved arrow in Fig. 2B). In addition there is an area of high SI on T2-weighted images (open arrow in Fig. 2B) in the region of the acetabular roof. The joint space is narrowed superiorly.

Diagnosis: Septic arthritis of the left hip with osteomyelitis of the acetabulum.

Discussion: Hip joint effusions are difficult to detect on plain radiographs unless they are large enough to produce lateral displacement of the femoral head. Ultrasound is the method of choice for detecting the presence of a hip joint effusion because of its availability, noninvasive nature, and low cost. However, MRI may occasionally be requested by the clinician to evaluate a painful hip.

The diagnosis of septic arthritis cannot be made with MRI based solely on the presence of a joint effusion. Trauma, AVN, and noninfectious arthritis can induce synovitis and cause an increase in the normal amount of joint fluid. The MR SI characteristics of the fluid in septic arthritis do not differ from other causes of joint effusion. However, septic arthritis is often associated with osteomyelitis and chondrolysis. As in this case, acute osteomyelitis produces an increase in the SI of the bone marrow on T2-weighted images. Acute destruction of the hyaline cartilage has been described in slipped femoral capital epiphysis, septic arthritis, Legg-Perthes disease, and idiopathic chondrolysis. The combination of joint effusion, hyperintense signal of the acetabulum, narrowing of the joint space, and clinical and laboratory findings strongly suggest the diagnosis of septic arthritis.

CASE 3

This 43-year-old woman had the insidious onset of right hip pain over 1 month. Conventional radiographs demonstrated a questionable degree of demineralization but were otherwise normal as was a CT scan. What are the differential considerations? (Fig. 3A. Coronal MR, TR/TE 500/20 msec. Fig. 3B. Coronal MR, TR/TE 2,000/80 msec.).

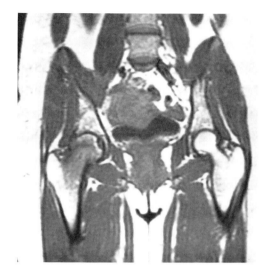 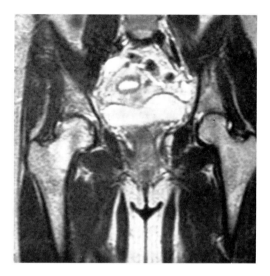

FIG. 3A.　　　　　　　　　　　　　　**FIG. 3B.**

Findings: There is a very large area of decreased signal present in the right femoral head, neck, and intertrochanteric regions, portions of which brighten on the long TR/TE sequence (Fig. 3A,B). Only a small part of the femoral head is spared. There is no evidence of bone destruction or soft tissue mass.

Diagnosis: Transient osteoporosis (TO) of the hip.

Discussion: The term *TO* refers to a clinical entity originally described as occurring in the third trimester of pregnancy but seen most often in middle-aged men. The disease is characterized by the insidious onset of hip pain, lasting for several months to 1 year. It is a self-limited disorder and therefore conservative treatment is recommended once the correct diagnosis is made. Whereas in male patients either hip may be involved, the left hip is almost always affected in female patients.

Patients with TO of the hip are often evaluated with multiple tests, including plain radiographs, bone scintigraphy, CT, MRI, and biopsy. Radiographs are often normal or reveal mild osteoporosis of the femoral head, with less involvement of the neck and acetabulum, and the joint space is maintained. Bone scintigraphy shows increased uptake in the region of the hip, and CT may demonstrate osteoporosis and joint effusion. Diffuse low SI on T1-weighted images and high SI on T2-weighted images involving the femoral head, neck, and intertrochanteric region, as in this case, are the characteristic findings on MRI, and joint effusions are invariably found. Ultimately, complete restoration of bone density and MRI changes occur (Fig. 3C).

Differential diagnosis must include osteomyelitis, septic arthritis, stress fracture, AVN, and tumor. Joint effusion can be an associated finding in all of these conditions. However, in the proper clinical setting there are some MRI features that can help in differentiating some of these entities from TO. Transient osteoporosis tends to involve the metaphysis and epiphysis of the femur. Stress fractures tend to be localized in the neck and are often seen as low SI lines or bands on all pulse sequences. The double-line sign, a low SI band surrounding a high SI line on T2-weighted images, has been considered pathognomonic of AVN and is not seen in TO. Joint aspiration may be necessary to exclude septic arthritis with osteomyelitis because it may be impossible to differentiate from TO on the basis of the MRI findings alone.

Regional migratory osteoporosis is closely related to TO of the hip. In regional migratory osteoporosis, joints of the lower limb are involved sequentially or occasionally simultaneously; it is the migratory nature that differentiates this condition from TO of the hip. The articulation nearest the diseased one is the next involved. Some authors consider these osteoporotic disorders as variants of Sudek atrophy.

The MRI manifestations of TO suggest an increase in the water content of the bone marrow (bone marrow edema) (Fig. 3D,E). The increased accumulation of radionuclide frequently found on scintigraphic studies probably reflects hyperemia and increased mineral turnover, which, in turn, can induce bone marrow edema. The term *transient bone marrow edema* has been suggested to replace TO because it better reflects the pathophysiologic events occurring in this entity and because osteoporosis is not uniformly present.

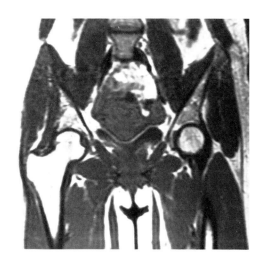

FIG. 3C. Transient osteoporosis of the hip (coronal MR, TE/TE 500/20 msec). This study, performed 3 months after Fig. 3A and 3B, demonstrates complete resolution of the bone marrow edema. The patient was asymptomatic at this time.

D

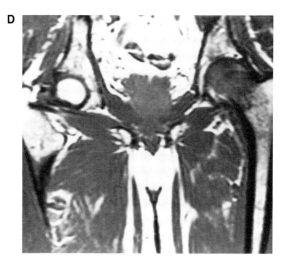

E

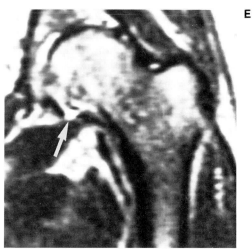

FIG. 3D,E. Transient osteoporosis of the hip. **D.** Coronal MR, TR/TE 300/15. **E.** Coronal MR, TR/TE 2,000/80 msec. The examination was made 2 months after the onset of symptoms in this 52-year-old man. The entire head and neck demonstrate signal abnormalities consistent with bone marrow edema. Portions of the lesions become quite bright. A small joint effusion (*arrow*) is present. The orthopedist performed a core biopsy because of his concern of early osteonecrosis. A follow-up study 2 months later was normal, and the biopsy specimen showed no evidence of necrosis.

CASE 4

This 45-year-old man has a long history of right hip pain. What is the diagnosis? What are the associated findings? (Fig. 4A. Axial MR, TR/TE 500/20 msec. Fig. 4B. Sagittal MR, TR/TE 2,000/80 msec.)

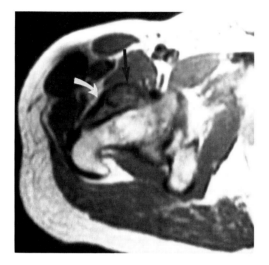

FIG. 4A.

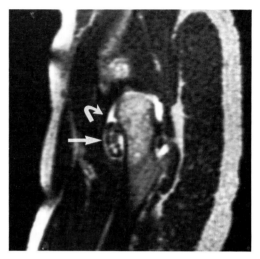

FIG. 4B.

Findings: Fig. 4A: There is a mass measuring about 2 cm in transverse diameter located anterior to the femoral neck (straight arrow) with a peripheral rim of low SI. The SI inside the mass is low, relative to the hyperintense fatty marrow of the femoral neck. The joint capsule is pushed anteriorly (curved arrow) by an effusion. The anterior cortical bone is normal. Fig. 4B: The mass (straight arrow) demonstrates slight increased SI related to cartilage component. The displaced capsule (curved arrow) is well delineated by high SI synovial fluid.

Diagnosis: Large osteochondral fragment secondary to primary synovial osteochondromatosis (SOC).

Discussion: Idiopathic SOC is a chronic, progressive, monoarticular disorder caused by metaplasia of the synovial membrane with formation of numerous cartilaginous intra-articular nodules. Synovial osteochondromatosis is more common in male than female subjects, and it is found between the third and fifth decades of life. The hip and the knee are most frequently involved, but any joint can be affected. Chronic pain and decreased range of motion are the most common clinical symptoms. Rarely, fracture and malignant transformation to chondrosarcoma can occur.

Radiologic findings include cartilaginous-type calcifications within the joint and, less frequently, bony erosions and secondary degenerative osteoarthritis (Fig. 4C). The intra-articular nodules range in size from a few millimeters to several centimeters (Fig. 4D,E). In general, the lesions are small and uniform in size. Occasionally these nodules do not calcify, and arthrography may be required to make the diagnosis. A joint effusion is a constant finding.

Magnetic resonance imaging is helpful in verifying the diagnosis and demonstrating the number, size, and location of the nodules, the presence of joint effusion, and the internal composition of the nodules.

Surgical removal and synovectomy is the definitive therapy, but complete synovectomy of the hip is difficult and recurrent disease is common.

Idiopathic synovial chondromatosis should be distinguished from "secondary" chondromatosis. In both diseases, a number of osteocartilaginous fragments may be present, but their etiologies are quite different. Primary SOC is a result of synovial metaplasia, with the production of small nodules that may calcify or ossify. The articular cartilage and bone are initially normal, although mechanical erosion of both can occur. In "secondary" SOC, primary cartilage disorders such as transchondral fracture or osteoarthritis result in fragmentation of cartilage and subsequent loose body formation. The number of fragments tends to be less than primary SOC, and cartilage and bone changes are often present early. To avoid confusion, it is probably best not to refer to the "secondary" form as SOC but rather as "osteocartilaginous loose bodies." Treatment for the secondary type is not synovectomy but rather removal of the bodies, abrasion of the articular surface, or possibly arthroplasty.

In this case, only a solitary nodule was present anteriorly. Surgical biopsy revealed SOC, but the presentation as a single nodule is quite unusual. The SI characteristics of the central area of the nodule, although nonspecific, are consistent with cartilaginous tissue. The peripheral rim of low SI reflects the heavy peripheral calcifications seen on plain films (Fig. 4C). Sundaram reported a case of SOC in which the SI was similar to that of the bone marrow on T1-weighted images, probably reflecting ossification and fatty marrow formation.

The differential diagnosis of the filling defects in a joint effusion must include pigmented villonodular synovitis, synovial hemangioma, lipoma arborescens, and synovial sarcoma.

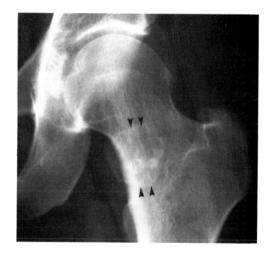

FIG. 4C. Loose body secondary to SOC. A sclerotic rimmed bony density (*arrowheads*) is projected over the left femoral neck in the patient in Fig. 4A and B.

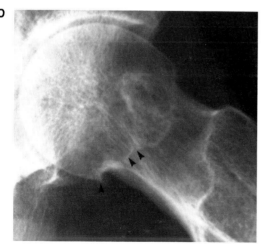
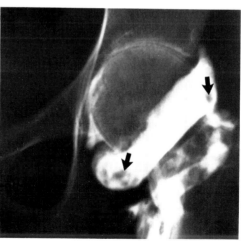

FIG. 4D,E. Primary synovial osteochondromatosis. **D.** Plain radiograph. **E.** Contrast arthrogram. This 25-year-old woman presented with the insidious onset of a monoarthritis over 1 year. Narrowing of the cartilage space, cyst formation, and an osteophyte at the head neck junction (*arrowheads*) are present on the conventional radiograph. Innumerable filling defects (*arrows*) representing the metaplastic nodules are present in the contrast pool. The patient underwent a synovectomy.

CASE 5

This is a 12-year-old girl with a history of intermittent right hip pain for about 6 months. Plain radiographs (not shown) obtained at the time of the MRI were normal. The patient was treated conservatively. What is the differential diagnosis? (Fig. 5A. Coronal MR, TR/TE 800/20 msec. Fig. 5B. Coronal MR, TR/TE 2,000/80 msec.)

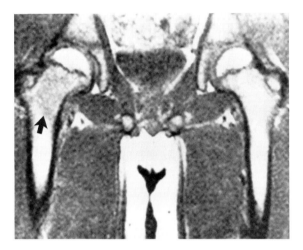 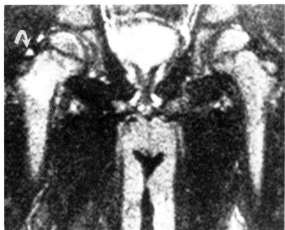

FIG. 5A.   **FIG. 5B.**

Findings: The T1-weighted image shows an area of low SI involving the right femoral neck (arrow in Fig. 5A) superiorly demarcated by the epiphyseal growth plate. The T2-weighted image demonstrates a hyperintense signal in the same area of the femoral neck and a small joint effusion (curved arrows in Fig. 5B). Follow-up radiograph of the right hip 1 year later shows sclerosis of the femoral neck medially with a small cortical nidus (arrow in Fig. 5C).

Diagnosis: Osteoid osteoma of the femoral neck with bone marrow edema.

Discussion: Osteoid osteomas are benign tumors seen most frequently in the second and third decades of life, with a male-to-female ratio of 3:1. The classic clinical presentation of pain worse at night than in the day, relieved by salycilates, is present in about one-fourth of the cases. Osteoid osteoma can involve almost any bone, but it is most frequently found in the femur and tibia. Typical radiographic findings include an area of sclerosis surrounding a central lucency or nidus measuring less than 1 cm. However, the appearance may vary depending on location in the skeleton.

Intramedullary osteoid osteomas elicit less sclerotic reaction than cortical or subperiosteal tumors. Pathologically, osteoid osteomas demonstrate a highly vascularized nidus surrounded by sclerotic bone formation.

The differential diagnosis of a small radiolucency surrounded by a rim of sclerosis includes a Brodie abscess, eosinophilic granuloma, and possibly stress fracture (Fig. 5D). In this case, the radiographic characteristics in the follow-up radiograph (Fig. 5C) are sufficient to make the specific diagnosis of osteoid osteoma. However, in the presence of normal plain radiographs, the MR findings, although nonspecific, can suggest the correct diagnosis. Decreased SI on T1-weighted images and increased SI on T2-weighted images within the medullary space reflect bone marrow edema, which can also be seen in a number of other conditions, including TO of the hip, occult intraosseous fracture, osteomyelitis, tumor, and AVN. The exact mechanism by which bone marrow edema developed in this case is unknown, but it can be postulated that reactive hyperemia and increased new bone formation led to the development of edema. Bone scintigraphy performed in cases of osteoid osteoma demonstrate an area of increased uptake in all three phases of the study, surrounding another central collection of higher accumulation of radionuclide (the double-density sign). The larger area of increased uptake reflects hyperemia and increased bone turnover, similar to what occurs in MRI.

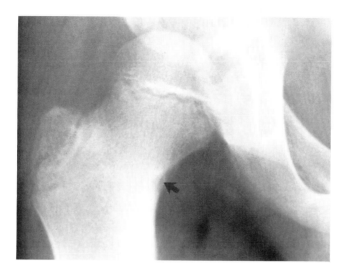

FIG. 5C. Osteoid osteoma. This plain radiograph was obtained on the same patient as in Fig. 5A and B one year after the MR. A small nidus (*arrow*) is clearly seen at this time, allowing the diagnosis to be confirmed.

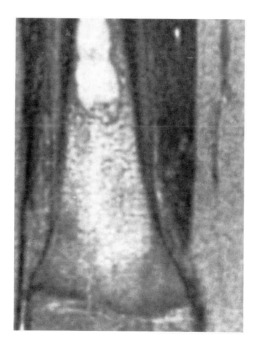

FIG. 5D. Brodie abscess (coronal MR, TR/TE 2,000/80 msec). There is diffuse edema of the medullary canal of the distal one-third of the femur, contrasting sharply with the low signal of the fat. A more discreet, well-circumscribed area of fluid is present within the edematous region. This proved to be an abscess.

CASE 6

This is a 63-year-old woman with recent onset of severe left hip pain. Axial sections of the left proximal femur, at the level of the ischial tuberosity, are shown in Fig. 6A (TR/TE 2,000/20 msec) and B (TR/TE 2,000/80 msec). A sagittal section through the left hip with the buttocks to the reader's right is shown in Fig. 6C (TR/TE 2,000/80 msec). The image shown in Fig. 6C was obtained with a rectangular surface coil placed in the left gluteal area. What is the diagnosis?

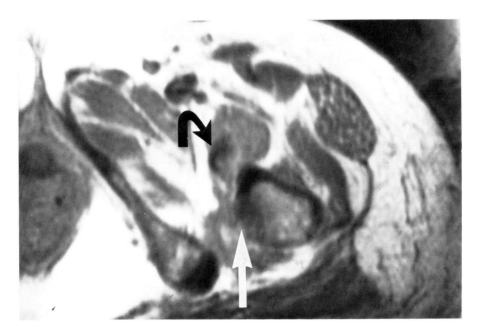

FIG. 6A.

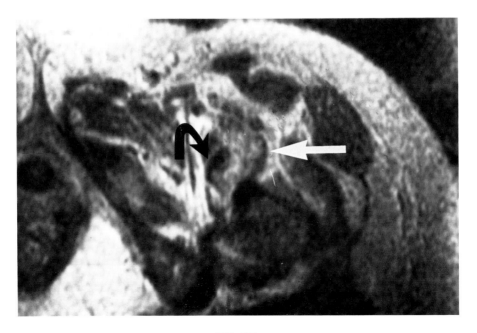

FIG. 6B.

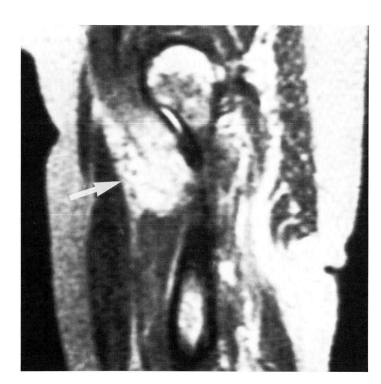

FIG. 6C.

Findings: An axial PD-weighted image (Fig. 6A) shows irregularity of the cortical bone at the level of the lesser trochanter (straight arrow in Fig. 6A), associated with an area of high SI on T2-weighted images anteriorly, in the region of the iliopsoas muscle (straight arrow in Fig. 6B). A low SI C-shaped structure is seen in the region of the iliopsoas muscle (curved arrows in Fig. 6A,B).

A sagittal T2-weighted image (Fig. 6C) demonstrates an ill-defined area of increased SI in the distal iliopsoas muscle at the level of its insertion upon the lesser trochanter (arrow in Fig. 6C).

Diagnosis: Metastatic avulsion of the lesser trochanter.

Discussion: Avulsion fractures in the area of the pelvis or proximal femora are frequently seen in young patients involved in sports. Clinically they present with pain and little external evidence of injury, and treatment is conservative. Avulsive injuries may result in secondary reparative changes in the host bone; these changes may simulate infection or even malignancy.

Avulsion of the lesser trochanter in adults is rare and, when present, frequently is related to a pathologic fracture, secondary to metastatic carcinoma. There is usually no history of trauma. Patients may relate the acute onset of pain to minimal activities such as coughing or walking. In this patient, biopsy of the lesion demonstrated metastatic adenocarcinoma. Prophylactic internal fixation of the femur was performed to prevent pathologic subtrochanteric fracture.

Plain radiographs demonstrate the avulsed fragment in most cases, but MRI may be required to exclude underlying pathology, as in this case. Metastatic foci were present in other areas of the pelvis and femora. Despite the low sensitivity of MRI in demonstrating cortical lesions, the small, cortical, low SI fragment can be seen in the region of the iliopsoas, surrounded by hyperintense edema and hemorrhage on T2-weighted images.

CASE 7

A 25-year-old woman had had right hip problems for years and finally sought medical care. An MRI was ordered to assess the problem. (Fig. 7A. Coronal MR, TR/TE 800/20 msec. Fig. 7B. Coronal MR, TR/TE 2,000/80 msec.) What is the probable diagnosis? What is the structure (arrow) in the joint, and what is its significance?

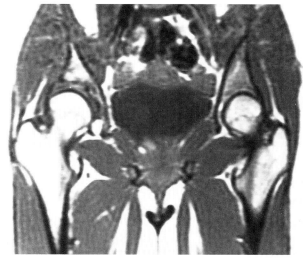

FIG. 7A.

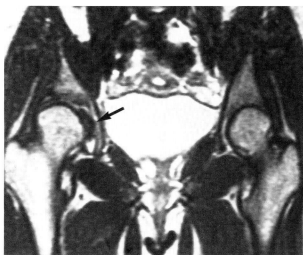

FIG. 7B.

Findings: The right hip is subluxed laterally and the right acetabulum is shallow (Fig. 7A). The hip is in valgus. The dark structure in Fig. 7B (arrow) represents a soft tissue mass, most likely a thickened pulvinar with both fatty and fibrous components. A plain radiograph of the patient is shown in Fig. 7C.

Diagnosis: Congenital dysplasia of the hip (CDH).

Discussion: The term *CDH* is preferable to congenital hip dislocation because it more correctly describes the basic pathologic process and because many affected hips are not dislocated. Congenital dysplasia of the hip refers to a deformity that affects *both* the proximal femur and the acetabulum that is a result of *in utero* and/or postnatal events. The term *dysplasia*, however, implies that the hip was malformed from the beginning; in fact, it is more appropriate to understand CDH as a congenital deformation.

The initiating event leading to CDH is not known and is probably multifactorial, including intrauterine positioning and capsular response to hormones. Anteversion of the neck of the femur, anterior acetabular deficiency, and external rotation-adduction contractures of the iliopsoas tendon and capsule all precede and predispose to subsequent dislocation.

There are several terms used frequently in discussions of CDH. The labrum is a fibrocartilaginous, intracapsular structure that in early life is extremely pliable. It normally covers the superolateral aspect of the femoral head. If it hypertrophies and inverts, it becomes an impediment to femoral head reduction and is defined as the limbus. The pulvinar is a mass of fibrofatty tissue in the acetabular notch; in the normal hip, it is quite small, but it may hypertrophy and become large enough to impede proper reduction.

The overwhelming majority of patients with CDH have an abnormality in which the hip is *positionally* unstable. Clinically, patients have flexion and adduction contractures, with the capsule being the most limiting structure. The most reliable methods for diagnosing CDH in the neonatal period are the Ortolani and Barlow maneuvers, but these examinations are positive for only a few days after birth. Both tests depend on the ability of the examiner to dislocate/locate the hip, a maneuver that leads to an audible clunk. Routine radiologic assessment is not dependable; in fact, *negative findings on radiographic examination do not exclude subluxation or dislocation.*

By 4 to 8 weeks of age, sufficient ossification of the chondro-osseous head and neck has occurred so that a number of plain radiographic signs may be present: (a) lateral and superior migration of the head and neck of the femur, (b) a shallow acetabulum, (c) development of a false acetabulum, and (d) delayed ossification of the femoral head.

Although arthrography may be extremely valuable in assessment of children with CDH, it should be limited to those patients who have failure of satisfactory reduction or who redislocate. Arthrography may demonstrate (a) inversion of the hypertrophied limbus between the femoral head and the acetabulum, (b) pulvinar hypertrophy, (c) thickening of the ligamentum teres, and (d) translocation of the iliopsoas tendon, imparting an hourglass deformity to the capsule (Fig. 7D,E).

The role of MR in assessment of the patient with CDH is not yet established, but anecdotal experience indicates that the major pathoanatomic lesions impeding reduction can be identified. While MR has a distinct advantage in that it is noninvasive, it does suffer from the inability to assess the hip *dynamically*, as can be done with arthrography.

In this case, the dark structure situated between the head and acetabulum probably represents the hypertrophied pulvinar, but there is no proof. A thickened ligamentum teres or a translocated iliopsoas tendon would have a different orientation. The labrum is not seen well, but it is not situated in the joint.

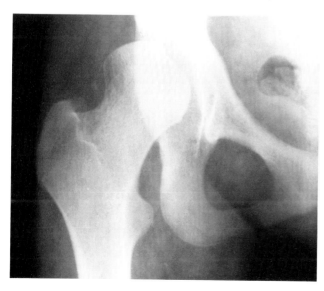

FIG. 7C. Congenital dysplasia of the hip. The plain radiograph of the patient in Fig. 7A demonstrates the shallow acetabulum characteristic of CDH.

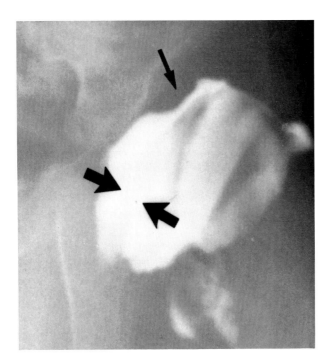

FIG. 7D. Congenital dysplasia of the hip. An arthrographic study demonstrates an inverted and thicked limbus (*thin arrow*) and a thickened ligamentum teres (*arrows*).

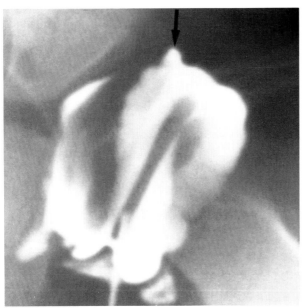

FIG. 7E. Normal infant hip. The normal limbus thorn (*arrow*) projecting above the contrast column indicates the normal everted position of the labrum just medial to it.

CASE 8

This is a 60-year-old woman with the onset of left hip pain over 1 week. Conventional radiographs of the left hip were normal. What is the diagnosis? (Fig. 8A. Coronal MR, TR/TE 2,000/20 msec. Fig. 8B. Coronal MR, TR/TE 2,000/80 msec.)

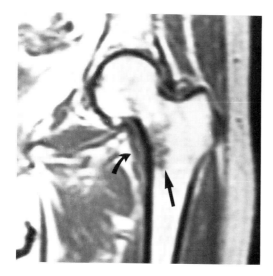

FIG. 8A.

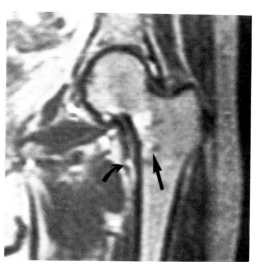

FIG. 8B.

Findings: The PD-weighted image (Fig. 8A) shows a broad-based area of low SI along the medial femoral cortex (straight arrow). The same area shows high SI on T2-weighted images (Fig. 8B). A band of low SI in the soft tissues, paralleling the femoral cortex, is noted medially on PD- and T2-weighted images (curved arrows). A follow-up radiograph is shown in Fig. 8C. (Reprinted with permission from Deutsch AL, Mink JH, and Waxman A. Occult fractures of the proximal femur: MR imaging. *Radiology* 1989;170:113–116.)

Diagnosis: Stress fracture of the femoral neck.

Discussion: The incidence of hip fractures in Caucasian women after the age of 50 increases dramatically, so that women in their eighth decade have a 40% chance of having sustained one or more hip fractures. A history of trauma is often missing, or, not infrequently, patients report the spontaneous onset of severe pain *after* which they stumble and fall. Radiographic examination is positive in the vast majority of cases, but in some patients the initial radiographs may be normal (Fig. 8D–F). Displaced fractures are readily diagnosed, but lack of displacement, decreased bone stock, and the inability of the patient to cooperate for optimal radiographs all result in some diagnostic uncertainty. In these cases, bone scintigraphy, conventional tomography, and CT may be used to establish the diagnosis of an occult fracture.

Bone scintigraphy is highly sensitive for early detection of stress and occult fractures in young individuals, but in the elderly, there may be a delay of up to 4 days between the time of the fracture and the appearance of increased activity on the bone scan. In addition, bursitis, Paget disease, and osteoarthritis, common findings in the elderly, can all produce increased activity on bone scintigraphy. Conventional tomography and CT have been shown to provide false-negative results in some cases.

Initial experience with MRI in the evaluation of occult and stress fractures is encouraging. Replacement of the normal hyperintense signal of the fatty marrow on T1-weighted images by low SI edema and hemorrhage related to the fracture has been seen during the first 25 hr of onset of the symptoms, although at this time, it is unknown if there may be a delay in MR findings, as there can be with scintigraphy. Areas of increased SI on T2-weighted images reflect bone marrow edema that may be found around the actual fracture line. In a recent study, MR correctly identified nine proximal femoral fractures that were not evident on conventional radiographs and correctly excluded fracture in 14 patients in whom a high clinical suspicion existed.

Some MRI features can help to differentiate occult intraosseous fractures from stress fractures. Occult intraosseous fractures are subchondral foci of impaction of bone with surrounding edema and hemorrhage. The term *occult* in this case has been used to indicate the universal finding of normal radiographs and lack of findings at arthroscopy. In fact, occult intraosseous fractures probably represent osteochondral fractures in which the elastic cartilage is only minimally deformed and in which the subchondral component dominates. These lesions are located in the epiphysis, whereas stress fractures are most often seen in the metaphysis. Occult intraosseous fractures are the result of a single traumatic event as opposed to a period of unusual physical activity. Periosteal elevation with fluid has been described only in stress fractures.

The case presented herein is best described as a stress response in bone. Poorly defined areas of bone marrow edema without a linear component probably represents edema/hyperemia and perhaps trabecular microfracture that precede actual skeletal failure.

This patient went home for several days and returned because of increasing pain. A follow-up MRI (and plain radiographs) demonstrated a completed fracture with a varus deformity of the femoral neck (Fig. 8C).

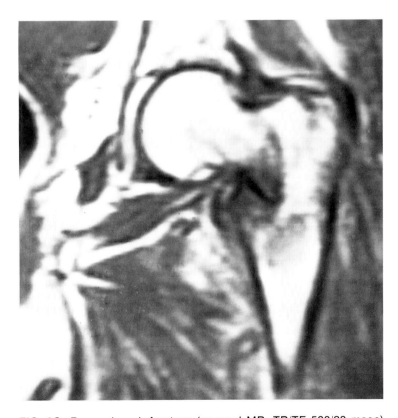

FIG. 8C. Femoral neck fracture (coronal MR, TR/TE 500/20 msec). The patient in Fig. 8A returned with increasing symptoms; the MR (and plain radiographs) demonstrated a varus deformity at the now-completed fracture site.

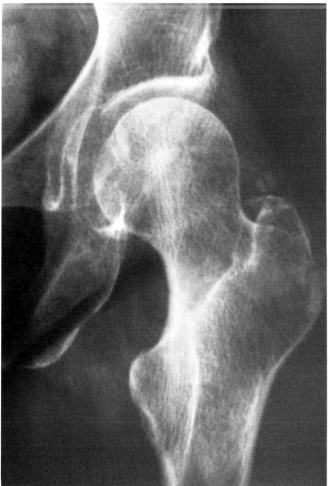

D

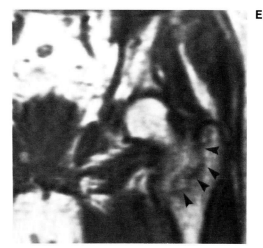

E

FIG. 8D,E. Radiographically occult intertrochanteric fracture. The conventional radiograph **(D)** is normal. **E.** Coronal MR, TR/TE 2,000/20 msec. The complete course of this intertrochanteric fracture is depicted as a line of decreased signal (*arrowheads*).

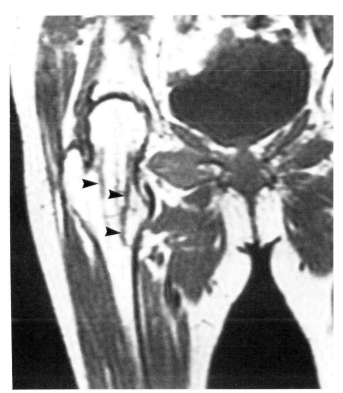

FIG. 8F. Old femoral fracture (coronal MR, TR/TE 300/15 msec). This elderly patient had had an intertrochanteric fracture years before that had been internally fixed by means of Ender rods. The rods were removed 2 years ago. Linear areas of decreased signal (*arrowheads*) are present but are oriented perpendicular to the expected plane of a fracture. The line represents the course of the rods. Interestingly, the old fracture is not visible, although the femur is in valgus.

CASE 9

This is a 28-year-old woman with a short history of left hip pain and palpable mass. Two coronal MRI images are shown in Fig. 9A and B (coronal MR, TR/TE 2,000/80 msec). What is the diagnosis?

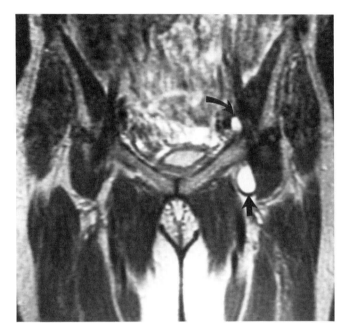

FIG. 9A.

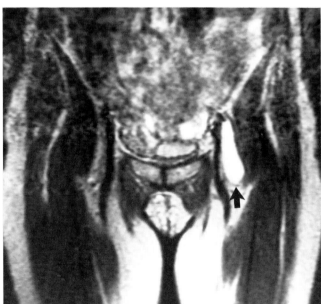

FIG. 9B.

Findings: There is an elongated fluid collection (straight arrows) with increased SI on these T2-weighted images, located medial to the iliopsoas muscle and lateral to the femoral vein and pectineus muscle. The fluid-filled mass extends superiorly above the region of the inguinal ligament (curved arrow). No joint effusion or other abnormalities are present.

Diagnosis: Iliopsoas bursitis.

Discussion: The iliopsoas bursa is the largest bursa of the hip joint and is present in more than 90% of adults. This bursa is located between the iliopsoas muscle anteriorly and laterally and the pectineus muscle and femoral vessels medially and anteriorly. Posteriorly, the bursa is in close relationship with a thin area of the joint capsule located between the iliofemoral and pubofemoral ligaments.

The iliopsoas bursa can become distended by fluid in a variety of inflammatory conditions, including rheumatoid arthritis, osteoarthritis, synovial chondromatosis, and pigmented villonodular synovitis. Occasionally the synovial fluid may become infected. In most cases of bursitis, there is communication with the hip joint through the normal anterior weak area.

Patients with distended iliopsoas bursae may present clinically with a palpable inguinal soft tissue mass. If associated inflammatory changes are present, the symptoms may include pain, sometimes referred to distant regions such as the abdomen, thigh, and knee. Paresthesias, neuropathy, urinary frequency caused by bladder compression, and edema of the extremity caused by compression of the femoral vessels have all been reported as complications of distended iliopsoas bursae.

Clinically, differential diagnosis includes soft tissue tumors, inguinal hernia, aneurysm, undescended testes, varices, and inguinal adenopathy (Fig. 9C). Preoperative diagnosis can be obtained by arthrography (if there is communication with the joint), direct puncture, and CT. In the case presented herein, MRI provided all the information necessary for preoperative evaluation, including the cystic nature of the mass, its relationship with the surrounding structures, and the lack of communication between the bursa and joint space.

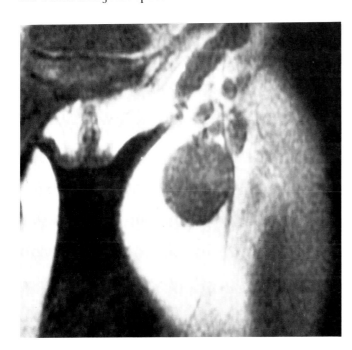

FIG. 9C. Inguinal lymphadenopathy (coronal MR, TR/TE 800/20 msec). Multiple rounded to ovoid masses are present in the left groin. The nodes are differentiated from an iliopsoas bursa by their number, shape, and lack of continuous extension above the inguinal ligament. This patient proved to have cat-scratch disease.

CASE 10

This is a 45-year-old woman who has had right hip pain of several weeks duration. What is the diagnosis, and what is the MR class (stage) of the disease? (Fig. 10A. Coronal MR, TR/TE 2,000/20 msec. Fig. 10B. Coronal MR, TR/TE 2,000/80 msec.)

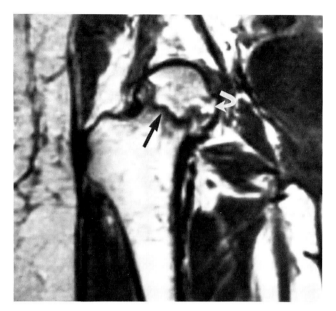

FIG. 10A.

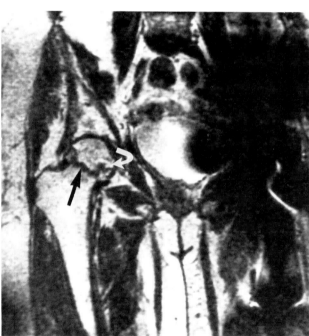

FIG. 10B.

Findings: There is a low SI band across the lower portion of the right femoral head on the PD-weighted image (Fig. 10A) and T2-weighted image (Fig. 10B) (straight arrows), surrounding a high SI band (curved arrows) (double-line sign). The upper portion of the femoral head is relatively isointense with the bone marrow of the femoral neck, establishing this hip as a class A. No joint effusion is present. A large area of signal void is seen in the region of the left hip (Fig. 10B), caused by a metallic total hip prosthesis.

Diagnosis: Avascular necrosis of the right femoral head.

Discussion: This patient has had a cadaveric renal transplant. Avascular necrosis (AVN) occurs in about 15% of patients undergoing renal transplant and steroid therapy. The femoral heads are the most frequently involved areas, although osteonecrosis can affect other bones of the skeleton.

Both a radiographic and an MR staging system of AVN have been proposed as a means of guiding medical and surgical therapy.

Ficat originally described stage I disease as a painful hip in which the radiographs were normal. In stage II disease, the radiographs demonstrate patchy sclerosis and lysis. A transition between stages II and III is characterized by the classic subchondral crescent; collapse of the subarticular bone is the hallmark of the stage III disease. By stage IV the joint is arthritic, and the acetabular side of the joint is involved. Most authors use the Cruess modification of Ficat's system, in which the appearance of the subchondral crescent is stage III, collapse of the head is stage IV, and stage V is narrowing and arthrosis of the joint. Stage O is truly the "silent hip." It is defined as an asymptomatic hip with normal radiographs in a high risk patient (e.g., a patient with AVN in the *opposite* hip). In one study, 17/27 silent hips had elevated intraosseous venous pressures and 11 of the 17 developed osteonecrosis.

Plain radiographs may be completely normal for 2 years following an acute ischemic event, and patients may be asymptomatic for prolonged periods of time. Bone scintigraphy has been capable of demonstrating changes within weeks of the onset of ischemia, but the sensitivity and specificity of bone scintigraphy are relatively low (in one study, bone scanning was 11% sensitive in the diagnosis of early AVN).

Magnetic resonance imaging has been proven to be highly accurate in detecting early changes of AVN with a sensitivity superior to that of bone scintigraphy and CT. One of the most frequent MRI signs of AVN is the so-called double-line sign. The exact nature of this sign is still under debate. Some authors believe it represents a chemical shift artifact, but others attribute the low SI band to the bone-forming layer of the reactive interface and believe the hyperintense band seen on T2-weighted images represents the ingrowing zone of granulation tissue and hypervascularity.

Four MRI patterns of AVN have been defined on the basis of signal characteristics. They *may* reflect the chronologic stage of disease. During early stages (class A pattern), high SI on T1-weighted images and low SI on T2-weighted images is seen within the area of necrosis of the femoral head. These SI changes are similar to those of normal fatty marrow (Fig. 10C,D). When subacute hemorrhage develops in a more advanced stage of AVN, high SI can be seen in both T1- and T2-weighted images. As more fluid accumulates within the necrotic segment of bone, a fluid-like pattern develops, with low SI on T1-weighted images (class C pattern). During the late stages of the disease, fibrosis and sclerosis predominate, demonstrating low SI on T1- and T2-weighted images (class D pattern). This pattern classification helps one to understand the pathologic events occurring in the necrotic segment of bone, but it is a simplified representation of these changes. Differentiation between viable and nonviable tissue cannot be made on the basis of SI changes seen on MRI.

The MRI classification correlates well with the severity of clinical symptoms and the radiographic stage of disease. Eighty-five percent of patients with class A (MR) had no or only mild symptoms, whereas 83% with stage D disease had moderate to severe symptoms. Eighty-three percent of patients with stage A disease had radiographic stage I or II disease, whereas 100% of patients with stage D disease had radiographic stage III, IV, or V. It is not known if the pattern of disease would predict which patients are most likely to benefit from a core decompression.

Large field-of-view MRI with a body coil is often used to evaluate patients suspected of having AVN because it is desirable to examine both hips. Small field-of-view surface coil technique provides improved assessment of the extent of the disease and associated findings, such as articular cartilage destruction, and presence of subchondral fractures. Gradient recalled sequences may more graphically depict articular cartilage fracture.

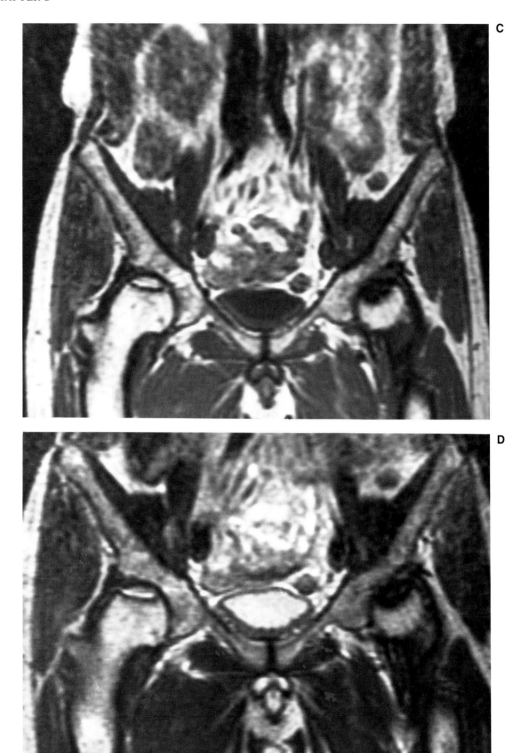

FIG. 10C,D. Osteonecrosis. **C.** Coronal MR, TR/TE 800/20 msec. **D.** Coronal TR/TE 2,000/80 msec. The right hip lesion corresponds to a class A hip with the lesion having a "fat" signal character. The left hip is a class D signal, remaining dark on both sequences. The conventional radiograph of the right hip was normal, but the patient had bilateral symptoms. (Right hip is radiographic stage I.)

CASE 11

An abnormality of the left femur was discovered on the radiographs of the pelvis obtained in this 18-year-old girl after she fell. An MRI of the pelvis and proximal femora was obtained in an attempt to assess and characterize the abnormality. What are the findings? What is the diagnosis? (Fig. 11A. Coronal MR, TR/TE 2,000/20 msec. Fig. 11B. Coronal MR, TR/TE 2,000/80 msec. Fig. 11C. Axial MR, TR/TE 2,000/80 msec.)

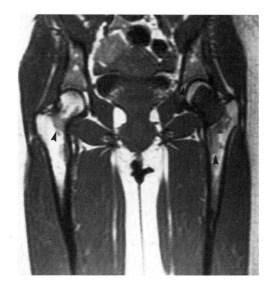

FIG. 11A.

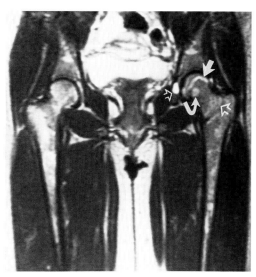

FIG. 11B.

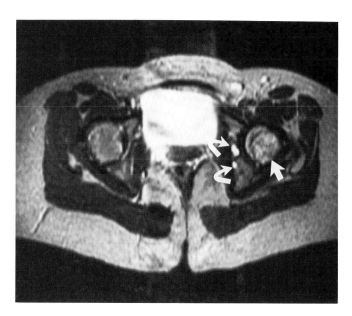

FIG. 11C.

Findings: The coronal T1-weighted image (Fig. 11A) demonstrates a well-demarcated area of low SI involving the left femoral neck, extending partially into the femoral head. A lower SI rim surrounds the lesion. Intermediate SI areas in the proximal left femoral diaphysis, right femoral neck, and pelvis represent hematopoietic marrow (arrowheads).

The coronal T2-weighted image (Fig. 11B) shows that most of the lesion involving the left femoral neck becomes relatively hypointense with the exception of a high SI band demarcating the proximal margin of the lesion (straight arrow in Fig. 11B), and a rounded area in the medial side of the neck (curved arrow in Fig. 11B). The most distal portion of the lesion (large open arrow in Fig. 11B) becomes isointense with the adjacent fatty marrow of the greater trochanter.

A second acetabular lesion is detected on the coronal T2-weighted image (small open arrow in Fig. 11B). The signal from this lesion is very intense, similar to the fluid-filled urinary bladder.

The axial T2-weighted image confirms the inhomogeneous hyperintensity of the proximal portion of the left femoral neck lesion (straight arrow in Fig. 11C) and shows two small fluid-like lesions in the acetabulum (curved arrows in Fig. 11C). The largest one corresponds to the cystic lesion seen in Fig. 11B.

Diagnosis: Fibrous dysplasia.

Discussion: Fibrous dysplasia is a developmental bone anomaly usually manifested in children and adolescents, characterized by slow-growing hamartomatous lesions composed mainly of fibrous tissue. Usually the disorder is monostotic, but it may involve multiple bones. The polyostotic variety may be associated with endocrine dysfunction (precocious puberty) and cafe-au-lait spots (McCune-Albright syndrome).

The disease is generally asymptomatic, and it is discovered incidentally when obtaining radiographs for unrelated reasons. The ribs, mandible, skull, and femurs are most frequently involved, but any part of the skeleton can be affected. Pathologically, fibrous dysplasia is composed of a fibrous collagenous matrix with irregular areas of woven bone that have variable degrees of mineralization.

The radiographic manifestations of fibrous dysplasia are varied and include radio-lucencies ("ground glass" appearance) (Fig. 11D), sclerosis, internal calcifications in areas of lucency, bone expansion, sclerotic rims, and bone deformity. The lucency or sclerosis seen on plain radiographs reflects the relative amounts of fibrous or osseous tissue composing the lesion.

The MRI appearance of fibrous dysplasia has not yet been reported, with the exception of an occasional anecdotal case. The reasons for the SI changes are not known but probably reflect the internal composition of the lesions. It has been shown that hypocellular fibrous tissue has a short T2, thus providing little or no signal on regular spin-echo T2-weighted pulse sequences. In addition, calcified tissue has long T1 and very short T2 relaxation times because of its relatively fixed molecular state. Therefore, both major components of fibrous dysplasia produce low SI on T1- and T2-weighted images.

Cartilaginous islands or nodules have been described in some cases of fibrous dysplasia. Cartilaginous tissue has more water content than fibrous tissue, and therefore high SI on T2-weighted images is evident (Fig. 11E,F). The presence of cartilaginous tissue would explain the relatively hyperintense signal seen in the femoral neck on T2-weighted images in this case.

Finally, fluid-filled areas have been described in fibrous dysplasia, probably representing degeneration and cellular necrosis. These cystic spaces would be responsible for the hyperintense lesions seen in the acetabulum.

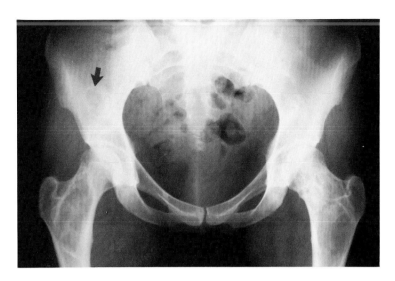

FIG. 11D. Polyostotic fibrous dysplasia. In another patient, "ground glass" type lesions are present in either proximal femur and the right ilium (*arrow*). The proximal extent of the left femoral lesion is difficult to determine.

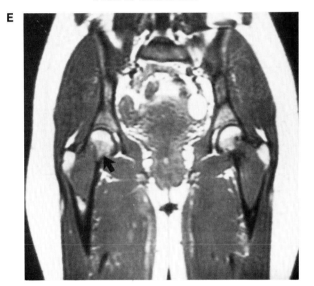

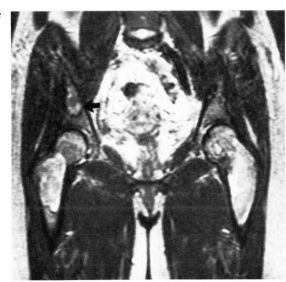

FIG. 11E,F. Polyostotic fibrous dysplasia (same patient as in 11D). **E.** Coronal MR, TR/TE 300/15 msec. **F.** Coronal MR, TR/TE 2,000/80 msec. Well-defined areas of intermediate signal compose both femoral lesions; the acetabular lesion is not well seen. The *arrow* in Fig. 11E points to the normal red marrow found in this location in a young adult. On the long TR/TE sequence, the femoral lesions have inhomogeneous signal increase, probably reflecting variable amounts of fibrous tissue and cartilage. The acetabular lesion (*arrow*) has a dark center and a brighter rim.

CASE 12

This is a 25-year-old man with sudden onset of buttock pain during water skiing. A coronal MRI through the ischial tuberosities is shown in Fig. 12A (TR/TE 500/20 msec). Axial MRI through the ischial tuberosities is shown in Fig. 12B (TR/TE 2,000/20 msec), and Fig. 12C (TR/TE 2,000/80 msec). Plain radiographs of the pelvis were normal. What are the findings?

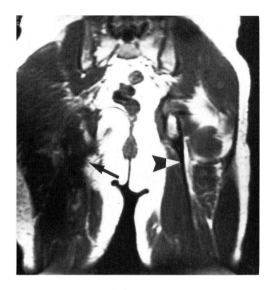

FIG. 12A.

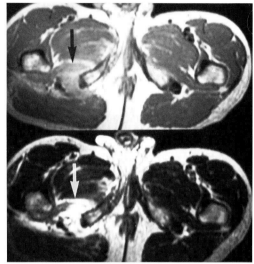

FIG. 12B,C.

Findings: Coronal T1-weighted images demonstrate absence of the right hamstring tendon (straight arrow in Fig. 12A) as compared with the left side (arrowhead). An ill-defined area of low SI is seen in the region of the hamstrings.

Axial PD- (Fig. 12B) and T2-weighted images (Fig. 12C) show increased SI in the region of the insertion of the hamstrings in the right ischial tuberosity (arrows in Fig. 12B and C).

Diagnosis: Avulsion of the hamstring muscles from the ischial tuberosity.

Discussion: The hamstring is composed of a group of muscles of the posterior compartment of the thigh, inserting on the ischial tuberosity. These muscles are the semitendinosus, the semimembranosus, the biceps femoris (long head), and the ischio-condylar portion of the adductor magnus. The function of the hamstring muscles includes flexion of the leg and extension of the thigh. The semimembranosus and semitendinosus muscles also contribute in part to medial rotation of the leg in flexion, and the biceps femoris contributes to lateral rotation of the leg.

Forceful contraction of the hamstring muscles can produce an avulsion of the ischial apophysis in young patients, before closure of the growth plate. In the adult patient, however, the weak area may not be at the epiphyseal plate but at the musculotendinous unit. In these cases the plain radiographs may be completely normal. Avulsion injuries followed by resorption of bone at the fracture site and reactive sclerosis can result in a complex radiographic image that has been confused with osseous malignancy.

Most hamstring injuries occur while running or sprinting and may be caused by poor muscle flexibility, inadequate muscle strength, dyssynergic muscle contraction, insufficient warm-up and stretching, awkward running style, and return to activity before complete rehabilitation following injury.

Clinical symptoms of avulsion of the hamstring muscles include sudden onset of buttock pain, difficulty on standing after trauma, palpable defect and tenderness in the region of the ischial tuberosity, and weakness of knee flexion. Treatment of hamstring avulsion includes rest and immobilization following injury, with a gradual return to activity. Large ruptures may require surgical intervention.

Tendon avulsion and rupture can be accurately diagnosed by ultrasound when the lesion is superficially located, but deep injuries such as hamstring avulsion are beyond the reach of ultrasound examination. MRI can provide useful information including confirmation of the clinical diagnosis and assessment of the extent of the lesion (Fig. 12D,E). Lack of visualization of the normal tendons of the hamstring muscles at the ischial tuberosity and inflammatory changes with high SI on T2-weighted images are the diagnostic findings in this case.

D

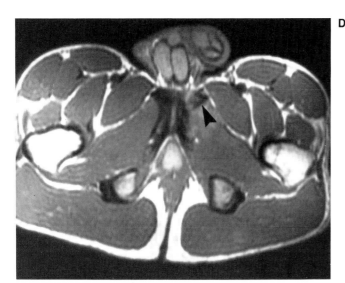

E

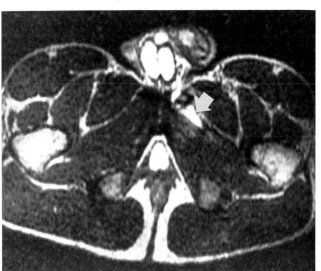

FIG. 12D,E. Traumatic avulsion of the pubic body (adductor avulsive injury). **D.** Axial MR, TR/TE 300/15 msec. A fragment (*arrowhead*) of the body of the left pubis is displaced. **E.** Axial MR, TR/TE 2,000/80 msec. There is considerable edema/hemorrhage (*arrow*) in the soft tissues adjacent to the avulsion. This injury required 4 weeks of injured reserve time for this professional football player.

CASE 13

This 30-year-old man with a chronic medical disorder presented with subacute left hip pain. Coronal sections of the pelvis and proximal thighs are shown in Fig. 13A (TR/TE 500/20 msec) and Fig. 13B (TR/TE 1,500/80 msec). What are the diagnoses?

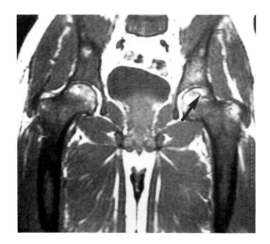 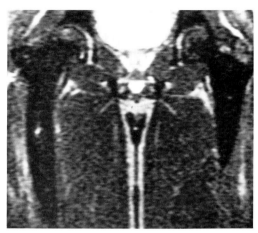

FIG. 13A.   **FIG. 13B.**

Findings: There is abnormal low SI of the bone marrow spaces of the pelvic bones (especially the right) and proximal femora on both the T1- and T2-weighted images. The only areas showing the expected high SI of fatty marrow on the T1-weighted image are portions of the femoral heads, the left greater trochanter, and small islands in either iliac wing. There is a low SI band across the left femoral head (arrow in Fig. 13A). The T2-weighted image (Fig. 13B) reveals a very striking decrease in signal from all marrow areas. There is a very small bright focus in the midfemoral shaft on the right.

Diagnosis: Sickle cell anemia (SCA) with hemosiderosis.

Discussion: In the fetus, the entire skeleton is devoted to hematopoietic production by red marrow, but soon after birth, replacement by fat commences. This conversion begins peripherally in the extremities and progresses centrally, but it may be somewhat inhomogeneous within a bone and asymmetric from side to side.

By the time an individual is 25 years old, the adult pattern of red marrow is achieved. Hematopoietic elements are confined to the skull, vertebrae, ribs, pelvis and sternum, and proximal femora and humeri. In the patient with SCA, several striking changes occur in the marrow space. The increased demands for red cell production results in reconversion of fatty to hematopoietic marrow, and marked red marrow expansion occurs. The slow sinusoidal circulation in hematopoietic marrow provides an ideal environment for sickling and ultimately marrow infarction. Repeated transfusions, accelerated marrow production, and infarction may all help produce excessive iron deposition in the reticuloendothelial system, leading to hemosiderosis.

The MR findings in patients with SCA depend on the severity and chronicity of the disorder. The most constant feature is reconversion of fatty to hematopoietic marrow. Focal and diffuse areas of mildly decreased signal on both short and long TR/TE sequences characterize reconversion. The appearance of this new marrow may be somewhat variable on T2-weighted images, depending on the ratio of cellular elements and tissue water. It may, therefore, be difficult to differentiate hyperplastic marrow from replacement disorders such as leukemia, and biopsy may be necessary. Marrow reconversion can be found in some *normal* adults in whom marrow biopsy has demonstrated hematopoietic hyperplasia (see Chapter 6, Case 55). In most cases, the reconversion affects the diaphysis and metaphysis, but if the demands for red cell production are sufficiently great, as in SCA, the epiphysis may participate as well.

Foci of decreased signal on T1-weighted images that brighten on long TR/TE sequences in patients with SCA are thought to represent areas of edema and presumed infarction (Fig. 13B,C), and the presence of these foci have correlated well with pain. Old infarctions with fibrosis show very low signal on both long and short TR/TE sequences.

Hemosiderin, an iron-containing protein, has an extremely short T2 value, producing a very dark appearance on both T1- and T2-weighted images.

Sickle cell anemia is a well-known predisposing condition for AVN, which can involve any epiphysis but is most frequent in the femoral heads. The low SI band across the left femoral head surrounding a relatively isointense region (straight arrow in Fig. 13A) represents the reactive interface surrounding a relatively recent area of AVN.

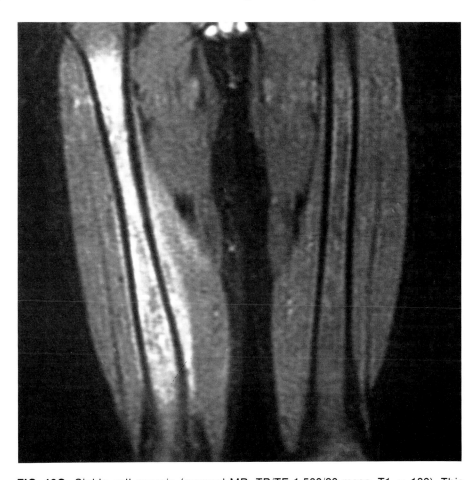

FIG. 13C. Sickle cell anemia (coronal MR, TR/TE 1,500/20 msec, T1 = 130). This patient with SCA had leg pain for 10 to 14 days. This inversion recovery sequence demonstrates that much of the right femur is bright, indicating an increase water content in the medullary canal. Similar findings are present in the soft tissues. The lack of bone destruction (after 10–14 days of symptoms), the normal conventional radiographs, nuclear medicine studies, and the clinical picture suggested bone infarction.

CASE 14

This is a 50-year-old woman with a 1-month history of left hip pain and fever. What are the findings and the differential diagnostic considerations? (Fig. 14A. Coronal MR, TR/TE 2,000/80 msec. Fig. 14B. Coronal MR, TR/TE 2,000/80 msec.)

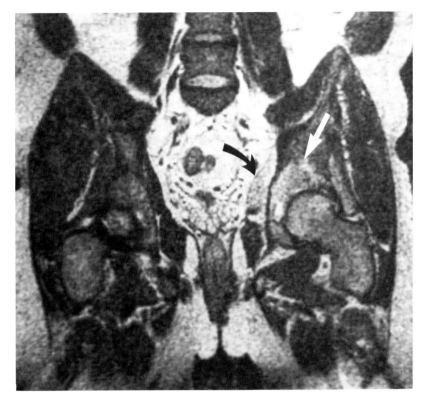

FIG. 14A.

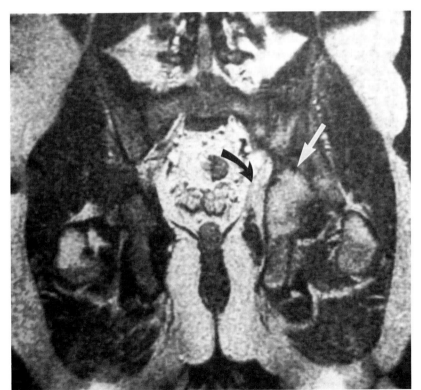

FIG. 14B.

Findings: There is high SI in the bone marrow of the left acetabulum (straight arrows in Fig. 14A,B) and in the obturator internus muscle (curved arrows in Fig. 14A,B). The left obturator internus muscle is thicker than the right one. Slight increase in SI is present in the left femoral head as compared to the right. There is no joint effusion.

Diagnosis: Acute osteomyelitis of the left acetabulum with edema or myositis of the left obturator internus muscle.

Discussion: Magnetic resonance imaging findings are nonspecific. Tumor and infection can produce the same increased SI on T2-weighted images within the bone marrow space. In this case, the preoperative diagnosis was based on the combination of the clinical and MRI findings. Without septic symptoms, tumors such as metastasis or myeloma should be included in the differential diagnosis.

Absence of a joint effusion makes the diagnosis of septic arthritis unlikely. The increased SI of the femoral head is probably due to reactive hyperemia, because no osteomyelitis of the femoral head was found at surgery.

The increased SI of the obturator internus muscle is due to edema.

Abscesses are well-defined, encapsulated fluid collections with dark rims and bright centers on long TR/TE sequences. Magnetic resonance has proven itself to be of great value in the assessment of musculoskeletal infections. Magnetic resonance can reliably differentiate bone from soft tissue infection and can distinguish soft tissue inflammation (phlegmon) from a drainable abscess. The latter use of MR is of special value in the diabetic foot. Radionuclide studies in such situations are completely nonspecific.

Bone scintigraphy demonstrated generalized increased uptake of the entire left hip area. The value of MRI in this case was to assess the exact extension of the infection in the bone and to exclude the presence of a drainable abscess or septic arthritis.

Inflammatory changes in the bone marrow precede bone destruction in acute osteomyelitis; therefore, plain radiographs and CT may be negative during the early stages of the infection. The main advantage of MRI over other diagnostic imaging modalities is its high sensitivity to small changes in the water content of the tissues. On the other hand, because inflammatory changes can occur in association with tumor, trauma, or infection, the specificity of MRI is low.

CASE 15

This is a 30-year-old woman with a history of right hip pain for about 6 months. She was involved in an auto accident, but no hip dislocation was detected at that time. What is the most likely diagnosis, and what is the value of MRI in this case? (Fig. 15A. Coronal MR, TR/TE 500/20 msec. Fig. 15B. Coronal MR, TR/TE 2,000/80 msec.)

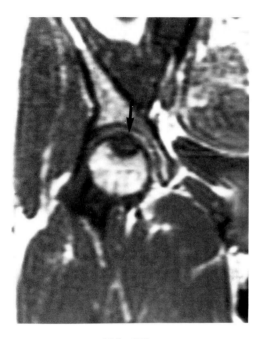 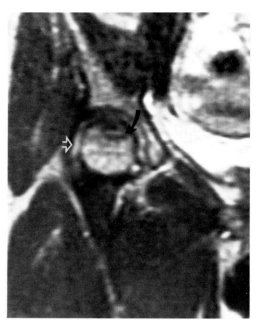

FIG. 15A. **FIG. 15B.**

Findings: There is a low SI segment of bone in the weight-bearing portion of the femoral head, seen on T1- and T2-weighted images. The subchondral cortex and cartilage in this abnormal area is bulging (straight arrow in Fig. 15A). The articular cartilage is thinner over the lesion than elsewhere. A lower SI band seen on both pulse sequences surrounds the lesion inferiorly. On T2-weighted images (Fig. 15B) there is a thinner hyperintense line immediately adjacent to the low SI band (curved arrow). A very small amount of fluid is seen laterally (open arrow).

Diagnosis: Osteochondritis dissecans (OCD).

Discussion: Osteochondritis dissecans or osteochondral fracture is a posttraumatic lesion involving the articular cartilage and subchondral bone. When the lesion involves only the articular cartilage, the term *transchondral fracture* is frequently used. Pathologically, the fragment is frequently necrotic, the cartilaginous layer covering the osteochondral fragment is thinned, and a healing response can be seen at the site of detachment. However, the lesion is clearly different from osteonecrosis.

Osteochondral fragments may remain in place or they may become detached. If the osteochondral fragment is separated from the bone, it may become a loose intraarticular body, it may be resorbed, or it may become imbedded in the synovium at a distant site. Revascularization of an *in situ* lesion may result in healing with complete anatomic restitution.

The knee and the ankle are the most frequent sites of OCD, but this lesion can occur in any traumatized articular surface. Plain radiographs, radionuclide studies, and arthrography have been used to establish the stability of the osteochondral fragment, but recently the value of MR in assessment of these lesions at the knee has been reviewed. Osteochondritis dissecans lesions may be described as loose, loose *in situ*, or stable. As judged by plain radiographs, those lesions less than 0.2 cm^2 were uniformly stable. A surrounding zone of sclerosis greater than 3 mm was indicative of a loose fragment. The most reliable MR signs of a loose fragment on MR is displacement of the fragment or interposition of a high signal interface, thought to represent joint fluid or possibly healing granulation tissue between the fragment and the host on long TR/TE sequences. Although most OCD fragments increase in signal on T2-weighted images, only loose fragments did not. The actual viability of the osteochondral fragment is difficult to evaluate with MRI, because SI changes do not exactly reflect the pathophysiologic events taking place.

The main differential diagnosis should be made with AVN. Occasionally the diagnosis may be difficult, based solely on the MRI findings. The presence of a "double-line" sign, with a band of decreased SI surrounding another hyperintense band, should favor AVN. However, as in this case, a partially detached osteochondral fragment can produce similar MRI findings. Localized thinning of the cartilage covering the lesion, as seen in this patient, would favor osteochondral fracture rather than AVN. Cartilage thinning occurs only in the late stages of AVN, at which time there is diffuse cartilage loss with associated degenerative changes. Indeed, in this case, the clinical history of auto accident and lack of predisposing conditions for AVN make the diagnosis of osteochondral fracture the most likely diagnosis.

A radiograph of the right hip (Fig. 15C) demonstrates the osteochondral fragment separated from the femoral head by a lucent line. The adjacent segment of the femoral head shows some reactive sclerosis (3 mm). This lesion was not operated. The zone of sclerosis on the plain radiograph and the bulging of the bone and cartilage suggest that this lesion was loose, but no fluid was present between the host and the fragment; therefore, this lesion was thought to represent a loose *in situ* fragment.

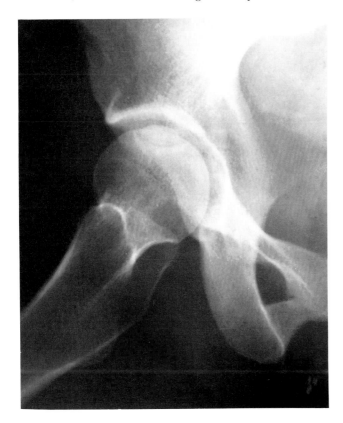

FIG. 15C. Osteochondritis dissecans of the hip. This frog-leg view of the hip demonstrates the "bulging" of the lesion appreciated on the MRI. A 3-mm zone of sclerosis is present around the lesion.

CASE 16

This is a 64-year-old woman with history of right hip pain for several weeks. What are the findings? What is the diagnosis? (Fig. 16A. Coronal MR, TR/TE 2,000/20 msec. Fig. 16B. Coronal MR, TR/TE 2,000/80 msec.)

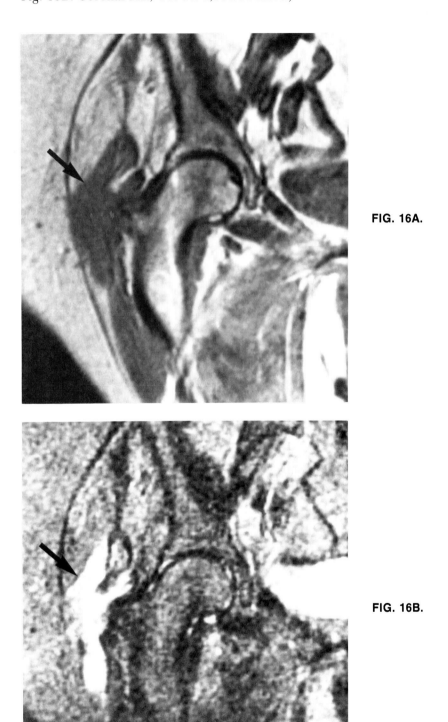

FIG. 16A.

FIG. 16B.

Findings: There is a lobulated fluid collection adjacent to the greater trochanter, with low SI on PD-weighted images and high SI on T2-weighted images (arrows in Fig. 16A,B). This fluid collection is well demarcated from the surrounding subcutaneous fat by a low SI line. A tiny joint effusion is present. There are no abnormal areas of SI changes in the bone marrow space of the right femur.

Diagnosis: Trochanteric bursitis.

Discussion: There are four bursae present around the hip joint: the iliopsoas, the iliopectineal, the ischiotrochanteric, and the trochanteric bursae. The iliopsoas bursa is the one that most commonly becomes inflamed and presents clinical symptoms. The ischiotrochanteric bursa is located around the lesser trochanter of the femur. Ischiotrochanteric bursitis has been described in patients with abnormal articulation between the lesser trochanter and the lateral aspect of the ischium in patients treated with total hip replacement for old congenital hip dislocation.

When septic arthritis is present, the infected joint effusion produces increased intraarticular pressure, leading to communication with one of the bursae around the hip, most frequently the iliopsoas bursa.

The trochanteric bursa often becomes infected or shows inflammatory changes following total hip replacement. It is present around the knots of the trochanteric wires. Communication with the hip joint space in postsurgical trochanteric bursitis is frequent and can be demonstrated with arthrography.

Trochanteric bursitis is also a relatively frequent cause of hip pain in patients without previous surgery or pathologic conditions of the hip. In these cases, communication with the joint space is rare, hence the absence of joint effusion in the patient presented herein.

The MRI appearance of bursitis is that of a fluid collection with low SI on T1-weighted images and high SI on T2-weighted images in the anatomical location of the bursa. The differential diagnosis with soft tissue abscess may be difficult. Infected bursa or soft tissue abscess should be considered when clinical and/or laboratory signs of infection are present. The SI changes seen on MRI are nonspecific. Infected and noninfected fluid collections have a similar MRI pattern. In this patient, percutaneous aspiration and culture of the fluid showed no infection.

CASE 17

This is a 21-year-old woman with a 5-day history of right hip and thigh pain, fever, and limitation of motion. Radiograph of the right hip and coronal MRI (TR/TE 2,000/ 80 msec) are shown in Fig. 17A and B, respectively. What is the diagnosis?

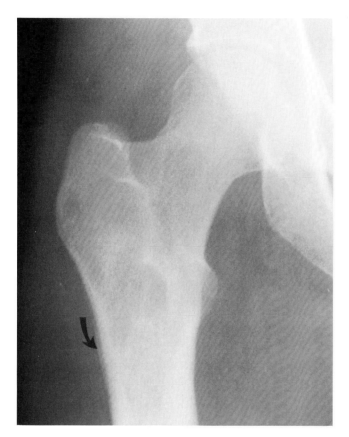

FIG. 17A.

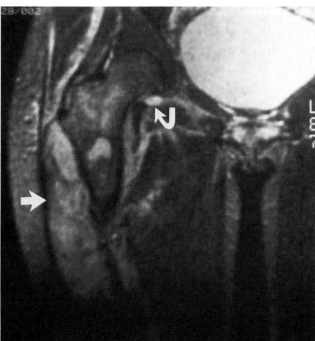

FIG. 17B.

Findings: Plain radiographs demonstrate a small periosteal reaction (curved arrow) in the lateral aspect of the femur. A faint lucency is seen within the medullary space of the proximal femur. Coronal MRI shows an area of high SI corresponding to the lucency seen on the plain film. This lesion extends to the outer cortex of the femur, in the region of the periosteal reaction seen on the plain radiograph. A small joint effusion is present (curved arrow in Fig. 17B). There is an area of ill-defined high SI in the neck of the femur. There is a large area of inhomogeneous high SI (straight arrow in Fig. 17B) in the soft tissues adjacent to the bone.

Diagnosis: Subacute osteomyelitis of the proximal femur (Brodie abscess), with draining sinus and a large abscess involving the vastus lateralis muscle.

Discussion: Magnetic resonance imaging has proven highly sensitive for the diagnosis of musculoskeletal infections. Increased water content of the infected tissue produces prolongation of the T1 and T2 relaxation times. This translates into low SI on T1-weighted images and high SI on T2-weighted images.

During the acute stage of the osteomyelitis, one can see ill-defined, hyperintense areas in the medullary space, on the T2-weighted images, similar to those seen in bone marrow edema. During the subacute phase of osteomyelitis, the intraosseous abscess is seen on MRI as a well-demarcated fluid collection of low SI on T2-weighted images and high SI on T2-weighted images. Finally, in the chronic stage of osteomyelitis, sequestrum formation can be seen as low SI tissue on all pulse sequences, within an area of high SI on T2-weighted images representing the intraosseous chronic infection.

The diagnosis of osteomyelitis can usually be made from plain radiographs, bone scintigraphy, and clinical findings. Magnetic resonance imaging, however, can provide information regarding the intra- and extramedullary extension of the infection.

Diagnosis of soft tissue infection, on the other hand, may be elusive even using sophisticated scintigraphic techniques. Soft tissue abscesses are seen on MRI as well-demarcated fluid collections with a peripheral rim of low SI representing the capsule. Edema and cellulitis have ill-defined margins.

In this patient, the abscess involving the vastus lateralis muscle was not detected by 3 phase technetium 99m bone scintigraphy and indium 111-labeled leukocyte scanning. At surgery, the abscess extended to the distal thigh.

The specific diagnosis of osteomyelitis cannot be made based on the MRI findings alone, because tumors and trauma can produce similar MR changes. Correlation with clinical and radiographic findings is essential for improved diagnostic accuracy.

CASE 18

Two different patients with the same diagnosis were being evaluated for possible surgical intervention. What is the diagnosis? What treatment is *probably* best for each? (Fig. 18A. Patient A, coronal MR, TR/TE 800/20 msec. Fig. 18B. Patient B, coronal MR, TR/TE 800/20 msec. Fig. 18C. Patient B, sagittal MR, TR/TE 2,000/80 msec.)

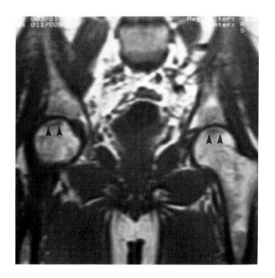

FIG. 18A.

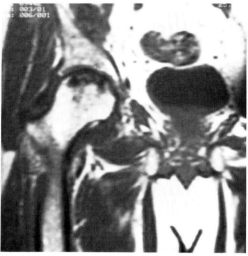

FIG. 18B.

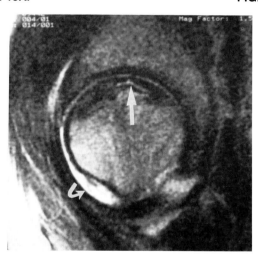

FIG. 18C.

Findings: In Fig. 18A, there are small crescent-shaped low SI areas in the upper portions of both femoral heads (arrowheads) without extension to the surface. In Fig. 18B and C, there is a large necrotic segment involving the weight-bearing surface. In the sagittal plane (Fig. 18C), the necrotic segment extends from the 9 o'clock position to the 2 o'clock position. The high intensity line within the fracture represents fluid or hemorrhage (straight arrow), and an effusion is present (curved arrow).

Diagnosis: Bilateral osteonecrosis.

Discussion: In the first patient (Fig. 18A) the diagnosis of osteonecrosis, or AVN, was confirmed by histologic material obtained during core decompression. The second patient had had a renal transplant and underwent an arthroplasty.

Most patients with AVN of the femoral head are symptomatic, but occasionally, no symptoms are present. The radiographic changes and the patient's symptoms are crucial in staging AVN. Stage I radiographic disease is defined as a normal radiograph in a symptomatic patient. An asymptomatic high risk patient with normal radiographs has been defined as stage O. The capability of MRI to depict small, early, and asymptomatic infarcts raises several questions: (a) Can MRI be used as a screening test for patients at high risk of developing AVN? and (b) If MRI is a good screening test, is there an effective therapy to change the natural course of the disease in patients with small asymptomatic infarcts?

At least two reports in the literature indicate that MRI is not a 100% sensitive test, because false-negative MRI cases have been found. In a group of patients who were at high risk for osteonecrosis, biopsy of 11 hips demonstrated evidence of AVN. Only six of the 11 hips had shown preoperative MR evidence of AVN (sensitivity = 46%). Bone scans and bone marrow scans in the same group had sensitivities of 11% and 50%, respectively, but in no case did the bone or bone marrow scan diagnose AVN when the MR was normal.

How early can MRI detect AVN following an ischemic insult is a question that has not yet been answered. This question is also compounded by the fact that AVN can occur as an acute event, following a fracture, or may be a summation of small ischemic episodes, as it is likely to occur in nontraumatic cases (i.e., steroid administration). Sampling errors at the time of the biopsy and lack of standardized pathologic criteria for AVN are also potential sources of error when evaluating MRI results. Despite the fact that MRI is the most sensitive examination for the detection of AVN, its use as a screening test cannot be recommended at the present because a negative MRI does not fully exclude AVN.

There is considerable controversy regarding the usefulness of core decompression. Conflicting data regarding short-term clinical improvement and alteration of the natural history exist. Core decompression is thought to decrease intramedullary pressures, improve venous return, and facilitate revascularization. Rabbits treated with steroids had femoral head blood flow measured pre– and post–core decompression. Before surgery, femoral blood flow gradually decreased with steroid use. These experimental data are consistent with the clinical observation of early conversion of hematopoietic to fatty marrow in adults with AVN. Such conversion is thought to result from alterations in blood flow. Following core decompression in the rabbits, however, femoral head blood flow improved.

Core decompression has been advocated for patients who have not yet suffered a subchondral fracture, e.g., patients with stage O, I, or II disease. It is accepted that 68% to 90% of patients with stage II or III disease, treated nonsurgically, will demonstrate progressive radiographic collapse. It was hoped that the core decompression would alter seemingly progressive hip destruction. Such patients may be stimulated to revascularize to prevent structural failure. The patient in Fig. 18A might be a candidate for core decompression, but the patient in Fig. 18B and C is not.

The results from *human* clinical studies of the utility of core decompression, however, have been quite mixed. One study suggests that, following core decompression, only 10% of hips demonstrated progression. Most recently, however, 66% of patients with stage II disease had radiographic evidence of progression, or increasing symptoms within 1 year of decompression. Only 33% of patients with stage I disease treated the same way had progression, but actually all were relieved of pain. Finally, core decompression is not without complications; major femoral fractures result from the stress-riser created in the femur.

There is only a minimal amount of data regarding the effect of core decompression on MR appearances of AVN, and the data are as mixed as the clinical results. In one study, seven of 18 patients treated with core decompression demonstrated a decrease in joint fluid, medullary edema, and the size of the lesion; progression of the disease occurred in only one case. However, in a group of eight patients with early AVN who were treated *nonsurgically*, follow-up MR showed no disease progression in six, and two demonstrated only increasing fatty marrow conversion.

The clinical implications of failure to detect early AVN are not known; it is quite possible that some cases will not progress. On the other hand, the efficacy of therapeutic procedures may be falsely negated by continued existence of risk factors (e.g., steroid administration). Clearly, the role of core decompression deserves much closer scrutiny.

Assessment of the exact location of the infarcted area in the sagittal plane may be useful to guide the surgeon considering transtrochanteric anterior rotational osteotomy (Sugioka). This procedure involves osteotomizing the proximal femur and rotating the femoral head forward (Fig. 18D). The necrotic segment, which nearly always affects the anterosuperior aspect of the head, is rotated inferomedially and is thus no longer weight-bearing. The amount of weight-bearing surface that is necrotic determines the feasibility of the procedure. The extensive involvement in Fig. 18B and C would indicate the patient is not a good candidate for rotational osteotomy. The "best" therapy for AVN is still not determined. Core decompression, rotational osteotomy, bone grafting curettage, sponge bone plasty, surface replacements, hemiarthroplasties, and ultimately total hip replacements represent the spectrum of procedures borne by patients with AVN.

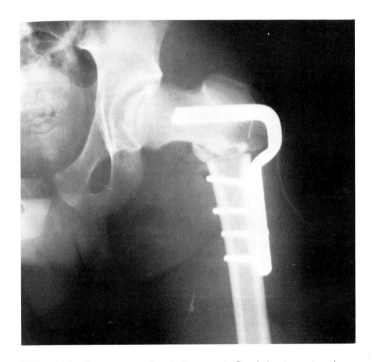

FIG. 18D. Osteonecrosis, status post–Sugioka transtrochanteric osteotomy. This young male patient had osteonecrosis (Perthes disease) of the left femoral head; a CT had demonstrated that only the anterior half of the head was affected. Because of his young age (10 years), an attempt was made to preserve bone stock. The osteotomy brought the unaffected portion of the femoral head into a weight-bearing position.

CASE 19

This is a 6-year-old boy. A coronal MR section through the hips is shown in Fig. 19A (TR/TE 500/20 msec). Sagittal sections through the right and left hip are shown in Fig. 19B and C, respectively (TR/TE 2,000/20 msec). What are the findings, and what information is most critical to transmit to the orthopedist?

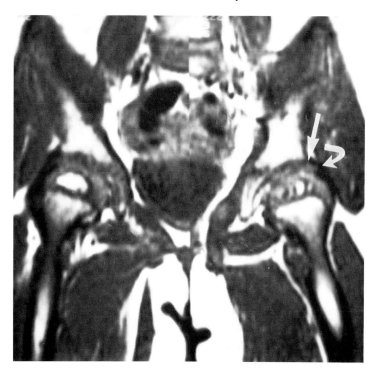

FIG. 19A.

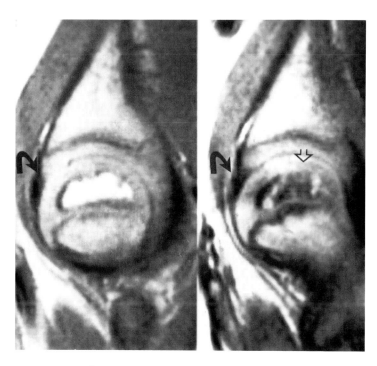

FIG. 19B.   **FIG. 19C.**

Findings: The coronal image demonstrates a normal configuration and SI of the right femoral head. The ossification center of the *left* femoral head is fragmented and has irregular areas of low SI, better seen in the sagittal section (Fig. 19C). The cartilaginous layer of the left femoral head is incompletely covered by the acetabulum (straight arrow). The fibrous labrum is seen in both coronal and sagittal sections (curved arrows).

Sagittal sections (Fig. 19B,C) demonstrate a distinct change in SI between the superficial and deep cartilaginous coverage of the femoral head, more conspicuous on the left (open arrow).

Diagnosis: Legg-Calve-Perthes (LCP) disease of the left femoral head.

Discussion: Although the causal mechanisms are not fully understood, the basic process in LCP disease is osteonecrosis. Ages 4 to 8 years, the time at which most cases of LPC disease present, is a "vulnerable" period in the development of the vascular supply to the femoral head.

Trauma, transient synovitis of the hip, intra-articular effusion, and perhaps a congenital predisposition may all be mechanisms that result in osteonecrosis. Conventional radiographic abnormalities include decreased size of the ossific nucleus, lateral displacement of the epiphysis (caused by hypertrophy of the cartilage along the medial side of the head), fragmentation, sclerosis, and flattening of the nucleus. Metaphyseal cysts resulting from a disturbance in endochondral ossifications and widening of the femoral neck are also frequently found.

Early in the course of LCP, the normally high SI of the femoral capital epiphysis is lost. The epiphysis usually is smaller than on the normal side, and it often becomes fragmented. At least one report indicated that MR demonstrated abnormal SI of the femoral head in a patient with LCP in whom the conventional radiographs and radionuclide bone scan were normal. This is not totally unexpected, as similar imaging findings have been reported in adult osteonecrosis. With revascularization, the femoral head signal may return to normal, but the bony deformity including flattening of the head and widening of the femoral neck usually persist.

There is still some controversy regarding the treatment of this disease, but most authors agree that conservative therapy is indicated if no significant flattening of the femoral head is present and if containment of the femoral head by the acetabulum can be achieved with the treatment.

Containment of the femoral head is calculated by determination of the acetabulum-head index (AHI) and acetabulum-head quotient (AHQ). In the former, the width of the acetabulum is divided by the width of the femoral head, and the result multiplied by 100. The AHQ compares the AHI of the affected hip with that of the opposite (normal) hip in the same patient:

$$\frac{\text{AHI (abnormal)}}{\text{AHI (normal)}} \times 100 = \text{AHQ}$$

The original description of this technique utilized the bony landmarks for assessment. Arthrography permits direct assessment of the articular cartilage and a more precise and anatomically correct evaluation. These highly desirable advantages can be achieved noninvasively with MR.

Because the articular cartilage covering the epiphyseal ossification center of the femoral head is not visible on conventional radiographs, measurements of femoral head containment may be misleading (Fig. 19D). The coronal MRI section in this case (Fig. 19A) shows that the left ossification center is nearly fully covered by the acetabulum, but a large segment of the articular cartilage is not covered (Fig. 19E).

Sagittal sections are ideal for assessment of congruity or lack of collapse of articular cartilage. Figure 19B and C demonstrate perfect congruity of the articular surfaces of the affected left femoral head and acetabulum.

Other findings occasionally seen in patients with LCP include acute chondrolysis, cartilage hypertrophy, and synovitis with pannus formation. All these abnormalities can be demonstrated with MRI.

An additional finding seen in this case is higher SI of the superficial layer of the femoral articular cartilage in relationship with the deep layer (Fig. 19C). The cause of this phenomenon is unknown, but it could be related to a greater degree of hydration of the superficial cartilaginous layers.

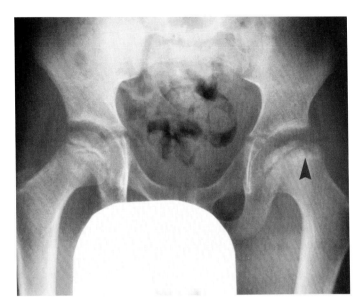

FIG. 19D. Legg-Perthes disease. The plain radiograph from the patient in Fig. 19A suggests virtual complete containment of the left femoral head (compared to the normal right side), whereas the MR indicates differently. The ossific nucleus is small and collapsed. Additional findings include widening of the left femoral neck and a metaphyseal cyst (*arrowhead*).

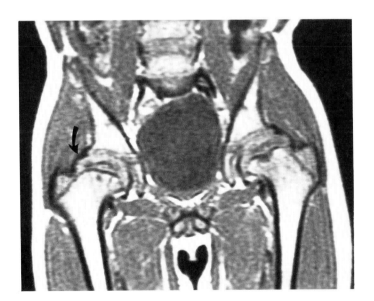

FIG. 19E. Legg-Perthes disease (coronal MR, TR/TE 600/20 msec). In another patient, both hips are affected; the ossification centers are small and irregular. The right hip is better covered by acetabular cartilage and the labrum (*arrow*) than is the left.

CASE 20

This 40-year-old man had a several month history of left hip pain. A coronal section of the pelvis and hips is shown in Fig. 20A (TR/TE 2,000/80 msec). Sagittal surface coil images of the left hip are shown in Fig. 20B (TR/TE 2,000/20 msec) and C (TR/TE 2,000/80 msec). What is the diagnosis?

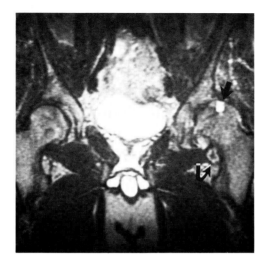

FIG. 20A.

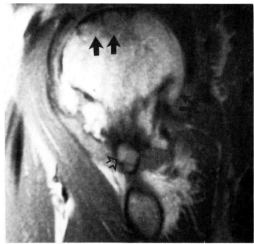

FIG. 20B.

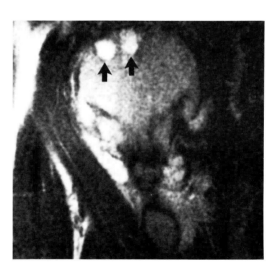

FIG. 20C.

Findings: The coronal section (Fig. 20A) demonstrates loss of the articular cartilage of the left hip joint, with marked narrowing of the joint space. There is also a small hyperintense lesion in the subchondral region of the left femoral head (straight arrow). There is a joint effusion. There is a low SI lesion adjacent to the medial aspect of the left femoral neck, surrounded by a high SI band (curved arrow).

Sagittal PD-weighted image (Fig. 20B) and T2-weighted image (Fig. 20C) show two subchondral cysts (straight arrows), osteophyte (curved arrow, Fig. 20B), and rounded lesions in the region of the femoral neck (open arrow, Fig. 20B).

Diagnosis: Osteoarthritis of the left hip with intra-articular loose bodies.

Discussion: Osteoarthritis of the hip is a relatively common disease in men. Occasionally, it can be seen in younger individuals and can be unilateral. Slipped femoral capital epiphysis during childhood has been suggested as possible etiology for "premature" osteoarthritis.

The pathologic changes occurring in the hip joint in osteoarthritis include narrowing of the joint space with superior migration, osteophyte formation, subchondral cysts, and sclerosis. Fragmentation of osteophytes and/or cartilage can lead to intra-articular loose bodies (Fig. 20D,E).

Radiographs accurately depict the articular changes occurring with osteoarthritis, but occasionally an MRI may be requested to exclude other conditions such as AVN, synovial chondromatosis, and inflammatory arthritis. The characteristic findings of AVN of the femoral head have been well-described and include areas of varying SI in the femoral head, surrounded by low SI rims, sometimes associated with high SI bands on T2-weighted images, the so-called double-line sign. These findings are different from those of osteoarthritis. Narrowing of the joint space and subchondral cysts are not a feature of AVN, unless secondary degenerative changes are present, in late stages of the disease.

The diagnosis of synovial chondromatosis is based on the presence of typical calcifications within the joint space on plain radiographs. Secondary degenerative changes and erosions can be found in advanced cases of synovial chondromatosis. In these cases, differentiation of advanced synovial chondromatosis from osteoarthritis might be difficult. Pigmented villonodular synovitis (PVNS) may also present with a pattern of "osteoarthritis" of the hip. Posterior and superior migration of the femoral head, presence of large osteophytes, and lack of marginal erosion would make the diagnosis of inflammatory arthritis unlikely.

Accurate evaluation of cartilage thickness, the cystic nature of the subchondral lesions, localization of the intra-articular loose bodies, and exclusion of AVN are features well demonstrated with MRI in this case.

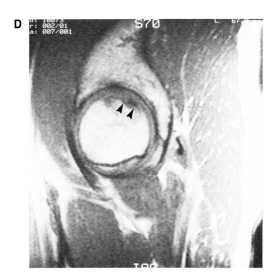 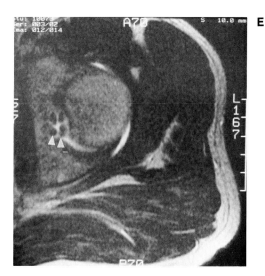

FIG. 20D,E. Osteochondral fracture with loose bodies. **D.** Sagittal MR, TR/TE 600/20. **E.** Axial MR, TR/TE 2,000/80. High resolution studies can be obtained with dual coils placed anterior and posterior to the hip. In Fig. 20D an osteochondral fracture (*black arrowheads*) is seen at the 2 o'clock position of the left femoral head. Multiple loose bodies are present (*white arrowheads*) surrounded by high signal joint fluid (Fig. 20E).

CASE 21

A 45-year-old man presented to his orthopedist with a 7-month history of severe left hip pain. Physical examination was normal except for mild pain at the extremes of hip motion. (Fig. 21A. Plain radiograph. Fig. 21B. Coronal MR, TR/TE 800/20 msec.) An abnormality, not shown in Fig. 21A and indicated by an arrow in Fig. 21B, did not increase in signal on a long TR/TE sequence.

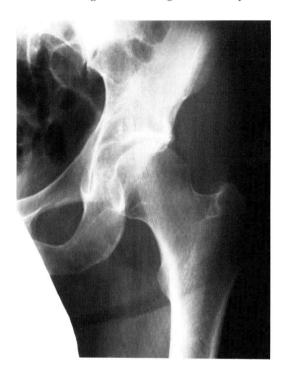

FIG. 21A.

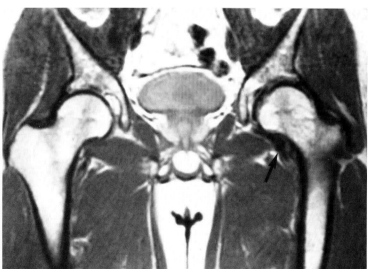

FIG. 21B.

Findings: There is an ovoid area of decreased signal just inferomedial to the left femoral neck (arrow). Its position immediately adjacent to bone suggests it is intracapsular, although without the presence of an effusion, it is difficult to be certain of this. There is no bone abnormality. The hip is displaced laterally.

Diagnosis: Pigmented villonodular synovitis.

Discussion: Although the abnormality was seen on the MR, its significance was not appreciated. The patient therefore underwent diagnostic arthrography and a CT scan (Fig. 21C, D), which demonstrated lobulated masses within the joint (arrows), especially along the medial femoral neck. The differential diagnosis included lipoma arborescens, synovial chondromatosis, synovial hemangioma, and multiple loose bodies. The diagnosis of PVNS was confirmed at the time of arthrotomy and subtotal synovectomy.

Pigmented villonodular synovitis is most frequently found in a 20- to 45-year-old patient who presents with several years of pain, limitation of motion, and swelling. All laboratory examinations are normal, and the joint fluid is almost always bloody. The knee is the most frequently involved large joint (80%), followed by the hip, ankle, small joints of the hands and feet, shoulder, and elbow. Rare instances of involvement of the temporomandibular joint and the facet joints of the spine have also been reported. Involvement of more than one joint is quite rare.

There are two morphologic forms of PVNS: the solitary nodule and the diffuse form, the latter of which usually occurs at the knee. Either form presents as a soft tissue mass, usually with a joint effusion. Calcification of the mass is notably absent, a distinguishing feature from synovial sarcoma. The involved joint space is normal in width, and osteoporosis is lacking. PVNS is frequently associated with bony cystic erosions, which are a result of direct pressure effect as well as a result of elevated intra-articular pressures. Such cysts are most commonly found in smaller articulations such as the wrist. At the hip, 37 of 40 (93%) cases had erosive/cystic disease.

There is scant data on the MR appearance of PVNS. The MR signal characteristics are quite variable and mixed, but this is explained by the variable composition of the mass. All PVNS lesions contain areas that are quite dark on both short and long TR/TE sequences, a reflection of the deposition of hemosiderin. Fat, also universally found in PVNS, has a bright signal on short and a less intense signal on long TR/TE sequences. Inflamed synovium, recent hemorrhage, and fluid are also found in these lesions and serve to produce inhomogeneous and variable signal foci. In this case, hemosiderin deposition and fibrosis predominated, and the lesion is quite dark.

The differential diagnosis includes hemophilic arthropathy, in which repeated hemorrhage could result in considerable hemosiderin deposition and synovial masses. Synovial chondromatosis may also have "dark-dark" appearance of synovial masses caused by calcification of the nodules. Rheumatoid arthritis and synovial hemangioma might be difficult to distinguish from PVNS. Interestingly, the coexistence of both rheumatoid arthritis and synovial hemangioma with PVNS has been reported.

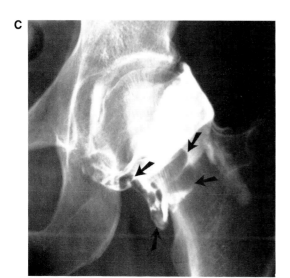

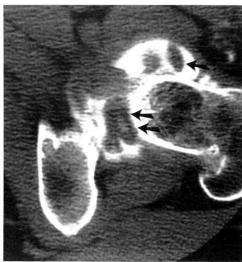

FIG. 21C,D. Pigmented villonodular synovitis. A diagnostic arthrogram and a CT scan reveal multiple masses in the joint (*arrows*), especially along the medial aspect of the femoral neck.

CASE 22

This is a 24-year-old female with a long history of bilateral hip pain. What is the differential diagnosis? (Fig. 22A. Coronal MR, TR/TE 600/20 msec. Fig. 22B. Sagittal MR of the right hip, TR/TE 2,000/20 msec. Fig. 22C. Sagittal MR of the right hip, TR/TE 2,000/80. Fig. 22D. Sagittal MR of the left hip, TR/TE 2,000/20 msec.)

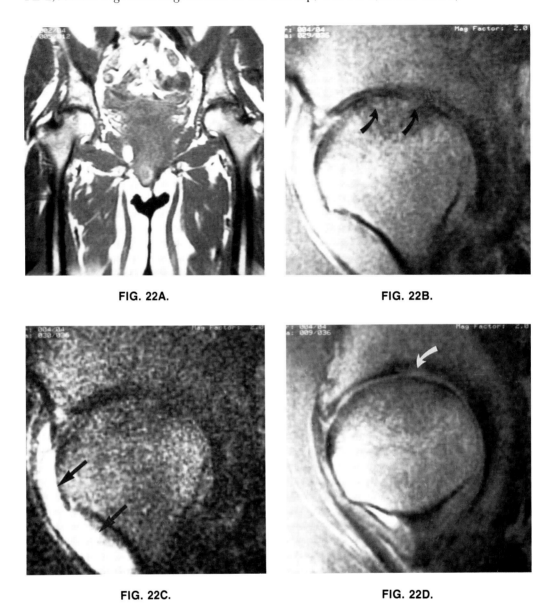

FIG. 22A.　　　　　　　　　　**FIG. 22B.**

FIG. 22C.　　　　　　　　　　**FIG. 22D.**

Findings:　Coronal T1-weighted images (Fig. 22A) demonstrates diffuse narrowing of the joint spaces bilaterally, with irregularity of the subchondral cortices of the acetabuli and femoral heads. Sagittal PD- and T2-weighted images through the hips (Fig. 22B–D) confirm the joint space narrowing and demonstrate cortical erosions (arrows) bilaterally. A joint effusion is also present on the right (Fig. 22C, large arrows). No osteophytes are seen. There is no evidence of a "double-line" sign. A normal sagittal MR is shown in Fig. 22E.

Diagnosis:　Rheumatoid arthritis.

Discussion: The main differential diagnoses in this case includes degenerative osteoarthritis, septic arthritis, AVN, and other noninfectious inflammatory hip diseases. The lack of bone production (osteophytes) practically excludes osteoarthritis.

Avascular necrosis could be seriously considered in this case, but there is absence of the characteristic "double-line" sign. In addition, there is severe cartilage destruction, a finding seen only in the late stages (stage V) of AVN in which there are associated degenerative changes, including osteophytes and subchondral sclerosis.

The presence of a joint effusion, bone erosions, and cartilage destruction are all findings of septic arthritis. The bilateral and almost symmetrical involvement of the hip joints, however, in this case makes the diagnosis of septic arthritis unlikely.

Finally, an inflammatory, noninfectious joint disorder such as rheumatoid arthritis or a seronegative spondyloarthropathy is the most likely possibility because of the axial migration of the femoral heads, diffuse symmetrical joint space narrowing, subcortical erosions and cysts, and joint effusions.

Additional MRI features of rheumatoid arthritis not seen in this case include pannus formation, whose SI is lower than the associated joint effusion, and tenosynovitis and tendon rupture.

Plain radiographs can accurately demonstrate the bone changes of rheumatoid arthritis, but occasionally, MRI may provide information regarding the extent and severity of cartilage damage, as well as the associated soft tissue lesions. Patients with rheumatoid arthritis may develop AVN secondary to steroid therapy.

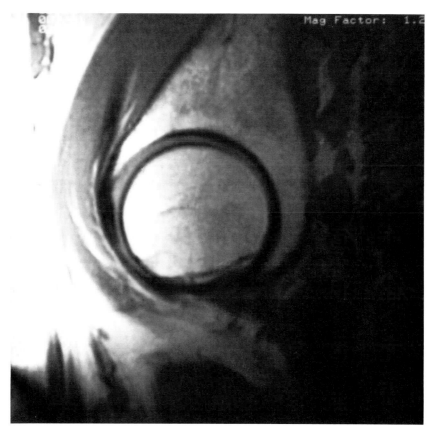

FIG. 22E. Normal hip (sagittal MR, TR/TE 2,000/20 msec). The articular cartilage is much thicker than in the patient with rheumatoid arthritis.

CASE 23

A 65-year-old woman presented to her orthopedist with a subcutaneous mass, thought to be a lipoma, in her *right* hip area. Plain radiographs of both hips were made. The physician ordered an MR. What are the findings? (Fig. 23A. Plain radiograph. Fig. 23B. Coronal MR, TR/TE 300/15 msec.)

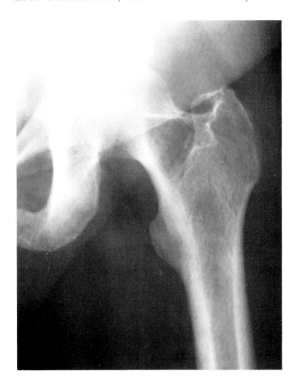

FIG. 23A.

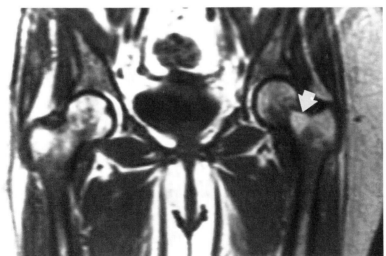

FIG. 23B.

Findings: A destructive lesion of the *left* femoral neck was seen on the plain radiograph (Fig. 23A). There is a linear calcification associated with the inferolateral extent of the lesion. The MR (Fig. 23B) demonstrates a high signal focus (arrow) in the femoral neck that is quite apparent because of the presence of adjacent hematopoietic marrow in the proximal femur. The lesion is isointense with the greater trochanter.

Diagnosis: Intraosseous lipoma.

Discussion: Intraosseous lipoma is an uncommon lesion of bone that is nearly always asymptomatic and is discovered accidentally. In a report of 61 cases of lipoma, the proximal femur was involved in 21 cases, the ilium in five cases, the ribs in four, and the calcaneus in five. In only a single case did the lesion undergo a pathologic fracture.

Intraosseous lipomas undergo involutional changes spontaneously, which result in alterations of the radiologic and pathologic features. These tumors have been classified into three stages. In stage I, there is no secondary necrosis, and the lesion consists of mature live lipocytes. In stage 2, there is partial necrosis, and in stage 3, there is near-complete necrosis of the tumor. In this latter stage one may find calcification, reactive ossification, and cyst formation.

Radiographically, a simple (stage 1 or 2) lipoma presents as a well-defined, lucent lesion with, perhaps, slight expansion of bone. Calcific and even ossific foci can be found within the mass. Stage 3 lipomas have a more complex appearance and resemble other calcifying lesions. An enchondroma is most easily confused with a stage 3 lipoma. Both are slow growing, calcified, and slightly expansible. Fibrous dysplasia, bone infarct, and osteoblastoma all share radiographic features with an ossifying lipoma.

In this case, there is no pathologic proof, as this was an incidental, asymptomatic discovery in an elderly woman. The radiographic and MR data were felt to be sufficiently diagnostic.

In most elderly individuals, this lesion probably would not have been seen on MR. The proximal femur is a site of red marrow and hematopoiesis. In young and middle-aged adults, the red marrow is an intermediate to dark geographic region, sparing the femoral epiphysis greater trochanters. As one gets older, the red marrow of the proximal femur is replaced by fat; this replacement is felt to occur secondary to alteration in local temperature and blood supply. Only 12% of normal patients older than 50 years have persistent hematopoietic marrow in the proximal femur as opposed to 95% of patients younger than 50. Therefore, if this patient had been like the other 88% of elderly individuals with fatty replacement in the proximal femur, the lipoma might not have been detectable on MR (Fig. 23C).

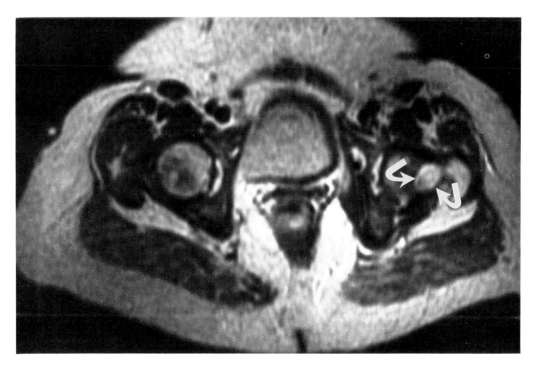

FIG. 23C. Intraosseous lipoma (axial MR, TR/TE 2,000/80 msec). The lesion (*arrows*) remains virtually isointense with the fatty marrow of the greater trochanter.

CASE 24

A 4-year-old boy presented with right hip pain and a limp. He had a low grade fever but was otherwise well. Are there any abnormalities in this study? What other MR examination might be valuable in the work-up of this patient? (Fig. 24A. Coronal MR, TR/TE 800/20 msec. Fig. 24B. Axial MR, TR/TE 2,000/80 msec.)

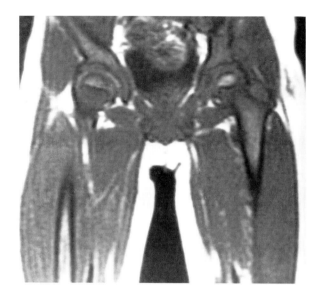

FIG. 24A.

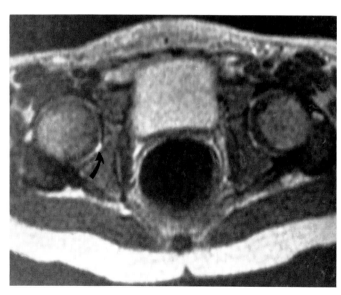

FIG. 24B.

Findings: The images of the right hip were normal (Fig. 24A). The femoral capital epiphysis is bright (the slight asymmetry is secondary to partial volume effect), and the marrow pattern is symmetric. The axial images (Fig. 24B) reveal a tiny amount of fluid in the right hip (arrow), but this is probably normal. The highest axial images (not shown) at the level of the midpelvis, demonstrated slight asymmetry of the psoas shadows. These findings led to an abdominal MR, demonstrating thickening and increased signal in the right psoas muscle, and increased signal in the L4 body (Fig. 24 C,D).

Diagnosis: Diskitis, L4/5 level.

Discussion: The psoas muscle arises from the transverse processes and bodies of T11, T12, L1, and L2 vertebrae and extends caudally in the retroperitoneum. At the iliac wing, the psoas joins with the iliacus muscle. They leave the pelvis over the acetabulum, just lateral to the femoral vessels and the femoral nerve. The iliopsoas tendon then crosses the femoral head in immediate juxtaposition to the hip joint capsule; it then inserts on the lesser trochanter. Injection studies confirm that the strong psoas fascia confines fluid collections arising within it.

The unique course of the psoas through the abdomen, pelvis, and groin provides a pathway for the spread of disease from one body area to another, permitting retroperitoneal processes to spread to the thigh, buttock, and hip.

In this patient, the initial symptoms were hip pain and a limp. Even following the MR, the patient had no back signs or symptoms (Fig. 24E). The clinicians caring for the patient elected not to perform a biopsy, and the patient was treated empirically with antibiotics.

It is important to remember that pediatric patients may present with signs and symptoms of musculoskeletal disease away from the site of the disorder.

C

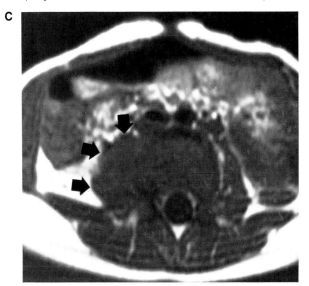

D

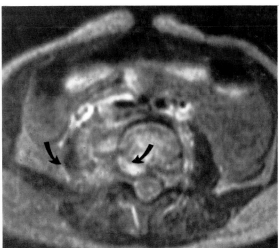

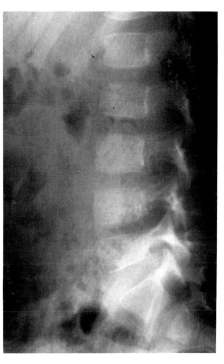

FIG. 24E. Diskitis L4/5. The plain radiograph demonstrates slight narrowing of the L4/5 disk space. The radiographic findings are quite minimal in this patient with symptoms for 2 weeks. This film was obtained after the MR suggested the diagnosis.

FIG. 24C,D. Psoas abscess secondary to L4/5 diskitis. **C.** Axial MR, TR/TE 300/15 msec. **D.** Axial MR, TR/TE 2,000/80 msec. A large mass is present in/around the right psoas muscle (*arrows*); the mass brightens on the long TR/TE sequences as does the involved L4 vertebra.

CONCLUSION

Magnetic resonance imaging of the hip joints is indicated in a variety of clinical situations. In many institutions, the most frequent indication is for evaluation of patients suspected of having AVN of the femoral heads. MRI has proven to be superior to other imaging techniques including plain films, bone scintigraphy, and CT for early detection of this disease. The so-called double-line sign is considered to be highly specific for AVN. Different patterns of SI in the area demarcated by the double line provide a rough estimate of the chronology of the infarction.

Despite the improved sensitivity and specificity of MRI, recent publications indicate that few cases of AVN may go undetected by MRI. If the patient is at high risk for AVN, the addition of bone scintigraphy to the general protocol will improve overall diagnostic sensitivity to this condition.

Tumoral, inflammatory, infectious, traumatic, and congenital or developmental conditions involving the hip can also be evaluated with MRI. The presence of a hip joint effusion can be accurately shown by MRI. Joint effusion with an associated abnormality of the acetabulum or femoral head should suggest the diagnosis of septic arthritis and prompt the clinician to joint aspiration and culture.

The use of surface coil technique combined with coronal and sagittal planes, and spin-echo T2-weighted images or fast scanning techniques, can provide dramatic depiction of the articular cartilage. This has been found to be extremely useful in the evaluation of the extent of AVN of the femoral head and in the assessment of cartilage damage in different conditions, such as Legg-Perthes disease, rheumatoid arthritis, and other inflammatory arthritides involving the hip.

The possibility of imaging the entire pelvis and hips using the body coil and large field of view provides a unique opportunity to study diffuse bone marrow disorders including bone marrow replacement, yellow to red marrow reconversion, and bone marrow edema.

The main disadvantage of MRI at present is its cost. Compared with other more invasive and less accurate diagnostic techniques, MRI is still a very expensive diagnostic test. Careful use of varied pulse sequences tailored to the specific clinical situation can help in cost reduction. The clinical implementation of fast scanning techniques in the hip will probably make MRI more competitive with other less expensive modalities.

Future developments in pelvis and hip MRI include improved software to decrease imaging time, three-dimensional Fourier transform with volumetric data acquisition, and improved surface coil design. Three-dimensional Fourier transform data acquisition is already being used in other areas of the body, such as the knee, spine, and brain. Thin true contiguous slices with improved signal-to-noise ratio, along with decreased imaging time, are the main advantages of three-dimensional Fourier transform over other MRI strategies. The possibility of performing secondary reconstructions in other planes and even surface reconstruction using the raw data once the patient has been scanned can produce significant savings in imaging time. The clinical applications of this technology are currently being developed in many institutions.

RECOMMENDED READING

Agre JC. Hamstring injuries. Proposed aetiological factors, prevention, and treatment. *Sports Med* 1985;2:21–23.

Aise AM, Martel W, Ellis JH, McCune WJ. Cervical spine involvement in rheumatoid arthritis: MR imaging. *Radiology* 1987;165:159–163.

Bassett LW, Gold RH, Reicher LR, Bennett LR, Tooke SM. Magnetic resonance imaging in the early diagnosis of ischemic necrosis of the femoral head. *Clin Orthop* 1987;214:237–248.

Beltran J, Burk JM, Herman LJ, et al. Avascular necrosis of the femoral head: early MRI detection and radiological correlation. *Mag Res Imaging* 1987;5:431–442.

Beltran J, Caudill JL, Herman LA, et al. Rheumatoid arthritis: MR imaging manifestations. *Radiology* 1987;165:153–157.

Beltran J, Herman LJ, Burk JM, et al. Femoral head avascular necrosis: MR imaging with clinical-pathologic correlation. *Radiology* 1988;166:215–220.

Beltran J, McGhee RB, Shaffer PB, et al. Experimental infections of the musculoskeletal system: evaluation with MR imaging and Tc-99m MDP and Ga-67 scintigraphy. *Radiology* 1988;167:167–172.

Beltran J, Noto AM, Herman LJ, et al. Joint effusions: MR imaging. *Radiology* 1986;158:133–137.

Beltran J, Noto AM, Herman LJ, Lubbers LM. Tendons: high-field strength, surface coil MR imaging. *Radiology* 1987;162:735–740.

Beltran J, Noto AM, McGhee RB, et al. Infections of the musculoskeletal system: high-field-strength MR imaging. *Radiology* 1987;164:449–454.

Berquist TH, Brown ML, Fitzgerald RH, May GR. Magnetic resonance imaging: application in musculoskeletal infection. *Magn Reson Imaging* 1985;3:219–230.

Bertin KC, Horstman J, Coleman SS. Isolated fracture of the lesser trochanter in adults: an initial manifestation of metastatic malignant disease. *J Bone Joint Surg* 1984;66:770–773.

Bloem JL. Transient osteoporosis of the hip: MR imaging. *Radiology* 1988;167:753–755.

Bluemm RG, Falke THM, des Plantes Z. Early Legg-Perthes disease (ischemic necrosis of the femoral head) demonstrated by magnetic resonance imaging. *Skeletal Radiol* 1985;14:95–98.

Boal DKB, Swenker EP. Infant hip: assessment with real time US. *Radiology* 1985;157:667–672.

Camp JF, Colwell CW. Core decompression of the femoral head for osteonecrosis. *J Bone Joint Surg* 1986;68A(9):1313–1319.

Coleman BG, Kressel HY, Dalinka MK, et al. Radiographically negative avascular necrosis: detection with MR imaging. *Radiology* 1988;168:525–528.

Deutsch AL, Mink JH, Waxman AD. Occult fractures of the proximal femur: MR imaging. *Radiology* 1989;170:113–116.

Dominguez R, Oh KS, Young LW, Goodman M. Acute chondrolysis complicating Legg-Calve-Perthes diseases. *Skeletal Radiol* 1987;16:377–382.

Editorial. Osteonecrosis. *J Bone Joint Surg* (A) 1986;68:1311–1312.

Feldman F. Tuberous sclerosis, neurofibromatosis, and fibrous dysplasia. In: Resnick D, Niwayama G, eds. *Diagnosis of bone and joint disorders*, 2nd ed. Philadelphia: WB Saunders, 1988;4003–4072.

Ficat RP. Idiopathic bone necrosis of the femoral head. Early diagnosis and treatment. *J Bone Joint Surg* (A) 1986;68:1313–1319.

Fletcher BD, Scoles PV, Nelson AD. Osteomyelitis in children: detection by magnetic resonance. *Radiology* 1984;150:57–60.

Fornage BD, Rifkin MD. Ultrasound examination of the tendons. *Radiol Clin North Am* 1988;26:87–107.

Genez BM, Wilson MR, Houck RW, et al. Early osteonecrosis of the femoral head: detection in high-risk patients with MR imaging. *Radiology* 1988;168:521–524.

Glickstein MF, Burk DL, Schiebler ML, et al. Avascular necrosis versus other diseases of the hip: sensitivity of MRI. *Radiology* 1988;169:213–216.

Gradwohl JR, Maillard JA. Cough induced avulsion of the lesser trochanter. *Nebr Med J* 1987;72:280–281.

Guerra J Jr, Armabuster TG, Resnick D, et al. The adult hip: an anatomic study. *Radiology* 1978;128:11–20.

Hayward I, Bjorkengren AG, Pathria MN, et al. Patterns of femoral head migration in osteoarthritis of the hip: a reappraisal with CT and pathologic correlation. *Radiology* 1988;166:857–860.

Helms CA, Hattner RS, Vogler JB. Osteoid osteoma: radionuclide diagnosis. *Radiology* 1984;151:779–784.

Hopson CN, Siverhus SW. Ischemic necrosis of the femoral head. *J Bone Joint Surg* (A) 1988;70:1048–1051.

Hungerford DS, Zizic TM. Alcoholism associated ischemic necrosis of the femoral head: early diagnosis and treatment. *Clin Orthop* 1978;130:144–153.

Ishikawa K, Kai K, Mizuta H. Avulsion of the hamstring muscles from the ischial tuberosity. A report of two cases. *Clin Orthop* 1988;232:153–155.

Kangarloo H, Dietrich RB, Taira RT, et al. MR imaging of bone marrow in children. *J Comput Assist Tomogr* 1986;10:205–209.

Kaplan SS, Stegman CJ. Transient osteoporosis of the hip. *J Bone Joint Surg* 1988;67A:490–492.

Kottal RA, Vogler JB, Matamoros A, et al. Pigmented villonodular synovitis: a report of MR imaging in two cases. *Radiology* 1987;163:551–553.

Kricum ME: Red-yellow marrow conversion: its effect on the location of some solitary bone lesions. *Skeletal Radiol* 1985;14:10–19.

Lang BA, Jergesen HE, Moseley ME, et al. Avascular necrosis of the femoral head: high-field-strength MR imaging with histologic correlation. *Radiology* 1988;169:517–524.

Lee JK, Yao L. Stress fractures: MR imaging. *Radiology* 1988;169:217–220.

Lequesne M. Transient osteoporosis of the hip: a nontraumatic variety of Sudek's atrophy. *Ann Rheum Dis* 1968;27:463–471.

Littrup PJ, Aisen AM, Braunstein EM, Martel W. Magnetic resonance imaging of femoral head development in roentgenographically normal patients. *Skeletal Radiol* 1985;14:159–163.

McKinstry CS, Steiner RE, Young AT, Jones L, Swirsky D, Aber V. Bone marrow in leukemia and aplastic anemia: MR imaging before, during and after treatment. *Radiology* 1987;162:701–707.

Mesgarzadeh M, Bonakdarpour A, Redeki PD. Case report 395. *Skeletal Radiol* 1986;15:584–588.

Mesgarzadeh M, Sapega AA, Bonakdarpour A, et al. Osteochondritis dissecans: analysis of mechanical stability with radiography, scintigraphy and MR imaging. *Radiology* 1987;165:775–780.

Milgram JW. Synovial osteochondromatosis: a histopathological study of thirty cases. *J Bone Joint Surg* (Am) 1977;59:792–801.

Milgram JW. Intraosseous lipomas: radiologic and pathologic manifestations. *Radiology* 1988;167:155–160.

Mink JH, Reicher MA, Crues JV. *Magnetic resonance imaging of the knee*. New York: Raven Press, 1987.

Mitchell DG, Joseph PM, Fallon M, et al. Chemical-shift MR imaging of the femoral head: an in-vitro study of normal hips and hips with avascular necrosis. *AJR* 1987;148:1159–1164.

Mitchell DG, Rao VM, Dalinka MK, et al. Femoral head avascular necrosis: correlation of MR imaging, radiographic staging, radionuclide imaging, and clinical findings. *Radiology* 1987;162:709–715.

Mitchell MD, Kundell HA, Steinberg ME, Kressel HY, Alavi A, Axel L. Avascular necrosis of the hip: comparison of MR, CT and scintigraphy. *AJR* 1986;147:76–81.

Modic MT, Pflanze W, Feiglin DH, Belhobek G. Magnetic resonance imaging of musculoskeletal infections. *Radiol Clin North Am* 1986;24:247–258.

Norman A, Steiner GC. Bone erosions in synovial chondromatosis. *Radiology* 1986;161:749–752.

Ogden JA. Congenital dysplasia of the hip. In: Resnick D, Niwayama G, eds. *Diagnosis of bone and joint disorders*, 1st ed. Philadelphia: WB Saunders, 1981:2452–2491.

Ogden JA. Legg-Calve-Perthes disease: detection of cartilaginous and synovial changes with MR imaging. *Radiology* 1988;167:473–476.

Olson DO, Shields AF, Scheurich CJ, Porter BA, Moss AA. Magnetic resonance imaging of the bone marrow in patients with leukemia, aplastic anemia, and lymphoma. *Invest Radiol* 1986;21:540–546.

Penkava RR. Iliopsoas bursitis demonstrated by computed tomography. *AJR* 1980;135:175–176.

Peters JC, Coleman BG, Turner ML, et al. CT evaluation of enlarged iliopsoas bursa. *AJR* 1980;135:392–394.

Phillips CD, Pope TL, Jones JE, et al. Nontraumatic avulsion of the lesser trochanter: a pathognomonic sign of metastatic disease? *Skeletal Radiol* 1988;17:106–110.

Rao VM, Fishman M, Mitchell DG, et al. Painful sickle cell crisis: bone marrow patterns observed with MR imaging. *Radiology* 1986;161:211–215.

Resnick D, Niwayama G. Degenerative diseases of extraspinal location. In: Resnick D, Niwayama G, eds. *Diagnosis of bone and joint disorders*, 2nd ed. Philadelphia: WB Saunders, 1988:1364–1479.

Rosenblum B, Overby C, Levine M, Handler M, Sprecher S. Monostotic fibrous dysplasia of the thoracic spine. *Spine* 1987;12:939–942.

Rush BH, Bramson RT, Ogden JA. Legg-Calve-Perthes disease: detection of cartilaginous and synovial changes with MR imaging. *Radiology* 1988;167:473–476.

Schuman WP, Castagno AA, Baron RL, Richardson ML. MR imaging of avascular necrosis of the femoral head: value of small-field-of-view sagittal surface-coil images.

Springfield DS, Enneking WJ. Surgery for aseptic necrosis of the femoral head. *Clin Orthop* 1978;130:175–185.

Spritzer CE, Dalinka MK, Kressel HY. Magnetic resonance imaging of pigmented villonodular synovitis: a report of two cases. *Skeletal Radiol* 1987;16:316–319.

Stafford SA, Rosenthal DI, Gebhardt MC, Brady TJ, Scott JA. MRI in stress fracture. *AJR* 1986;147:553–556.

Steinbach LS, Schneider R, Goldman AB, Kazam E, Ranawat CS, Ghelman B. Bursae and abscess cavities communicating with the hip. Diagnosis using arthrography and CT. *Radiology* 1985;156:303–307.

Sugioka Y. Transtrochanteric anterior rotational osteotomy of the femoral head in the treatment of osteonecrosis affecting the hip: a new osteotomy operation. *Clin Orthop* 1978;130:191–202.

Sugioka Y. Transtrochanteric rotational osteotomy in the treatment of idiopathic and steroid induced femoral head necrosis, Perthes disease, slipped capital femoral epiphysis, and osteoarthritis of the hip. *Clin Orthop* 1984;184:12–23.

Sundaram M, McGuire MH, Fletcher J, et al. Magnetic resonance imaging of lesions of synovial origin. *Skeletal Radiol* 1986;15:110–116.

Sundaram M, McGuire MH, Schajowicz F. Soft-tissue masses: histologic basis for decreased signal (short T2) on T2-weighted MR images. *AJR* 1987;148:1247–1250.

Sweeney JP, Helms CA, Minagi H, et al. Widened teardrop distance: a plain film indicator of hip joint effusion in adults. *AJR* 1987;149:117–119.

Tang JSH, Gold RH, Bassett LW, Seeger LL. Musculoskeletal infection of the extremities: evaluation with MR imaging. *Radiology* 1988;166:205–210.

Tehranzadeh J. The spectrum of avulsion and avulsion-like injuries of the musculoskeletal system. *Radiographics* 1987;7:945–974.

Totty WG, Murphy WA, Ganz WI, Kumar B, Daum WJ, Siegel BA. Magnetic resonance imaging of the normal and ischemic femoral head. *AJR* 1984;143:1273–1280.

Vogler JB, Murphy WA. Bone marrow imaging. *Radiology* 1988;168:679–693.

Wang GJ, Dughman SS, Reger SI, et al. The effect of core decompression on femoral head blood flow in steroid-induced avascular necrosis of the femoral head. *J Bone Joint Surg* 1985;67A(1):121–124.

Weiss KL, Beltran J, Lubbers LM. High-field MR surface coil imaging of the hand and wrist. II. Pathologic correlations and clinical relevance. *Radiology* 1986;160:147–152.

Wilson AJ, Murphy WA, Hardy DC, Totty WG. Transient osteoporosis: transient bone marrow edema? *Radiology* 1988;167:757–760.

Yao L, Lee JK. Occult intraosseous fracture: detection with MR imaging. *Radiology* 1988;167:749–751.

Yulish BS, Lieberman JM, Newman AJ, et al. Juvenile rheumatoid arthritis: assessment with MR imaging. *Radiology* 1987;165:149–152.

CHAPTER 6

The Knee

Jerrold H. Mink and Andrew L. Deutsch

There has been an explosion of interest in and use of MR for the assessment of musculoskeletal disorders. In many centers, MR has replaced knee arthrography. The advantages of MR include its noninvasive nature, lack of known side effects or ionizing radiation, ability to image in any plane, and lack of operator dependence. Magnetic resonance is additionally capable of diagnosing and evaluating periarticular (e.g., neoplasms, collateral ligaments) and subchondral abnormalities (e.g., stress fracture, osteonecrosis), lesions that are not accessible to the arthroscopist. Because the pre–MRI diagnosis is often not known or incorrect, it is not always possible to tailor an MR examination to a specific clinical problem. We have worked to develop a knee imaging protocol that would permit rapid, complete, and accurate assessment of virtually any knee disorder.

Regardless of the MR system used, surface coils, preferably those with send and receive functions, are mandatory to achieve images with high signal-to-noise ratio. The images must be acquired in an acceptable period of time to reduce patient motion and improve throughput so that costs can be contained. The standard examination of the knee is performed with the patient supine and the affected knee in 10 to 20° of external rotation and full extension, a position that optimizes visualization of the anterior cruciate ligament (ACL). It is essential that both coronal and sagittal sequences be performed on each patient. The coronal series is performed first. The repetition time (TR) is 600 to 800 msec, depending on the number of slices needed to cover the knee; the echo time (TE) is 20 msec; and the image sections are 5 mm, with no interslice gap. A 16-cm field of view (FOV) and a 128 × 256 matrix acquisition provides spatial resolution of 1.25 mm in the phase-encoded direction and 0.6 mm in the frequency-encoded direction. The sagittal sequence is prescribed from the coronal. Again, a 16-cm FOV and 5-mm section thickness with no interslice gap is selected. A spin-echo multiecho sequence (TR/TE 2,000/20, 80 msec) provides intermediate and mildly T2-weighted images. This protocol can be completed in 7 min of imaging time; we allow 25 min of "room time" for each patient, but the majority of patients are examined in 15 min.

In an effort to save time, a variety of shorter protocols have been proposed. Because the sagittal is the "primary" imaging plane, some have suggested eliminating the coronal sequence. This would represent a savings of 1½ min of imaging time. We have, however, found the coronal plane to be of value in a number of important instances: (a) Bucket handle tears may be oriented such that the peripheral rim has a triangular, albeit small, configuration, and the significance of the lesion is not appreciated. The "handle" is readily identified in the coronal plane in the intercondylar area, but partial volume effect of the dark fragment with the dark cortical bone of the vertical edge of the femoral condyle in the sagittal plane can cause confusion. (b) In the sagittal plane, the proximal extent of the ACL and its site of attachment to the femur volume average and often produce a "mass," suggesting an ACL tear. The coronal plane eliminates such partial volume effects. (c) The free edge of the midzone of either meniscus may not be optimally assessed in the sagittal plane because visualization of the edge depends on its relation to the 5-mm section thickness. The coronal plane allows unimpeded visualization of small edge tears and degeneration.

Performing a short TR/TE sequence rather than a multiecho long TR/TE sagittal sequence would save approximately $2\frac{1}{2}$ min of imaging time, but we believe considerable diagnostic information would be lost. Our own data suggest that the sensitivity of MR for assessment of the ACL is approximately 93% if T1-weighted sequences are used; accuracy is 91%. By utilizing long TR/TE sequences, the sensitivity improves to 100%, and the accuracy improves to 95%. Detection of osteochondral fractures, loose bodies, menisco-capsular separations, and extra-articular fluid collections (meniscal cysts and ganglia) is facilitated on images in which the synovial fluid is bright.

Axial images are not routinely utilized but are reserved for detailed examination of the patellofemoral joint, at which time dynamic imaging is usually performed. Although chondromalacia is best assessed in the axial plane, our referral base requests MR to determine if there is *another*, surgical cause for the patient's anterior knee pain; chondromalacia is usually evident clinically before MR abnormalities, and the overwhelming majority of patients with chondromalacia are not operated. Evaluation of bony and soft tissue neoplasms and infections and patellar dislocation with osteochondral fracture usually necessitate axial imaging.

We do not routinely utilize gradient recalled images (GRASS, General Electric Medical Systems, Milwaukee, WI), which are often billed as "fast scans." Such sequences are not faster than the spin-echo protocols detailed above, and we find the decreased contrast resolution in the soft tissues and medullary bone to be unacceptable. There are two instances in which we have found GRASS sequences to be of value when combined with a standard knee exam. In patients who have a high likelihood of having articular cartilage abnormalities [e.g., postoperative knee, osteochondritis dissecans (OCD)], we have utilized GRASS scan with a three-dimensional Fourier transform. By utilizing a TR/TE of 55/15 msec and a flip angle of 80°, we have maximized differentiation of articular cartilage and synovial fluid, a distinction not usually achieved with standard GRASS (2-D) sequences. We do not concur with early published data suggesting that three-dimensional GRASS can replace spin-echo imaging.

Patients who have deep venous thrombosis of the calf veins may present with enlargement of the calf and/or pain. In suspected cases, we perform a "MR-venogram" utilizing a TR/TE of 33/13 msec and a flip angle of 45°. This same protocol is added to the standard assessment of patients with musculoskeletal neoplasms. The GRASS sequences optimize visualization of the vascular bundle and permit easier detection of vascular invasion/thrombosis.

All images from the coronal and sagittal sequences are filmed utilizing a 15-on-1 format. These images provide a global view of all the periarticular structures in a long gray-scale setting. Additionally, to provide the best assessment of meniscal abnormalities, the intermediate weighted sagittal sequence is magnified 1.5 times and the images filmed on a 9-on-1 format. The contrast settings for these "meniscal window" images are optimized for detection and assessment of intrameniscal signal. Although it is true that most meniscal tears are seen or suspected on long gray-scale images, we have encountered innumerable instances in which a suspected grade 2 abnormality was in fact a definite tear on the meniscal windows. For meniscal assessment, the window widths are extremely narrow and the window level is adjusted so that slight inhomogeneities in the signalless meniscus can be seen. If the level is too dark, the meniscus will be black, and tears will be overlooked; if very light, any signal can be made to look like a tear. Many of the images in this chapter are deliberately displayed at "meniscal settings" to demonstrate an abnormality more optimally.

CASE 1

Two patients present with symptoms of internal derangement. What is your diagnosis? (Fig. 1A. Sagittal MR, TR/TE 2,000/20 msec. Fig. 1B. Sagittal MR, TR/TE 2,000/20 msec.)

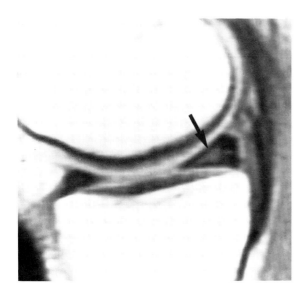

FIG. 1A.

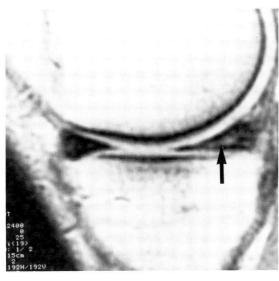

FIG. 1B.

Findings: There is a "cast-like" focus of intrameniscal signal that is nonlinear and irregular in morphology (Fig. 1A, arrow). This signal does not extend to an articular surface. In Fig. 1B a horizontal focus of linear signal is present within the posterior horn of the medial meniscus (arrow). It extends to the menisco-capsular junction but does not extend to a meniscal articular surface.

Diagnosis: Intrameniscal signal (grade 1 and grade 2). No evidence of a meniscal tear.

Discussion: Both patients in the present case demonstrate increased intrameniscal signal that does not clearly extend to an articular surface. The presence of such signal led to initial difficulty in the interpretation of MR images with regard to predicting the presence of an arthroscopically demonstrable meniscal tear. The histologic correlates of abnormal intrameniscal signal have been subject to detailed anatomic/pathologic investigation. Both "degenerative changes" as well as meniscal tears are manifest by areas of increased intrameniscal signal on MR.

In an attempt to separate minor degenerative abnormalities from arthroscopically detectable and significant tears, a grading system of intrameniscal signal based on its morphology has been developed. The most popular system divides intrameniscal signal into three principal grades. Grade 1 signal is globular and does not extend to either the superior or inferior articular surface. Histologically, menisci in which grade 1 signal is seen demonstrate foci of mucinous, hyaline, or myxoid degeneration in areas of chondrocyte deficiency. The findings are frequently seen in asymptomatic individuals and may represent a response to mechanical stress and loading with increased production of mucopolysaccharide ground substance. Grade 2 intrameniscal signal is primarily a linear signal within the meniscus that again does not extend to an articular surface. In nearly all cases, grade 2 signal arises within the substance of the posterior horn of the medial meniscus at the meniscosynovial junction; it is oriented in the midplane of the meniscus without a vertical component. Histologically these foci are characterized by more extensive bands of mucinous degeneration bordering hypocellular regions of the meniscus. Microscopic areas of collagen fragmentation may be observed, although no distinct fibrocartilagenous separation is present. Grade 2 signal abnormality represents a continuum of progressive degeneration. Patients with grade 2 signal may or may not be symptomatic, although the histologic stage has been described as a precursor to frank tears.

Grade 3 signal is defined as intrameniscal signal that unequivocally extends to an articular surface (Fig. 1C). Grade 3 has been further subdivided into grade 3A, which is linear intrameniscal signal abutting an articular margin, and grade 3B, which has irregular morphology. An articular surface is defined as the upper, lower, or free edge of the meniscus. The surface of the meniscus that is attached to the capsule is not considered an articular surface. Grade 3 signal reflecting a meniscal tear is commonly seen on more than two contiguous 5-mm sections. In small radial tears of the lateral meniscus, however, the abnormal signal may characteristically be seen on only one section along the free edge (see Case 4). Histologically, frank fibrocartilagenous separation is identified in menisci demonstrating grade 3 signal along with extensive regions of mucinous degeneration and chondrocyte death. Grade 3B lesions have been associated with more extensive degenerative change in the surrounding meniscus than the more linear grade 3A tears. Menisci containing linear grade 3A signal may appear initially near normal at gross inspection. Extensive probing at the time of arthroscopic evaluation may be required to delineate these tears.

Horizontal linear areas of increased signal that do not communicate with the articular surface are not uncommonly visualized in the menisci of pediatric patients. These areas likely correspond to perforating arteries and veins in the more highly vascularized meniscus of a child and do not have the same clinical significance as the areas of mucinous degeneration observed in the avascular meniscus of the adult. Apparent intrameniscal signal may be mimicked by partial volume effects (Fig. 1D). This is most commonly observed on the initial sagittal section that intersects the periphery of the meniscus. As a consequence of the concave capsular attachment, a peripheral section through the meniscus could partial volume average extracapsular structures, which would appear to project within the meniscus. This should not prove particularly difficult to differentiate from a meniscal tear, and coronal images will be within normal limits. Intrameniscal signal may also be seen at the central ligamentous meniscal attachments. Signal in the "meniscal horns" adjacent to the intercondylar notch should be critically evaluated and not be interpreted as indicative of a meniscal tear unless clearly depicted on more than one section. The signal relates to the junction of the meniscus with its central attachment. Isolated tears of the meniscus in these sites are highly unusual.

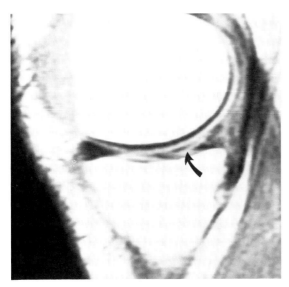

FIG. 1C. Grade 3 signal (sagittal TR/TE 2,000/20 msec). An oblique line of high signal is seen within the posterior horn of the medial meniscus. The line intercepts the tibial articular surface and represents an oblique tear (*arrow*).

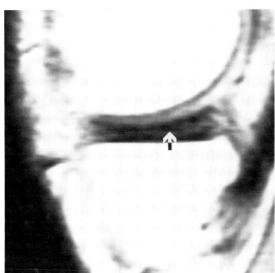

FIG. 1D. Pseudotear medial meniscus (sagittal TR/TE 2,000/20 msec). On this extreme peripheral sagittal section through the medial meniscus, an apparent line of intrameniscal signal is demonstrated (*arrow*). In actuality, this represents a partial volume effect related to the concave capsular meniscal attachment and not true intrameniscal signal.

CASE 2

A 26-year-old man presented to his physician with mild knee joint discomfort. To what structure do the arrows point? (Fig. 2A. Coronal MR, TR/TE 800/20 msec.)

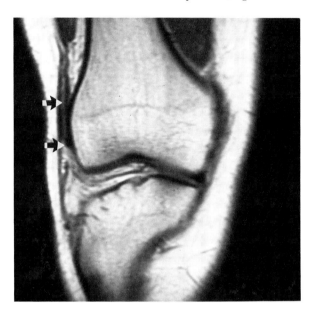

FIG. 2A.

Findings: The arrows outline a sharply circumscribed, elongated soft tissue density with signal characteristics suggestive of fluid. Just lateral to it is the iliotibial band, extending parallel to the femur from the buttock to the anterolateral tibia.

Diagnosis: Small joint effusion in the lateral recess of the suprapatellar bursa.

Discussion: The patient undergoing MR is supine and the leg is fully extended. Fluid is, therefore, distributed differently than it is during conventional radiography when the patient is in a lateral decubitus position. The distribution of fluid during MR is determined by the degree of knee flexion, the degree of external/internal rotation, and the amount of fluid present.

Fluid normally assumes the most dependent portion of the knee joint that is available. When the knee is slightly flexed, the posterior capsule and the gastrocnemii muscles relax and fluid accumulates behind the femoral condyles and the posterior cruciate ligament (PCL) (Fig. 2B). If, however, the knee is in full extension, the gastrocnemii displace the fluid into the intercondylar region and ultimately into the suprapatellar bursa (Fig. 2C).

The suprapatellar bursa has an inverted-U appearance, with the medial and lateral reflections actually being more dependent than the antefemoral component (Fig. 2D). Fluid will therefore first be seen in these recesses rather than in the radiographic suprapatellar bursa. These reflections medially and laterally are best depicted on coronal rather than sagittal images. Because the knee is usually externally rotated during MR, small amounts of fluid will preferentially fill the lateral rather than the medial recess. In this case, the distended lateral gutter is seen as a fluid-filled structure just medial to the iliotibial band.

The iliotibial band is one of the four main lateral supporting structures of the knee depicted on MRI; the other three are the popliteus tendon, the fibular collateral ligament (FCL), and the biceps femoris tendon. The iliotibial band inserts on the anterolateral tibia at a site known as the Gerdy tubercle. Occasionally, it may be torn during extreme varus forces acting at the knee.

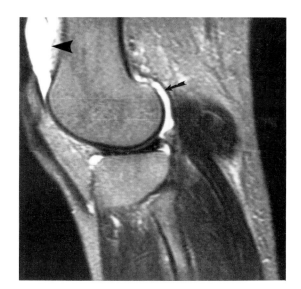

FIG. 2B. Effusion (sagittal MR, TR/TE 2,000/80 msec). The knee is slightly flexed so joint fluid accumulates behind the condyle (*arrow*). A suprapatellar effusion (*arrowhead*) is present.

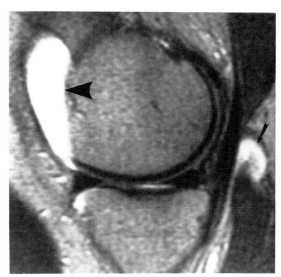

FIG. 2C. Joint effusion (sagittal MR, TR/TE 2,000/80 msec). In spite of the fact that a large joint effusion is present (*arrowhead*), there is no fluid behind the femoral condyle because the leg is in full extension. A small popliteal cyst is present (*arrow*).

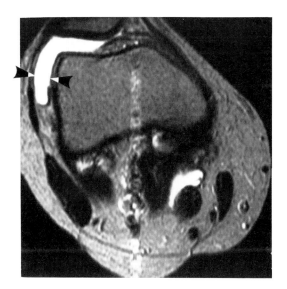

FIG. 2D. Joint effusion (axial MR, TR/TE 2,000/80 msec). The medial and lateral (*arrowheads*) reflection of the suprapatellar bursa present an inverted-U appearance and are more dependent than the prefemoral component. Because the knee is externally rotated, the lateral gutter fills preferentially.

CASE 3

A 30-year-old woman underwent an MR because of a palpable 2-cm mass in her calf. What MR findings suggest the correct diagnosis? (Fig. 3A. Coronal MR, TR/TE 800/20 msec. Fig. 3B. Axial MR, TR/TE 2,000/80 msec.)

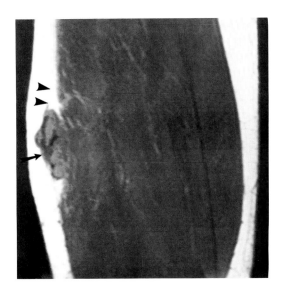

FIG. 3A.

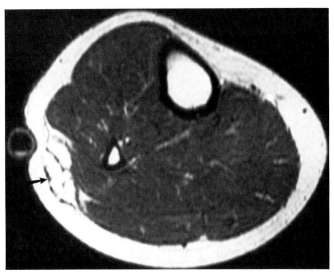

FIG. 3B.

Findings: There is a 2-cm mass (Fig. 3A, arrow) in the peroneal muscles on the lateral side of the midcalf just deep to the fascia. The fat (arrowheads) that projects just above and below the lesion was actually part of the tumor at examination. The long TR/TE sequence results in the lesion (Fig. 3B, arrow) becoming bright. A marker is present along the skin laterally.

Diagnosis: Hemangioma (intramuscular).

Discussion: Vascular malformations are among the most common congenital abnormalities in humans. Although usually benign, they may result in considerable deformity or produce life-threatening complications. A variety of terms have been applied to these lesions depending on their histology and clinical course. They have been known as angiomas, hemangiomas, arteriovenous malformations, and angiolipomas; collectively, they are best referred to as hemangiomas. Within the musculoskeletal system, they are most commonly found in the skull and thoracic vertebrae.

Long TR/TE sequences are optimal for detection and evaluation of the size and extent of hemangiomas. Within soft tissues, the lesions appear as serpiginous areas of high signal intensity when compared to the surrounding muscle and fat because of its long T2 value, reflecting the predominance of cystic spaces in the lesion. Although hemangiomas are generally homogeneous, various components of fat, vascular thrombosis, phlebolith formation, and fibrosis can lead to inhomogeneities. The appearance of hemangiomas on a T1-weighted image is variable. They may be identified by distortion of the surrounding soft tissues only, because many are isointense with skeletal muscle. However, many lesions also show an *increased* signal on T1-weighted images as well as T2-weighted images. This finding is rather unique among soft tissue tumors and is thought to be due primarily to the variable amount of adipose tissue that composes the lesion. Slow flow through large vessels and pooling of blood in dilated sinuses can also result in increased signal on short TR/TE sequences. The vascular channels are also thought to be responsible for the serpiginous character of the tumor.

A soft tissue mass in the skeletal muscle of a young individual should raise the possibility of a hemangioma. Differential considerations might include a lipoma and soft tissue sarcoma. A lipoma would have signal characteristics identical to subcutaneous fat. A sarcoma would best be distinguished by its rounded and irregular margin. Sarcomas do not have increased signal on short TR/TE sequences (Fig. 3C,D). Hemorrhage or necrosis within a sarcoma might produce signal characteristics similar to hemangiomas but the well-demarcated serpiginous pattern would not be present.

Vertebral hemangiomas also have a distinctive MR appearance (Fig. 3E,F). The intraosseous component, composed in part of adipose tissue, is bright on short TR/TE sequences. The extraosseous component of a vertebral hemangioma, however, has little fat and therefore is not bright on short TR/TE sequences. Both intra- and extraosseous components, however, have long T2 values.

C
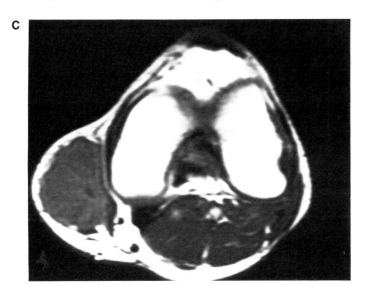
D
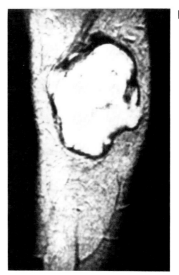

FIG. 3C,D. Fibrosarcoma. **C.** Axial MR, TR/TE 800/20 msec. **D.** Sagittal MR, TR/TE 2,000/80 msec. A soft tissue mass with irregular but rounded margins is seen in the subcutaneous fat of the leg. This sarcoma becomes inhomogeneously brighter on the long TR/TE sequence. The lesion has thick, irregular walls as opposed to the thin walls of a hemangioma.

E
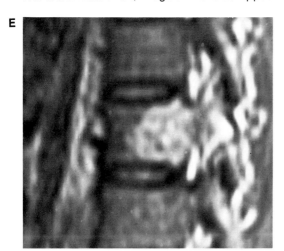
F
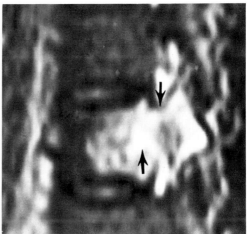

FIG. 3E,F. Vertebra hemangioma. **E.** Sagittal MR, TR/TE 700/20 msec. **F.** Sagittal MR, TR/TE 2,000/80 msec. The vertebra body is quite bright on this T1-weighted sequence. Portions of the tumor increase in signal intensity (*arrows*) while other areas change rather minimally on the long TR/TE sequence.

CASE 4

The patient is a 46-year-old woman who fell while attempting forcibly to pull open a drawer that was stuck. She experienced lateral joint line symptoms. Is the meniscus abnormal? What are the common patterns of meniscal tear that involve the lateral meniscus? How do they differ from each other? (Fig. 4A. Sagittal MR, TR/TE 2,000/20 msec. Fig. 4B. Coronal MR, TR/TE 800/20 msec.)

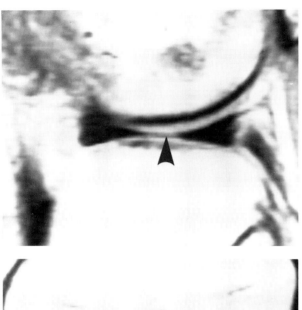

FIG. 4A.

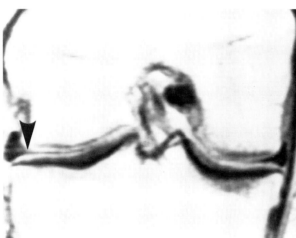

FIG. 4B.

Findings: On this sagittal image through the lateral meniscus (Fig. 4A), there is a focus of increased signal intensity seen along the free edge (arrowhead). The coronal view (Fig. 4B) confirms increased signal and blunting of the meniscal tip (arrowhead). The medial meniscus demonstrates a normal triangular contour and is without signal.

Diagnosis: Radial (parrot beak) tear lateral meniscus.

Discussion: Radial or transverse tears occur across the body of the meniscus and extend from the inner edge to the periphery. These tears may be complete or incomplete and are most common in the lateral meniscus particularly at the junction of the body and posterior horn. The occurrence of radial tears in the lateral meniscus is principally related to its more circular shape and smaller radius of curvature. External rotation of the femur on the tibia with the knee flexed displaces the posterior horn of the lateral meniscus toward the center of the joint. Forceful extension of the knee in this position of rotation does not cause a longitudinal tear (in contrast to the resultant injury to the medial meniscus) but rather straightens or elongates the lateral meniscus. This places the greatest strain along the inner concave margin of this more circular meniscus and tears it transversely. An incomplete radial tear will at times extend in an anterior and posterior direction. This creates both vertical and horizontal components in an oblique plane at the free edge of the meniscus and is commonly referred to as a "parrot beak" tear.

Radial tears can be quite small and therefore subtle in their MR appearance. As stated they typically occur at the junction of the posterior horn and body of the lateral meniscus and may be seen on only one contiguous 5-mm MR image. They could go readily undetected if not specifically looked for on sagittal images. Additionally, increased signal along the free edge related to partial volume effects must be distinguished from true tears. High resolution coronal images of the meniscus may allow better evaluation of the location and extent of these tears than sagittal images alone. Small foci of signal localized to the tip of the free edge of the lateral meniscus may only be associated with visual fraying of the free edge at arthroscopy and evidence of fibrillation of the margin on histology. It may be extremely difficult to differentiate between simple fraying of the free edge and a small parrot beak tear by MR.

Closely related to the radial tear is the oblique tear, which is a full thickness tear through the body of the meniscus without any evidence of a horizontal cleavage plane (Fig. 4C). Like the radial tear, the oblique tear begins from the inner edge of the meniscus and extends obliquely to the posterior or anterior horn. If it extends posteriorly, it is called a posterior oblique tear, which has a posterior base and an anterior apex. The tear is produced by sudden straightening of the meniscus with a consequent strain placed on the thin, concave, unattached inner edge. This strain precipitates the oblique tear, which commonly involves the medial meniscus, in distinction to the radial tear, which involves the body of the lateral meniscus. Flap tears are the same as oblique tears, but these occur in a degenerated meniscus that already contains the horizontal cleavage split. The flaps may involve either the superior or inferior half of the meniscus and are referred to as superior flap or inferior flap tears, respectively (Fig. 4D). Meniscal tears are considered complex when there are several tears, each in different places, such as a flap tear, a horizontal tear, and possibly a coexistant radial or longitudinal tear. These usually occur in the degenerated meniscus that already has a preexisting horizontal cleavage split (see Fig. 16A).

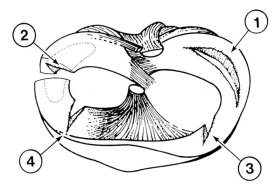

FIG. 4C. Schematic diagram of meniscal tears. This diagram depicts the menisci as they appear lying on the tibial articular surface. The four principal types of meniscal tears are illustrated. Within the posterior horn of the medial meniscus, a minimal to mildly displaced complete vertical tear is illustrated (*1*). Occupying the posterior horn and midzone of the lateral meniscus is a horizontal cleavage tear separating the meniscus into superior and inferior halves (*2*). Within the anterior horn of the medial meniscus, an oblique tear is demonstrated (*3*). A radial tear is noted, extending from the free edge to the periphery of the anterior horn of the lateral meniscus (*4*).

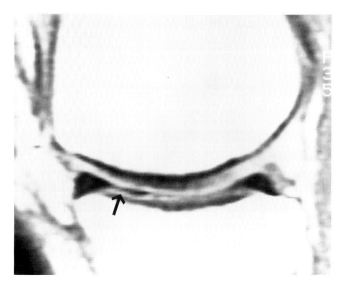

FIG. 4D. Flap tear (sagittal TR/TE 2,000/20 msec). On this sagittal section through the medial meniscus, the typical appearance of a flap tear (*arrow*) is demonstrated.

CASE 5

This 62-year-old woman had the sudden onset of knee pain when she stepped off a curb. What is the diagnosis, and what are the associated MR findings? (Fig. 5A. Sagittal MR, TR/TE 800/20 msec.)

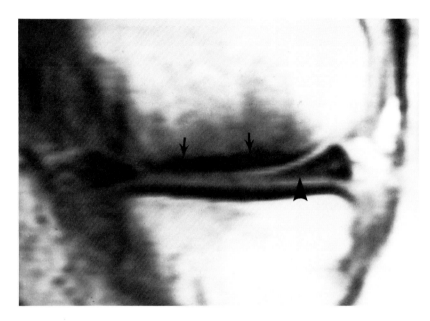

FIG. 5A.

Findings: The MR (Fig. 5A) demonstrates a crescentic area of markedly decreased signal (arrows) in the immediate subchondral bone of the femur. There is a less well-defined and much larger area of diminished signal throughout much of the medial femoral condyle, probably representing edema. A linear focus of increased signal (arrowhead) seen in the posterior horn of the medial meniscus did extend to the free meniscal edge on other images.

Diagnosis: Spontaneous osteonecrosis (SON) of the medial femoral condyle (with an associated meniscal tear).

Discussion: Spontaneous osteonecrosis is a well-recognized entity occurring in elderly individuals, which usually results in collapse of the weight-bearing surface of the medial femoral condyle. Women are affected three times more often than men. Patients report the very sudden onset of pain that they can relate to a specific event on a specific day. The acute onset of symptoms leads to confusion of SON with fracture and meniscal tear. Physical examination reveals localized tenderness of the medial joint line and an effusion. A similar clinical and radiographic syndrome has been reported in individuals with SON of the medial tibial plateau (Fig. 5B).

The cause of SON is unknown. Proposed mechanisms include an alteration in the blood supply to the condyle or repeated trauma. There has been much discussion as to whether meniscal tears, found in 40% of cases, produce osteonecrosis, follow it, or are fortuitously found. Plain radiographs are notoriously delayed in depicting osteonecrosis. Bone scans are extremely sensitive in detection, but they are nonspecific and may be abnormal secondary to meniscal injury, fracture, and osteoarthritis, conditions that may be clinically confused with SON. Particular advantages of MR include the ability to assess the condition of the overlying articular cartilage, the meniscus, and the capability of determining the presence and precise location of loose bodies.

Osteonecrosis secondary to systemic causes, such as occurs with steroid use, has an identical MR appearance but differs from SON in that the former is frequently bilateral (50%), often involves the lateral condyle (60%), and may involve multiple joints. Osteochondritis dissecans is differentiated from SON by its location (on the nonweight-bearing lateral side of the medial femoral condyle), the younger age group affected, and the insidious rather than abrupt onset of pain. Treatment of osteonecrosis includes avoidance of weight bearing early in the course; tibial osteotomy and even arthroplasty, as was ultimately necessary in this case, may be necessary after articular collapse (Fig. 5C).

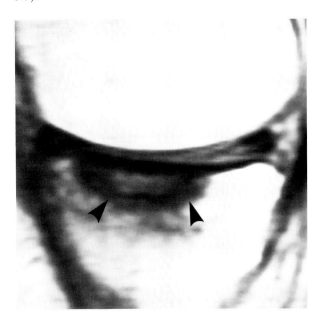

FIG. 5B. Spontaneous osteonecrosis of the tibia. Sagittal MR, TR/TE 800/20 msec. An osteonecrotic lesion of the medial tibial plateau (*arrowheads*) is present.

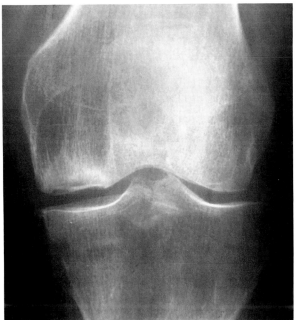

FIG. 5C. Spontaneous osteonecrosis. The medial femoral condyle has collapsed 2 months after the initial MR.

CASE 6

The patient is a 28-year-old woman. To what structures do the arrowheads in Fig. 6A point? Is the contour normal? How is the appearance of this structure influenced by changes in the degree of knee flexion? (Fig. 6A. Sagittal MR, TR/TE 2,000/20 msec.)

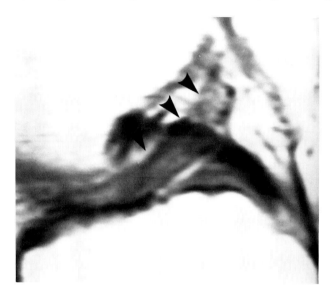

FIG. 6A.

Findings: On this sagittal section through the lateral aspect of the intercondylar notch (Fig. 6A), both the ACL and PCL are identified. The anterior contour of the ACL is moderately concave, suggesting ligamentous laxity secondary to mechanical insufficiency.

Diagnosis: "Lax" ACL.

Discussion: The ACL is an intra-articular yet extrasynovial structure. It originates from the nonarticular area anterior and lateral to the tibial spine, 23 mm from the anterior tibial cortex. At this site it is in immediate continuity to the central attachments of the anterior horn of the lateral meniscus. The tibial attachment of the ACL is wider and stronger than the femoral attachment. The ACL passes beneath the transverse meniscal ligament and courses obliquely upward and backward along the lateral aspect of the intercondylar notch to insert with a broad attachment over the posterior medial surface of the lateral femoral condyle.

The width of the ACL averages 11.1 mm, and its length averages 31 to 38 mm. The ACL is generally described as consisting of two principal fiber bundles: (a) a wider and stronger anteromedial bundle (AMB), and (b) a smaller posterolateral band (PLB). Each band is composed of a collection of individual fascicles. Although a majority of the fibers of the ACL are longitudinally oriented, some spiral from one attachment site to the other in a slight outward (lateral) direction. The spiral orientation is significant in allowing winding and unwinding of the ligament along its long axis, thus providing tension on some portion of the ligament throughout the range of motion. The ACL is noted to decrease nearly 50% in length at 90° of knee flexion.

Theoretically the ACL should become tense in extension, when the two points of attachment are maximally separated, and relaxed in flexion, when the attachment points are more closely approximated. During flexion, however, part of the potential slack is reduced by internal rotation of the tibia in relation to the femur. This allows counterclockwise twisting and subsequent tightening of the ligament. In arthroscopic assessment of the ACL in anesthetized patients, the ligament has been identified to be lax between 45 and 90° of flexion. Beyond that point, tension has been noted to increase gradually with increasing degrees of knee extension. Knowledge of this dynamic anatomy is helpful in understanding the MR appearance of the ACL as well as the mechanism of ACL injuries and the basis for clinical testing of the ligament. The ACL is most commonly injured with external rotation and abduction with the knee flexed or with hyperextension with the leg internally rotated. The normal ACL appears taut on MR because the examination is performed with the knee in extension. The degree and significance of laxity of the ACL as depicted on knee MR must be interpreted with knowledge of the degree of knee flexion, if any, with which the examination was accomplished (Fig. 20C).

On MR, the ACL most commonly appears as a solid dark continuous band of varying thickness or as two or three separate fiber bundles extending from the medial aspect of the lateral femoral condyle to the anterior tibial plateau (Fig. 6C). The identification of "individual" bundles is most often recognized on coronal scans near the femoral attachment and on sagittal studies near the tibial attachment. Visualization of the ACL on MR is optimized when the leg is positioned in mild (10–20°) external rotation. The normal ACL should be well seen on sagittal images in more than 90% of adequately positioned cases. When positioning patients for MR evaluation of the knee, it is preferrable to underrotate rather than overrotate the knee. Coronal images can be of particular value, especially in evaluating the femoral attachment of the ACL. In our experience, when there is uncertainty with regard to the integrity of the ACL as depicted on the sagittal study and the ACL as seen on the coronal images appears normal, invariably no tear has been present.

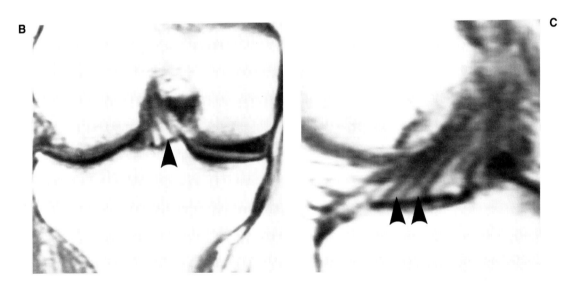

FIG. 6B,C. Anterior cruciate ligament fiber bundles. **B.** Coronal TR/TE 800/20 msec. On this coronal section individual fiber bundles of the ACL are also well demonstrated (*arrowhead*). The femoral attachment of the ACL is often well evaluated in the coronal plane. **C.** Sagittal TR/TE 2,000/20 msec. On this magnified view from a sagittal knee MR sequence, three separate fiber bundles of the ACL can be resolved (*arrowheads*). Note the straight anterior margin of this normal ACL. The examination was performed with the knee in extension.

CASE 7

A 25-year-old man suffered an injury to the knee during a football game. He felt the knee "come out of joint," but it relocated itself immediately. There was rapid swelling of the knee. Can one comment on the nature of the joint effusion, and what are its implications? To what does the arrow in Fig. 7A point? (Fig. 7A. Sagittal MR, TR/TE 2,000/20 msec. Fig. 7B. Sagittal MR, TR/TE 2,000/80 msec.)

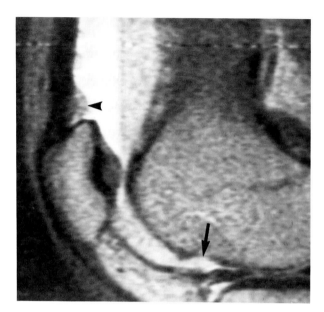

FIG. 7A.

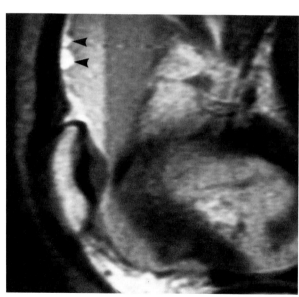

FIG. 7B.

Findings: The intermediate echo image demonstrates a large joint effusion in the suprapatellar bursa. A fluid/fluid level is present in the center of the bursa. Several small globules, bright on the shorter TR/TE sequence and darker on the longer, are floating on top of the fluid. An osteochondral defect is seen in the anterolateral femoral condyle (arrow).

Diagnosis: Lipohemarthrosis secondary to patellar dislocation and an osteochondral fracture.

Discussion: The composition of a joint effusion may occasionally be apparent on MR. Simple synovial effusions are homogeneous and approximately isointense with muscle on short TR/TE sequences. The amount of protein in joint fluid is generally 1% to 4% weight/volume, an amount insufficient to shorten the T1 value of the fluid enough to increase its signal intensity on short TR/TE sequences. High protein concentrations may rarely be encountered in inflammatory disorders or very rarely in posttraumatic disorders. Therefore, long TR/TE sequences on which effusions become bright are generally necessary to allow one to use simple effusions as a biologic contrast agent.

Almost without exception, an effusion occurring in the first 4 to 6 hr following trauma is a hemarthrosis. Such effusions generally indicate the presence of an ACL tear (72%), osteochondral fracture, or peripheral meniscal tear. Nontraumatic lesions such as pigmented villonodular synovitis, intra-articular tumors, and hemophilia may also produce a hemarthrosis. On MR, grossly bloody effusions are certainly bright on long TR/TE sequences. If the patient is supine for a sufficient length of time before the MR study, as in this case, the cellular elements of the hemarthrosis may settle so that a serum/cell level may be depicted on sagittal MR images. The cellular component remains dark on short and long TR/TE sequences, but the supernatant increases in signal on long TR studies.

Tibial plateau fractures may first be suspected by the content of a joint aspirant. The presence of fat in a joint aspiration obliges physicians to search diligently for fractures, but because fat is normally present in and around ligaments as well as in the synovium, it is possible no fracture will be found. The presence of fat and *bone spicules* in a joint effusion, however, is evidence of an intra-articular fracture.

The presence of fat in a bloody effusion creates the most complex MR appearance. Because fat is less dense than joint fluid, it floats on the effusion (Fig. 7C). On MR, the fat appears as small globules floating on the joint fluid. Like fat elsewhere, it is rather bright on T1-weighted images and is darker on long TR/TE sequences. In this case, all of the components of a lipohemarthrosis are depicted in this patient who dislocated his patella and suffered an osteochondral fracture of the anterolateral femoral condyle.

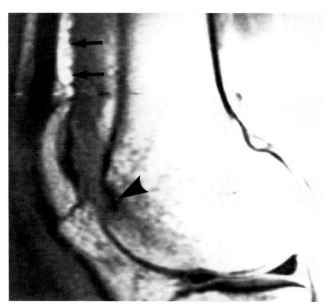

FIG. 7C. Lipohemarthrosis (sagittal MR, TR/TE 800/20 msec). This young man fell on his flexed knee and had the rapid onset of an effusion. The orthopedist tapped the fluid and found it to be bloody but without fat. The MR demonstrates fat globules (*arrows*) floating on the effusion (no fluid/fluid level is present). An impacted fracture of the anterolateral femoral condyle is seen (*arrowhead*), with a surrounding zone of edema.

CASE 8

A 27-year-old professional basketball player experienced sudden knee pain while decelerating. He was unable to extend his leg fully on examination. What is your diagnosis? (Fig. 8A. Sagittal MR, TR/TE 2,000/20 msec. Fig. 8B. Sagittal MR, TR/TE 2,000/80 msec.)

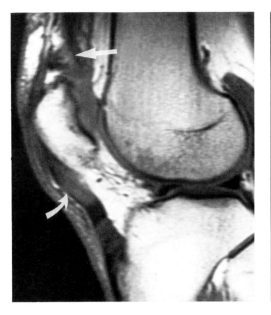

FIG. 8A.

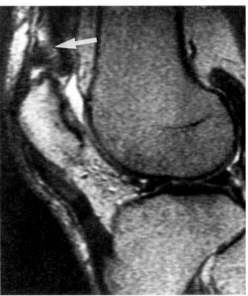

FIG. 8B.

Findings: On the proton-density image (Fig. 8A), there is evidence of complete disruption of the quadriceps tendon, starting at the level of the upper pole of the patella and extending cephalad (straight arrow). The T2-weighted image (Fig. 8B) demonstrates extensive increased signal representing edema and interstitial hemorrhage (arrow). There is an abnormal tilt of the patella. The patellar tendon demonstrates a well-defined ovoid area of intratendonous signal, which may relate to a partial tear or prior injury (Fig. 8A, curved arrow).

Diagnosis: Quadriceps tendon tear.

Discussion: The quadriceps muscle serves as the principal extensor of the knee and is composed of four discrete parts. The rectus femoris arises from the anterior inferior iliac spine, crosses the hip joint, and influences both knee and hip function. The three vastii, named medialis, intermedius, and lateralis, arise from the shaft of the femur. All four muscles converge to form a common tendon that crosses the knee joint and attaches to the tibial tuberosity via the patella. The quadriceps tendon is composed of three lamina, which can normally be resolved on MR scans (see Fig. 50B). The superficial layer is contributed by the rectus femoris, the middle layer from the tendons of the vastus lateralis and medialis, and the deep layer from the vastus intermedius. The tendon inserts into the patella with a central component that passes anterior to the patella and contributes to the patellar tendon. The patellar tendon extends from the inferior border of the patella to the tibial tubercle. The patella tendon should be approximately equal in length to the patella. Patella alta (high patella position) and patella baja (low patella position) are readily demonstrated on sagittal sections (Fig. 8C–E).

Extensor mechanism injuries are seen both in deconditioned (often elderly) individuals and in highly trained athletes. Tears of the mechanism may occur from either direct or indirect trauma. The disruption may occur at any one of multiple sites including (a) muscle-tendon junction, (b) intratendenous (quadriceps or patella), (c) tendo-osseous junction (supra- or infrapatellar), and (d) tendo-tubercle insertion. Tears of the quadriceps mechanism may be partial or complete. A number of medical conditions are known to predispose to tendon rupture including hyperparathyroidism, chronic renal disease, gout, systemic lupus erythematosus (SLE), and diabetes mellitus. In elderly individuals, the tear most commonly involves the quadriceps tendon approximately 2 cm superior to the upper pole of the patella. The tear characteristically begins within the central anterior aspect of the tendon and then extends both medially and laterally with increased force. A typical history is that of an elderly individual who, while descending stairs or jumping down, sustains severe anterior knee pain, commonly resulting in a fall. In the athlete, quadriceps tears often occur with running, with the lead foot on the ground and the individual decelerating. This results in the full body weight of the individual adding to the force of the foot strike with the resultant stress causing tendon rupture. Patellar tendon tears in distinction to those involving the quadriceps are more often identified in younger individuals. Avulsion in these younger patients may occur at either the tendo-tubercle or infrapatellar insertion. The spectrum of quadriceps tendon injuries are well depicted by MR (Fig. 8D,E). Complete disruption is characterized by tendon discontinuity that is readily demonstrable. In complete quadriceps tears there may be associated redundancy of the patellar tendon. Hemorrhage and edema are present within and surrounding the tendon. Partial tears are characterized by focal areas of augmented signal intensity often accompanied by focal tendon thickening (Fig. 8D). No complete gap is identified, and this finding is utilized to differentiate partial from full thickness tears. Axial images are often helpful in precisely defining the muscle groups involved and in accomplishing the distinction between complete tears with separation from partial tears with atrophy (Fig. 8E). A retracted muscle bundle may typically demonstrate higher signal intensity than the remainder of the muscle. Chronic quadriceps injuries may be associated with marked muscle atrophy and fatty infiltration (see Chapter 9, Fig. 3C,D).

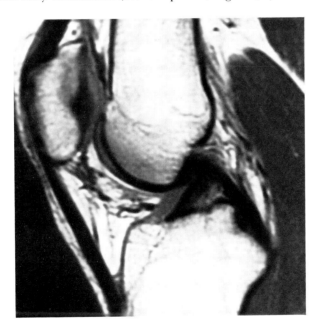

FIG. 8C. Patella alta (sagittal TR/TE 2,000/20 msec). The patella is high in position with increase in length of the patella tendon to patella ratio. Patella alta is associated with an increased incidence of patella tracking abnormalities and chondromalacia.

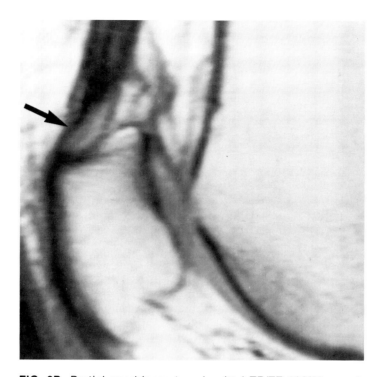

FIG. 8D. Partial quadriceps tear (sagittal TR/TE 800/20 msec). Magnified view demonstrates a focus of increased signal intensity within the distal quadriceps tendon (*arrow*). No complete gap is present, and the findings are compatible with a partial tear.

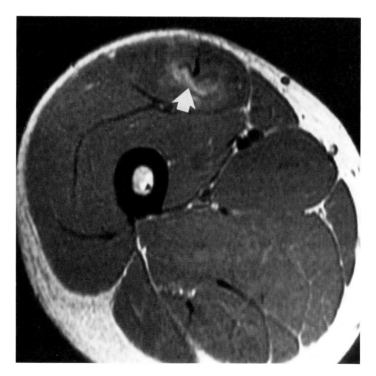

FIG. 8E. Partial tear rectus femoris (axial TR/TE 2,000/20 msec). There is a stellate focus of increased signal intensity corresponding to a partial tear of the rectus femoris muscle proximal to the origin of the quadriceps tendon (*arrow*).

CASE 9

A 75-year-old man presented to his physician with a 2-week history of pain in the right leg and enlargement of the right thigh. Plain radiographs (Fig. 9A) were made before an MRI (Fig. 9B,C). The patient then underwent an open biopsy. What are the differential diagnoses? (Fig. 9A. Plain radiograph. Fig. 9B. Coronal MR, TR/TE 800/20 msec. Fig. 9C. Axial MR, TR/TE 3,000/80 msec.)

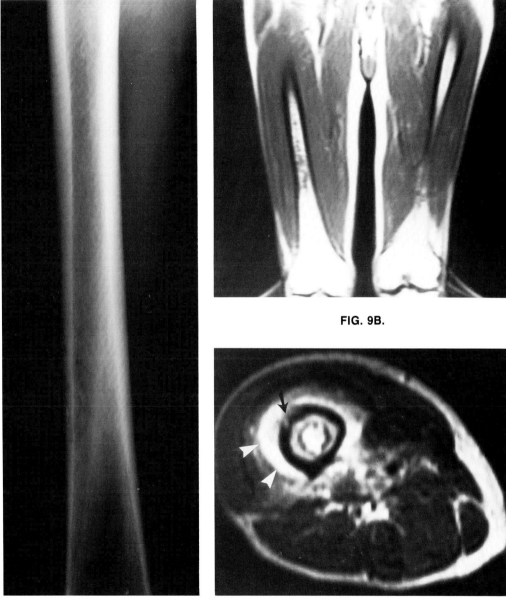

FIG. 9B.

FIG. 9A.   **FIG. 9C.**

Findings: There is a long destructive lesion of the midfemur and distal femur that involves the medullary bone over a considerable length (Fig. 9A,B). The axial image (Fig. 9C) depicts fluid in the medullary canal and soft tissues around the femoral shaft (arrowheads). A cortical defect is seen (arrow). No discrete mass is identified.

Diagnosis: Osteomyelitis of the femur (*Staphylococcus aureus*).

Discussion: The diagnosis of osteomyelitis of bone is often made from a combination of clinical studies, positive blood cultures, and compatible radiographs. In this case, the patient had no identifiable risk factors, no fever, and only a slightly elevated white count. The MRI was made to assess the lesion, which was thought possibly to be a primary sarcoma of bone.

During acute osteomyelitis, an increase in intramedullary water resulting from the inflammatory exudate, hyperemia, and edema results in an intermediate signal on T1-weighted images. The relatively long T1 and T2 of inflamed tissue contrasts sharply with the short T1 and T2 of the normal medullary fat (Figs. 9D,E). In this case, the lesion is rather poorly defined and the surrounding soft tissues are edematous, producing high signal on long TR/TE sequences. As the infection becomes more established, the bony margins become better defined, producing a thick sclerotic border, which is dark at all imaging parameters. Violation of the cortex results in an abscess continuous with the medullary origin of the infection.

In a chronic phase, sequestra and involucra can be found as band-like or linear areas of markedly decreased signal surrounded by pus. With healing and resolution of the infection, the bone marrow signal may return to normal or actually increase on T1-weighted images because of replacement of the intermediate signal hematopoietic elements by fat. A similar phenomenon has been noted in patients who have had radiation therapy to the spine for malignant disease.

There is early evidence to indicate that MR is quite efficient in the detection of bone and soft tissue infections. The ability of MR to detect bone infection is equivalent to that of radionuclide (RN) studies, but MR is more specific; the greater anatomic detail, axial plane imaging capability, and greater spatial resolution permit differentiation of osteomyelitis from soft tissue infection with a 95% efficiency for MRI versus 84% for RN. This is of great importance, especially in the diabetic foot, in which RN scans are difficult to interpret. The presence of a bone infection would, however, alter medical/surgical management. Magnetic resonance is also capable of distinguishing cellulitis from a drainable abscess (Fig. 9F,G), a distinction of considerable surgical import. Radionuclide imaging has no role in assessment of cellulitis versus abscess.

The differential diagnosis in this case must include a primary malignant tumor of bone as well as a metastasis. Sarcomata in general have well-defined intramedullary and extramedullary components. They tend to be inhomogeneous, but signal characteristics might be similar to those of infection. Plain radiographs and clinical parameters are helpful in making the distinction. Metastases to bone usually do not have a significant extraosseous component.

D

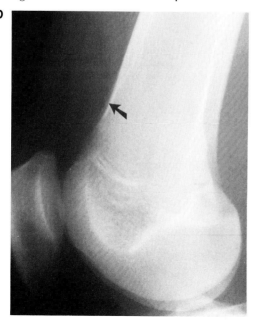

E

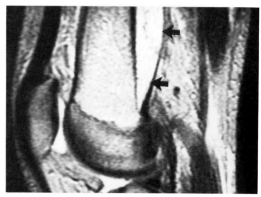

FIG. 9D,E. Osteomyelitis, femur. This plain radiograph **(D)** is normal except for a very tiny area of periosteal response (*arrow*). The MR **(D)** (sagittal TR/TE 2,000/80 msec) demonstrates that the entire marrow space of the femoral metadiaphysis is bright. Additionally, a subperiosteal fluid collection is seen along the posterior femur **(E,** *arrows*).

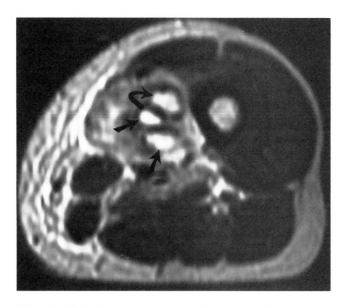

FIG. 9F. Soft tissue abscess (axial MR, TR/TE 2,500/80 msec). The signal of the vastus medialis and adductors is abnormal, and several loculations of fluid with thick walls are seen (*arrows*). This patient with AIDS had an abscess drained.

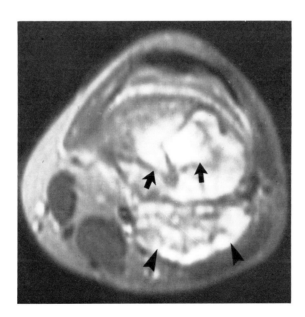

FIG. 9G. Osteosarcoma of femur (axial MR, TR/TE 2,000/80 msec). A large mass occupies the medullary canal of the femur; it has violated the cortex (*arrows*) and extended into the soft tissues (*arrowheads*). There is no significant edema.

CASE 10

A 37-year-old man sustained an acute knee injury. Physical examination revealed locking. What is your diagnosis? What should the normal relationship in size and shape be between the anterior and posterior horns of the medial meniscus? How is this injury produced and what are the new developments in treatment? (Fig. 10A. Sagittal MR, TR/TE 2,000/20 msec.)

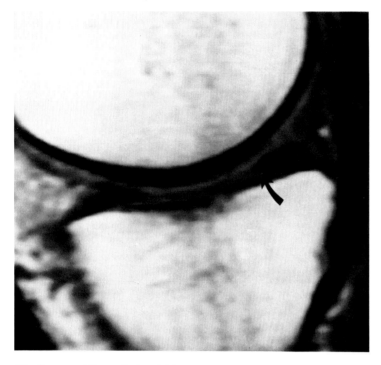

FIG. 10A.

Findings: There is focal blunting and deformation of the posterior horn of the medial meniscus (Fig. 10A, curved arrow). The meniscus is smaller in size than expected. No abnormal intrameniscal signal is present.

Diagnosis: Displaced bucket handle tear of the medial meniscus.

Discussion: There are two principal MR criteria for detection of meniscal abnormalities: (a) abnormal signal, and (b) abnormal shape. Much attention has been directed toward the grading system utilized to evaluate intrameniscal signal (see Case 1). Meniscal morphology, however, as in conventional arthrography, must also be critically evaluated. Menisci that are too big, too small, and/or asymmetric in configuration must be regarded with suspicion.

Both menisci are 3 to 5 mm in height. The average width of the anterior horn of the medial meniscus is 6 mm and that of the posterior horn is 12 mm. The lateral meniscus is C-shaped with an approximate width of 10 mm throughout. Because the middle zone of each meniscus is approximately 10 mm, when utilizing 5-mm contiguous sections, there are nearly always two MR images through the midzone. The more central of the two typically falls at or quite near the free edge. Central to the free edge, separate triangular-shaped anterior and posterior horns are demonstrated for two to three sections. On the medial side, the posterior horn is always larger in height and breadth than the anterior horn. Focal alteration of meniscal size is always abnormal, regardless of whether there is increased signal present. The situation is classically encountered with vertical longitudinal (bucket handle) tears in which the plane of separation and displacement of the "handle" occurs through the region of abnormal meniscal signal, resulting in a signalless and often triangular meniscal rim, which can appear surprisingly normal. Detection of these subtle MR findings requires appreciation that the meniscus is "too small" and/or a fragment is displaced into the intercondylar notch. Coronal views can be essential to the diagnosis of displaced meniscal tears (Fig. 10B–D).

Vertical longitudinal tears are the classic type of tears seen secondary to acute trauma in the young athlete and most commonly involve the medial meniscus. During flexion and extension of the knee, the menisci move anteriorly and posteriorly, respectively. With maximum knee flexion, the posterior horns of the menisci are compressed between the posterior aspects of the tibial and femoral condyles. Internal rotation of the femur on the tibia with the knee in flexion will force the posterior horn of the medial meniscus toward the center of the joint space. With sudden knee extension the meniscus may be momentarily entrapped, with the resulting traction causing a longitudinal tear. The more mobile lateral meniscus, as a consequence of its less extensive capsular attachment, escapes this entrapment and longitudinal tear. A second postulated theory to explain the mechanism of this tear is the combination of knee flexion with forced valgus. With this mechanism, the medial joint space is widened and the meniscus intrudes into the center of the joint. With reextension, the opposing femoral and tibial condyles grasp the now entrapped meniscus. The meniscus is crushed, creating a longitudinal tear with the inner fragment often displaced into the intercondylar notch.

If the initial longitudinal tear occurs exclusively in the posterior horn, the meniscus will return to a more normal position. With a more extensive tear that extends anteriorly beyond the collateral ligament, the meniscus commonly becomes entrapped between the two condyles and produces locking. With the most extensive tear, the inner fragment may be displaced into the center of the joint (intercondylar), and no mechanical locking will occur. These injuries are invariably accompanied by synovial effusions, which result from injury to the synovium, capsule, or ligaments. There has been increasing interest in repair of acute vertical meniscal tears and menisco-capsular separations (see Case 56). Longitudinal follow-up of a group of patients treated either conservatively or by arthroscopic repair has demonstrated persistence of abnormal signal in most, but not all, patients (Fig. 10E,F) considered clinically stable. This persistence of signal likely reflects known histologic differences in repair fibrocartilage as compared to the virgin meniscus.

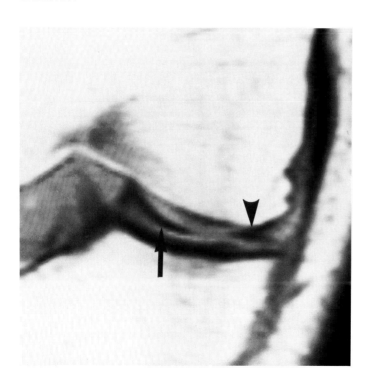

FIG. 10B. Displaced bucket handle tear (coronal TR/TE 2,000/20 msec). This section is from the coronal sequence obtained on the patient illustrated in Fig. 10A. The peripheral remnant is small and relatively without signal (*arrowhead*). A large, centrally displaced fragment is present (*arrow*). Coronal sections can be critical to accurate assessment of displaced longitudinal tears.

C

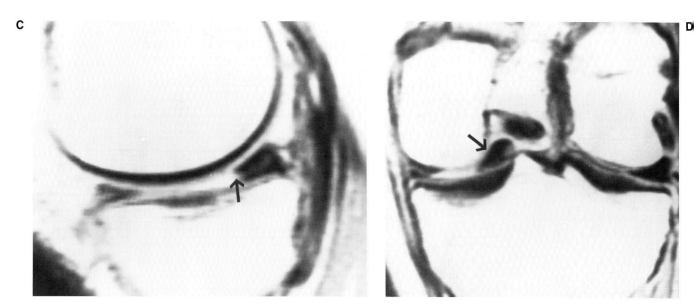

D

FIG. 10C,D. Displaced bucket handle tear. **C.** Sagittal TR/TE 2,000/20 msec. On this sagittal section through the medial meniscus there is marked blunting of the tip of the posterior horn (*arrow*). **D.** Coronal TR/TE 800/20 msec. A massively displaced fragment (*arrow*) is seen extending toward the intercondylar notch. The patient experienced intermittent knee locking.

E

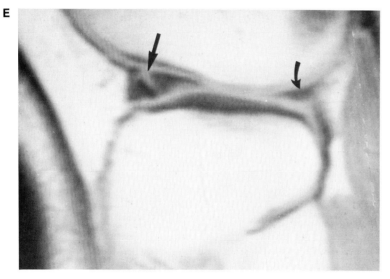

FIG. 10E,F. Meniscal repair. **E.** Sagittal TR/TE 2,000/20 msec. The posterior horn of the lateral meniscus is markedly foreshortened but without significant intrameniscal signal (*curved arrow*). The anterior horn is normal in size and configuration but contains grade 3 signal (*arrow*). A displaced vertical (bucket handle) tear was evident on other sections. **F.** Sagittal TR/TE 2,000/20 msec. The patient underwent repair of this tear with the displaced fragments apposed with suture material. A follow-up MR demonstrates a more normal size and configuration of the posterior horn. The abnormal signal within the anterior horn has largely been resolved.

F

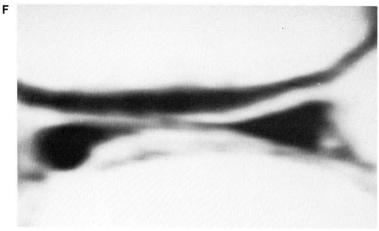

CASE 11

A 24-year-old novice skier fell forward, catching the inside edge of her left ski in the snow. She felt a sudden "pop" and experienced immediate pain. What is your diagnosis? (Fig. 11A. Sagittal MR, TR/TE 2,000/20 msec. Fig. 11B. Sagittal MR, TR/TE 2,000/80 msec.)

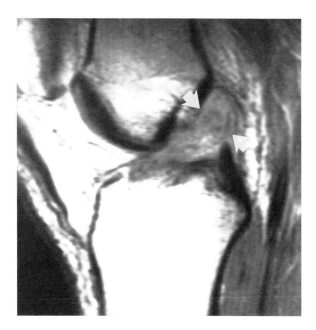

FIG. 11A.

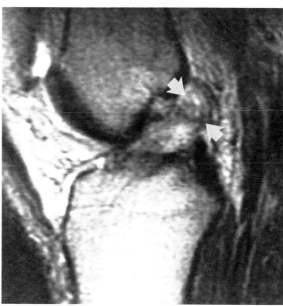

FIG. 11B.

Findings: On the proton-density image obtained through the lateral aspect of the intercondylar notch (Fig. 11A), the mid to proximal ACL is obscured by surrounding increased signal (arrows). The findings are better delineated on the T2-weighted image (Fig. 11B), where a prominent focus of increased signal intensity representing edema and hemorrhage is well seen within the substance of the torn ligament (arrows).

Diagnosis: Acute complete interstitial tear of the ACL.

Discussion: Alpine skiing has produced a virtual epidemic of ACL injuries. Changes in equipment, technique, and possibly terrain have all been implicated as contributing factors. In particular the modern high stiff boot and tight bindings with no upward release on the toe piece, while dramatically decreasing the incidence of injuries to structures below the knee, have added to the risk of injury to the ACL. The technique of utilizing the stiff ski tails to unweight through avalement ("sitting back") causes the force generated by the quadriceps contraction and stiff boot to combine to contribute to forward displacement of the tibia on the femur, often sufficient to rupture the ACL without a fall. In the present case, the patient suddenly caught the inside edge of the anterior portion of her ski, causing external rotation and valgus stress to the knee.

The ACL is one of the principal stabilizers of the knee functioning as the first line of defense against hyperextension. The ACL is also the most frequently torn knee ligament. In patients sustaining acute knee trauma sufficient to produce hemarthrosis, an accompanying ACL tear is reported in more than 70% of cases. The majority of ACL tears are initially interstitial. The rupture is most often (90%) within the main substance of the ligament and toward the proximal (femoral) side. Less commonly, the ACL is avulsed from bone: 7% from the femoral attachment and 3% from the tibial attachment. Tears with bony or chondral fractures classically demonstrate tense joint effusions with gross fat in the aspirate as compared with interstitial tears. These findings may be demonstrable with MR. Tears with attached bone usually occur in younger individuals. In these latter injuries, the displaced fragment off the tibial eminence often extrudes itself from beneath the transverse ligament. Recognition of the position of the fragment has direct management implications, as the fracture may not be anatomically reducible utilizing the arthroscope.

There are two principal mechanisms of an ACL injury. "Isolated" ACL tears result from hyperextension of the knee with the leg internally rotated. This type of tear need not be associated with contact, and it can occur with jumping injuries (basketball). The second principal mode of injury occurs with external rotation of the tibia with the knee flexed and a valgus force. These tears have a high association with concurrent bony and meniscal injury. In our review of 25 consecutive ACL tears (as judged by MR), 23 were found to have a lateral compartmental bony injury. Eighteen of the 25 patients (72%) with an ACL tear had a bone bruise (medullary trabecullar microfracture), and five (20%) had an impaction-type osteochondral fracture directly over the anterior horn of the lateral meniscus (Fig. 11C). Injuries to the medial capsulo-ligamentous complex are frequently seen in association with ACL tears. Meniscal tears are also often commonly associated with ACL tears. Many of these meniscal tears are vertically oriented and located within the posterior horn of the medial meniscus often at the

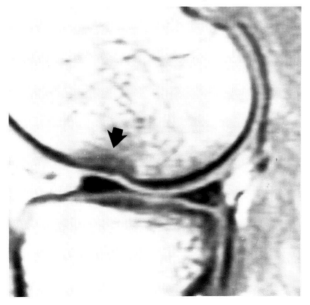

FIG. 11C. Osteochondral impaction fracture (sagittal TR/TE 2,000/20 msec). A well-circumscribed ovoid-shaped area of diminished signal intensity (*arrow*) is seen within the subchondral bone immediately underlying an impaction fracture of the articular cartilage. This site directly over the anterior horn of the lateral meniscus is characteristic of this injury when associated with ACL tears.

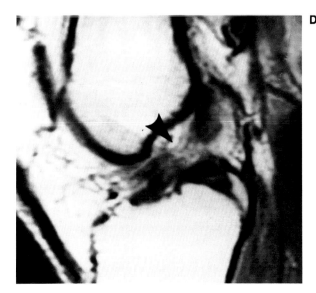

D

FIG. 11D,E. Partial ACL tear. **D.** Sagittal TR/TE 2,000/20 msec. A small "mass" of intermediate signal intensity is seen near the proximal attachment of the ACL (*arrowhead*). The section is through the lateral aspect of the intercondylar notch and does not represent partial volume with the lateral femoral condyle. **E.** Sagittal TR/TE 2,000/80 msec. This section demonstrates increased signal intensity within the mass representing edema/hemorrhage (*arrowhead*). The tendon does not appear entirely discontinuous, a finding of value in distinguishing acute partial from complete tears. A partial tear was identified at surgery.

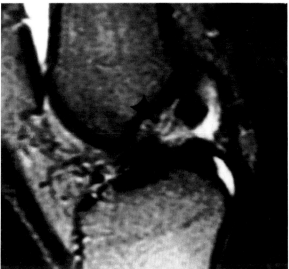

E

menisco-capsular junction. This site is known to be the most difficult for arthroscopic detection, particularly in patients with a "tight" medial compartment. In many cases the meniscal tear may have more immediate mechanical effects than the ACL tear. Injuries of the posterior capsule, particularly lateral capsular disruption, may be seen in association with ACL injuries. Plain radiographs in this setting may demonstrate a lateral tibial avulsion fracture (Segond fracture) indicative of capsular tear (see Case 22). Evidence of capsular disruption on MR is depicted by fluid extravasation from the joint into surrounding soft tissue structures.

The MR appearance of an ACL tear depends on the age of lesion and the degree of disruption. In an acute tear, the ACL is either clearly discontinuous or demonstrates a serpiginous or grossly concave anterior margin. Acute partial or complete tears may demonstrate a mass of intermediate signal on T1-weighted images, most often at the proximal end of the tendon, with or without an associated identifiable discontinuity of the tendon itself (Fig. 11D,E). The "mass" consists of fluid, hemorrhage, and acute synovitis, which often increases in signal on T2-weighted images. Rapid synovial overgrowth is characteristic of acute ACL injuries and likely relates to the blood supply of the ligament. Differentiation between a complete and partial disruption may be more difficult but generally can be accomplished. Minimal degrees of attenuation, fraying, and chronic partial tears are unreliably detected by MR and indeed, usually require extensive probing at the time of arthroscopy for detection.

CASE 12

A runner suffered a twisting injury to the knee and the rapid onset of a hemarthrosis. What is the diagnosis? What related chronic disorder may occur as a result of the same mechanism? (Fig. 12A. Sagittal MR, TR/TE 2,000/20 msec. Fig. 12B. Sagittal MR, 5 mm more medial than A, TR/TE 2,000/80 msec.)

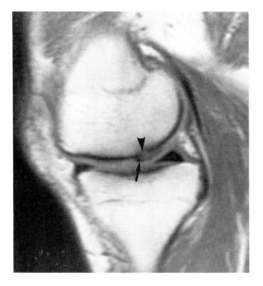

FIG. 12A.

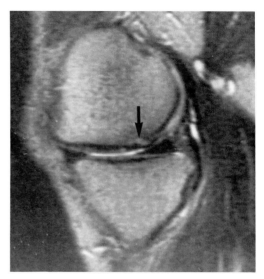

FIG. 12B.

Findings: The articular cartilage over the posterior medial femoral condyle has a band of decreased signal (Fig. 12A, arrow) within it. Additionally, the continuity of the subchondral cortical bone is violated (arrowhead). A small amount of fluid is present (arrow) in the defect in Fig. 12B.

Diagnosis: Minimally displaced osteochondral fracture of the medial femoral condyle.

Discussion: Whereas any intra-articular abnormality can result in a knee joint effusion, the rapid posttraumatic onset of swelling suggests a hemarthrosis, which is most frequently secondary to a cruciate or collateral ligament injury, osteochondral fracture, or a very peripheral meniscal tear. ACL tears are found in 72% of patients presenting with an acute posttraumatic hemarthrosis.

Osteochondral fractures may be described as impacted or displaced, with the latter more likely resulting from shear forces. The most common type of displaced osteochondral fracture occurs as a result of patellar dislocation/relocation, at which time the fracture occurs on the medial patellar facet or the anterior lateral femoral condyle (Fig. 12C). Most authors believe that osteochondritis dissecans (OCD) is a posttraumatic disorder occurring as a result of chronic impingement of the tibial spines on the lateral aspect of the medial femoral condyle.

Intra-articular fractures may involve calcified or uncalcified cartilage or subchondral bone. In adults, the fracture occurs at the "tide mark," the junction between calcified and uncalcified cartilage. Purely chondral fractures are virtually impossible to diagnose by plain radiography (Fig. 12D). Osteochondral injuries in adolescents, however, usually involve the subchondral bone. Although conventional spin-echo techniques may well reveal a chondral defect, especially when an effusion is present, a variety of gradient refocused techniques may prove better in the evaluation of articular cartilage.

In this case, a minimally displaced fracture was identified at surgery, and it was reseated and internally fixed.

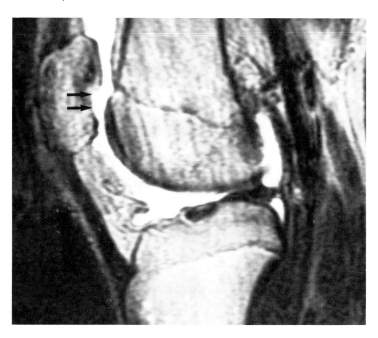

FIG. 12C. Osteochondral fracture (sagittal MR, TR/TE 2,000/80 msec). A large osteochondral fracture defect (*arrows*) is seen near the lower pole of the patella in this patient who had a recent patellar dislocation.

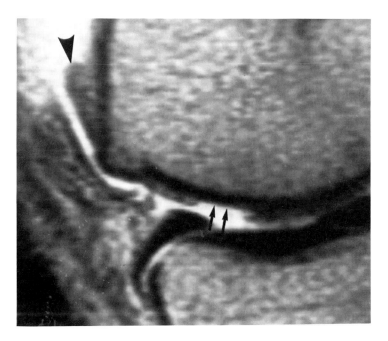

FIG. 12D. Chondral fracture (sagittal MR, TR/TE 2,000/80 msec). A chondral defect (*arrows*) is seen in the anterior medial femoral condyle, resulting in a discontinuity of the articular cartilage. The loose fragment (*arrowhead*) is seen in the suprapatellar bursa.

CASE 13

A 36-year-old man complained of pain and medial joint line tenderness. What is your diagnosis? What is the functional anatomy of the medial collateral ligament (MCL)? How are ligamentous injuries classified? (Fig 13A. Coronal MR, TR/TE 2,000/80 msec.)

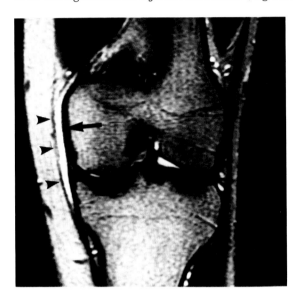

FIG. 13A.

Findings: The normal homogeneous high signal intensity subcutaneous fat is infiltrated by edema and hemorrhage. A band of high signal intensity fluid parallels the superficial aspect of the tibial collateral ligament (TCL) (Fig. 13A, arrowheads). The TCL is of normal thickness, is not displaced from the underlying cortical bone, and appears intact (arrow).

Diagnosis: Grade 1 MCL tear.

Discussion: The MCL has classically been divided into a superficial and deep portion. The superficial component, the TCL, arises from the medial femoral epicondyle and inserts onto the medial aspect of the tibia approximately 5 cm below the joint line. The TCL is composed of anterior fibers that are longitudinally oriented parallel to the long axis of the ligament and middle to posterior fibers that become more obliquely and posteriorly inclined and eventually blend with the deep capsular ligament at the posterior medial corner of the knee. The condensation of the deep capsular ligament and oblique posterior extension of the TCL is referred to as the oblique posterior ligament, which functions independently of the TCL. The TCL is separated from the medial capsular ligament and attached meniscus by a bursa that allows for movement between the two (Fig. 13B). The ligament itself is best depicted on coronal spin-echo images and is seen as a thin low signal band on both T1- and T2-weighted images. The ligament blends imperceptibly with the low signal tibial and femoral cortices. The capsular ligament attaches firmly onto the middle third of the meniscus, securing the meniscus to the femur by the menisco-femoral ligament and to the tibia by the menisco-tibial (coronary) ligament. The deep layer of the MCL can occasionally be demonstrated on high resolution coronal images as can the intervening bursa (Fig. 13B). Controversy exists over which specific structures of the medial capsulo-ligamentous complex are principally responsible for resisting abnormal medial opening and external rotation of the tibia on the femur. Although the posterior oblique ligament is thought to function as the principal restraint according to some investigators, most recent studies suggest that the TCL is the principal static medial stabilizer of the knee.

As a consequence of the normal valgus position of the knee, the medial ligamentous structures are more vulnerable to injury than those on the lateral side. Most major ligamentous injuries are caused by contact and severe valgus stress often associated with external rotation ("clipping injury"). The deep capsular ligament fibers tear initially. With increased valgus stress, the TCL and ultimately the ACL disrupts. A torn TCL implies that the capsular ligament is already torn. Peripheral meniscal tears and meniscal capsular separations may also be seen. Peripheral tears of the medial meniscus in the posteromedial corner may include a tear of the posterior oblique ligament.

Ligamentous injuries are commonly subclassified into three clinical grades: (1) disruption of only a few fibers, (2) disruption of up to 50% of the ligamentous fibers, and (3) complete disruption. The diagnosis is primarily clinical, based on history and physical examination, the latter of which often necessitates general anesthesia to accomplish because the rapid development of muscle spasm, soft tissue edema, and hemorrhage preclude optimal assessment. Radiography is principally utilized to assess for avulsion fractures indicative of capsular rupture or cruciate injury (tibial spine). Arthroscopy is utilized in many cases, although in major ligament injuries the possibility of extravasation of the arthroscopic irrigation fluid through a capsular tear often precludes performance of the examination. With arthroscopy, the assessment of collateral ligament injuries is indirect, with detection of peripheral meniscal separation, synovial disruption, and/or hemorrhage, the principal indicators of significant ligamentous damage. As optimal therapy is based on as comprehensive and accurate assessment of the extent of injury as possible, the value of MR in noninvasively evaluating the acutely traumatized knee is evident. A sprain of the TCL (clinical grade 1), as demonstrated in this case, results in edema and hemorrhage in the soft tissue, which can be well defined by coronal MR sequences. The TCL itself is essentially normal in thickness, completely continuous (intact), and not displaced from the underlying bone (Fig. 13C). No evidence of peripheral meniscal tear, menisco-capsular separation, or ACL injury was evident in this patient.

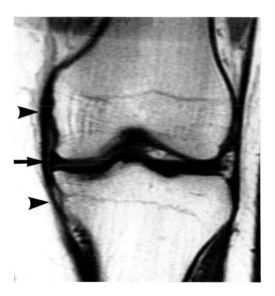

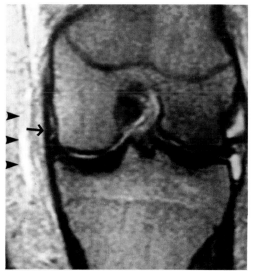

FIG. 13B. Normal tibial collateral ligament (TCL) (coronal TR/TE 800/20 msec). The TCL is seen as a well-defined low signal intensity band extending from the epicondyle of the femur to the proximal tibia (*arrowheads*). A small collection of fat (*arrow*) may separate the TCL from the medial meniscus. This should not be confused with an abnormality.

FIG. 13C. Grade 1 MCL sprain (coronal TR/TE 2,000/80 msec). On this coronal image obtained from an acute knee series, there is evidence of increased signal intensity representing edema within the soft tissues (*arrowheads*). The TCL is intact, normal in thickness, and not displaced from underlying bone (*arrow*).

CASE 14

A 55-year-old woman had the insidious onset of *left* knee pain following a *right* total hip replacement. What, if any, is the relationship between the MR image of the left knee, and the surgical procedure on the right hip? (Fig. 14A. Coronal MR, TR/TE 800/20 msec.)

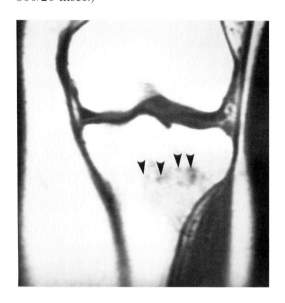

FIG. 14A.

Findings: There is a poorly defined, inhomogeneous area of decreased signal (arrowheads) in the lateral tibial metaphysis, extending to the lateral cortex. There is no linear component nor is there any evidence of violation of the cortex. The soft tissues are normal.

Diagnosis: Stress response, manifest by bone marrow edema.

Discussion: *Stress fracture* is the term applied to the failure of the skeleton to withstand repeated submaximal loads. These fractures have been classically divided into two types. Insufficiency fractures have been found in patients with decreased bone stock secondary to metabolic, inflammatory, or neoplastic processes. The stresses placed on these skeletons are not excessive, but the affected bones are unable to cope with even normal loads. Fatigue fractures, on the other hand, occur in normal bones that are subjected to stresses to which they are unaccustomed. Both types of fractures can occur in individuals who undergo surgical procedures that produce imbalanced stresses on their bones. Stress fractures may be found in the metatarsals following bunion surgery, the clavicles following radical neck dissections, and in the lower extremities following hip surgery (as in this case).

Regardless of the etiology, the stress response begins as a focal area of bone remodeling with resorption of the lamellar bone. There is a vulnerable period at which time the foci of resorption may be transformed into microfractures if the stresses persist. The fracture begins as a small crack in the cortex, ahead of which subcortical infraction and presumably edema can be identified. Appropriate modification of behavior may result in healing without the appearance of fracture, but persistent stresses during the vulnerable period lead to structural failure.

Two MR patterns of stress response have been described. The globular type, depicted above, presents as an amorphous, poorly defined area of signal loss on short TR/TE sequences (Fig. 14B). Within the lesion are smaller foci that are quite dark at all imaging parameters, although the surrounding zone may brighten on long TR/TE sequences.

The lesions are metaphyseal and adjacent to a cortical margin. In this instance, clinical improvement occurred by the time of the follow-up MR (Fig. 14C). The histopathologic correlate of the signal abnormality is unknown, but hyperemia, edema, and hemorrhage may contribute.

A classic linear stress fracture presents as a serpiginous area of decreased signal perpendicular to an adjacent cortex. In the active phase, the fracture is surrounded by a zone of (presumed) edema. Occasionally, extensive cortical, periosteal, and soft tissue involvement may result in a pattern suggestive of malignancy or infection.

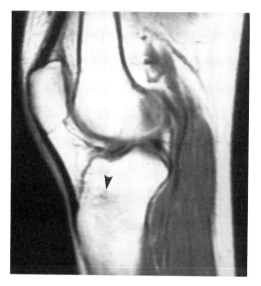

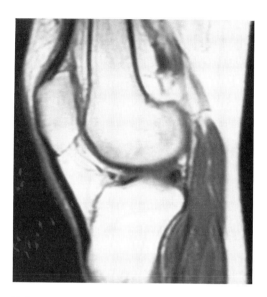

FIG. 14B. Stress response (sagittal MR, TR/TE 800/20 msec). The metaphyseal lesion (*arrowhead*) is in the anterior half of the tibia. This is the sagittal image made at the same time as Fig. 14A.

FIG. 14C. Stress response (sagittal MR, TR/TE 800/20 msec). A follow-up study at 2 months, at which time the patient was asymptomatic, reveals MR resolution of the lesion.

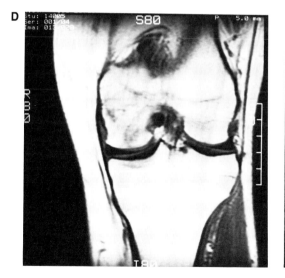

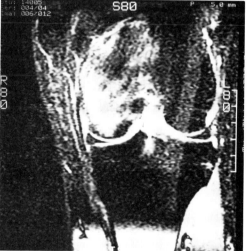

FIG. 14D,E. Transient osteoporosis. (**D.** Coronal MR, TR/TE 500/20 msec. **E.** Coronal MR, TR/TE/TI 1,500/20/160 msec.) This 45-year-old male presented to his physician with a two week history of severe knee pain without an injury. No other joints were involved. The T1 image (**D**) reveals patchy areas of decreased signal in the medial femoral condyle; the short inversion recovery image (**E**) much more graphically demonstrates the extensive edema. Within six weeks, the patient's symptoms had improved at the same time that his MR had nearly returned to normal.

CASE 15

A 45-year-old man with known lymphoma was referred for an MRI because he had an abnormal staging bone scan. He was asymptomatic. What is the most likely differential diagnosis and how will this lesion affect his therapy? (Fig. 15A. Anteroposterior radiograph of the left leg. Fig. 15B. Coronal MR, TR/TE 800/20 msec. Fig. 15C. Sagittal MR, TR/TE 2,000/80 msec.)

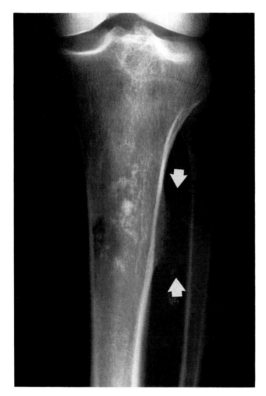

FIG. 15A.

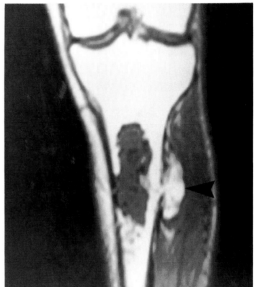

FIG. 15B.

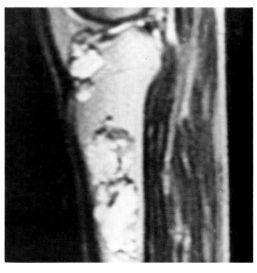

FIG. 15C.

Findings: There are multiple lobulated and confluent areas of abnormal signal extending from the tibial plateau caudally to the midtibia. The lesions uniformly become bright on the T2-weighted images. Also present is an exostosis of the proximal posterolateral tibia seen on both the plain radiograph (Fig. 15A, arrows) and the coronal MR (Fig. 15B, arrowhead).

Diagnosis: Multiple benign enchondromata and a benign osteochondroma of the tibia.

Discussion: Magnetic resonance has proven itself extremely efficient in its ability to detect a number of disorders of bone marrow on short TR/TE sequences by virtue of replacement of the normally bright medullary fat (short T1) by processes that almost universally have longer T1 values. However, the changes in the MR may be quite nonspecific; tumor, infection, and fracture may have similar characteristics, and T1 and T2 values have not proven useful in distinguishing benign from malignant disease. For the most part, *characterization* of suspected bone neoplasms has not been accomplished by MR; rather, the plain radiograph often is the most useful tool in this regard. Depiction of a fluid–fluid level within a bone lesion on MR may suggest an aneurysmal bone cyst or echinococcal disease. Angiomas are serpiginous masses that contain fat and therefore may have portions of the lesion that are bright on short TR/TE images. Simple cysts and intraosseous lipomata also have characteristic MR appearances.

In this case there are multiple lesions oriented along the long axis of the tibia. They have an MR appearance that has been described as being specific for hyaline cartilage origin tumors (chondrosarcoma, osteochondroma, and enchondroma): the matrix portion of the tumor demonstrates a homogeneous, high signal intensity (relative to muscle) in a *lobular* pattern on long TR/TE images. The high water content, trapped within the interstices of the matrix, presumably produces the very high signal. High signal may be seen in bone cysts, osteomyelitis, and a variety of primary and secondary malignant neoplasms, but none of these entities has a lobular character. It is the lobular pattern, therefore, that does appear to be unique to hyaline cartilage origin lesions. Nonhyaline cartilage origin tumors (chondroblastoma, clear cell chondrosarcoma, and synovial chondromatosis) demonstrate isointensity or even hypointensity relative to muscle on long TR/TE sequences and in general have an amorphous, as opposed to lobular, configuration. It must be emphasized that there are no MR criteria that allow one to distinguish benign from malignant cartilage reliably. The standard radiographic film signs of cortical breakthrough, soft tissue mass, etc., remain critical adjuncts to MR data.

Osteochondromata are hyaline cartilage origin tumors that may have a cartilagenous cap with enchondral ossification at their bases and a well-circumscribed periphery. The thickness of the cartilage cap as measured by CT may bear some relationship to the likelihood of benignancy or malignancy, but the accuracy of MR in assessing the cartilagenous cap has not been critically examined. In this case, the cartilagenous cap was approximately 2 mm in thickness and thus one might predict that this is a benign osteochondroma (Fig. 15D).

This patient had no symptoms referable to his leg, but because of the extent of the lesion, abnormal bone scan, and the endosteal erosion, the question of malignancy arose and the patient underwent an open biopsy with curettage of multiple lesions. The pathologic specimen was completely benign and diagnostic of enchondroma.

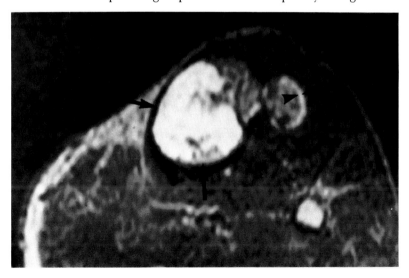

FIG. 15D. Enchondroma and osteochondroma (axial MR, TR/TE 2,000/80 msec). The medullary canal is filled with bright hyaline cartilage (*arrows*). A thin cartilaginous cap (*arrowheads*) is present over the exostosis. The medullary canal of the exostosis as well as the lateral portion of the tibia is filled with fat.

CASE 16

A 58-year-old man presents with clinical findings suggestive of internal derangement. What is your diagnosis? Is this a common type of meniscal abnormality? What is the significance of the findings? (Fig. 16A. Sagittal MR, TR/TE 2,000/20 msec.)

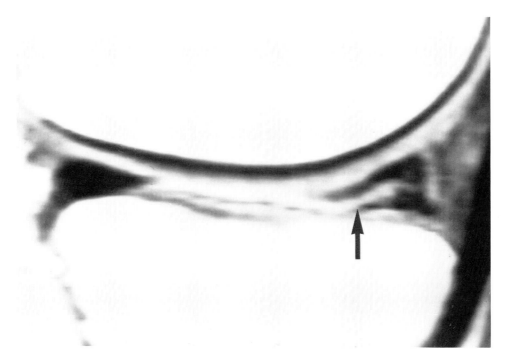

FIG. 16A.

Findings: Sagittal section through the posterior horn of the medial meniscus (Fig. 16A) demonstrates a linear focus of intrameniscal signal that courses horizontally through the posterior horn (arrow), separating the meniscus into superior and inferior halves.

Diagnosis: Horizontal (fish mouth) tear of the posterior horn of the medial meniscus.

Discussion: Horizontal tears roughly parallel the superior and inferior surfaces of the meniscus and separate the meniscus into two halves. The horizontal tear, also referred to as a cleavage tear, is one of the most commonly encountered meniscal lesions and generally occurs in older people and is considered degenerative in nature. This is in distinction to vertical tears, which are classically associated with acute trauma. The posterior half of the medial meniscus or midsegment of the lateral meniscus are the most common sites of horizontal tears. These tears commonly communicate with the margin of the meniscus at the free edge. At arthroscopy, the meniscal free edge may resemble the tips of a fish's scales, with the tear itself suggesting the appearance of the mouth of a fish, hence the common designation of fish mouth tear. Following repeated trauma, a horizontal tear may extend in an oblique manner, ultimately becoming long enough to impinge itself repeatedly in the joint. This may produce constant pain instead of mechanical symptoms.

As a consequence of age-related changes, the elasticity and resistance of the meniscus is gradually reduced, predisposing the meniscus to injury. During flexion and extension, the femur moves with the superior half of the meniscus, and the tibia moves with the inferior half, creating a mechanical shear stress within the central meniscus. Histologically, myxoid material preferentially accumulates along the "middle perforating bundle," a collagen bundle that roughly divides the meniscus into superior and inferior halves. With increasing degeneration, a rent forms parallel to the middle perforating bundle corresponding to the intrasubstance tear first described by Smillie. Central extension results in the common horizontal cleavage tear. Many of these tears detected on MR represent confined or closed intrasubstance cleavage tears. Diagnosis of these closed meniscal tears may not be possible by visual inspection alone and frequently require extensive probing at the time of surgery to allow their detection (Fig. 16B). It is important to note that a horizontal cleavage lesion is a part of the degenerative aging process and that these tears may or may not be symptomatic. Further injury to a degenerative meniscus may result in development of a flap or complex tear (Fig. 16C).

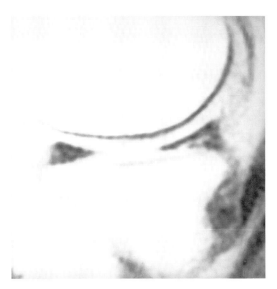

FIG. 16B. Sagittal TR/TE 2,000/20 msec. Magnified high contrast image of the medial meniscus demonstrates an oblique line of high signal intensity within the posterior horn of the medial meniscus that extends to the tibial articular surface (grade 3 signal). At arthroscopy, the surgeon reported that no abnormality was obvious on visual inspection and that based on the MR report only after extensive probing did he "fall into" the tear.

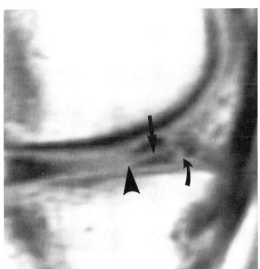

FIG. 16C. Complex tear (sagittal TR/TE 2,000/20 msec). This sagittal section demonstrates a complex tear involving the posterior horn of the medial meniscus. These tears commonly develop as extensions from preexisting horizontal tears. This tear has prominent horizontal (*arrow*) and vertical (*curved arrow*) components with associated blunting of the meniscal tip (*arrowhead*).

CASE 17

This young woman twisted her knee while skiing and had the immediate onset of swelling of her knee approximately 24 hr before this MR. What is the major MR finding, and what associated abnormality might be seen on other MR images from this study? (Fig. 17A. Sagittal MR, TR/TE 2,000/20 msec. Fig. 17B. Sagittal MR, TR/TE 2,000/80 msec.)

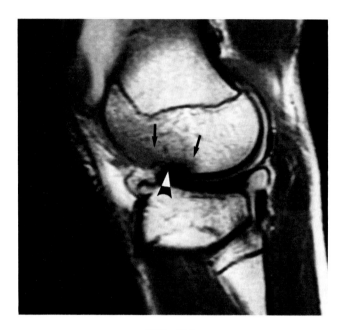

FIG. 17A.

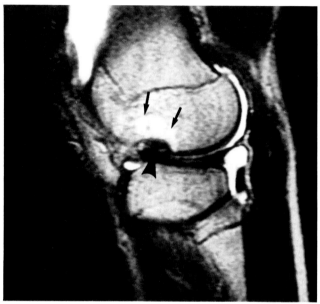

FIG. 17B.

Findings: There is a poorly defined zone of decreased signal intensity (Fig. 17A,B; arrows) in the medullary bone of the femur immediately above the anterior horn of the lateral meniscus. Additionally, the contour of the cortical bone and articular cartilage over the lesion is depressed (arrowheads). The focus of subchondral depression is dark on both images.

Diagnosis: Impacted osteochondral fracture of the lateral femoral condyle, associated with a tear of the ACL.

Discussion: There are multiple mechanisms of injury to the chondral surfaces. Dislocation/relocation of the patella is the most common and results in injury to the patella or the anterior nonweight-bearing surface of the lateral femoral condyle. Rotatory motions in which the tibial spine impacts the femur may cause injury to the medial femoral condyle; this mechanism is thought by some to be responsible for the production of OCD. Finally, a direct blow to the patella may result in forces being transmitted to the condyle with a resultant impaction injury.

The lesion described above is commonly associated with an ACL tear. In a study of 25 acute, complete ACL tears, this particular impaction fracture was found in five cases. Eighteen patients in the same group suffered lateral compartment bone bruises. Thus 23 of 25 knees with ACL injuries suffered lateral compartment bony and cartilaginous injuries. External rotation of the femur over a fixed tibia and foot, a common mechanism of ACL tears, produces compressive forces on the lateral side of the knee and is thought to be the etiology of the injury. Perhaps this lateral impaction fracture and a bone bruise may differ only in the magnitude of the force applied.

The impaction injuries occurring with ACL tears are generally not significantly depressed. Some lesions have not been identified at arthroscopy, and none have been treated. Young cartilage is quite elastic, and compressive forces in the lateral compartment may result in significant *subchondral* injury with "rebound" of the overlying elastic cartilage, therefore accounting for the lack of an arthroscopically detectable lesion.

Interestingly, the bony defect in the lateral femoral condyle in patients who have acute ACL tears occurs precisely in a normal concavity of the condyle. This concavity accepts the anterior horn of the meniscus when the knee is at full extension, although the cartilagenous and cortical contour may be slightly concave. In both the normal situation and the osteochondral fracture, the signal characteristics of the cartilage and the subchondral bone distinguish normal from abnormal (Fig. 17C,D).

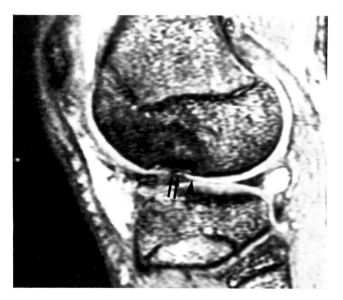

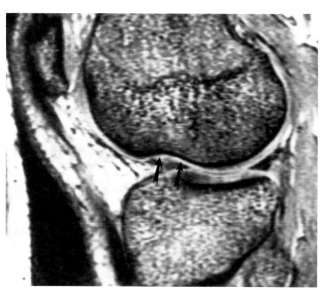

FIG. 17C. Impacted osteochondral fracture (sagittal MR, TR/TE 55/15 msec). This GRASS sequence on the same patient as in Fig. 17A graphically depicts the altered signal of the articular cartilage (*arrows*), which contrasts sharply with the normally bright cartilage (*arrowheads*). The subchondral defect, however, is not optimally seen.

FIG. 17D. Normal lateral femoral condyle (sagittal MR, TR/TE 55/15 msec). Although the contour of the femoral condyle and cartilage are concave (*arrows*), the normal signal of the cartilage and lack of a subchondral lesion indicate this is normal.

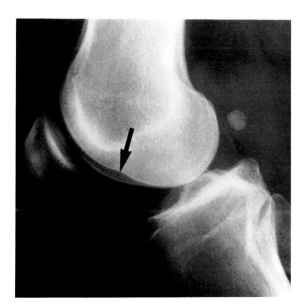

FIG. 17E. Normal lateral femoral condyle. The normal concavity of the lateral femoral condyle is seen (*arrow*). A similar concavity may be seen medially.

CASE 18

A 22-year-old ballet dancer landed off balance following a jump and felt a sudden "pop." His knee buckled but he did not fall. He could not continue with the performance. What single event can explain all the findings? (Fig. 18A. Sagittal MR, TR/TE 2,000/20 msec. Fig. 18B. Sagittal MR, TR/TE 2,000/20 msec. Fig. 18C. Coronal MR, TR/TE 2,000/80 msec. Fig. 18D. Axial MR, TR/TE 2,000/20 msec.)

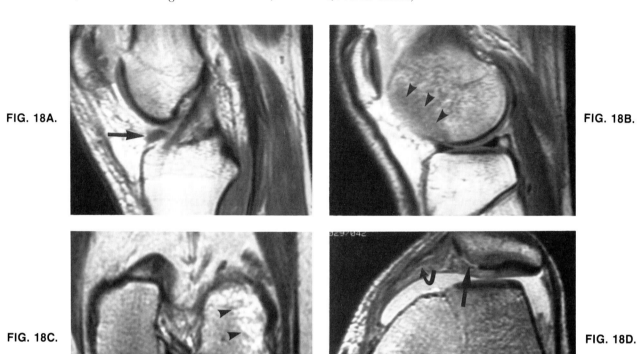

FIG. 18A.

FIG. 18B.

FIG. 18C.

FIG. 18D.

Findings: Figure 18A reveals the ACL to be entirely intact. Immediately anterior to the tibial attachment of the ACL is an ovoid focus of low signal intensity suspicious for an intra-articular osteochondral body (arrow). On a sagittal section through the lateral femoral condyle (Fig. 18B), a hemispheric zone of decreased signal intensity (arrowheads) is seen within the subchondral bone. On the T2-weighted coronal image (Fig. 18C), an extensive region of mildly increased signal intensity is seen within the lateral femoral condyle most consistent with an extensive bone bruise (arrowheads). On the axial section (Fig. 18D), an osteochondral defect is seen within the patellar articular cartilage along the median ridge and medial facet (arrow). The medial retinaculum is wavy in contour, and there is extensive surrounding edema (curved arrow).

Diagnosis: (a) Lateral dislocation of the patella with an osteochondral impaction fracture involving the medial facet of the patella and lateral femoral condyle. (b) Loose intra-articular osteocartilagenous bodies. (c) Disruption of the medial patellar retinaculum.

Discussion: Traumatic dislocation of the patella may result from a direct blow or, as in this case, secondary to severe rotary stress imposed on the weight-bearing knee. The clinical presentation may at times be difficult to differentiate from acute ACL disruption. Before the use of advanced imaging techniques and diagnostic arthroscopy, 10% of patients undergoing arthrotomy for suspected ACL tear turned out to have patellar subluxation with osteochondral injury.

Patellar dislocation is often accompanied by traumatic disruption of the medial retinacular supporting structures or by avulsion of a small medial fragment of patella at the site of retinacular insertion. The MR appearance of retinacular injuries varies with the extent of disruption. With lateral dislocation, the medial patellar ligament is torn from the intramuscular septum directly beneath the vastus medialis obliquus. Increased signal intensity, representing edema and/or hemorrhage, is seen along the medial aspect of the retinaculum on T2-weighted images (see Fig. 37D,E). A discontinuous or wavy contour to the retinaculum itself is commonly evident.

Osteochondral injuries as seen in this case are commonly associated with patellar dislocation. The medial facet of the patella and lateral femoral condyle are most commonly involved, and injury occurs at the time of spontaneous reduction. Intra-articular osteocartilagenous fragments may result and are common in the patellofemoral space as well as the lateral gutter adjacent to the popliteal tendon. Axial sections are ideal for assessment of the patellar articular cartilage defects. The curved contour of the lateral femoral condyle is often best evaluated on sagittal sections. Hemarthrosis is common and the presence of lipohemarthrosis is diagnostic of osteochondral fracture with intra-articular extension.

Acute dislocations can predispose the knee to repeated episodes. A number of structural conditions have been associated with recurrent patellar subluxation/ dislocation. As a consequence of the inclination of the femur, the quadriceps muscle does not pull in a direct line with the patella. The angle formed between the two is valgus and is known clinically as the Q angle. This angle results in a natural tendency toward lateral displacement of the patella. The medial retinacular supporting structures serve to counteract the lateral deviating vector resulting from the Q angle. An abnormality/deficiency of the vastus medialis complex can predispose to patellar instability. Osseous conditions implicated as contributing to patellar instability include decreased height of the lateral femoral condyle, shallow or dysplastic trochlear sulcus, and flattening of the patellar facets and in particular the Wyberg Type 3 patellar configuration. These conditions are well evaluated both by conventional MR as well as the recently developed kinematic technique utilizing rapid scan sequences (see Chapter 10).

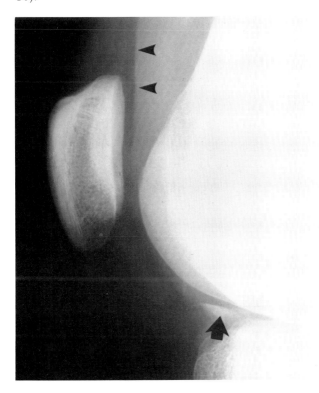

FIG. 18E. Osteochondral fracture. This plain radiograph is of the patient in Fig. 18A. The *large arrow* depicts the osteochondral patellar fracture, and the *arrowheads* show a lipohemarthrosis.

CASE 19

This long distance runner has had mild knee pain for several years. An MR was normal except for the lesion depicted by the arrow. What is the diagnosis, and what is the surgical significance? (Fig. 19A. Sagittal MR, TR/TE 2,000/80 msec. Fig. 19B. Coronal MR, TR/TE 800/20 msec.)

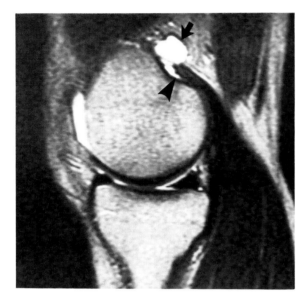

FIG. 19A.

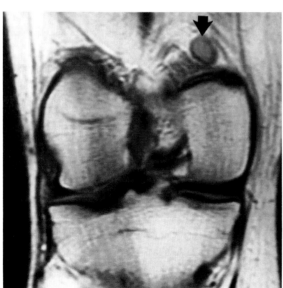

FIG. 19B.

Findings: A mass (Fig. 19A,B; arrows) is seen just above the expected insertion of the medial joint capsule into the bone. A fluid-filled stalk (Fig. 19A, arrowhead) extends under the origin of the gastrocnemius, where it presumably connects to the joint.

Diagnosis: Ganglion.

Discussion: A variety of fluid collections can occur around the knee, the most common of which is a popliteal cyst. Large dissecting medial meniscal cysts may occasionally be mistaken for popliteal cysts. Other cystic masses around the knee include angiomas, popliteal artery aneurysms, and distended anserine, prepatellar, and infrapatellar bursae.

Ganglia most commonly present as masses near a joint to which they invariably have a connection (Fig. 19C). The communication may be a patent one, as in this case, or a fibrous stalk that may be difficult to detect. Ganglia have been reported deep in the vastus lateralis and semimembranous muscles, rather far from their joint of origin. It is critical that the capsular origin of these lesions be recognized at surgery because failure to resect it will result in recurrence.

Although the distinction between a ganglion and a meniscal cyst may seem a matter of semantics, it must be remembered that a ganglion is *not* associated with a meniscal tear, and therefore an intra-articular procedure is not necessary. Definitive treatment of a meniscal cyst invariably requires a meniscectomy.

Ganglia are often septated and may have daughter cysts communicating with the main mass. On MR, ganglia are best visualized on long TR/TE sequences, which result in high signal because of its fluid nature. They may be appreciated on short TR/TE sequences by virtue of their mass effect or possibly by the bone erosion that occasionally occurs (Fig. 19D).

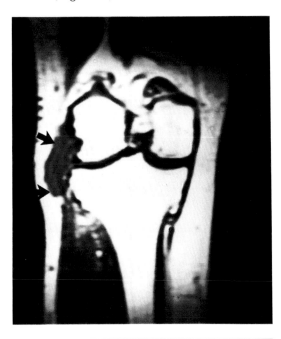

FIG. 19C. Ganglion of the knee joint (coronal MR, TR/TE 800/20 msec). An elongate ganglion is seen arising at the joint line and extending cephalad and caudad (*arrows*).

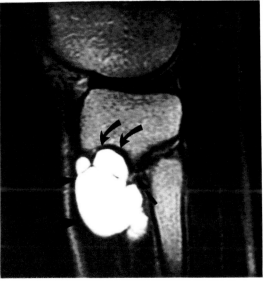

FIG. 19D. Ganglion of the tibiofibular joint (sagittal MR, TR/TE 2,000/80 msec). A large septated fluid mass (*arrowheads*) is seen just anteroinferior to the proximal tibiofibular joint, to which it appears connected by a thin stalk (*arrow*). The top of the mass has slightly eroded the tibial shaft (*curved arrows*).

CASE 20

A 32-year-old man had experienced a knee injury in a motorcycle accident several years ago. He did not seek medical attention at that time. He now presents with "popping and clicking" within his knee and feels that his knee "gives out." What is your diagnosis? Is the PCL normal? How is the contour of the PCL influenced by changes in knee flexion? (Fig. 20A. Sagittal MR, TR/TE 2,000/20 msec.)

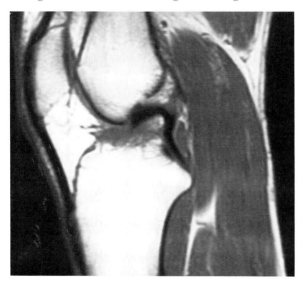

FIG. 20A.

Findings: The PCL is intact but demonstrates an abnormal contour with a mildly buckled appearance (Fig. 20A). No abnormality of the subchondral bone in the region of the tibial insertion of the PCL is identified to suggest avulsion. The ACL is not visualized.

Diagnosis: Hyperbuckled PCL secondary to chronic ACL insufficiency.

Discussion: The normal PCL is generally the most easily depicted structure within the knee utilizing MR. Originating from a fan-shaped attachment on the posterior aspect of the lateral surface of the medial femoral condyle, it courses posteriorly and distally across the joint to attach to the extreme posterior intercondylar (nonarticular) region of the tibia. Although this tibial attachment extends for several millimeters on the posterior surface of the tibia, the functional attachment of the ligament is proximal. The PCL is narrowest at its midportion, fanning out proximally on the femur and to a lesser degree on the tibia. The fascicles comprising the PCL are divided into an anterior portion, from which the majority of the ligament is comprised, and a posterior segment, which is much smaller and courses obliquely across the joint. The posterior aspect of the ligament frequently blends with the posterior joint capsule.

The PCL is longer, wider, and stronger than the ACL and is a fundamental stabilizer of the knee in flexion, extension, and internal rotation. The PCL controls the movement of the medial femoral condyle during flexion and prevents backward displacement of the tibia on the femur. The fibers of the PCL twist on their longitudinal axis in a clockwise direction during flexion and unwind in extension. This movement is analogous to but directly opposite to that of the ACL. The contour of the PCL as depicted on knee MR will be affected by the degree of extension in which the examination is accomplished. The dominant anterolateral bundle tightens on flexion and relaxes in extension; the smaller posterolateral bundle acts in the opposite direction.

The morphology of the course of the PCL is important in diagnosis and follows the dynamic anatomy. In the extended position, in which most knee MR examinations are performed, the PCL demonstrates a smooth, mildly convex posterior curve (Fig. 20B). With mild flexion the ligament appears more taut and thinner in overall girth (Fig. 20C). A minimally "buckled" appearance to the PCL may be seen normally. A sharply buckled or S-shaped PCL is pathologic. The differential diagnosis for a hyperbuckled PCL includes (a) a distal PCL tear, (b) avulsion of the tibial insertion of the PCL, and (c) complete ACL tear. In the latter situation, the lack of anterior restraints allows anterior translation of the tibia in the relation to the femur (MR drawer sign), with consequent shortening and buckling of the PCL. The abnormal course of the PCL may also be appreciated on coronal images, where a section through the sharply inclined posterior limb allows significantly more enhanced visualization of the PCL than normal.

The meniscofemoral ligament, once described as the "third cruciate ligament," is intimately associated with the PCL. An accessory ligament, it extends from the posterior horn of the lateral meniscus to the lateral aspect of the medial femoral condyle close to the attachment of the PCL. The ligament may divide to pass anteriorly and/or posteriorly to the PCL. The anterior division (ligament of Humphry), when present, can vary considerably in size and may be up to one-third the size of the PCL or larger. The posterior meniscofemoral ligament (ligament of Wrisberg) lies behind the PCL and extends from the posterior aspect of the lateral meniscus to the medial condyle of the femur. The ligament of Wrisberg varies in size and in some specimens is half the diameter of the PCL. The presence of these structures within the knee is quite variable. They may be confused with loose bodies within the joint (see Case 29). One or the other is present in approximately 70% of knees examined, although both ligaments are uncommonly present in the same knee.

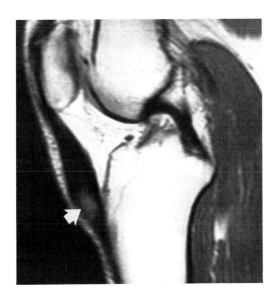

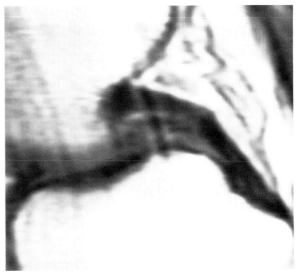

FIG. 20B. Normal PCL (sagittal TR/TE 2,000/20 msec). The normal PCL demonstrates a smooth convex posterior contour with the knee examined in flexion. Note the increased signal intensity and widening of the infrapatellar tendon; findings compatible with patellar tendonitis (*arrow*).

FIG. 20C. Taut PCL (sagittal TR/TE 800/20 msec). The PCL appears "stretched" secondary to the degree of knee flexion in which the study is performed.

CASE 21

A 21-year-old woman presented with knee pain out of proportion to the mild changes of patellofemoral dysfunction that were present on physical examination. A plain radiograph and diagnostic arthroscopy were essentially normal. What is the differential diagnosis on this MR (made 6 weeks following the arthroscopy), and why might the arthroscopy have been normal? (Fig. 21. Sagittal MR, TR/TE 800/20 msec.)

Findings: The MR demonstrates virtual complete replacement of the medullary canal of the femur. There are several small masses that are clearly extraosseous (arrows).

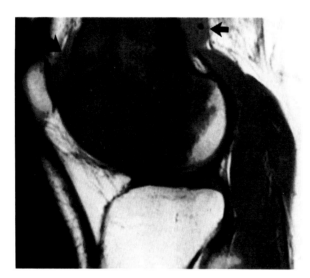

FIG. 21.

Diagnosis: Lymphoma.

Discussion: Lymphoma of the musculoskeletal system may involve bone primarily or secondarily. The latter pattern is generally part of a widely disseminated process involving lymph nodes. In this case, the abnormal MR led to a bone scan demonstrating multiple skeletal lesions. The ilium was biopsied and led to the diagnosis of histiocytic lymphoma.

Lymphoma, like leukemia, metastases, and primary tumors, are distributed throughout the skeleton within the red marrow either by originating there (leukemia, myeloma) or by being secondarily deposited in the red marrow because of its rich vascular supply. Marrow replacement disorders are exquisitely well seen on short TR/TE sequences. Their generally long T1 value contrasts with the short T1 values of fat within the medullary canal. The appearance of T2-weighted images, however, is variable, depending on the degree of cellularity of the tumor, its water content, and the amount of necrosis and fibrosis that is present. T1 and T2 values do not generally permit characterization of lesions, and most importantly they *do not reliably separate benign from malignant disease.*

The differential diagnosis in this case must include all infiltrative marrow disorders including primary sarcoma, metastatic tumor, infection, and Gaucher disease. The several features that do suggest lymphoma, however, are (a) the involvement of a very large segment of bone and (b) little if any soft tissue component. The latter, however, is variable, but primary sarcomas virtually always have extensive extraosseous disease.

Lymphoma is one of many lesions that originate in the subarticular bone (osteonecrosis, stress fractures) and, in such a location, are not detectable by arthroscopy. The ability of MR to detect extra synovial lesions makes it an invaluable tool for both the radiologist and orthopedist investigating knee disorders.

CASE 22

The patient is a 36-year-old man who sustained an acute knee injury while playing basketball. (Fig. 22A. Coronal MR, TR/TE 600/20 msec. Fig. 22B. Sagittal MR, TR/TE 2,000/20 msec.)

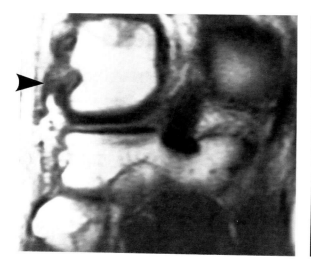

FIG. 22A.

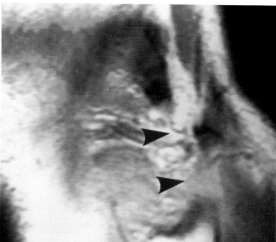

FIG. 22B.

Findings: On the coronal section (Fig. 22A), there is complete interruption of the fibular collateral ligament (FCL) with the proximal ligament seen coiled near its attachment site (arrowhead). On the sagittal image obtained through the fibular head (Fig. 22B), there is complete disorganization of the lateral ligaments. The normal conjoined tendon is absent (arrowheads).

Diagnosis: Tear of the FCL.

Discussion: The ligamentous anatomy of the lateral aspect of the knee is quite complex but knowledge of the anatomy is essential to recognition and understanding of injuries associated with lateral instability. The lateral soft tissue structures can be divided into anterior, middle, and posterior subdivisions.

The lateral capsular ligament (LCL) is the deepest layer of the lateral supporting structures and extends from the patella to the PCL. The anterior subdivision of the lateral capsulo-ligamentous complex consists of that section of the LCL that extends from the patella and patellar tendon to the anterior portion of the iliotibial band. The iliotibial band is an extracapsular structure that is superficial to the LCL. It inserts proximally on the epicondylar tubercle of the femur and distally on the Gerdy tubercle of the anterior tibia. The middle section of the lateral ligament is composed of the iliotibial band and LCL deep to it and extends posteriorly to the FCL. The middle section of the LCL is strong and has attachments both to the femur and tibia.

The posterior third consists of LCL and extracapsular ligaments, which form a single functional unit called the arcuate complex. The components of the complex are the FCL, arcuate ligament, and tendoaponeurotic unit formed by the popliteus muscle. The FCL is an extracapsular structure that originates from the lateral epicondyle of the femur directly anterior to the insertion of the gastrocnemius and passes directly beneath the lateral knee retinaculum to form a cord-like structure that joins with the biceps femoris tendon to insert into the head of the fibula as a conjoined tendon. The popliteus tendon passes under the FCL to perforate the posterior horn of the lateral meniscus and join its muscle belly on the posterior tibia. The arcuate ligament forms a triangular sheet that diverges upward from the fibular styloid. The lateral limb of the arcuate is dense and strong and is attached to the femur and popliteal tendon. The weaker medial limb curves over the popliteus muscle and is attached to the posterior horn of the lateral meniscus. The posterior third receives dynamic reinforcement from the biceps femoris and popliteus muscles as well as from the lateral head of the gastrocnemius.

Injuries of the lateral compartment are less common than those on the medial side of the joint. The injuries, however, may be more disabling because the lateral ligaments are under maximum tension when the leg is extended during walking. Anterolateral rotary instability results from traumatic disruption of the middle third LCL and/or injury to the ACL. Tears of the iliotibial tract may also be associated. Posterolateral rotary instability occurs when the arcuate complex is injured. The ACL may be torn in association and adds to the severity of the instability when both the PCL and arcuate complex are injured. The result is a combined PCL and posterolateral instability. Utilizing conventional radiographs, injuries to the lateral compartment are identified by widening of the lateral joint space and by depiction of a small avulsion from the tibial insertion of the LCL. This avulsion is referred to as the Segond fracture and is an indicator of extensive ligament damage. A Segond fracture results from traumatic detachment of the meniscosynovial portion of the middle third of the LCL. The fragment is usually small, minimally displaced, and arises not from the plateau but 2 to 10 mm distal to the articular surface superior and posterior to the Gerdy tubercle. On radiographs it is best seen on anteroposterior or tunnel views (Fig. 22C).

The normal iliotibial band, representing the thickened portion of the fascia lata, is identified on peripheral MR images as a delicate longitudinal low signal intensity line coursing longitudinally in the high signal intensity subcutaneous fat of the lateral thigh. The FCL and its common insertion with the biceps tendon on the head of the fibula are depicted on coronal images (Fig. 22D) and occasionally on extreme lateral sagittal views (Fig. 22E). As in other articulations, the normal ligaments are of low signal intensity on all pulse sequences. Disruption of these ligamentous structures is manifest by their complete absence or interruption of their normal contour (Fig. 22F). A wavy appearance or localized fluid within or around them is characteristic of lateral compartmental injuries. Medial compartmental bony injuries may be seen secondary to varus stress that disrupts the lateral compartment. Capsular disruptions may be recognized by fluid extravasation into the surrounding soft tissues and is often best depicted tracking from the joint superficial to the popliteus muscle and tendon.

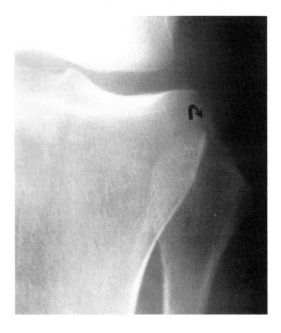

FIG. 22C. Segond fracture. Magnification view from an anterior-posterior radiograph demonstrates a small vertically oriented fragment along the lateral tibial plateau, originating just below the articular surface (*arrow*). This is a classic appearance of a Segond fracture.

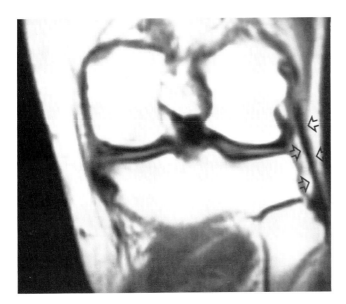

FIG. 22D. Normal FCL (coronal 600/20 msec). The normal appearance of the FCL as it courses from the head of the fibula to the lateral epidondyle of the femur is well demonstrated. The ligament is thin and of uniform low signal intensity (*open arrows*).

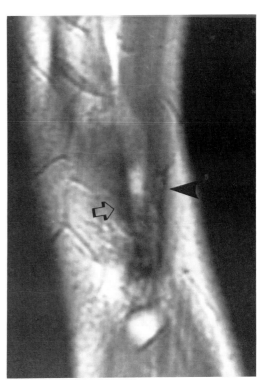

FIG. 22E. Sagittal TR/TE 2,000/20 msec. The biceps femoris tendon (*open arrow*) merges with the fibular collateral ligament to insert on the head of the fibula as a conjoint tendon (*arrowhead*).

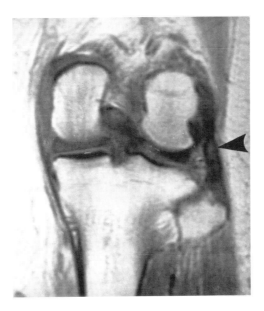

FIG. 22F. Partial FCL tear (coronal TR/TE 2,000/20 msec). Section demonstrates thickening of the FCL with focal area of increased signal intensity within its midsubstance representing a partial tear (*arrowhead*).

CASE 23

A 25-year-old woman fell down stairs and had severe pain in her knee. Except for diffuse tenderness on the tibial side of the joint and an effusion, no other abnormalities were identified. Plain radiographs were normal. After 2 weeks of conservative care, the patient was still not able to bear weight without pain, and she was sent for an MR. What is the diagnosis? For the lesion below, what are the potential benefits of MR? (Fig. 23A. Coronal MR, TR/TE 800/20 msec.)

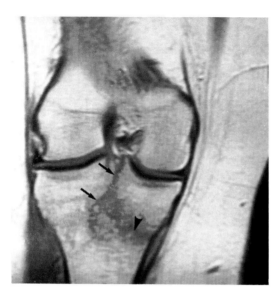

FIG. 23A.

Findings: There is a line of decreased signal (arrows) extending obliquely from the intercondylar area across the lateral tibial metaphysis to the lateral tibial cortex. Another limb of the lesion extends transversely to the medial femoral cortex (arrowhead). There is considerable edema around the lesion.

Diagnosis: Tibial plateau fracture (Hohl type I).

Discussion: Magnetic resonance, by virtue of its unmatched capabilities to detect changes in bone marrow composition, may be capable of detecting plateau fractures earlier than conventional radiographs.

Tibial plateau fractures result from excessive axial loading of the knee, combined with a valgus strain, during which time the lateral femoral condyle is driven into the plateau. The degree of comminution, depression, and displacement are the factors that determine operative versus nonoperative therapy. In general, types I, II, and III (undisplaced, local compression, split compression), characterized by depression of less than 8 mm and without significant displacement, can be treated nonoperatively. Plain radiography underestimates these critical features; CT and planar tomography have been used adjunctively. The multiplanar capabilities of MRI may facilitate three-dimensional perception and evaluation of such fractures and permit assessment of the collateral and cruciate ligaments and the menisci (Fig. 23B).

In addition to assessing known fractures, MR may have a role in the detection of such fractures that are radiographically occult (Fig. 23C,D). In one study 13 tibial plateau and distal femoral fractures were found on MRs that were normal (11 cases) or only slightly suspicious (two cases) for fracture on plain radiographs. Magnetic resonance has also been utilized in the detection of radiographically occult fractures of the *hip* and was found to be more sensitive than conventional radiographs.

There is no data regarding the use of MR to assess fracture healing. In this case, the follow-up study, performed when the patient was asymptomatic (Fig. 23E,F), revealed resolution of the presumed zone of the edema around the fracture, but the fracture line itself, evident on plain radiographs, was well seen on MRI. Mere depiction of a fracture line cannot be used as an indicator of nonunion.

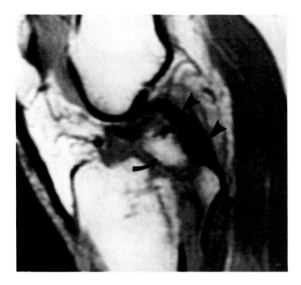

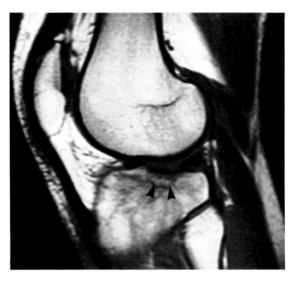

FIG. 23B. Tibial plateau fracture (sagittal MR, TR/ TE 800/20 msec). This fracture was clearly evident on the plain radiographs, and because the fragments were not significantly displaced, nonoperative therapy was contemplated. The MR demonstrated that a large fragment (*arrows*) was attached to the posterior cruciate ligament (*arrowheads*). A single screw secured the fragment and reestablished the integrity of the cruciate mechanism. The other large fragment (*curved arrow*) was reduced without internal fixation.

FIG. 23C. Depressed lateral tibial plateau fracture (sagittal MR, TR/TE 800/20 msec). A small segment of the middle of the lateral tibial plateau (*arrowheads*) is depressed 2 to 3 mm.

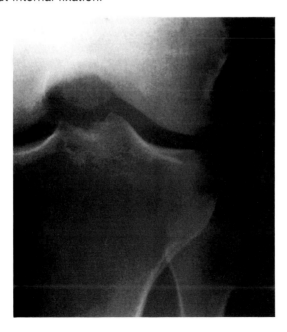

FIG. 23D. The plain radiograph made at the same time as the MR in Fig. 23C is normal; the patient was treated conservatively.

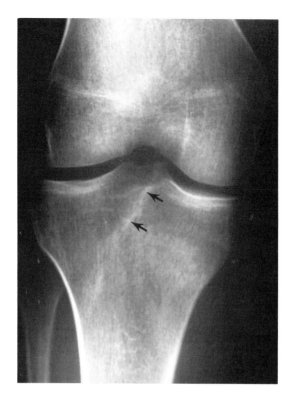

FIG. 23E. A follow-up plain radiograph, made at the same time as the MR in Fig. 23B, demonstrates a sclerotic fracture line (*arrows*).

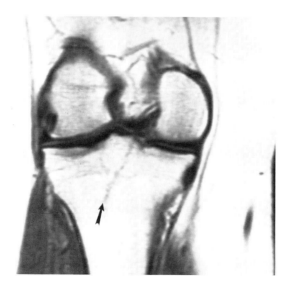

FIG. 23F. Plateau fracture (MR, TR/TE 2,000/20 msec). An MR made 2 months after Fig. 23A reveals resolution of the edema, but the fracture line persists (*arrow*).

CASE 24

The patient is a 47-year-old man with history of prior meniscal surgery. The patient has had recent knee symptoms and is referred to exclude meniscal retear. What is your diagnosis? What problems are unique to assessment of the postoperative meniscus? (Fig. 24A. Sagittal MR, TR/TE 2,000/20 msec. Fig. 24B. Sagittal MR, TR/TE 2,000/80 msec.)

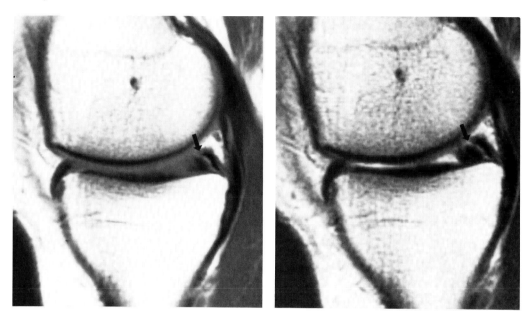

FIG. 24A.   **FIG. 24B.**

Findings: The patient is status post–moderately extensive meniscal resection. On the intermediate echo (Fig. 24A), the meniscal tip appears blunted, although it is difficult to distinguish meniscus from contiguous synovial fluid. There is an apparent high signal intensity line intersecting the posterior horn of the meniscus and apparently extending to an articular surface (arrow). On the T2-weighted image (Fig. 24B), the interface between meniscus and synovial fluid is better defined, and the meniscal tip, although rounded, appears smooth. The high signal intensity line persists (arrow) and suggests retear.

Diagnosis: Stable posterior horn remnant of the medial meniscus.

Discussion: Patients in whom previous meniscal surgery has been performed represent a heterogeneous group. The postoperative meniscus will demonstrate an altered morphology directly dependent on the degree of surgical intervention. We have found it useful to divide postoperative patients into three groups. The first group is composed of patients in whom relatively minimal meniscal resection has been previously performed with the meniscus appearing "near normal" in overall configuration. In these patients the presence of unequivocal linear grade 3 signal predicts the presence of a meniscal retear with a degree of accuracy comparable to that established in the virgin meniscus. Additionally, minimal blunting of the meniscal tip may be allowable as a postsurgical finding in these patients. It is critical, however, to determine that prior meniscal surgery has indeed been performed. Prior arthroscopy should not be equated with prior meniscal surgery. Evaluation of patients in whom a more extensive meniscal resection has been performed but in whom a qualitatively substantial remnant remains present is considerably more problematic. These patients comprise the second group. The presence of grade 3 signal within these menisci is a far less accurate predictor of meniscal tear than in the virgin meniscus and is associated with a significantly increased incidence of false-positive examinations. The patient illustrated in this case was diagnosed prospectively as having a meniscal retear by MR. At arthroscopy the meniscus was reported stable.

The third group of menisci are those in whom extensive prior meniscal resection has been performed and in whom only a small peripheral rim remains present. These menisci are far less problematic for MR evaluation than the prior group. Diffuse increased signal often extending to and involving an articular surface is not uncommonly identified in these small remnants and has not proven to be an accurate discriminator of meniscal stability. Extensive degeneration of the articular cartilage is a universal finding identified at arthroscopy in this group of patients. It is arguable how significant the determination of extensive degeneration or retear of these remnants is in patient management. Attention should be directed toward assessment of the articular cartilage and in detection of potential loose osteocartilagenous bodies in these patients. T2-weighted images or gradient refocused sequences may be of particular value in this regard (Fig. 24C,D).

There are several possible explanations to account for the discrepant findings between MR and arthroscopy in this group. A basic tenet of present arthroscopic surgery is restraint in meniscal resection. The challenge for the orthopedic surgeon is in balancing the need for debridement with the need for preservation of tissue. In some patients, particularly with degenerative meniscal tears, the surgeon will excise any excessive flaps but may elect to preserve some shaggy meniscal tissue. Furthermore, complete resection of a horizontal tear is not often considered necessary. This underscores the difficulty in MR morphological analysis of the postoperative meniscus for stability. If the surgeon, in an effort to preserve the peripheral rim, elects to leave the nonsignificant horizontal component of a complex tear within the remnant, it will, without exception, contribute to false-positive MR diagnosis of retear. Secondly, in patients with a prominent horizontal cleavage or "fish mouth" tear, the line of abnormal signal frequently extends from the articular surface of the meniscus to the capsular margin. The recognized extent of the tear at arthroscopy is often less than that of the full extent of the abnormal signal. When the meniscus is trimmed to the base of the tear, it becomes theoretically possible to create from the remaining intrasubstance (grade 2) signal an appearance that could mimic a grade 3 tear (Fig. 24E). A third consideration with regard to discordant MR-arthroscopy findings in the postoperative meniscus is the fact that this has principally occurred in the posterior horn of the medial meniscus. This parallels the experience in the nonoperated knee. Given the acknowledged operator dependence of arthroscopy and the known difficulty in arthroscopic assessment of the posterior horn of the medial meniscus, it is possible that in some cases false-positive MR findings could actually represent false-negative arthroscopies.

C   D

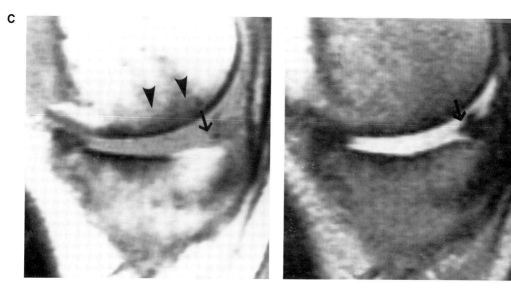

FIG. 24C,D. Group 3 meniscus. The value of T2-weighted images. **C.** Sagittal TR/TE 2,000/20 msec. There has been marked meniscal resection with only a small posterior horn remnant. On this intermediate weighted sequence, the borders of the meniscus are difficult to differentiate from surrounding synovial fluid (*arrow*). There are extensive changes within the subchondral bone (*arrowheads*) indicative of the degree of degenerative change so commonly identified in these patients. **D.** Sagittal TR/TE 2,000/80 msec. On this T2-weighted image, the contour of the meniscus is better evaluated (*arrow*). There has been marked meniscal resection. No significant intrameniscal signal was identified. The case reflects a stable group 3 meniscal remnant with osteoarthritic changes within the knee.

A   B

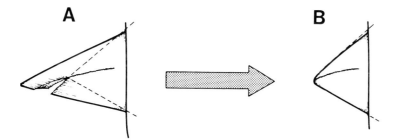

FIG. 24E. Schematic diagram. This diagram demonstrates the potential conversion of intrasubstance grade 2 signal to appear to mimic a grade 3 tear. The extent of the arthroscopically demonstrable meniscal tear is less than the degree of abnormal signal seen on the MR (A). Following meniscal resection (*dotted lines*), the postoperative meniscus (B), which is clinically stable, could contain intrasubstance signal that now appears to extend toward the articular surface and could thus be confused with grade 3 signal indicative of a meniscal tear.

CASE 25

A 32-year-old woman underwent evaluation for chronic knee complaints. She had previously sustained a major knee injury following a skiing accident. What is your diagnosis? What is your level of confidence? What is the natural history of ACL injuries? What is the accuracy of MR in evaluation of the ACL? (Fig. 25A. Sagittal MR, TR/TE 2,000/20 msec.)

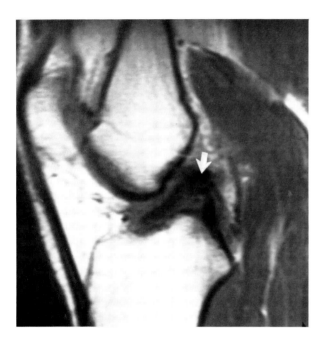

FIG. 25A.

Findings: On this section through the lateral intercondylar notch (Fig. 25A), the femoral attachment of the ACL is absent. The distal ACL is low lying in position and follows an abnormal course. The ACL appears adherent to the midportion of the PCL (arrow).

Diagnosis: Chronic tear of the ACL.

Discussion: Anterior cruciate ligament injuries are characterized pathologically by rapid synovial overgrowth and angioblastic invasion. Within 2 weeks of an ACL tear, arthroscopic inspection reveals a contracted ligament that is covered by a synovial cap. Destructive changes already present do not allow for mobilization and primary repair. With time there is further smoothing of the ACL stump, revascularization, and a yellow-white synovial covering encases the ligament. The natural history of an ACL injury is usually attachment to the PCL. The torn ACL falls on the PCL and gains a secondary source of attachment and blood supply. A surgical repair technique modeled after this natural mechanism with surgical attachment of the ACL to the PCL has been described by Wittek. In cases in which there has been a violent disruption of the ligament, there may be no ACL tissue arthroscopically demonstrable. On MR, the chronically torn ACL often appears completely absent; no structure is identified in the lateral intercondylar notch (Fig. 25B). Occasionally, a low signal intensity band, representing the remnant of the ACL, is seen to lie on the tibial plateau (Fig. 25C). A chronically torn ACL attached to the PCL may be recognized on MR, as in the patient illustrated. This type of insufficient ACL can be recognized by its low lying position, its lack of a proximal attachment, and an abnormal course through the knee.

A number of clinical methods have been established to assess the integrity of the major knee ligaments and to evaluate the degree of instability associated with ligamentous injuries. The Lachman test is one of the principal clinical methods utilized to test for ACL integrity. With the knee in 20 to 30° of flexion, the examiner applies a manual anterior displacement force to the proximal calf and perceives the tibial displacement both manually and visually and "endpoint stiffness" manually. Increased anterior excursion on this test is pathognomonic of a torn ACL. The degree of laxity is graded in relation to the patient's normal knee. Although the Lachman test is considered specific, reliable, and only minimally painful for the patient, a recent study found it only 89% sensitive in a group of patients with arthroscopically proven ACL tears. The efficacy of MR in detecting abnormalities of the ACL has been the subject of multiple critical evaluations. In the largest series to date, MR was found to be 95% accurate in assessment of ACL tears; in patients in whom T2-weighted images were performed, the accuracy improved to 97%. Several smaller studies have reported accuracy rates greater than 95% for MR assessment of ACL injuries.

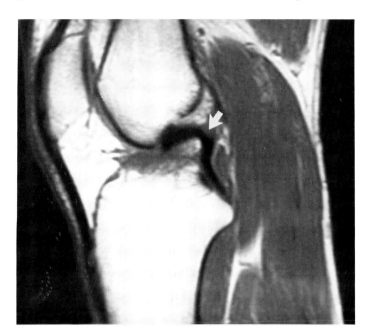

FIG. 25B. Chronic ACL tear (sagittal MR, TR/TE 2,000/20 msec). There is complete absence of the ACL, which should normally be visualized on this section through the lateral intercondylar notch. The tibia is translated anteriorly in relation to the femur, a finding indicative of ACL insufficiency. The PCL is mildly buckled (*arrow*).

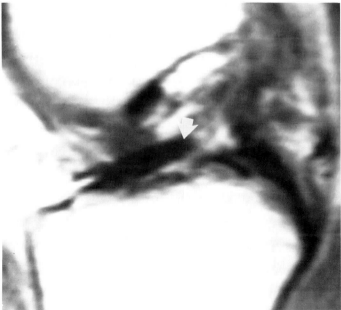

FIG. 25C. Complete ACL tear (sagittal MR, TR/TE 2,000/20 msec). The ACL is completely torn and detached from the proximal femur. Only a small remnant of the distal ACL is seen to lie freely on the intercondylar notch (*arrow*).

CASE 26

A 22-year-old woman noticed a firm tender mass on the medial side of her knee. She was told 3 years prior that she had an exostosis in the area. At the time of the MR examination 1 week later, however, the mass had decreased in size. What might have been the transient lesion, and what might be another reason for ordering an MR in this individual? (Fig. 26A. Oblique coronal MR, TR/TE 800/20 msec. Fig. 26B. Axial MR, TR/TE 2,000/80 msec.)

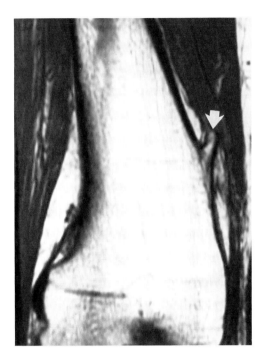

FIG. 26A.

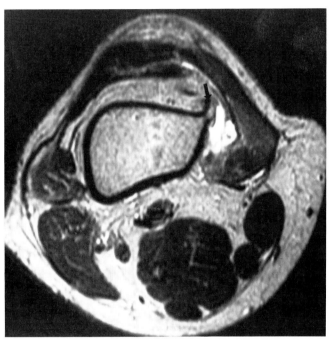

FIG. 26B.

Findings: A typical osteochondroma is seen arising from the medial aspect of the supracondylar region of the femur (Fig. 26A, arrow). Figure 26B reveals fluid behind the lesion (arrow). There is no demonstrable cartilagenous cap.

Diagnosis: Benign osteochondroma with probable rupture of an associated bursa.

Discussion: Osteochondromata are relatively common benign tumors of bone arising from the region of the physis in a child and "migrating" away from the growth plate. The lesion arises as a focal herniation of the growth plate, which becomes oriented at 90° to the growing bone. Therefore, osteochondromata uniformly point away from their site of origin and are never epiphyseal. In the immature skeleton, the osteochondroma is composed of bone and cartilage. The bony portion has both medullary bone, which blends with that of the bone of origin, and a cortical shell. The cartilagenous portion is analogous to a physis. It is relatively thick in the immature skeleton but becomes thin and smooth in the adult and stops growing at the time of closure of the nearest physis. Radiographic growth of an osteochondroma (in an adult), pain, loss of a cortical margin, development of a mass and increasing scintigraphic activity are signs *suggesting* malignant transformation. The rate of malignant change of osteochondromata is unknown because the prevalence of these benign lesions in the general population is also unknown, but it is believed that for a solitary osteochondroma, malignant transformation occurs in much less than 1% of cases. In patients with multiple hereditary exostoses, however, the rate rises to 10% to 20%. In an effort to distinguish benign exostoses from exostotic chondrosarcomas, an attempt was made to measure the thickness of the cartilagenous cap directly by means of CT. Benign lesions had smooth caps with a maximum thickness of 3 cm (range, 0.1–3 cm with an average of 0.6 cm), and chondrosarcomas had a minimal cartilage thickness of 2 cm, with most having cartilage thicker than 3 cm (range, 1.5–12 cm with a mean of 5.9 cm). It is apparent that there is a wide range of overlap, and these values thus can only serve as guidelines. Magnetic resonance should exceed the ability of CT to measure the cartilagenous cap correctly. The low cellularity and rather high water content of the hyaline cartilage that composes the cap produces high signal on long TR/TE sequences, and therefore the thickness and contour of the cartilage should be easy to discern.

Bursa formation in association with osteochondromata has been well described. Awareness of this possibility is important because formation of such a fluid-filled mass can result in an alteration of the shape of the exostosis, an increase in its surrounding soft tissues, and an increase in pain. Bursae have a synovial lining, usually containing mucinous fluid, and may contain fibrin and loose bodies as well. It is these cartilaginous bodies that radiographically suggest increasing perilesional calcification (a sign of malignant degeneration).

Several other complications of exostoses are worthy of mention. Skeletal deformity, especially limb shortening, is not infrequent. The lesion may directly compress the spinal cord or peripheral nerves. Trauma to nearby arteries, especially in the popliteal space, may result in formation of a pseudoaneurysm (Fig. 26C). Finally osteochondromata may fracture or infarct, both of which may result in an increase in pain.

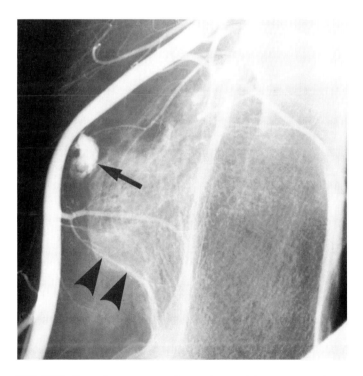

FIG. 26C. Pseudoaneurysm. An angiographic sequence demonstrates a small false aneurysm (*arrow*) intimately related to a large osteochondroma (*arrowheads*) of the posterior distal femur. (Courtesy of Donald Resnick, MD.)

CASE 27

The patient is a 46-year-old mason complaining of chronic knee pain. What is your diagnosis? (Fig. 27A. Sagittal MR, TR/TE 2,000/80 msec.)

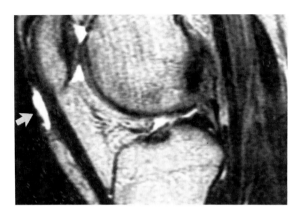

FIG. 27A.

Findings: There is an ovoid, relatively well-circumscribed high signal intensity fluid collection located anterior to the proximal patellar tendon (Fig. 27A, arrow). The collection most closely corresponds to the superficial infrapatellar bursa.

Diagnosis: Superficial infrapatellar bursitis.

Discussion: Twelve bursae regularly occur in the region of the knee. Under certain circumstances, any of these bursae may become infected or inflamed and produce symptoms. Bursal inflammation (bursitis) is characterized by pain, which characteristically is aggravated by motion and at times is worse at night. Point tenderness is commonly depicted on physical exam.

There are three bursae related to the patellar tendon: (a) prepatellar, (b) superficial infrapatellar, and (c) deep infrapatellar. The prepatellar bursa is located between the skin and outer surface of the patella and commonly extends from the midplane of the patella to cover the upper half of the patellar tendon. Its superficial location subjects the bursa to repeated trauma. Prepatellar bursitis is most commonly seen in those whose occupation involves much kneeling, as in the patient illustrated in this case. This condition is also referred to as "housemaid's" knee and has been reported as well in overenthusiastic lovers (*genu amoris*). The diagnosis is usually easy to establish with local tenderness, swelling, and inflammation as symptoms. Pain most often results from tension of the skin in extreme knee flexion or by direct pressure. Treatment is usually conservative but occasionally may require aspiration. Chronic bursitis may be more recalcitrant and require surgical excision of the bursa. Prepatellar bursitis is not uncommonly encountered on knee MR, often as an unexpected finding.

The findings in this case are localized to the superficial infrapatellar bursa. The deep infrapatellar bursa lies between the patellar tendon and the upper part of the tuberosity of the tibia (Fig. 27B). It is small and separated from the synovial membrane of the knee by the infrapatellar fat pad (Hoffa fat pad). Deep infrapatellar bursitis is typically associated with overactivity and may result from friction with the upper tibia. It rarely results from direct trauma as it is well protected by the overlying patella and underlying fat pad. Tenderness can be illicited by pressure behind the infrapatellar tendon. Treatment is supportive and consists of heat, immobilization, and aspiration (Fig. 27C–E). Also to be considered in the differential diagnosis is tendonitis of the patellar tendon. This is typically associated with focal widening of the tendon, with associated areas of increased signal within the tendon.

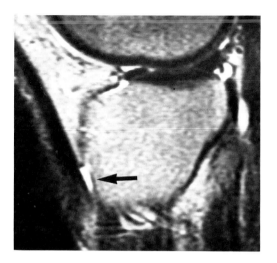

FIG. 27B. Fluid in infrapatellar bursa (sagittal MR, TR/TE 2,000/80 msec). This figure demonstrates a small collection of fluid (*arrow*) within the infrapatellar bursa. This degree of fluid is commonly encountered normally on knee MR. It denotes the position of the infrapatellar bursa.

FIG. 27C–E. Infrapatellar bursitis. **C.** Sagittal TR/TE 2,000/20 msec. A circumscribed intermediate signal intensity collection is seen in the region of the infrapatellar bursa (*arrow*). **D.** Sagittal TR/TE 2,000/80 msec. The collection increases significantly in signal intensity on the T2-weighted image (*arrow*). The findings are compatible with infrapatellar bursitis. **E.** Sagittal TR/TE 2,000/20 msec. Follow-up exam performed 2 months after original study. There has been total resolution with no abnormality in the infrapatellar bursa (*arrow*).

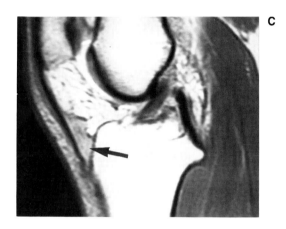

C

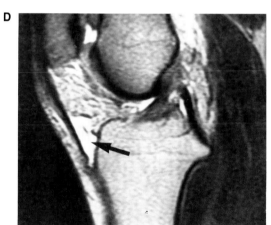

D

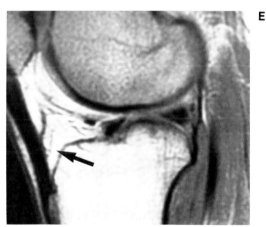

E

CASE 28

A 28-year-old man suffered pain in his knee immediately following an injury. What is the important MR finding and what is its clinical significance? (Fig. 28A. Coronal MR, TR/TE 2,000/20 msec. Fig. 28B. Coronal MR, TR/TE 2,000/80 msec.)

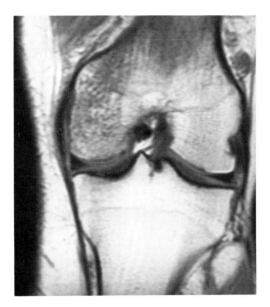

FIG. 28A.

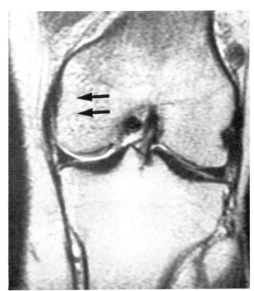

FIG. 28B.

Findings: There is a poorly defined inhomogeneous area of abnormal signal seen in the medial femoral metaphysis (Fig. 28A). The lesion brightens somewhat on the long TR/TE sequence (Fig. 28B, arrows). The cortical bone and overlying articular cartilage are normal.

Diagnosis: Bone bruise.

Discussion: Bone bruises were first described in association with major injuries to the contralateral collateral ligament. In the initial description, these lesions, thought to be areas of edema and hemorrhage, were found in five of eight patients who had significant MCL injury. In a more detailed analysis of acute bone fractures, bone bruises were found in five of five severe MCL injuries. Additionally, lateral bone bruises were identified in 18 of 25 *acute* ACL tears without collateral ligament injuries. Acute ACL tears commonly occur during external rotation of the femur over a fixed tibia and foot. Such action increases the compressive forces on the lateral side of the knee and most likely is the etiology of a bone bruise. There is no available material to determine the pathologic correlate of the MR image, but given the traumatic nature of the lesion, it appears reasonable that edema, hyperemia, hemorrhage, and perhaps trabecular disruption are present.

Most bone bruises are on the lateral side of the knee joint in association with ACL or MCL tears (Fig. 28C). *Medial* bone bruises have been reported in association with lateral collateral ligament tears and as a result of direct trauma (Fig. 28A,B). In one study, six bruises involved the tibia, 17 the femur, three involved both bones, and one bruise occurred in the fibula (Fig. 28D).

Although direct proof is lacking, there is some evidence to suggest that a bone bruise is a benign abnormality: (a) several lesions have been followed to rather rapid clinical and MR resolution, (b) there has been no reported progression of the lesion, and (c) there is no arthroscopic correlate (as expected in this *subchondral* abnormality). Although a bruise may possibly be a cause of symptomatology when it is an isolated abnormality, it is difficult to assess its significance if there is an associated ACL or MCL tear in which permanent disability might ultimately occur. Bone bruises may be etiologically related to impaction types of osteochondral fractures; in fact, bruises and osteochondral fractures may be manifestations of similar forces but of differing magnitudes.

On MRI, bone bruises appear as poorly defined areas of decreased signal in the subcortical bone. By definition, the cartilage surface and cortical bone are intact. On long TR/TE sequences, a bone bruise demonstrates inhomogeneous signal increase. The lesion involves the epiphyseal region but may occasionally extend well into the meta-diaphysis.

In this case, the patient had fallen on his knee. He did not suffer any twisting injury. Although no MR follow-up was available, the patient was asymptomatic within 1 month.

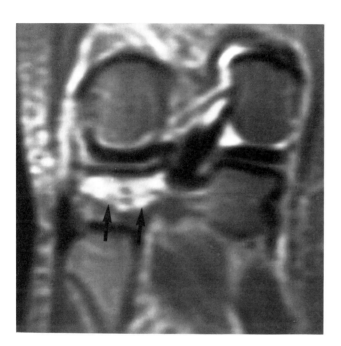

FIG. 28C. Bone bruise (coronal MR, TR/TE 2,500/80 msec). A band of markedly increased signal (*arrows*) is present in the subcortical bone of the tibial plateau in this patient with MCL tear seen best on other images.

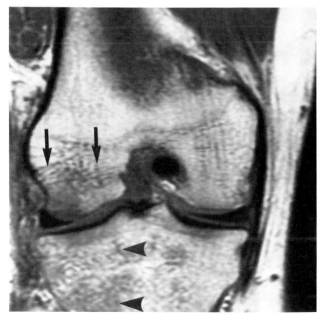

FIG. 28D. Bone bruise (coronal MR, TR/TE 2,000/20 msec). A bone bruise is present in both the lateral femoral condyle (*arrows*) and lateral tibial plateau (*arrowheads*). The patient suffered a twisting injury and had a completely torn ACL.

CASE 29

This 42-year-old woman frequently experienced mild discomfort in the knee after running; physical examination was completely normal. The patient was to enter a marathon and requested an MRI to ensure that there were no intra-articular abnormalities. This study was completely normal except for the two lesions depicted by the arrows in Fig. 29A and B. What advice might the orthopedist give to the runner? (Fig. 29A. Sagittal MR, TR/TE 2,000/20 msec. Fig. 29B. Sagittal MR, 5 mm medial to Fig. 29A, TR/TE 2,000/20 msec.)

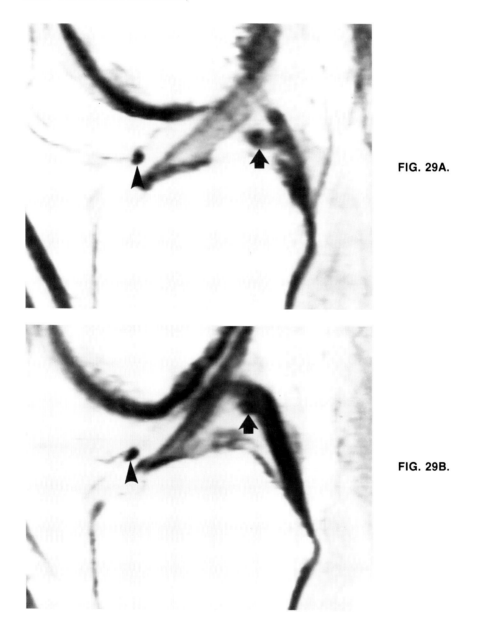

FIG. 29A.

FIG. 29B.

Findings: There is an ovoid density (arrow) seen in the concavity of the PCL that is more cephalad on the more medial image (Fig. 29B). Another rounded structure is seen just above the lateral tibial plateau anterior to the ACL (Fig. 29A,B; arrowheads).

Diagnosis: Normal knee MRI; the posterior "abnormality" related to the PCL is the ligament of Humphry. The more anterior lesion is the transverse meniscal ligament.

Discussion: A dark round structure can often be seen related to the anterior portion of the proximal PCL on one or more routine sagittal images. Although there is no pathologic support in this case, the "mass" probably represents the ligament of Humphry. This structure, best seen on coronal images, is a band of tissue that secures the posterior horn of the *lateral* meniscus to the *medial* femoral condyle (menisco-femoral ligament). It is actually composed of two components, one of which lies just anterior (ligament of Humphry) and the other posterior (ligament of Wrisberg) to the PCL (Fig. 29C,D). The anterior limb is the one that is more consistently seen on sagittal images. Atypical masses in the intercondylar notch may represent meniscal fragments (Fig. 29E).

The transverse meniscal ligament, a dark, rounded structure that secures the anterior horns of the menisci to each other, is routinely seen on knee MRI studies. It is best identified by its shape and its constant location. It can be followed from one meniscus to the other on sequential sagittal images.

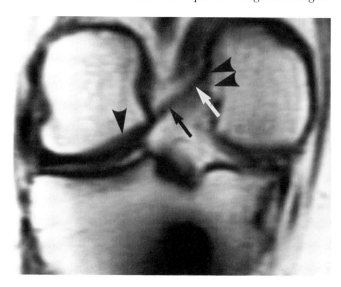

FIG. 29C. Ligament of Wrisberg (coronal MR, TR/TE 800/ 20 msec). This posterior coronal image demonstrates the ligament of Wrisberg (*arrows*), extending from the posterior horn of the lateral meniscus (*arrowhead*) and rising across the posterior knee joint to attach to the postero-medial femoral condyle (*double arrowhead*). A metal artifact is present in the tibia.

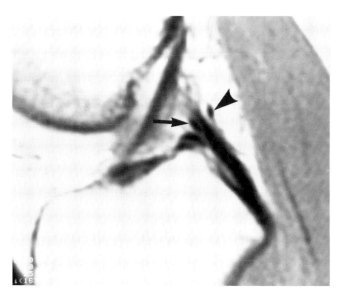

FIG. 29D. Ligaments of Humphrey and Wrisberg. Both arms of the menisco-femoral ligament can be seen anterior (*arrow*) and posterior (*arrowhead*) to the PCL.

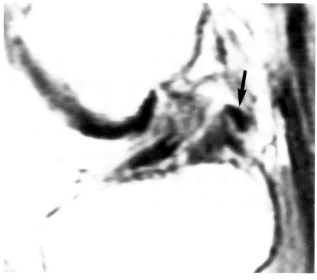

FIG. 29E. Meniscal fragment (sagittal MR, TR/TE 2,000/20 msec). A triangular structure (*arrow*) is seen in the posterior intercondylar area, but it is quite large and does not have the relationship to the PCL that characterizes the ligament of Wrisberg. A displaced lateral meniscal fragment, seen best on coronal images, was removed at surgery.

CASE 30

A 32-year-old professional football player experienced an acute hyperextension injury. What is your diagnosis? How common is this injury? What are some associated findings? (Fig. 30A. Sagittal MR, TR/TE 2,000/20 msec.)

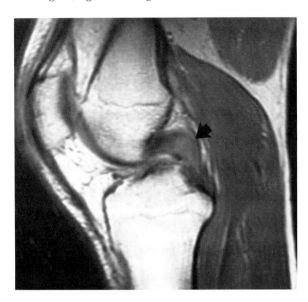

FIG. 30A.

Findings: There is diffuse increased signal intensity replacing the normally homogeneous low signal intensity of the PCL (Fig. 30A). The ligament is focally widened (arrow). No osseous avulsion is present.

Diagnosis: Posterior cruciate ligament tear (interstitial type).

Discussion: Acute and chronic tears of the PCL most commonly appear as foci of increased signal intensity, replacing the normally homogeneously dark ligament on T1-weighted images. The ligament is often focally widened in caliber, but the tear is characteristically less mass-like than those involving the ACL. With complete tears a gap can be identified separating the two ends of the ligament (Fig. 30B). In subacute tears, focal areas of signal relating to areas of hemorrhage may be identified. Chronic tears or fibrous scar may manifest little abnormal signal and may only be demonstrable by subtle contour changes. Bony avulsions secondary to PCL injury occur most commonly at the tibial insertion (Fig. 30C). A potential problem in diagnosis occurs when a torn or avulsed PCL remains closely apposed to the tibia. A clue to the correct diagnosis can be demonstrated in the immediate subchondral bone, which will often demonstrate diminished signal intensity secondary to edema.

The PCL is most typically injured following a posteriorly directed force against the flexed knee (motor vehicle accident), resulting in posterior displacement of the tibia in relation to the femur. Hyperextension injuries are a second major mechanism for PCL tears. In the acute setting, clinical diagnosis is often made difficult by a tense hemarthrosis and marked associated muscle spasm. In patients with "dashboard" injuries, associated fractures of the femoral diaphysis are common.

Disruption of the PCL can be "isolated" but often is associated with other serious knee injuries. Principal among these are posterior lateral capsular disruptions (Fig. 30D) and tears of the arcuate ligament complex. Accurate diagnosis of the extent of injury is essential as isolated tears of the PCL can be treated nonoperatively, whereas PCL tears associated with capsular tears are generally treated surgically. Arthroscopy in the acute setting is often precluded as a consequence of concern for irrigation fluid extravasation from a torn capsule. Additionally, the PCL is not optimally visualized from the typical anterolateral arthroscopic approach. With an intact ACL, only the femoral attachment of the PCL is routinely well seen. A posteromedial portal may be required to see the tibial attachment of the PCL, one of the most common sites of injury. Often the synovial investment covering the cruciate ligament needs to be opened to allow identification of the tear. These considerations underscore the contribution that MR can make in assessment of individuals with this injury complex.

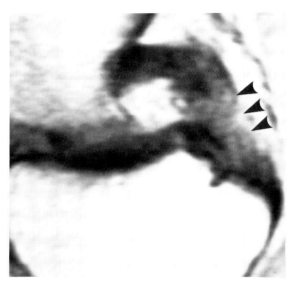

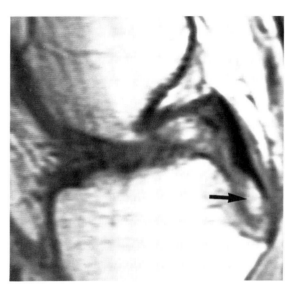

FIG. 30B. Posterior cruciate ligament tear (sagittal TR/TE 2,000/20 msec). The PCL is entirely discontinuous with only a small segment of the femoral attachment intact. An extensive mass representing edema, hemorrhage, and synovial proliferation replaces the normal low signal intensity tendon (*arrowheads*).

FIG. 30C. Posterior cruciate ligament tear (sagittal TR/TE 2,000/20 msec). There is complete disruption of the distal PCL attachment from the tibia. A small bony fragment is evident (*arrow*). There is associated signal diminution relating to edema within the immediate subchondral bone.

FIG. 30D. Posterior capsular tear. (sagittal TR/TE 2,000/80 msec). Irregular and serpiginous foci of increased signal representing extravasated fluid from the joint is seen tracking posterior to the popliteus tendon and muscle (*arrowheads*).

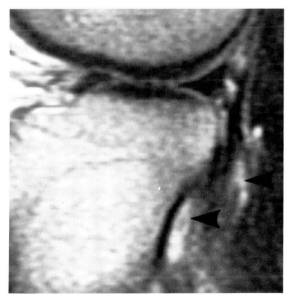

CASE 31

A 30-year-old man presented to his orthopedist with chronic knee discomfort and a small effusion. What is the diagnosis, and what is the critical information to transmit to the orthopedist regarding this study? (Fig. 31A. Coronal MR, TR/TE 800/20 msec. Fig. 31B. Sagittal MR, TR/TE 2,000/80 msec.)

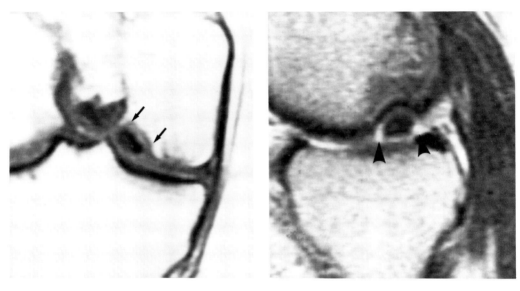

FIG. 31A.   **FIG. 31B.**

Findings: The coronal image (Fig. 31A) reveals a typical OCD lesion (arrows) measuring 1.5 cm. It is surrounded by a thin zone of decreased signal. The sagittal spin-echo sequence (Fig. 31B) reveals fluid anterior (arrowheads) and posterior to the fragment.

Diagnosis: Osteochondritis dissecans with a partially loose fragment.

Discussion: Osteochondritis dissecans is a painful condition of the epiphysis of a child or young adult in which a segment of articular cartilage and underlying bone may become detached from its site of origin. The knee and elbow are the most frequently affected joints. Within the knee, the lateral aspect of the medial femoral condyle (80%), the lateral femoral condyle, and the patella are affected in decreasing order of frequency. Several etiologies for OCD have been proposed, but most authors favor acute and/or chronic trauma.

The management of OCD is critically dependent on the mechanical stability of the lesion. It may be described as: (a) loose *in situ*, (b) grossly unstable, or (c) stable (Fig. 31C). The grossly loose fragment is one in which the overlying articular cartilage is violated and the lesion is partially or completely detached from the bone. The loose *in situ* lesion is ballotable (at surgery), but the articular cartilage is intact. A recent report analyzed the ability of plain radiography, scintigraphy, and MR to determine the mechanical stability of OCD lesions of the knee. All lesions less than 0.2 cm were stable, whereas those greater than 0.8 cm were loose. Lesions with a thick sclerotic rim (greater than 3 mm) were loose, but 42% of lesions with thin rims were also loose. The most reliable MR sign of loosening was obvious displacement of the fragment, but fluid interposed between the lesion and the host bone was seen in all displaced lesions. The bright interface on long TR/TE sequences is presumed to represent synovial fluid traversing the cartilagenous defect. The signal intensity of the fragment itself increased in all patients with stable fragments and 10 of 12 patients with loose fragments. In this case, the fluid interface between the fragment and the femur, the large size (1.5 cm), and the lack of high signal in the lesion all suggest an unstable fragment.

Two-dimensional and three-dimensional Fourier transform techniques may be a more optimal method of demonstrating the articular cartilage abnormality occurring in OCD. The patient in Fig. 31D had a large displaced fragment removed 1 year before the MR; the base was drilled in an attempt to stimulate healing. An MR was ordered to assess the degree of repair. These three-dimensional GRASS images demonstrate that the large defect has been filled in, presumably with fibrocartilage (although no histology is available in this case).

Magnetic resonance may have great value in longitudinal assessment of the patient with OCD. One must be careful not to mistake a normal variant for OCD (Fig. 31E,F). Preoperatively, the stability of the lesion can be reliably assessed. The natural history can be noninvasively determined. If surgery is necessary, the results can be depicted without need for a "second-look" operation.

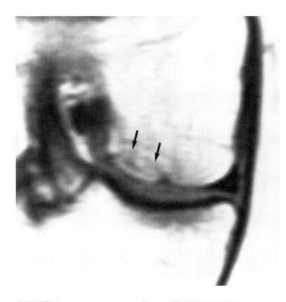

FIG. 31C. Osteochondritis dissecans, stable fragment (coronal MR, TR/TE 800/20 msec). An ovoid fragment is nearly inseparable from the host (*arrows*) without a fluid interface.

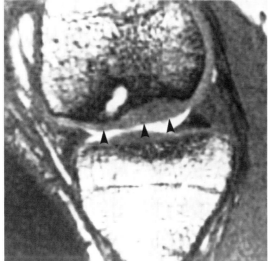

FIG. 31D. Osteochondritis dissecans (sagittal MR, TR/TE 55/15 msec). One year after the base of a large loose fragment was drilled, an MR revealed that the defect filled in with a tissue (*arrowheads*) with signal characteristics similar to cartilage. The surface is somewhat irregular but is intact.

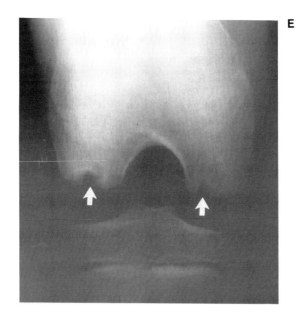

E

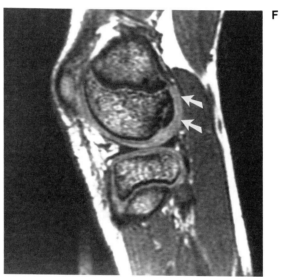

F

FIG. 31E,F. Normal ossification centers. **E.** Plain radiograph. **F.** Sagittal MR, TR/TE 55/15 msec. This 11-year-old boy had been followed for 2 years with the diagnosis of OCD of the medial and lateral condyles bilaterally (*arrows*, Fig. 31E). Because the lesions were posterior, they were felt to represent irregular but normal ossification centers; an MR confirmed that the articular cartilage (*arrows*, Fig. 31F) is smooth and intact. The "lesion" is not optimally seen.

CASE 32

This 30-year-old woman fell, striking her knee on a hard surface. Besides an effusion, is there an abnormality present? What might one do to improve detection of intra-articular lesions? (Fig. 32A. Coronal MR, TR/TE 500/20 msec.)

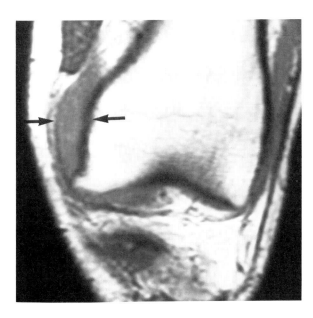

FIG. 32A.

Findings: On this T1-weighted image (Fig. 32A), there is a very faint area of inhomogeneity in the medial reflection of the suprapatellar bursa (arrows), which is distended with fluid. The lesion is difficult to identify confidently but is easily seen on the T2-weighted image (Fig. 32B).

Diagnosis: Osteochondral loose body.

Discussion: Loose bodies in the knee most commonly arise from OCD or a result of patellar dislocation in young adults and from foci of severe osteoarthritis in the older population. There has been considerable discussion as to the appearance of loose fragments on MR. When the lesion is ossified, it has medullary bone in its center and has a dark cortical rim on at least one side. This lesion is generally bright on short TR/TE sequences and becomes darker on longer TR/TE sequences (Fig. 32C,D). Some loose bodies are heavily calcified but do not have a medullary center; they are dark at all imaging sequences. Finally, a purely chondral fragment may have signal characteristics not too dissimilar from joint fluid on T1-weighted images and may be seen only on T2-weighted sequences (as in this case). In our experience, most loose bodies have had mixed characteristics, and in cases of multiple loose bodies, each may have different components.

Long TR/TE sequences are invaluable in both detection of a suspected loose body and confirmation of its intra-articular position. In this case, the lesion is nearly impossible to see on a short TR/TE study, but it is readily detected on the more T2-weighted image. Routine use of T2-weighted images has been shown to improve detection of osteochondral defects, abnormal plicae, joint effusions, ACL injuries, meniscal and popliteal cysts, muscle injuries, and menisco-capsular separations.

Nodular synovitis, as may be seen in patients with an inflammatory arthritis, may present an appearance quite similar to loose bodies, but synovial nodules are not generally dependent within the joint. Pigmented villonodular synovitis and exostotic spurs may be other sources of confusion with loose bodies. Displaced meniscal fragments are loose bodies, and when one encounters a single fragment in an otherwise normal joint, one must suspect the menisci as the site of origin. Finally, air introduced inadvertently into the joint during arthrocentesis may mimic a heavily calcified fragment (Fig. 32E), and extra articular fat may be mistaken for a loose body (Fig. 32F,G).

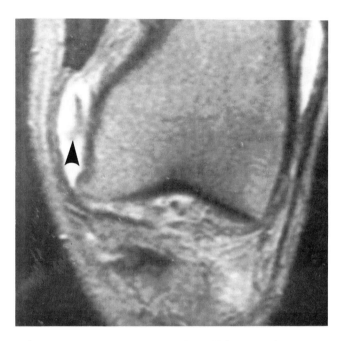

FIG. 32B. Loose body (coronal MR, TR/TE 2,000/80 msec). The chondral fragment (*arrowhead*) present in Fig. 32A is much more easily seen on this long TR/TE sequence.

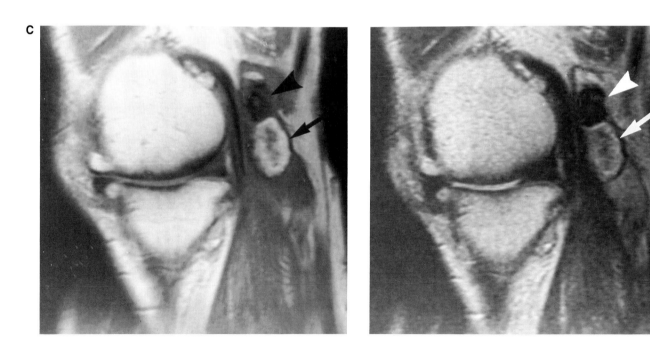

FIG. 32C,D. Loose bodies in a popliteal cyst. **C:** Sagittal MR, TR/TE 2,000/20 msec; **D:** Sagittal MR, TR/TE 2,000/80 msec. An osteochondral fragment, composed largely of medullary bone (*arrow*), is bright on this intermediate sequence. A heavily calcified fragment (*arrowhead*) is dark. The ossific fragment, composed of fat, becomes darker on this long TR/TE sequence (*arrow*). The calcified fragment (*arrowhead*) is dark at all imaging sequences.

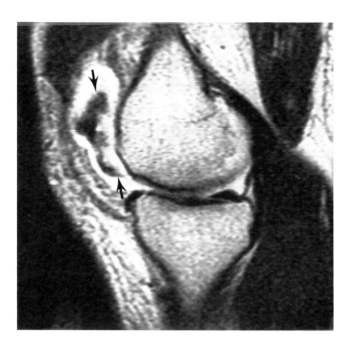

FIG. 32E. Air bubbles/clotted blood (sagittal MR, TR/TE 2,000/80 msec). Large dark areas (*arrows*) are present in this hemarthrosis, which was tapped 12 hr prior. No loose fragments were present. Although blood does not normally clot in a joint, tiny fragments of foreign matter introduced during arthrocentesis may provide a nidus for clot formation.

F

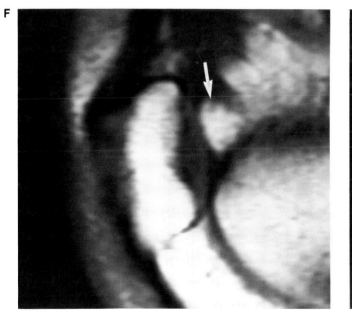

G

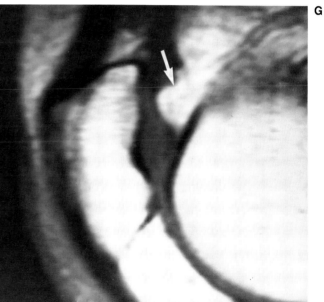

FIG. 32F,G. Pseudo loose body (sagittal MR, TR/TE 800/20 msec). The bright rounded density "in" the suprapatellar bursa (*arrows*) represents periarticular fat. On the more lateral image **(G),** the continuity of the lesion with periarticular fat is easily seen. (From Mink JH, Reicher MA, and Crues JV: *MRI of the knee.* Raven Press, New York, 1987.)

CASE 33

The patient is a 46-year-old woman with complaint of anterior knee pain accentuated when walking downhill. What is your diagnosis? What are the differential considerations for this MR appearance? (Fig. 33A. Sagittal MR, TR/TE 2,000/20 msec.)

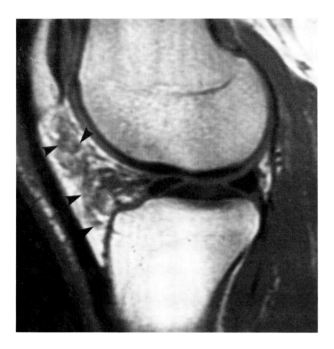

FIG. 33A.

Findings: The normally homogeneous high signal intensity infrapatellar fat pad is diffusely infiltrated by multiple serpiginous and confluent foci of low signal intensity (Fig. 33A, arrowheads). The fat pad is minimally increased in size. The patellar tendon is normal.

Diagnosis: Hoffa disease.

Discussion: The term *Hoffa disease* has been variably utilized to refer to traumatic changes within the infrapatellar fat pad. Hoffa initially implicated traumatic and inflammatory changes of the infrapatellar fat pad as a clinical entity at the turn of the century. The condition is characterized by painful enlargement of the fat pad, which is often accompanied by an effusion and limitation of function of the knee. Clinical symptoms are nonspecific. The complaints are usually produced or increased by strong tension on the patellar tendon, particularly when walking downhill. In contrast to meniscal lesions, tenderness in the anterior joint space does not change with knee flexion. It is generally agreed that Hoffa disease as a primary independent condition is extremely uncommon (Fig. 33B).

In present usage, the syndrome refers to hypertrophy of the infrapatellar fat pad most commonly seen in younger patients following trauma. The enlarged fat pad may then become impinged between the femur and tibia when the flexed knee is suddenly extended. The entrapment of the hypertrophic villi of the fat pad between the condyles can mimic other causes of internal derangement including meniscal tear. On physical exam, swelling on either side of the patellar tendon and vague tenderness to deep pressure may be identified. Quadriceps atrophy has been reported. Enlargement of the fat pad has also been associated with edema in the premenstrual water retention syndrome. Bleeding into the fat pad can occasionally lead to calcifications. Radiographic findings include: (a) calcifications that range from small flecks to dense deposits, (b) absence of the normal lucency of the infrapatellar fat pad, and (c) pressure erosion on the inferior pole of the patella.

The principal differential diagnostic consideration would be that of scarring within the fat pad. Such scarring invariably is seen in patients postarthroscopy and most typically presents as a smaller stellate focus of diminished signal intensity within the fat pad (Fig. 33C). The presence of scarring can indeed be utilized as an excellent MR marker of prior arthroscopic surgery. Overall enlargement of the fat pad is not present with postoperative scarring. Calcified hematomas within the fat pad could mimic the low signal intensity findings seen with both Hoffa disease and infrapatellar fat pad scarring.

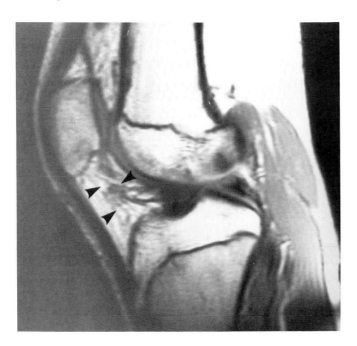

FIG. 33B. Hoffa disease (sagittal TR/TE 2,000/20 msec). Image of a second surgically confirmed case of Hoffa disease in an adolescent. The findings are less extensive than the previously illustrated cases and demonstrate the spectrum of changes that may be observed in this condition (*arrowheads*).

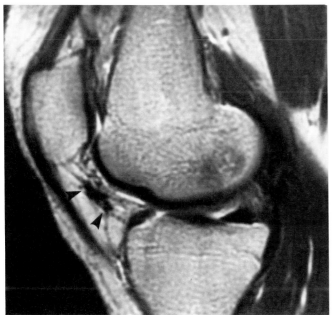

FIG. 33C. Postoperative scarring (sagittal TR/TE 2,000/20 msec). A stellate focus of low signal intensity is seen within the infrapatellar fat pads (*arrowheads*). Note the similarity to the previously illustrated cases. Scarring within the fat pad is an invariable accompaniment of prior arthroscopic surgery.

CASE 34

A 13-year-old boy had a long-standing knee disorder for which he had sought medical care on a number of occasions. He was eventually referred for an MRI. What is the diagnosis? (Fig. 34A. Sagittal MR, TR/TE 2,000/20 msec. Fig. 34B. Sagittal MR, TR/TE 2,000/20 msec.)

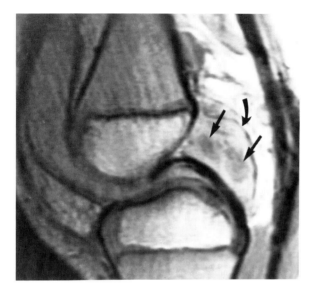

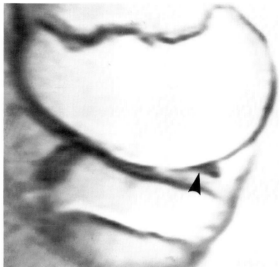

FIG. 34A.   **FIG. 34B.**

Findings: A midline sagittal image (Fig. 34A) demonstrates a distended suprapatellar bursa and posterior capsule (curved arrow). Two rounded, nondependent densities (arrows) are seen posteriorly within the fluid. A highly magnified, high contrast sagittal image through the periphery of the medial compartment (Fig. 34B) demonstrates a very small posterior horn of the medial meniscus (arrowhead); the anterior horn is essentially absent.

Diagnosis: Juvenile rheumatoid arthritis (JRA).

Discussion: Juvenile rheumatoid arthritis is a systemic inflammatory condition in which joint disease is usually the most prominent manifestation. Many of the tissues involved in the inflammatory condition, the synovium, cartilage, and menisci, are invisible to conventional radiography yet may be demonstrated on MR. The osseous structures may eventually develop erosive disease, and changes secondary to hyperemia, growth retardation, and premature epiphyseal closure are readily depicted on conventional radiographs.

Early synovial disease is the most common manifestation of JRA and is manifest by an effusion and irregularity and thickening of the synovial margin, especially where it covers the fat pad (Fig. 34C). Pannus appears as nodular areas of decreased signal (on most imaging parameters) that project into the fluid. These changes are best demonstrated on T2-weighted images. Many patients with inflammatory arthritis have popliteal cysts, and some have massive cystic lesions about the knee in atypical locations (Fig. 34D).

The joint instability that occurs in children with JRA may be secondary to several causes. Hyperemia results in soft tissue laxity, but the abnormally small menisci found in 13 of 20 patients with JRA may also contribute. Meniscal changes in these patients vary from slightly small menisci to ones that are virtually absent (Fig. 34E). A number of proposed mechanisms for the meniscal change include destruction secondary to release of proteolytic enzymes, nutritional strangulation secondary to pannus (menisci are nourished by joint fluid), or part of a pattern of generalized growth retardation. Small menisci in a chronically inflamed joint have been seen by us in one patient with a nonspecific synovitis and in two adults with typical adult-type rheumatoid disease.

In the normal child's knee, the articular cartilage has a homogeneously intermediate signal greater than the subjacent cortical bone and overlying meniscus. In a study of 20 affected knees, Senac found that 14 patients showed cartilage loss, whereas plain radiographs showed joint space narrowing in only nine of the same patients.

Bone infarction may occur in JRA and has been noted on MR in patients with normal radiographs. The infarction may be metadiaphyseal; when epiphyseal, subarticular fracture and loose body formation may occur.

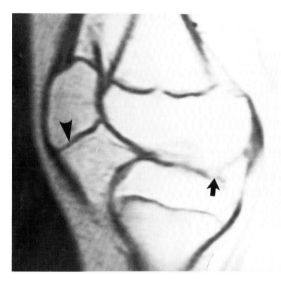 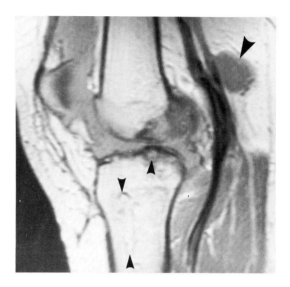

FIG. 34C. Juvenile rheumatoid arthritis (sagittal MR, TR/TE 2,000/20 msec). The epiphyses are enlarged, the lateral meniscus is virtually absent (*arrow*), and the inferior pole of the patella (*arrowhead*) is hypoplastic.

FIG. 34D. Juvenile rheumatoid arthritis (sagittal MR, TR/TE 2,000/20 msec). Metaphyseal and subchondral infarcts (*arrowheads*) are present in this patient. A para-articular cyst (*large arrowhead*) is seen in the popliteal space.

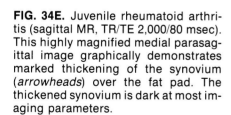

FIG. 34E. Juvenile rheumatoid arthritis (sagittal MR, TR/TE 2,000/80 msec). This highly magnified medial parasagittal image graphically demonstrates marked thickening of the synovium (*arrowheads*) over the fat pad. The thickened synovium is dark at most imaging parameters.

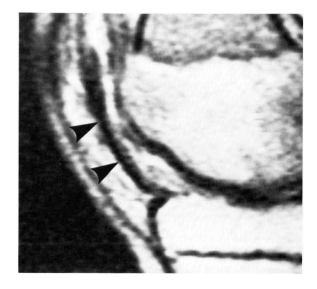

CASE 35

The patient is a 38-year-old man being evaluated for suspected internal derangement. The patient's symptoms are retropatellar and in the region of the anterior horn of the lateral meniscus. What is your diagnosis? (Fig. 35A. Sagittal MR, TR/TE 2,000/20 msec. Fig. 35B. Sagittal MR, TR/TE 2,000/20 msec.)

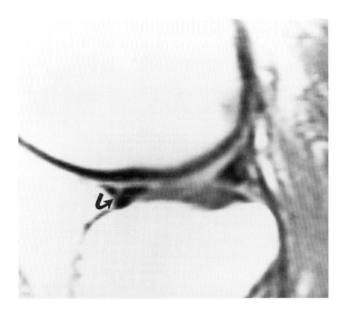 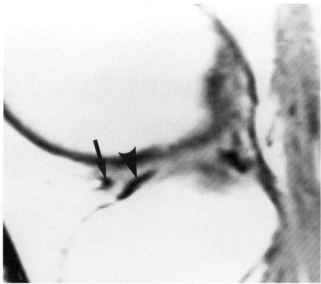

FIG. 35A.   **FIG. 35B.**

Findings: An oblique line of intermediate signal intensity is seen and appears to course obliquely across the anterior horn of the lateral meniscus and suggests a meniscal tear (Fig. 35A, arrow). In actuality the line represents the interface between the transverse ligament and meniscus. On the more central section (Fig. 35B), the transverse ligament (arrow) is further separated from what is now the rhomboid-shaped central attachment of the anterior horn of the lateral meniscus (arrowhead). This appearance can simulate a displaced meniscal fragment.

Diagnosis: Anterior horn lateral meniscus pseudo-tear related to the transverse ligament.

Discussion: The anatomy of the central attachment of the anterior horn of the lateral meniscus may simulate the appearance of a meniscal tear on MR as in the present case. The lateral meniscus is nearly circular in configuration and covers a greater percentage of the tibial plateau than its medial counterpart. Triangular in cross section, its width is constant over most of its length. The anterior horn of the lateral meniscus is secured to the tibial intercondylar eminence, ACL, and anterior horn of the medial meniscus by a fibrous band known as the transverse ligament. The transverse ligament arises as a rounded band of tissue from the anterior superior convex margin of the lateral meniscus immediately above and anterior to the rhomboid central attachment (Fig. 35C). It courses between the Hoffa fat pad and the tibial attachment of the ACL to insert ultimately onto the anterior superior corner of the anterior horn of the medial meniscus. In approximately 30% of knee MR examinations, a line of intermediate signal intensity is seen coursing obliquely through the anterior superior corner of the anterior horn of the lateral meniscus. This line simulates the appearance of a tear with the "peripheral fragment" representing the transverse ligament separated from the rhomboid central attachment of the meniscus as seen in this case. The complete extent of the transverse ligament can be demonstrated in 15% of MR examinations and is often well seen on axial sections. It has recently been suggested that the lateral inferior genicular artery, which is closely applied to the meniscus as it wraps around the knee, may also simulate the appearance of a tear of the anterior horn of the lateral meniscus.

The central ligamentous attachments of the menisci may also cause diagnostic confusion with regard to meniscal tears and in particular with regard to abnormalities involving the anterior horn of the lateral meniscus. The ligamentous attachment of the anterior horn of the lateral meniscus can usually be seen on one or two images (5 mm) prior to its insertion on the tibia and can be confused with tears of the fibrocartilagenous meniscus itself. Whereas the meniscal horn is essentially an equilateral triangle with apex pointing transversely across the tibial plateau, the central ligamentous attachment is characteristically more rhomboid and is oriented obliquely upward. The transition point between the meniscus and ligamentous attachment is the site of origin of the transverse ligament; tissue central to the transverse ligament origin is not fibrocartilage. Signal within a central attachment should thus not be mistaken for a meniscal tear (Fig. 35D). It may also be noted that isolated tears of the anterior horn of the lateral meniscus are unusual. Central ligamentous detachments may occur but are also exceedingly uncommon.

POSTERIOR

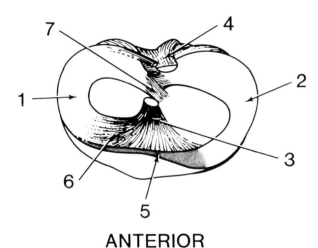

ANTERIOR

FIG. 35C. Schematic diagram of the transverse ligament and central ligamentous attachments. The articular surface of the tibia is visualized from above. The site of origin of the transverse ligament from the transition point between the anterior horn of the lateral meniscus and and its central attachment is demonstrated. Tissues central to the transverse ligament origin is not fibrocartilage. (*1*) lateral meniscus, (*2*) medial meniscus, (*3*) ACL, (*4*) PCL, (*5*) transverse ligament, (*6*) central attachment anterior horn lateral meniscus, (*7*) central attachment posterior horn.

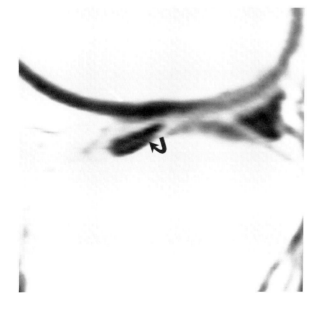

FIG. 35D. Anterior horn, central attachment (sagittal TR/TE 2,000/20 msec). On this section an apparent focus of grade 3 signal traverses the "anterior horn" of the lateral meniscus (*arrow*). In actuality this section is central to the origin of the transverse ligament and represents the more rhomboid-shaped central attachment. Signal within the ligamentous central attachments is common and should be differentiated from a meniscal tear.

CASE 36

Following a skiing injury, this man had the rapid onset of a joint effusion and pain. What is the major pathologic lesion, and what is the significance of the finding depicted by the arrow? (Fig. 36A. Sagittal MR, TR/TE 2,000/80 msec. Fig. 36B. Sagittal MR, TR/TE 2,000/80 msec.)

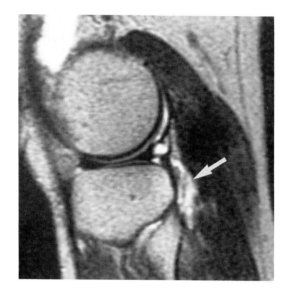 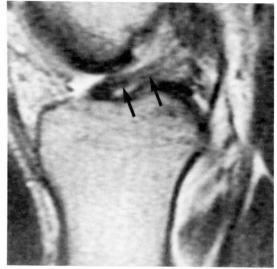

FIG. 36A.   **FIG. 36B.**

Findings: Figure 36A demonstrates fluid posterior to and within the popliteus muscle beneath the tendon (arrow). In Fig. 36B, a complete tear of the ACL is seen with the ligament laying on the tibia (arrows). Fluid is present behind the tibia and within the popliteus muscle.

Diagnosis: Tear of the ACL with an associated tear of the posterolateral capsule.

Discussion: The popliteus tendon is an extra-articular structure. Its anteromedial surface is convered by a synovial lined sheath that normally communicates with the knee joint (Fig. 36C,D). The presence of fluid within the sheath, or the presence of a small amount of fat adjacent to the sheath at the point at which the tendon penetrates the posterolateral meniscus has been described as a pseudo-tear of the meniscus. In this case, fluid is *posterior* to the popliteus tendon and extends down the tendon and around the muscle. Although there is no pathologic proof in this case, the lesion is thought to represent a capsular tear. Such tears have been seen in more than 50% of acute ACL tears (Fig. 36E). The natural history of the capsular disruption and its contribution to anterolateral instability of the knee is unknown.

Several other abnormalities could produce an MR pattern such as this. Rupture of the popliteus tendon is a rare and difficult-to-diagnose injury that is associated with a tear of its sheath (Fig. 36F). A peripheral meniscal tear can similarly produce a capsular disruption. One must look carefully at the meniscal fasicles; in individuals with a joint effusion, the fasicles can usually be resolved on long TR/TE sequences.

In this instance, the patient had an ACL tear by MR and arthroscopy, but no lateral meniscal tear was present and no other abnormalities were found at surgery.

C

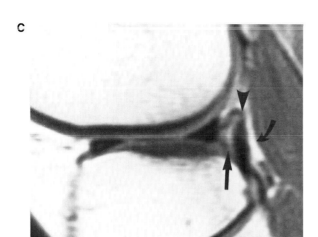

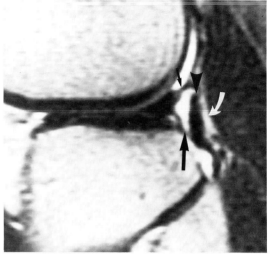

D

FIG. 36C,D. Normal popliteal tendon. **C.** TR/TE 2,000/20 msec; **D.** TR/TE 2,000/80 msec. A small amount of fluid (*arrow*) is present anterior to the popliteal tendon (*arrowhead*), but the tendon is extrasynovial. Only fat (*curved arrows*) is present posterior to the tendon. The superior (*small arrow*) and inferior (*large arrow*) fasicles are well seen on the second echo image.

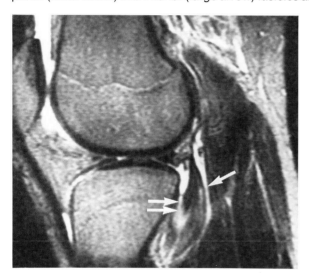

FIG. 36E. Torn capsule (sagittal MR, TR/TE 2,000/80 msec). Fluid is present posterior (*arrow*) and anterior (*double arrow*) to the popliteus muscle/tendon; it dissects along the muscle.

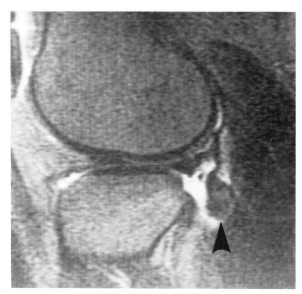

FIG. 36F. Torn popliteal muscle/tendon (sagittal MR, TR/TE 2,000/80 msec). The popliteal muscle/tendon appears retracted (*arrowhead*), and fluid is present around the "mass." There is no proof in this case.

CASE 37

The patient is a 28-year-old football player who sustained a direct injury to the medial aspect of the left knee. Where is the high signal intensity located? What is the anatomic function of this structure, and how is it related to patellar tracking? What is the nature of the fluid collection, and what factors account for the decreased signal within the center of this collection? (Fig. 37A. Coronal MR, TR/TE 2,000/80 msec.)

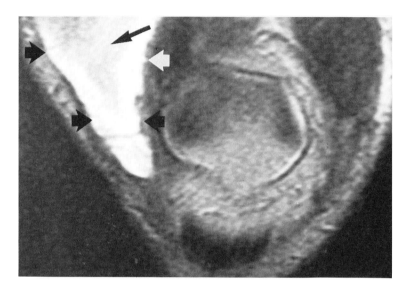

FIG. 37A.

Findings: On this T2-weighted image, a large high signal intensity fluid collection representing edema and hemorrhage is seen within the vastus medialis muscle (arrows). The medial retinaculum is predominantly intact. The signal intensity within the center of the fluid collection is slightly decreased (long arrow).

Diagnosis: Extensive tear of the vastus medialis with intramuscular hematoma.

Discussion: The injury in the present case resulted secondary to a direct blow incurred during a professional football game. An extensive hematoma within the vastus medialis was identified and was surgically evacuated. The medial retinaculum was predominantly intact. The diminished signal intensity within the center of the hematoma on the coronal T2-weighted image relates to the magnetic susceptibility effects of deoxyhemoglobin within intact red blood cells. This finding is most evident at high field strengths. The appearance of blood on MR is quite complex and is influenced by multiple factors including location, configuration, age of the injury, and field strength of the imaging system.

The vastus medialis muscle serves to counteract the naturally occurring lateral deviating vector (K vector) that results from the normal valgus position of the knee (Q angle) (see Case 18). The lowest fibers of the vastus medialis arise from the tendon of the adductor magnus and pass almost horizontally to the common tendon and medial border of the patella. This part of the muscle is referred to as the vastus medialis obliquus (VMO) and is particularly important in patellofemoral alignment. Abnormalities of the VMO predispose to patellar instability. The tendency toward lateral subluxation of the patella is increased by VMO weakening, high insertion of the VMO on the patella, and a stretched VMO and medial retinaculum secondary to dislocation and injury.

The vastus medialis (as does the vastus lateralis) contributes a fibrous expansion that is critical to the maintenance of normal patellar alignment and motion. The expansions, known as the medial and lateral retinacula, are attached to the margins of the patella and patella tendon and extend backward on both sides as far as the corresponding collateral ligament and downward to the condyles of the tibia (Fig. 37B,C). Injuries to the retinaculum are associated with major knee trauma and patellar dislocation and are well characterized by MR (Fig. 37D,E).

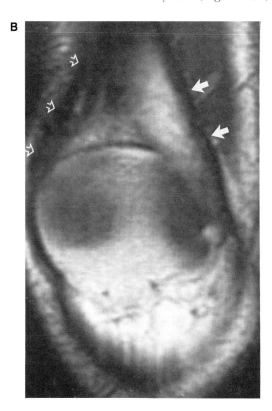

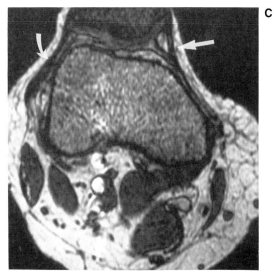

FIG. 37B,C. Normal retinaculum. B. Coronal TR/TE 800/20 msec. On this magnified coronal section the medial (*arrows*) and lateral retinaculum (*open arrows*) are seen as they insert on the patella. C. Axial ''GRASS'' TR/TE 80/13 theta 80. The typical Y-shaped configuration of the medial retinaculum is demonstrated (*arrow*). The lateral retinaculum is straight (*curved arrow*).

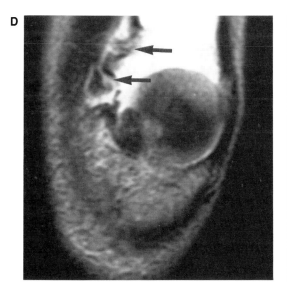

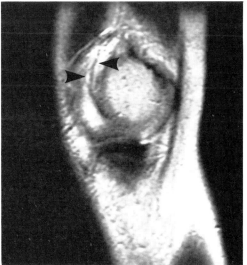

FIG. 37D,E. Medial retinaculum tear. D. Coronal TR/TE 2,000/80 msec. A far anterior coronal image from an acute knee series demonstrates high signal intensity fluid within the suprapatellar bursa. The medial retinaculum demonstrates an abnormal wavy appearance consistent with a retinacular disruption (*arrows*). E. Sagittal TR/TE 2,000/80 msec. On this sagittal section through the medial aspect of the knee, the torn retinaculum is again demonstrated (*arrowheads*).

CASE 38

This is a 36-year-old woman with medial joint line pain. Is this meniscus normal? What are some causes of small menisci? (Fig. 38A. Sagittal MR, TR/TE 800/20 msec.)

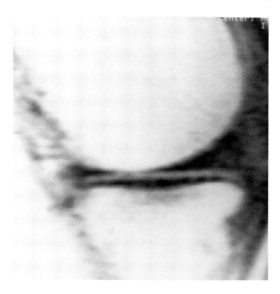

FIG. 38A.

Findings: The meniscus is small in overall size with a disproportionately small anterior horn. The meniscal contours are normal, and no significant intrameniscal signal is noted (Fig. 38A). The possibility of a vertical tear (bucket handle) with centrally displaced fragment must be strongly considered.

Diagnosis: Normal small meniscus.

Discussion: In addition to intrameniscal signal, morphological alterations in size and contour are also a characteristic of meniscal tears. The peripheral rim ("bucket") of the vertical longitudinal tear may be devoid of signal and recognized solely on the basis of focal diminution in size (see Fig. 10A). Intrasubstance tears of the posterior horn may be present and manifest as a decrease in height so that it is equal to or less than that of the anterior horn (Fig. 38B). The sharpness of the tip of the free edge of the meniscus must also be critically evaluated; a blunted free edge may be the only manifestation of a meniscal tear, and often coronal images are required for optimal evaluation (see Fig. 4A,B).

Detection of focal alteration in meniscal size is usually less problematic than assessment of global (symmetric) diminution in size. Meniscal dimensions (see Case 10) provide a baseline for evaluation and comparison with the opposite extremity and can be helpful in assessment of normal variations (hypoplastic menisci).

The differential diagnosis of the small meniscus includes: (a) displaced ("bucket handle") tear, (b) prior partial meniscectomy, (c) inflammatory arthritis, and (d) normal variation. In displaced bucket handle tears, the remaining periphery ("bucket") demonstrates an asymmetric triangular shape and/or apical angle. Coronal images are essential for depiction of the displaced fragment ("handle"). A low signal scar in the fat pad may provide an important clue as to the etiology of the small meniscus. With partial meniscectomies, the meniscal flap is commonly resected, and the remaining peripheral portion is trimmed into a roughly triangular shape. Following a complete meniscectomy, fibrous tissue may proliferate from the highly vascularized meniscal capsular region. The regenerated tissue may resemble the meniscus in shape but is only 2 to 4 mm in width (Fig. 38C). Patients with JRA have been observed to have uniformly small menisci (Fig. 38D). It has been speculated that this results from "starvation" of the menisci related to the markedly thickened accompanying synovium interfering with normal meniscal nutrition obtained from the synovial fluid. Small menisci have also been described in adults with rheumatoid and nonspecific synovitis, in whom arthroscopy revealed the meniscus to be enveloped in a markedly thickened synovial membrane (Fig. 38E). The meniscus in the patient illustrated in this case was small but otherwise normal on arthroscopic inspection. No tear was identified, and no intra-articular pathology was identified to account for the patient's symptoms.

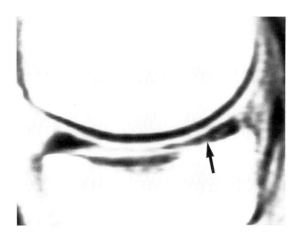

FIG. 38B. Flattened meniscus (sagittal MR, TR/TE 2,000/20 msec). A small focus of grade 3 signal is seen along the undersurface of the meniscus (*arrow*). The posterior horn is markedly decreased in height representing meniscal collapse secondary to a tear. The normal posterior horn should be taller than the anterior horn of the medial meniscus.

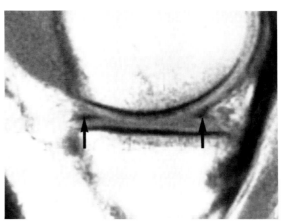

FIG. 38C. Post menisectomy (sagittal TR/TE 2,000/20 msec). The patient is status post–subtotal menisectomy. The peripheral remnant is small but roughly triangular (*arrows*). The contour of postoperative menisci are often best assessed on T2-weighted images, which provide greater contrast between the meniscus and synovial fluid (see Fig. 24C,D). There is mild decreased signal within the subchondral bone of the femoral condyle, a finding reflecting degenerative arthritis commonly seen in postoperative knees.

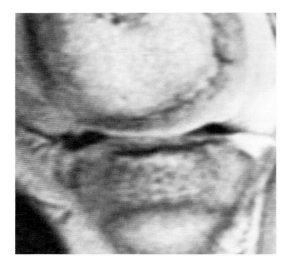

FIG. 38D. Juvenile rheumatoid arthritis (sagittal TR/TE 2,000/20 msec). There is enlargement of the femoral ossification center and osseous erosions. The meniscus is small.

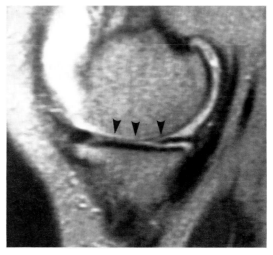

FIG. 38E. Adult rheumatoid arthritis (sagittal TR/TE 2,000/80 msec). In this 28-year-old woman with established rheumatoid arthritis, there is marked thinning and destruction of articular cartilage particularly well demonstrated involving the medial femoral condyle (*arrowheads*). The meniscus, particularly the posterior horn, is diminutive in size.

CASE 39

A 28-year-old man had a 3-week history of gradual onset of pain in his left knee. On physical examination he had a small joint effusion and lateral joint line pain. Diagnostic arthroscopy was normal. How would a preoperative MRI (as opposed to this postoperative study) have helped in assessment of the patient? (Fig. 39A. Coronal MR, TR/TE 800/20 msec.)

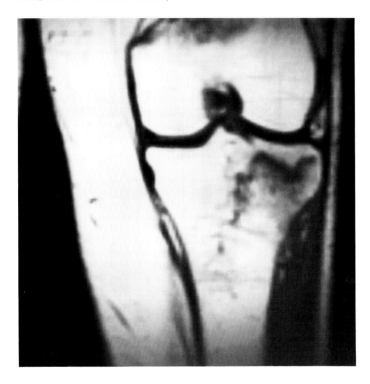

FIG. 39A.

Findings: There is a long serpiginous line of decreased signal seen in the lateral tibial metaphysis surrounded by a zone of presumed edema. The lesion abuts the lateral tibial cortex but does not extend to the articular surface.

Diagnosis: Stress fracture, linear type.

Discussion: This lesion is a classic fatigue fracture occurring in a "weekend warrior" who stressed his skeleton beyond the load it was accustomed to handling. At the time of presentation, the patient had a normal plain radiograph. Because of the presumed clinical diagnosis of a meniscal tear, he underwent arthroscopy as the first examination. When no lesion was found and his pain persisted, the patient was sent for the above MRI, which revealed this clearly extra-articular abnormality (Fig. 39A). A follow-up radiograph (Fig. 39B) several weeks later confirmed the presence of a lesion with a topography identical to that seen on the MRI.

A stress fracture is the final chapter in a series of events in bone more properly termed *a stress response*. As discussed in Case 14, remodeling of lamellar bone occurs initially during abnormal stresses. Subcortical infraction with edema occurs in advance of the microfractures. It is the leading edge of edema/hemorrhage/hyperemia that probably accounts for the nonlinear component of abnormal signal found in stress fracture (Fig. 39C).

Although it is generally accepted that MR is more sensitive to changes in the composition of marrow than is plain radiography, there are no data available regarding the comparative value of MRI and scintigraphy in detection of stress fractures. Bone scanning, although highly sensitive, lacks specificity; OCD, meniscal injury, osteoarthritis, and osteonecrosis might all produce a positive bone scan. One author has advocated MRI over bone scanning for suspected hip fractures because of the improved specificity and greater degree of anatomic information obtainable from the MR examination.

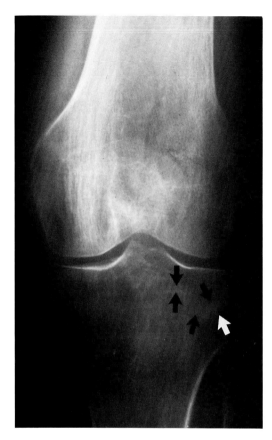

FIG. 39B. Stress fracture. Three weeks following the MR, a repeat plain radiograph confirmed the presence of a cancellous stress fracture (*arrows*).

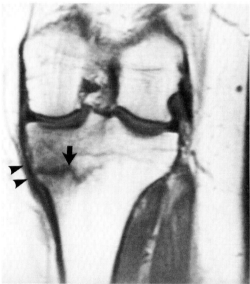

FIG. 39C. Stress fracture (coronal MR, TR/TE 700/20 msec). A metaphyseal stress fracture (*arrow*) is present. There is thickening and presumed edema (*arrowheads*) of the medial soft tissues.

CASE 40

A 36-year-old woman presents with recent knee injury and evidence of internal derangement. What is your diagnosis? What factors contribute to this appearance? What MR features are characteristic of an ACL tear? (Fig. 40A. Sagittal MR, TR/TE 2,000/20 msec.)

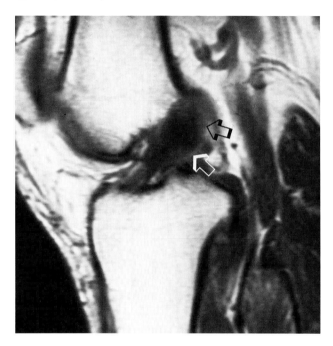

FIG. 40A.

Findings: On this sagittal section (Fig. 40A), the tibial attachment of the ACL is well demonstrated and appears normal. There is an apparent ovoid-shaped low signal intensity "mass" seen in the region of the femoral attachment of the ACL (open arrows).

Diagnosis: Anterior cruciate ligament pseudo-tear.

Discussion: Acute partial and complete ACL tears are often associated with a "mass" of moderate signal intensity on T1-weighted images in the region of the proximal attachment site of the tendon to the femoral condyle. The appearance of a torn ACL may be mimicked, as in this case, by a partial volume effect with the lateral femoral condyle, producing a "pseudo-mass." T2-weighted images may be of value in allowing the distinction of pseudo-mass from a true tear to be accomplished. With the partial volume pseudo-mass, there is generally no signal change on T2-weighted images in distinction to the true ACL tear in which signal may increase secondary to fluid and hemorrhage within and around the torn ligament (Fig. 40B,C). Additionally, demonstration of an intact ligament on the coronal sequence may allow a high degree of certainty in excluding a true ACL tear. The absence of a significant hemarthrosis may represent a helpful secondary sign.

An additional potential diagnostic difficulty in evaluating the ACL is the occasional failure to visualize a completely normal ligament. This may cause confusion with a chronic ACL tear. An additional spin-echo sequence at 3-mm increments or a gradient refocused sequence at 1.5-mm increments may be reassuring in accomplishing an accurate diagnosis (Fig. 40D,E). In evaluating both the ACL "pseudo-tear," as well as in excluding chronic ACL tear, it is critical to obtain a section through the true lateral intercondylar notch. This section is characterized by an oblique low signal line (Fig. 40F), and a normal ACL should be seen on this section.

B 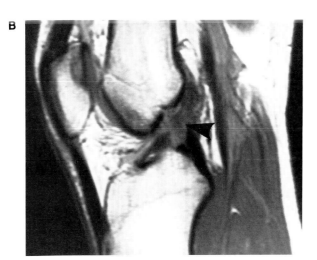 C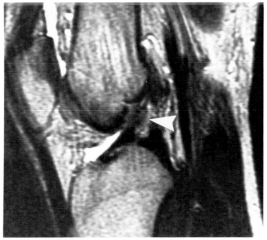

FIG. 40B,C. Anterior cruciate ligament tear with "true" mass. **B.** Sagittal TR/TE 2,000/20 msec. This section obtained parallel to the lateral intercondylar notch demonstrates a "mass" of intermediate signal intensity near the expected femoral attachment of the ACL (*arrowhead*). **C.** Sagittal TR/TE 2,000/80 msec. In contrast to an ACL "pseudo-tear," the "mass" demonstrates focally increased signal on the T2-weighted image representing edema and hemorrhage (*arrowhead*). T2-weighted image can be of particular value in assessment of ACL injuries.

D 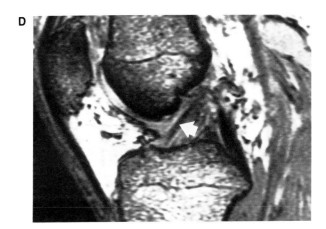 E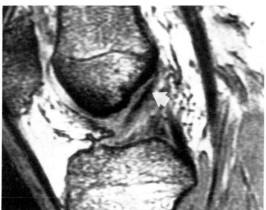

FIG. 40D,E. Normal ACL. Sagittal TR/TE 55/13 theta 80 msec. These two images are from a three-dimensional volume acquisition utilizing a gradient refocused echo technique (GRASS) to demonstrate the ACL to advantage (*arrows*). This sequence provides 64 contiguous sections and can be of value in assessment of the ACL when diagnostic uncertainty exists on the routine spin-echo sequence.

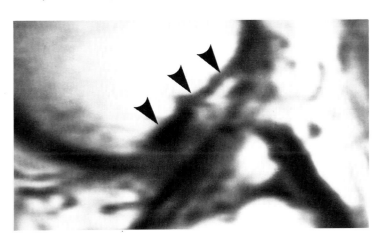

FIG. 40F. Sagittal TR/TE 2,000/20 msec. A true sagittal section through the lateral intercondylar notch is characterized by an oblique low signal line that should be visualized throughout the section (*arrowheads*). The normal ACL should be seen on this section. Compare this image with Fig. 40A in which the proximal extent of the lateral intercondylar notch is imaged obliquely and contributes to the appearance of a "pseudo-tear."

CASE 41

This 25-year-old woman had had knee pain for several months. Physical examination revealed an effusion and a palpable snap with knee flexion. The patient was referred for an MRI (Fig. 41A). What is the nature and significance of the lesion indicated by the arrows? (Fig. 41A. Sagittal MR, TR/TE 2,000/80 msec.)

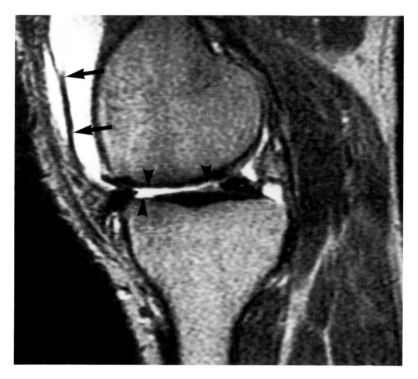

FIG. 41A.

Findings: A thick dark band (arrows) runs vertically in the suprapatellar bursa, surrounded by bright synovial fluid. There is a tear of the posterior medial meniscus and extensive loss of articular cartilage (arrowheads) over the femoral condyle.

Diagnosis: Thickened medio-patellar plica.

Discussion: There are three identifiable ridges of synovium within the knee joint capsule thought to represent residua of the membranes that divided the knee into separate compartments in the fetus. These folds, known as plicae, are found in up to 80% of adults. The infrapatellar fold, or ligamentum mucosum, is the most commonly encountered and extends from the fat pad up into the intercondylar notch just anterior to and paralleling the ACL. Its only significance is that it may rarely divide the knee into separate medial and lateral compartments. The suprapatellar plica crosses the bursa from front to back near the superior pole of the patella. In most cases, it is fenestrated, but rarely it completely divides the upper portion of the bursa from the remainder of the joint. There is considerable controversy as to whether a suprapatellar plica can produce symptomatology.

It is the medio-patellar plica that has most commonly been implicated as a cause of knee joint pathology. It runs vertically along the medial wall of the capsule from the suprapatellar plica above to the fat pad below. The incidence of symptomatic plica in the population is unknown, although it is probably small. When the leg is in full extension, the medio-patellar plica lies along the medial femoral condyle, just posterior to the patella. With flexion and external tibial rotation, the plica becomes wedged between these two bones; it is in this position that it is vulnerable to trauma transmitted through the patella. Edema, thickening, fibrosis, and synovitis result. Once thickened, the plica can produce chondromalacia of the patella, either by direct erosion or by tethering of the quadriceps mechanism (Fig. 41B). The snapping knee syndrome is due to the plica snapping against the anterior medial femur. Pain may result from compression of the plica or by traction on the fat pad where multiple nerve endings are present. Some authors suggest that a thickened plica is the result of an intra-articular disorder rather than a cause. Although most patients respond to conservative care, some plicae require surgical intervention. A normal plica can be depicted on MRI when excessive joint fluid is present, and mere demonstration of such a band is not necessarily pathologic (Fig. 41C). T2-weighted sagittal or axial images can be used to demonstrate both the plica and the changes that may be induced in the medial femoral condyle and patella.

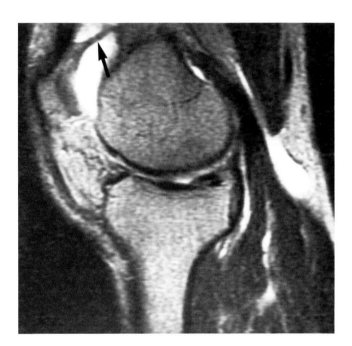

FIG. 41B. Thickened suprapatellar plica (sagittal MR, TR/TE 2,000/80 msec). An abnormal fold of synovium is seen extending from the front to the back of the bursa (*arrow*) just above the patella.

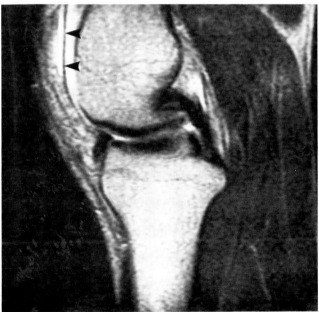

FIG. 41C. Normal plica (sagittal MR, TR/TE 2,000/80 msec). A normal medio-patellar plica is depicted (*arrowheads*). The thickness should be contrasted with the abnormal plica in Fig. 41A.

CASE 42

The patient is a 26-year-old football player who sustained a "clipping injury." What is your diagnosis? How are these injuries classified? What are the important associated abnormalities that should be looked for? (Fig. 42A. Coronal MR, TR/TE 2,000/20 msec. Fig. 42B. Coronal MR, TR/TE 2,000/80 msec.)

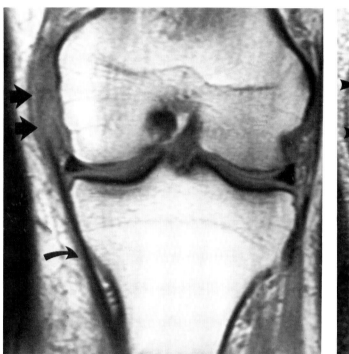

FIG. 42A.

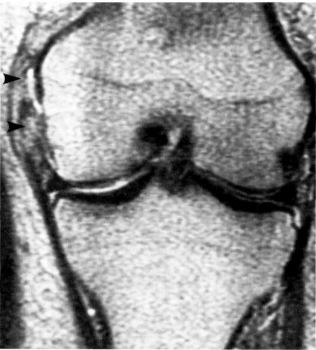

FIG. 42B.

Findings: Increased distance is seen between the medullary and subcutaneous fat. There is complete disruption of the proximal (femoral) attachment of the TCL (Fig. 42A, arrows). The distal (tibial) attachment is normal (curved arrow). On the T2-weighted image (Fig. 42B), there is mildly augmented signal intensity secondary to extra-articular edema and hemorrhage (arrowheads).

Diagnosis: Grade 3 MCL tear.

Discussion: Tears of the medial ligamentous structures result from valgus stress on the knee and follow a predictable progression of injury. With grade 1 tears, only a limited number of deep capsular fibers are torn, and the ligament appears normal in thickness and contour on MR. Extracapsular soft tissue edema and hemorrhage are well demonstrated and manifest on MR by decreased signal on T1-weighted images and increased signal on T2-weighted images. In grade 3 tears, both the deep capsular ligament and superficial ligament (TCL) are completely disrupted. Complete TCL rupture is associated with marked thickening, discontinuity, and serpiginous contours of the affected ligament on MR. Increased distance is seen between the high intensity signal from the medullary and subcutaneous fat. The site of the tear can frequently be localized precisely, particularly on T2-weighted images, which we routinely obtain in all cases of acute knee trauma for best depiction of collateral ligament injuries. Grade 2 injuries have features of both grade 1 and grade 3 tears and are less precisely characterized by MR. We commonly classify injuries as either grade 1 or grade 2/3. Tears of the MCL may be "isolated" or more commonly associated with a number of related injuries. The nature of the interface between the foot and the surface often determines the nature and extent of knee pathology. For example, isolated MCL injuries are relatively common in hockey players. The interface between skate and ice is more forgiving than between shoe and turf or ski and snow. The skate may become trapped in a valgus sense but without the rotary deceleration of the lower limb so common in skiing. Among the most commonly encountered associated injuries with high grade MCL tears are bone bruises (medullary trabecular microfractures) (Fig. 42C). When seen in association with MCL tears, these lesions may involve the tibia, femur, or both sides of the joint. Bone bruises typically involve the epiphyseal region and a portion of the metaphysis and demonstrate a geographic area of signal loss on T1-weighted images that increases in signal on T2-weighted images (Fig. 28A,B). Tears of the ACL are also commonly associated with both medial collateral and lateral compartmental bony injuries. Tears of the medial meniscus, in particular, longitudinal tears of the posterior horn, and peripheral tears including meniscal capsular separations are also often identified.

Many complete MCL tears may be satisfactorily managed conservatively. The critical determinant for operative versus nonoperative management is the integrity of the posterior oblique capsular ligament. When this ligament is intact, the natural history of the nonoperatively treated MCL is quite satisfactory. When the posterior oblique ligament is involved, a more aggressive approach with surgical MCL reconstruction may be elected, particularly in the elite athlete. On arthroscopy, involvement of the posterior oblique ligament is recognized by the presence of a menisco-capsular separation, disruption of the synovium, or hemorrhagic blush of the synovium. Meniscocapsular separations and peripheral meniscal tears involving the posterior horn of the medial meniscus can be detected utilizing MR (see Case 56).

An apparently thickened and abnormal-appearing medial collateral ligament may be seen in patients with medial compartmental osteoarthritis (Fig. 42D). This may simulate the appearance of a chronic MCL injury. The findings relate to buckling of the ligament secondary to varus deformity. The absence of edema and the continuity of the tibial collateral ligament allow distinction from acute tear, although a chronic injury could have such an appearance. Pellegrini-Stieda calcification, the classic radiographic manifestation of prior MCL injury, may be demonstrated on MR scans, although its appearance at times may be subtle (Fig. 42E).

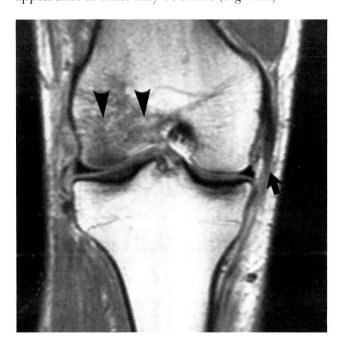

FIG. 42C. Bone bruise (coronal TR/TE 2,000/20 msec). The mid to distal TCL is discontinuous (*arrow*). There is extensive decreased signal within the subchondral bone of the lateral femoral condyle (*arrowheads*). The findings are typical of marrow edema seen with a bone bruise.

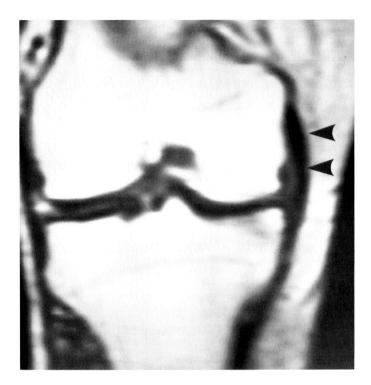

FIG. 42D. Medial compartment osteoarthritis (sagittal TR/TE 800/20 msec). There is generalized thickening of the TCL, particularly in its proximal portion (*arrowheads*). There is medial compartment joint space narrowing with osteoarthritis. The findings can be difficult to distinguish from a chronic tear.

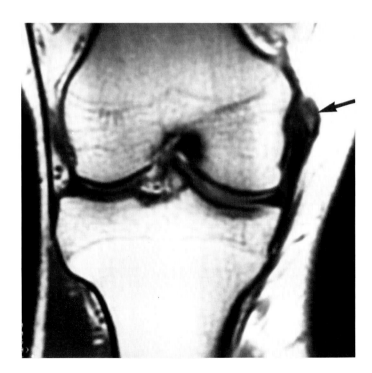

FIG. 42E. Pellegrini-Stieda (coronal TR/TE 800/20 msec). There is thickening of the proximal TCL compatible with prior injury. There is a focus of high signal intensity likely reflecting an area of ossification and correlating with a radiographically evident Pellegrini-Stieda lesion (*arrow*).

CASE 43

The patient is a 36-year-old man with intermittent symptoms of knee "popping." He was sent to exclude internal derangement. What is your diagnosis? What is the accuracy of knee MR? Are false-positive or false-negative interpretations more of a diagnostic problem? What are some factors responsible for disagreement between MR and arthroscopy? (Fig. 43A. Sagittal MR, TR/TE 2,000/20 msec.)

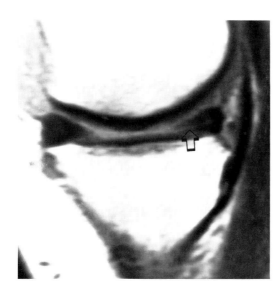

FIG. 43A.

Findings: On this sagittal section through the posterior horn of the medial meniscus (Fig. 43A), a well-defined line of high signal intensity courses through the posterior horn. The line appears to extend to the tibial articular surface (arrow). The findings are highly suggestive of an oblique tear.

Diagnosis: Minimal free edge fibrillation but otherwise normal meniscus at arthroscopic inspection.

Discussion: The present case is highly suggestive of a meniscal tear by presently established MR criteria. An oblique line of high signal courses through the posterior horn of the medial meniscus and extends to an articular surface corresponding to a grade 3A meniscal abnormality. Menisci demonstrating these MR findings may initially appear normal to gross inspection and must often be extensively probed to delineate the tear. In this patient the arthroscopist noted that the medial compartment was "tight," limiting visualization of the posterior horn. This raises the possibility that this false-positive MR could represent a false-negative arthroscopy. Documenting the abnormality on an orthogonal sequence (coronal) can allow added confidence in diagnosis. The majority of meniscal tears can be detected on both sagittal and coronal sections.

Multiple studies of varying size have compared the agreement of MR with arthroscopy. In the largest series to date, preoperative MR examinations of 459 virgin menisci were compared with arthroscopic findings. The overall agreement of MR and arthroscopy in meniscal evaluation was 93%. Of interest, in a subgroup in which the surgical procedures were performed by a single knee subspecialist arthroscopist, the MR-arthroscopy agreement rate was even higher (95%). Several other large series have confirmed this extremely high agreement rate of MR and arthroscopy.

False-positive MR examinations have been a greater source of error than have false-negative examinations. A number of factors may contribute to this problem. The

CASE 44

A 25-year-old woman had swelling and pain in her knee for several years. What is the diagnosis? (Fig. 44A. Plain anteroposterior radiograph. Fig. 44B. Sagittal MR, TR/TE 700/20 msec. Fig. 44C. Sagittal MR, TR/TE 1,500/60 msec.)

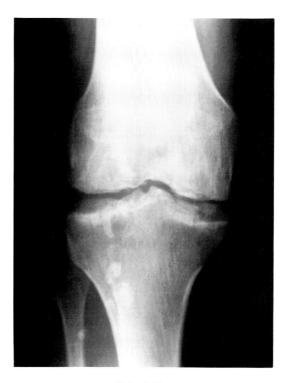

FIG. 44A.

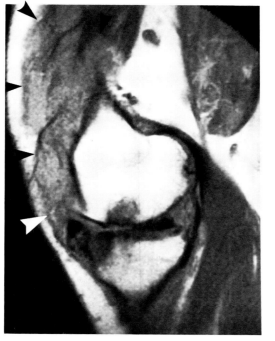

FIG. 44B.

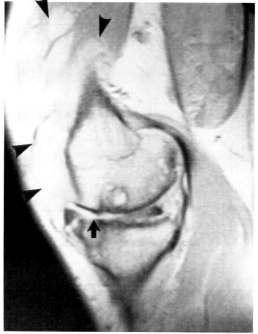

FIG. 44C.

CASE 43

The patient is a 36-year-old man with intermittent symptoms of knee "popping." He was sent to exclude internal derangement. What is your diagnosis? What is the accuracy of knee MR? Are false-positive or false-negative interpretations more of a diagnostic problem? What are some factors responsible for disagreement between MR and arthroscopy? (Fig. 43A. Sagittal MR, TR/TE 2,000/20 msec.)

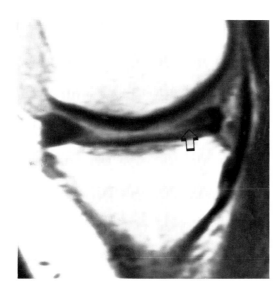

FIG. 43A.

Findings: On this sagittal section through the posterior horn of the medial meniscus (Fig. 43A), a well-defined line of high signal intensity courses through the posterior horn. The line appears to extend to the tibial articular surface (arrow). The findings are highly suggestive of an oblique tear.

Diagnosis: Minimal free edge fibrillation but otherwise normal meniscus at arthroscopic inspection.

Discussion: The present case is highly suggestive of a meniscal tear by presently established MR criteria. An oblique line of high signal courses through the posterior horn of the medial meniscus and extends to an articular surface corresponding to a grade 3A meniscal abnormality. Menisci demonstrating these MR findings may initially appear normal to gross inspection and must often be extensively probed to delineate the tear. In this patient the arthroscopist noted that the medial compartment was "tight," limiting visualization of the posterior horn. This raises the possibility that this false-positive MR could represent a false-negative arthroscopy. Documenting the abnormality on an orthogonal sequence (coronal) can allow added confidence in diagnosis. The majority of meniscal tears can be detected on both sagittal and coronal sections.

Multiple studies of varying size have compared the agreement of MR with arthroscopy. In the largest series to date, preoperative MR examinations of 459 virgin menisci were compared with arthroscopic findings. The overall agreement of MR and arthroscopy in meniscal evaluation was 93%. Of interest, in a subgroup in which the surgical procedures were performed by a single knee subspecialist arthroscopist, the MR-arthroscopy agreement rate was even higher (95%). Several other large series have confirmed this extremely high agreement rate of MR and arthroscopy.

False-positive MR examinations have been a greater source of error than have false-negative examinations. A number of factors may contribute to this problem. The majority of false-positive examinations have occurred within the posterior horn of the medial meniscus, the most difficult area to evaluate arthroscopically and one frequently requiring extensive probing particularly to exclude the confined intrasubstance cleavage tear. Diagnosis of these closed tears often may not be accomplished by visual inspection alone. Arthroscopy is highly operator-dependent; manual dexterity, three-dimensional perception, and surgical experience are all reported to be critical to its success. Surgeons who have practiced arthroscopy extensively for more than 4 years can expect to double the accuracy rate obtained in the first 4 years. Additionally, surgeons who judge themselves as knee subspecialist arthroscopists reported a higher order of agreement of MR with surgical findings than do general orthopedic surgeons. These several factors suggest that many of the MR-arthroscopy discrepancies may represent false-negative arthroscopies.

Difficulty in assessment of meniscal signal intensity by the interpreting radiologist also undoubtedly accounts for false-positive examinations. Minimal areas of signal confined to the free edge or meniscal apex may represent only fraying or fibrillation and must be distinguished from arthroscopically significant tears (Fig. 43B). Linear signal that equivocally extends to an articular surface must not be overinterpreted as representing a significant tear. Grade 3 signal that diminishes in intensity as it approaches an articular surface may represent a confined tear, and this possibility should be recognized. Confirmation of the extension of intrameniscal signal to an articular surface on an orthogonal view should be sought to minimize false-positive interpretations (Fig. 43C–F).

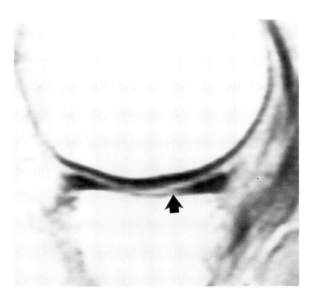

FIG. 43B. False-positive MR (sagittal TR/TE 2,000/20 msec). On this sagittal section through the posterior horn of the medial meniscus, there is minimal increased signal within the meniscal apex (*arrow*). The apex of the anterior horn demonstrates a sharp tip without signal. Although the present findings may be seen with small tears, only free edge fibrillation deemed clinically insignificant by the arthroscopist was identified in this case.

C 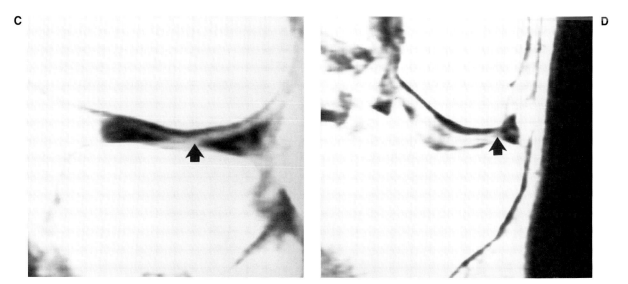 D

FIG. 43C,D. Free edge tear. Value of coronal images. **C.** Sagittal TR/TE 2,000/20 msec. On this sagittal section through the free edge of the lateral meniscus there is minimal increased signal that could represent either a small radial tear or simply mild fibrillation (*arrow*). **D.** Coronal TR/TE 800/20 msec. On the coronal section there is unequivocal blunting of the free edge confirming the presence of an arthroscopically significant tear (*arrow*).

E 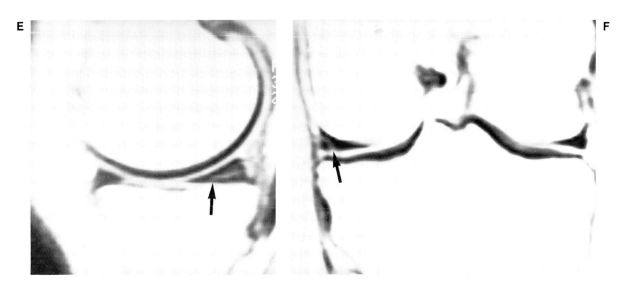 F

FIG. 43E,F. Oblique tear. Value of coronal images. **E.** Sagittal TR/TE 2,000/20 msec. On this magnified high contrast image through the posterior horn of the medial meniscus, an oblique line of intrameniscal signal is identified (*arrow*). The line does not clearly extend to the articular surface and would be interpreted as grade 2 (no tear). **F.** Coronal TR/TE 800/20 msec. On the coronal section, the line is seen to be steeply inclined and clearly extends to the tibial articular surface (*arrow*). This extension was missed on the sagittal views and would have resulted in a false-negative examination.

CASE 44

A 25-year-old woman had swelling and pain in her knee for several years. What is the diagnosis? (Fig. 44A. Plain anteroposterior radiograph. Fig. 44B. Sagittal MR, TR/TE 700/20 msec. Fig. 44C. Sagittal MR, TR/TE 1,500/60 msec.)

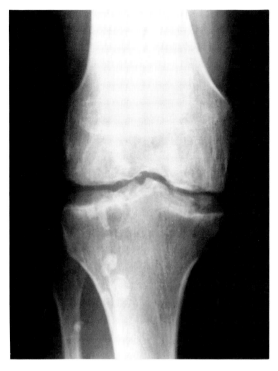

FIG. 44A.

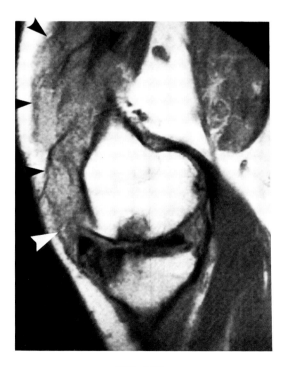

FIG. 44B.

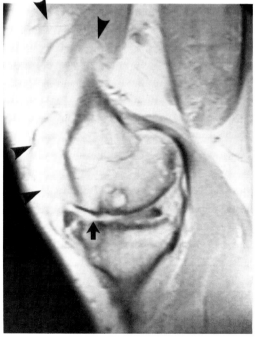

FIG. 44C.

Findings: The plain radiograph (Fig. 44A) reveals deformity and marked irregularity of the articular surfaces of the femur and tibia with subchondral cyst formation. Multiple phleboliths are present near and probably in the joint. The MR (Fig. 44B,C) demonstrates a mass involving the joint cavity, extending anterior and medial to the femur (arrowheads). There is loss of articular cartilage (Fig. 44C, arrow).

Diagnosis: Synovial hemangioma.

Discussion: Synovial hemangiomas most frequently involve the knee, although these benign tumors have been reported at the elbow, ankle, and a variety of tendon sheaths. They usually occur in adolescents with girls being affected more frequently than boys.

There are two pathologic types of synovial hemangioma, each having a different radiographic appearance. The localized pedunculated mass may produce mechanical difficulties at the knee, but it is easy to excise. It may be difficult to distinguish from the localized form of pigmented villonodular synovitis.

The lesion in this case represents the diffuse involvement of the synovial membrane. Surgically, these lesions are much more difficult to resect, and they tend to recur. Associated visceral or cutaneous vascular lesions may be found in 40% of patients. Both the Klippel-Trenaunay syndrome (varicose veins, soft tissue and bony hypertrophy, and cutaneous hemangiomas) and the Kasabach-Merritt syndrome (capillary hemangiomas and purpura) have been associated with an arthropathy similar to that in this case. The articular changes are quite similar to those reported in hemophilia. The pathogenesis is most likely repeated hemorrhage resulting in hemosiderin deposition, fibrosis of subsynovial tissue, and villous hypertrophy. Similar pathologic changes have been seen in hemophilia and instances of posttraumatic hemarthrosis. In this case, there is soft tissue swelling/mass, osteoporosis, epiphyseal overgrowth, deformed irregular articular surfaces with a beaten silver appearance, widening of the intercondylar notch, and squaring of the patella. Arthrography may demonstrate multiple confluent synovial masses, and arteriography may demonstrate the hypervascular tumors. All reported synovial hemangiomas are associated with phleboliths, and therefore conventional radiography is often sufficient for the diagnosis. MR, however, is of value in depicting the intra- and extra-articular extent of the tumor when surgical extirpation is contemplated (Fig. 44D).

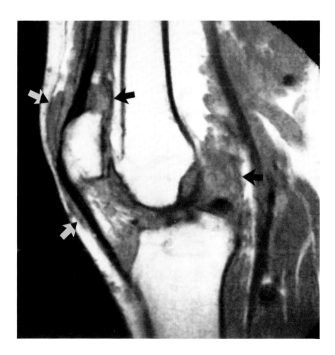

FIG. 44D. Synovial hemangioma (sagittal MR, TR/TE 500/20 msec). This less magnified view demonstrates involvement of the prepatellar, suprapatellar, and popliteal space soft tissues (*arrows*) by the tumor. (Courtesy of Ron Reinke, MD.)

CASE 45

A 25-year-old healthy woman had had a prior partial medial menisectomy. Since then, she has been bothered by recurrent joint effusions and a mass in the back of the knee. She had the sudden onset of pain and tenderness in her calf, which was red and swollen. What is the MR finding, and what is the differential diagnosis? (Fig. 45A. Sagittal MR, TR/TE 2,000/80 msec. Fig. 45B. Axial MR, TR/TE 2,000/80 msec.)

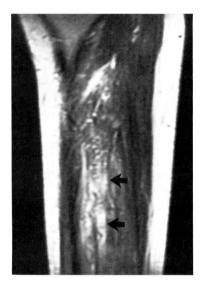

FIG. 45A.

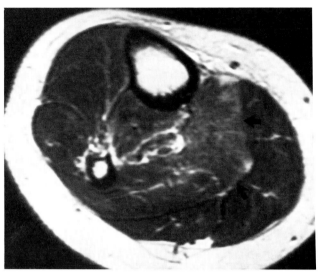

FIG. 45B.

Findings: There is increased signal in and around the soleus muscle (Fig. 45A,B; arrows) and the neurovascular bundle seen in both planes, but the muscle itself is intact. No discrete mass is identified. The abnormality is confined to the deep posterior compartment.

Diagnosis: Rupture of a popliteal cyst with the development of the pseudo thrombophlebitis syndrome.

Discussion: The pseudo thrombophlebitis syndrome refers to a *clinical* picture resembling thrombophlebitis, which occurs in patients in whom the venous system is patent. Rupture of a popliteal cyst with escape of fluid has been shown to produce symptoms identical to those in patients with acute venous obstruction.

Any condition associated with a chronic joint effusion and increased intra-articular pressures, such as an inflammatory arthritis, can lead to communication of the knee joint with the semimembranosis-gastrocnemious bursa. Progressive distention of the bursa can result in *dissection* of the cyst up or down the leg and subsequent presentation as a mass in the thigh or the ankle (Fig. 45C). The fluid remains encysted in a synovial membrane. With actual *rupture* of the cyst, the extravasated joint fluid escapes into the soft tissues. This inflammatory fluid is highly irritating and produces the symptoms seen with venous obstruction (pain, swelling, redness).

A venogram performed just before this MRI was normal; the patient was treated with anti-inflammatory drugs with a good clinical response. Two days later, however, her pain suddenly worsened and the swelling increased. A repeat MRI was unchanged, but a repeat venogram showed extensive deep venous clot (Fig. 45D). The effect of extravasated joint fluid from a popliteal cyst producing sufficient irritation to the calf veins to result in thrombosis might be termed the "pseudo-pseudo thrombophlebitis syndrome." It is conceivable that the initial venographic study induced subsequent thrombosis, a recognized complication of venography. Another alternative explanation is that the patient, being relatively confined to bed, experienced thrombosis secondary to inactivity. Regardless, it should be understood that the coexistence of venous thrombosis and ruptured popliteal cysts is well recognized.

The term *compartment syndrome* refers to a condition in which patients present with pain out of proportion to their injury (usually minimally displaced tibial fractures), sensory deficits, and intact pulses. Elevated compartment pressures above 30 mm Hg establish the diagnosis. Treatment consists of fasciotomy. On MR, the affected limb is increased in girth. Edema is present diffusely within a given compartment and is manifest by an increase in signal on long TR/TE sequences (Fig. 45E). In one instance, the veins proximal to an affected compartment appeared dilated.

The differential diagnosis of deep muscular edema must include thrombophlebitis, direct trauma, compartment syndrome, rupture of a popliteal cyst, cellulitis, and osteomyelitis.

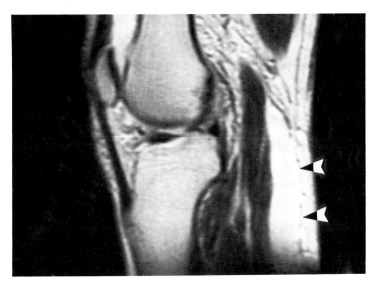

FIG. 45C. Dissecting popliteal cyst (sagittal MR, TR/TE 2,000/80 msec). A subcutaneous popliteal cyst (*arrowheads*) has dissected down the upper one-third of the leg. The fluid is sharply margin-ated because it is contained within the synovial membrane. This dissecting cyst appears different than when the cyst is ruptured, and no wall is identified.

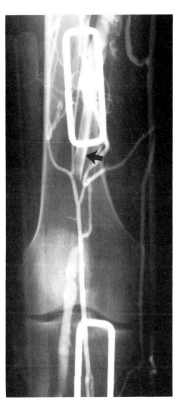

FIG. 45D. Deep venous throm-bosis. The second venogram demonstrates extensive deep venous clot (*arrow*).

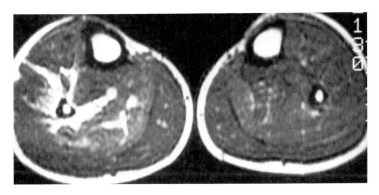

FIG. 45E. Compartment syndrome (axial MR, TR/TE 2,000/80 msec). The lateral compartment and to a lesser degree the gas-trocnemius muscle and the region of the neurovascular bundle demonstrate a diffuse increase in signal. This patient suffered trauma to his leg without fracture.

CASE 46

The patient is a 36-year-old man undergoing evaluation for internal derangement. To what structure does the arrowhead point? What is the significance of the findings in this individual? (Fig. 46A. Sagittal MR, TR/TE 2,000/20 msec.)

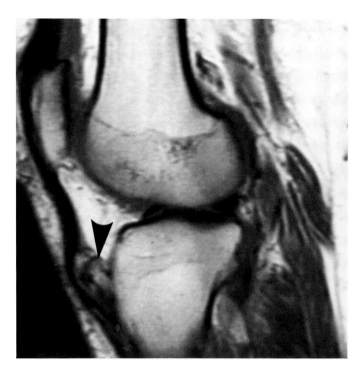

FIG. 46A.

Findings: There is extensive irregular ossification in the region of the anterior tibial tubercle (arrowhead). The findings extend into and merge with the attachment of the patellar tendon.

Diagnosis: Osgood-Schlatter disease (old).

Discussion: Osgood-Schlatter disease occurs in adolescence and is seen three times more commonly in boys than girls. Clinically, the condition is characterized by local pain and tenderness in the region of the anterior tibial tubercle that is frequently aggravated during activity and ameliorated with rest. Soft tissue swelling and firm masses may be palpated in the involved region. The etiology of the condition is believed to relate to a traumatically induced disruption of a portion of the patellar tendon from the developing tibial tuberosity with subsequent avulsion of fragments of cartilage and underlying bone.

There are three classic radiographic signs essential to the diagnosis of this condition: (a) soft tissue swelling anterior to the tibial tubercle, (b) thickening and irregularity of the patellar tendon, and (c) fragmentation and irregularity of the tibial tubercle. Magnetic resonance findings identified to date in this condition include: (a) localized diminution in signal intensity from medullary bone underlying the tuberosity, (b) focal thickening and slight increased signal involving the patellar tendon, and (c) localized soft tissue swelling anterior to the tibial tubercle.

The patient illustrated in this case had a prior history of Osgood-Schlatter disease. The extensive changes within the tibial tuberosity are compatible with this prior history. In the absence of this history, however, the present findings need to be differentiated from multiple ossification centers of the tibial tuberosity that may be seen as variations in normal ossification. It should be noted that a bursa may form between the residual ununited ossicles and the underlying tibial tuberosity, which may give rise to symptoms of bursitis in some individuals.

A similar condition also seen in adolescents occurs at the proximal end of the patellar tendon. It is referred to as Sinding-Larsen-Johansson disease (Fig. 46B). This condition is characterized clinically by local pain and tenderness on palpation and radiographically by separation and fragmentation of the lower pole of the patella associated with soft tissue swelling and occasionally calcifications in the region of the patellar tendon. The findings are well demonstrated on soft tissue technique radiography. As with Osgood-Schlatter disease, it is believed that the lesion is related to a traction phenomenon in which contusion or tendonitis in the proximal attachment of the patellar tendon can be followed by calcification and ossification. During the active phase of the disease, the presence of overlying soft tissue swelling allows differentiation of this condition from normal multiple ossification centers in the lower pole of the patella. The natural duration of the disease is approximately 3 to 12 months. Fragmentation of the inferior pole of the patella has also been associated with spastic paralysis. This is also believed to result from a traction phenomenon. Tendinitis of the patella tendon may also commonly be observed in athletes, in whom the term "jumper's knee" has been applied (Fig. 46C). In this condition, patellar subluxation or dislocation has been advanced as one of the common mechanisms of injury.

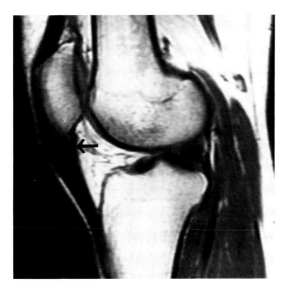

FIG. 46B. Sinding-Larsen-Johansson disease (sagittal TR/TE 800/20 msec). This section signal demonstrates a well-defined ossicle relating to the inferior pole of the patella (*arrow*). The findings are consistent with a prior history of this condition.

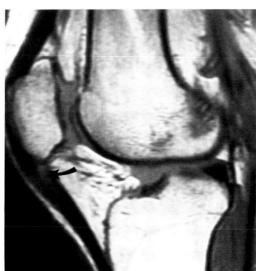

FIG. 46C. Jumper's knee (sagittal TR/TE 2,000/ 29 msec). This section demonstrates a focus of increased signal within a widened patellar tendon (*arrow*). The location immediately caudal to the lower pole of the patella is characteristic for this condition.

CASE 47

A 35-year-old man presented to an orthopedist with a tender mass on the lateral side of the knee. There are two related MR findings. What are they? (Fig. 47A. Coronal MR, TR/TE 800/20 msec. Fig. 47B. Sagittal MR, TR/TE 2,000/80 msec.)

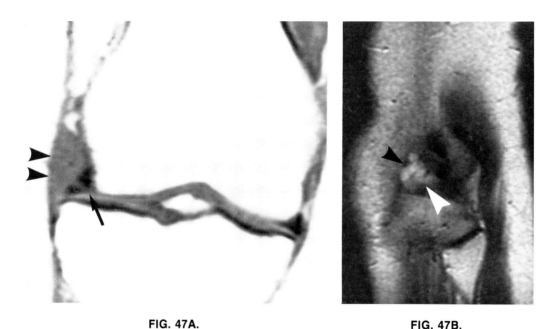

FIG. 47A.   FIG. 47B.

Findings: A magnified high contrast image shows a grade 3 signal (arrow) extending in a horizontal plane in the lateral meniscus in Fig. 47A. A fluid-filled mass (arrowheads) is seen just lateral and anterior to the lateral meniscus from which it is inseparable. The extreme lateral sagittal image (Fig. 47B) demonstrates the cystic nature of the mass (arrowheads).

Diagnosis: Meniscal cyst secondary to a horizontal cleavage tear of the lateral meniscus.

Discussion: Meniscal cysts are fluid collections that accumulate in the parameniscal region of either meniscus, although they are four times more common laterally than medially. In a detailed study of the histopathology of meniscal cysts, Ferrer-Roca found that cysts were invariably associated with horizontal meniscal tears, which extended into the parameniscal soft tissues. On the lateral side of the knee, the cysts tend to remain confined to the joint line. They insinuate themselves between the FCL and iliotibial band or occasionally dissect into the Hoffa fat pad. Medially, cysts frequently dissect along soft tissue planes. If they occur posterior to the MCL, they expand in an unrestricted manner and may ultimately present as a mass at some distance from the joint. Rarely, a meniscal cyst may erode bone. Recurrence of a cyst is common after surgery if the relationship to an underlying meniscal tear goes unrecognized.

A ganglion presents as a fluid-filled mass invariably communicating with a nearby joint but with no connection to the underlying meniscus. Therefore, although meniscal resection is not necessary in definitive treatment of ganglia, the capsular origin must be recognized and resected or recurrence will occur.

On MRI, both meniscal cysts and ganglia present as masses that are bright on long TR/TE sequences; T2-weighted images also improve detection of the communication between the joint and the mass. In patients with meniscal cysts, the associated meniscal tears are best seen on short TR/TE or long TR/short TE sequences (Fig. 47C). A double-echo sequence is therefore optimal (TR/TE 2,000/20,80 msec) when examining the knee, because the sensitivity of long TR/TE sequences in detecting meniscal tears is low.

Differential considerations of a meniscal cyst include popliteal cysts, angiomas, and distended bursae (e.g., anserine, semimembranosis-gastrocnemius, infrapatellar), ganglia, popliteal artery aneurysms, and venous varices (Fig. 47D–F).

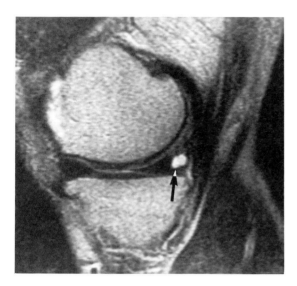

FIG. 47C. Meniscal cyst (sagittal MR, TR/TE 2,000/80 msec). A fluid-filled mass (*arrow*) is present behind the posterior medial meniscus.

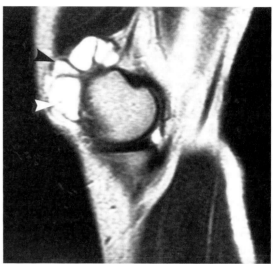

FIG. 47D. Meniscal cyst (sagittal MR, TR/TE 2,000/ 80 msec). This septated cyst (*arrowheads*) presented anterior to the femur.

E

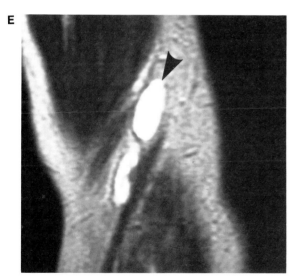

F

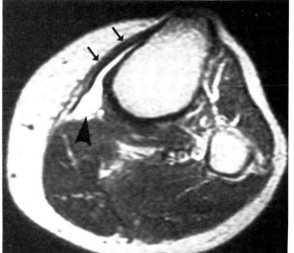

FIG. 47E,F. Anserine bursitis. **E.** Sagittal 2,000/80 msec; **F.** Axial MR, TR/TE 2,000/80 msec. An elongated, septated fluid-filled mass (*arrowhead*) is situated just deep to the tendons of the pes anserines (semitendinosis, gracilis, and sartorius) (*arrows*) and the femur. This could represent a meniscal cyst, but no tear was present.

CASE 48

The patient is a 39-year-old man with symptoms of internal derangement and lateral joint line pain. Is a meniscal tear present? What anatomic structures account for the findings in this case? (Fig. 48A. Sagittal MR, TR/TE 800/20 msec.)

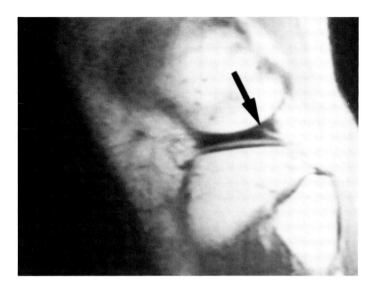

FIG. 48A.

Findings: On this sagittal section through the body of the lateral meniscus (Fig. 48A), a high signal intensity line (arrow) intersects the meniscus posteriorly. The line appears to extend to the articular surface and suggests a tear.

Diagnosis: Pseudo-tear of the posterior horn of the lateral meniscus.

Discussion: The anatomic relationships of the posterior horn of the lateral meniscus are complex, and this unique anatomy has been a well-recognized source of interpretive difficulty for arthrographers in evaluation of the lateral meniscus. The principal difficulty relates to the unique relationship of the popliteal tendon and its tendon sheath to the joint capsule and lateral meniscus. Knowledge of this complex anatomy is critical to correct interpretation of knee MR.

In approximately 15% of MR examinations, an oblique line of augmented signal intensity can be demonstrated to course along the posterior horn of the lateral meniscus in an anterosuperior to posterior inferior direction as demonstrated in this case. This line, which extends to the superior articular surface of the meniscus, closely mimics a meniscal tear. In actuality, the "line" represents fat interposed between the collapsed popliteal tendon sheath and the meniscus and is seen as a line of increased signal on T1-weighted images.

The popliteal tendon arises from the lateral femoral condyle below the origin of the FCL and courses downward, posteriorly, and medially to join its muscle belly on the posterior tibia. The tendon is extracapsular and it is bordered anteromedially by a sheath that communicates with the joint space. In distinction to the medial meniscus, which is firmly attached to the joint capsule throughout, the popliteal tendon violates the menisco-capsular junction of the lateral meniscus, creating two struts or fascicles that function as the peripheral meniscal attachments. The intricate anatomy of the fascicles is not routinely demonstrated on standard T1-weighted sequences but can be demonstrated to advantage on T2-weighted images, particularly when joint fluid is present (Fig. 48B,C).

The "pseudo-tear" (popliteal tendon sheath) of the posterior horn of the lateral meniscus predictably courses in a 30 to 45° superior anterior to posterior inferior direction. The peripheral fragment (popliteal tendon) can be correctly recognized by sequentially following it downward onto the posterior tibia. True tears of the posterior horn generally have a less constant and typically more horizontal course (Fig. 48D).

B
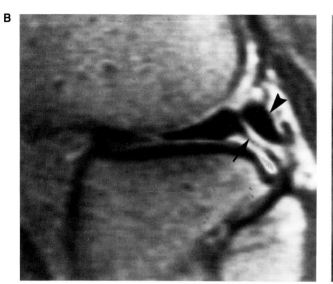
C
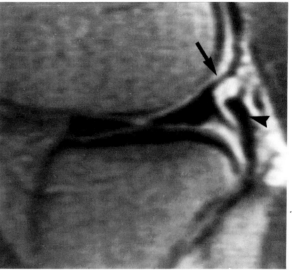

FIG. 48B,C. Popliteal tendon relationships (sagittal TR/TE 2,000/80 msec). **B.** On this extreme sagittal section, the popliteal tendon (*arrowhead*) is seen violating the menisco-capsular junction. Only the inferior fasicle is present (*arrow*). **C.** On a more central section, the tendon (*arrowhead*) has moved caudally and the superior fasicle (*arrow*) is now the peripheral attachment of the posterior lateral meniscus.

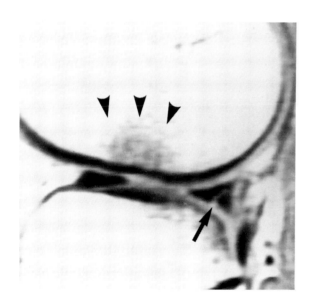

FIG. 48D. True tear posterior horn lateral meniscus (sagittal TR/TE 2,000/20 msec). This section demonstrates an oblique line of high signal intensity coursing in the opposite direction to that of "pseudo-tear." The line intercepts the tibial articular surface and represents an undersurface tear. A focus of low signal intensity, better defined on different window settings, represents an area of osteoarthritis within the lateral femoral condyle (*arrowheads*).

CASE 49

A 48-year-old man presented with a firm mass in the popliteal space. He gave a history of some discomfort around the knee, but the mass itself was painless. Is specific therapy really necessary? What is the diagnosis? (Fig. 49A. Sagittal MR, TR/TE 800/20 msec.)

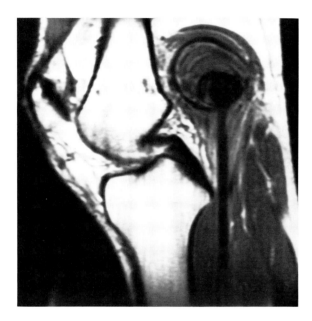

FIG. 49A.

Findings: A rounded mass in the popliteal space is inseparable from the popliteal artery (Fig. 49A).

Diagnosis: Popliteal artery aneurysm (PAA).

Discussion: Popliteal artery aneurysms are a serious medical and surgical problem. They represent the most common peripheral aneurysm and are bilateral in up to 60% of cases. Trauma and atheromatous disease are the most common causes of PAA. Nearly all patients, like the one above, have a nonpulsatile mass at presentation. Popliteal artery aneurysms carry a high incidence of complications including distal embolization; in 10% to 15% of cases, popliteal venous obstruction occurs at the point at which the vein normally crosses from a lateral to a medial position relative to the artery.

Recently, there has been interest in utilizing MR to assess the patency of the venous system in the leg (Fig. 49B). In a series of 17 extremities examined by both MR and standard contrast venography, diagnostic agreement between the two examinations occurred in all but one case, in which MR overestimated the extent of the clot (Fig. 49C).

Magnetic resonance-venography has several advantages over contrast studies. The procedure is fast, requiring only 20 min. The potential danger of contrast-induced thrombophlebitis is eliminated. Finally, both extremities can be examined at the same time, a particular advantage when one is searching for a source of emboli.

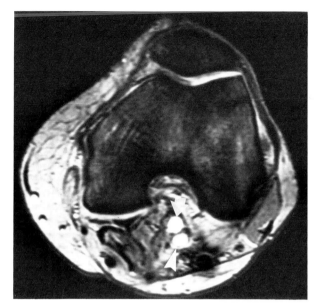

FIG. 49B. Normal popliteal vessels (axial MR, TR/TE 33/13 msec). Gradient refocused sequences are exquisitely sensitive to flow, and in this case the bright signal with the popliteal artery *(arrow)* and vein *(arrowhead)* is produced by the flowing blood.

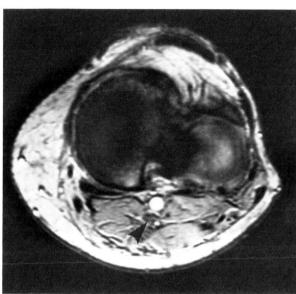

FIG. 49C. Popliteal venous occlusion (axial MR, TR/TE 33/13 msec). A filling defect is found in the popliteal vein *(arrowhead)* in this patient with clinically suspected venous obstruction.

CASE 50

The patient is a 32-year-old woman with complaints of aching pain over the front of the knee, particularly aggravated when climbing steps. What is your diagnosis? How are these lesions classified? (Fig. 50A. Axial MR, TR/TE 55/17 theta 80 msec.)

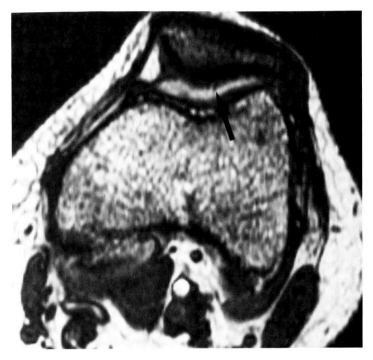

FIG. 50A.

Finding: An axial 1.5-mm section from a three-dimensional gradient refocused volume acquisition (Fig. 50A) demonstrates marked diminution in thickness of the patellar cartilage, particularly well seen along the lateral facet. Fine linear low intensity lines along the superficial articular cartilage surface can be defined (arrow). These corresponded to areas of "crabmeat" fibrillation seen on arthroscopic examination.

Diagnosis: Chondromalacia patella.

Discussion: The patella is a sesamoid bone contained within the quadriceps tendon. It functions both to protect the femoral articular surface as well as to increase the efficiency of the quadriceps mechanism by virtue of the fulcrum effect. By increasing the distance of the extensor mechanism from the axis of flexion-extension motion, the force of extension can be increased 30% to 50%. A considerable compressive force is generated at the patellofemoral joint and can reach 3.5 times body weight during stair descent. The patellar articular surface is divided into medial and lateral facets, with the medial facet often subdivided into a smaller (odd) facet and a larger medial facet proper. The lateral facet is usually the largest, although considerable variations in size and shape of the patella are recognized and have been categorized by Wyberg. The cartilage covering the articular surface of the patella is the thickest of any articulation and measures nearly 6 mm in depth (Fig. 50B). This compares to the cartilage covering the femoral surface of the trochlea, which measures 2 to 3 mm.

The term *chondromalacia* is commonly utilized, although a universally accepted definition has not yet been agreed on. Multiple factors have been implicated in the etiology of chondromalacia. Trauma has been considered a primary causative factor, and it is believed that the process can be initiated by either a sudden traumatic event causing contusion of the articular surface or by repetitive minor trauma (microtrauma) as can be seen with patellofemoral malalignment abnormalities. Inherent metabolic abnormalities have also been implicated in the etiology of this entity. The pathoanatomy described differs for young and older patients. In younger individuals the degenerative process is referred to as basal degeneration and results from the initial disruption of the vertical fibers that connect the superficial to the deep layer of articular cartilage. There is initial softening of the cartilage that can be detected arthroscopically by probing the affected area. With progression of the disease the superficial layer of the cartilage separates further from the deeper layer with the formation of a blister. With continued progression the blister can eventually rupture, leading to ulceration and fragmentation of the articular cartilage. The cartilagenous fragments attached to the subchondral bone have a "crabmeat" appearance as seen on arthroscopy. In the end stage there is complete loss of articular cartilage, with exposure of bone in some areas giving the appearance of a crater surrounded by a rim of frayed cartilage. In older individuals the process of degeneration takes place more gradually and starts in the superficial layer rather than in the deep layer. There is initial fibrillation of the superficial cartilage with later fissure formation. The latter stages are grossly similar to those of basal degeneration as the fissures enlarge, extend to the subchondral bone, and the fragmented articular cartilage separates exposing bone and causing crater formation.

The patellar articular cartilage can be demonstrated on routine sagittal spin-echo imaging, although the axial plane is most optimal for cartilage assessment. On T1-weighted images, the patella cartilage demonstrates a homogeneous intermediate signal. The cartilage should be uniform in thickness and normally demonstrates a sharply defined and gently convex contour on sagittal sequences (Fig. 50B). On T2-weighted images, the cartilage maintains intermediate signal; contrast between cartilage and synovial fluid increases, although decreased signal-to-noise ratio degrades image quality. Gradient refocused echo sequences and in particular three-dimensional volume acquisitions show considerable promise in evaluating articular cartilage (Fig. 50C). Magnetic resonance has been reported capable of detecting irregularity and swelling of the articular cartilage, although this has not been a subject of critical evaluation. Focal areas of decreased signal within the patella articular cartilage are frequently seen in both symptomatic and asymptomatic individuals, and their significance is uncertain. Well-defined areas of low signal on T1-weighted images at the cartilage surface may correspond to blister formation (arthroscopic grade 2). The fibrillation of cartilage recognized arthroscopically (grade 3) as "crabmeat" appearance may be recognized on MR by conspicuous loss of the normal sharp cartilage interface on T1-weighted images. Focal thinning and ulceration of the articular cartilage is readily demonstrated on MR but correlates with more advanced degrees of degeneration. This is often seen in association with abnormalities of subchondral bone. Abnormal patellar tracking, which may be etiologically related to many cases of chondromalacia, is best evaluated utilizing a dynamic MR technique (see Chapter 10). Axial MR images may also be utilized for calculation of the congruence and patella tilt angles, commonly utilized indices of lateral patella subluxation.

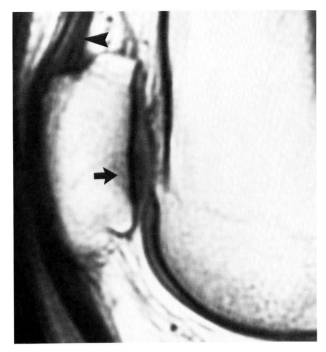

FIG. 50B. Normal patella cartilage (sagittal TR/TE 800/20 msec). A magnified view demonstrates the thickness and gentle convex contour of the patella cartilage in a normal patient. A focal area of low signal intensity within the lateral facet is commonly identified (*arrow*). Note the normal trilaminar appearance of the distal quadriceps tendon (*arrowhead*).

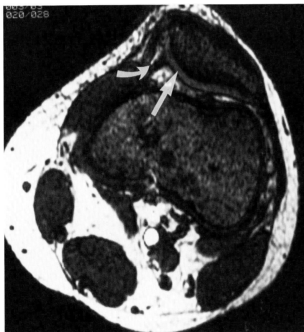

FIG. 50C. Normal patellar cartilage (axial gradient recalled echo, TR/TE 55/17, theta 80 msec). This image is from a study performed on a 14-year-old child. The medial (*arrow*) and lateral patellar facets are identified. The articular cartilage demonstrates a homogeneous intermediate signal intensity on this pulse sequence. Also note the normal Y-shaped configuration of the medial patella retinaculum (*curved arrow*).

CASE 51

A 45-year-old man presents with medial joint line tenderness following a knee injury. The McMurray test is positive. What is your diagnosis? What is the significance of the findings? (Fig. 51A. Sagittal MR, TR/TE 2,000/20 msec. Fig. 51B. Sagittal MR, TR/TE 2,000/20 msec. Fig. 51C. Sagittal MR, TR/TE 2,000/20 msec. Fig. 51D. Coronal MR, TR/TE 800/20 msec.)

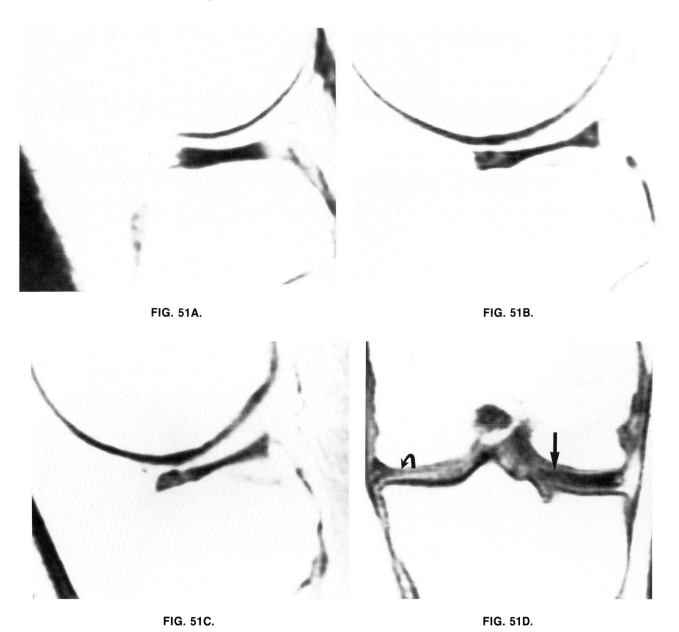

FIG. 51A.

FIG. 51B.

FIG. 51C.

FIG. 51D.

Findings: On the three consecutive 5-mm sagittal images (Fig. 51A–C), the normally expected tapering of the midzone of the meniscus is absent. The black "bow tie" is seen on all three sections. On the most central section (Fig. 51C), there is no separation into the expected separate anterior and posterior horns. The coronal section (Fig. 51D) demonstrates a nontapering "slab-like" appearance to the meniscus with the central tip extending toward the intercondylar notch (arrow). The tip of the medial meniscus is blunted with evidence of a small flap tear (curved arrow).

Diagnosis: (a) Discoid lateral meniscus. (b) Flap tear medial meniscus.

Discussion: Discoid menisci are morphologically enlarged menisci and can be readily depicted on MR. Several different schemes have been proposed for classification of discoid menisci and depend on the degree of fibrocartilage between the femoral condyle and tibial plateau ("complete or incomplete"), as well as the presence of an intact menisco-femoral ligament (Wrisberg ligament-type). Discoid menisci are overwhelmingly lateral but may occasionally be identified involving the medial meniscus (Fig. 51E,F). The middle zone of each normal meniscus is approximately 10 mm in size. Utilizing 5-mm contiguous sections, only two sagittal slice locations through the body of the normal lateral meniscus should demonstrate the normal black bow tie. Central to the midzone/free edge, separate triangular-shaped anterior and posterior horns should be demonstrated for two to three sections. Three or more contiguous 5-mm sagittal sections demonstrating the bridge of tissue between the anterior and posterior horns is highly suggestive of discoid meniscus.

Discoid menisci may also have increased inferior superior dimensions, and an abnormally thickened bow tie can suggest a discoid lateral meniscus. Failure to demonstrate the normal rapid tapering of the meniscus as one moves from the periphery to the center of the joint also suggests the diagnosis. Coronal images may be of particular value in diagnosis. The width of the normal lateral meniscus is 10 to 11 mm. The majority of discoid menisci are manifest by widths that exceed 20 mm, and often the meniscus can be seen extending into the intercondylar notch. Asymmetric discoid menisci may present with abnormally enlarged anterior or posterior horns or an enlarged body with normal horns.

The "snapping knee syndrome" is a classical, although uncommon, presenting feature of discoid lateral meniscus. With flexion/extension a snapping sound is heard, often associated with lateral joint line pain/tenderness. Discoid menisci demonstrate an increased incidence of tears related to altered joint mechanics. In any child presenting with suspected internal derangement, the presence of a discoid meniscus must be strongly suspected. Intrasubstance tears of discoid menisci have been likened to the central openings of a pita bread (Fig. 51G,H). Magnetic resonance does not appear to suffer the limitations with regard to diagnosis of tears involving discoid menisci as has been reported with arthrography. Symptomatic or torn discoid menisci are often treated with partial meniscectomy and contouring of the dimensions to a near normal size and configuration.

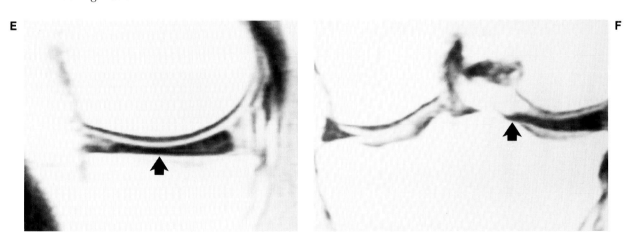

FIG. 51E,F. Discoid medial meniscus. **E.** Sagittal TR/TE 2,000/20 msec. On this sagittal section through the midplane of the medial tibial plateau, the meniscus would normally be expected to demonstrate separate anterior and posterior horns. Instead a continuous bridge of meniscus is present (*arrow*). **F.** Coronal TR/TE 800/20 msec. The large minimally tapering "slab-like" medial meniscus is seen to extend into the intercondylar notch (*arrow*). Medial discoid menisci are extremely uncommon.

G

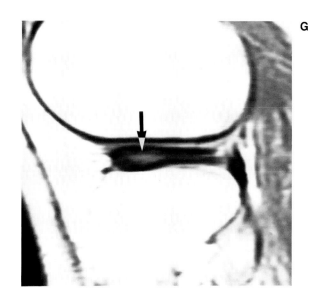

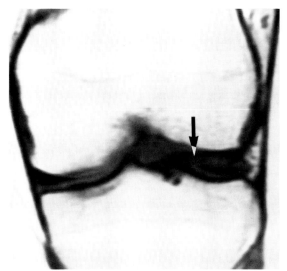

H

FIG. 51G,H. Discoid meniscus with intrasubstance tear. **G.** Sagittal TR/TE 2,000/20 msec. The lateral meniscus is grossly enlarged with diffuse increased intrasubstance signal, resembling the center of a pita bread (*arrow*). **H.** Coronal TR/TE 800/20 msec. The enlarged meniscus extends into the intercondylar notch (*arrow*). An intrasubstance tear was identified at surgery and the meniscus was contoured to a near normal size and configuration.

CASE 52

A 16-year-old boy presented to an orthopedist with a 6-month history of knee pain. On physical examination, his symptoms were localized to the patellofemoral joint. What is the lesion depicted by the arrow in Fig. 52A, and what is the significance of the smooth contour of the structure outlined by arrowheads? (Fig. 52A. Sagittal MR, TR/TE 800/20 msec.)

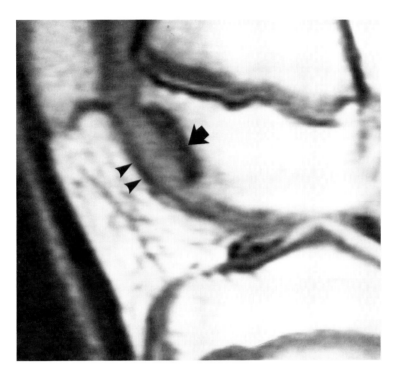

FIG. 52A.

Findings: There is a semicircular area of decreased signal in the anterolateral femoral condyle just distal to the open growth plate. The articular cartilage, a band of intermediate signal paralleling the bone, is smooth and intact.

Diagnosis: Lucent articular condylar lesion (OCD).

Discussion: Cayea first described the above lesion, which he attributed to chronic trauma, representing either an osteochondral fracture or OCD. It occurs in athletic male subjects 14 to 27 years old who present with chronic knee pain and it must be considered in the differential diagnosis of patellofemoral disorders. Routine radiography may not demonstrate the lesion; it is often not appreciated on tunnel, skyline, or anteroposterior views, and often a vague lucency is seen on the lateral film. Specialized views such as the Merchant or mountain views are necessary to demonstrate the segment of the femur proximal to that which is viewed in tangent on a skyline view (Fig. 52B). Scintigraphy has been reported to be abnormal in patients with a lucent articular condylar lesion. In the two patients studied by arthrography, the articular cartilage was normal in one and hypertrophied in the other. There are no data available upon which to make treatment recommendations because the natural history of this form of OCD is unknown, but assumedly complete resolution or free fragment generation could occur. Magnetic resonance imaging can noninvasively assess the integrity of the cartilage and assess the presence or absence of loose bodies.

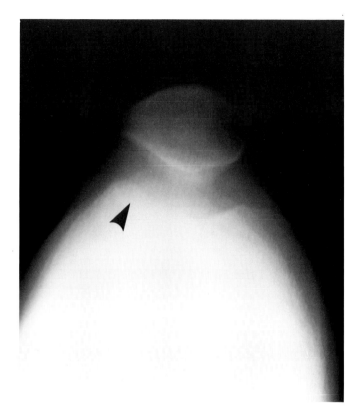

FIG. 52B. Osteochondritis dissecans. A Merchant view of the distal femur demonstrates a difficult-to-see concavity to the contour of the anterior aspect of the lateral femoral condyle (*arrowhead*).

CASE 53

A 37-year-old man with anterolateral knee instability. What procedure has been performed on this patient? Is the reconstruction intact? What types of treatment are available for ACL injuries? (Fig. 53A. Sagittal MR, TR/TE 2,000/20 msec. Fig. 53B. Sagittal MR, TR/TE 2,000/80 msec.)

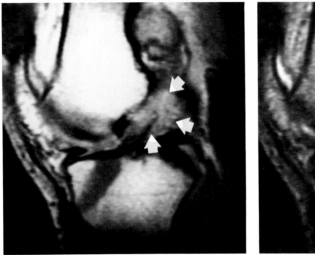
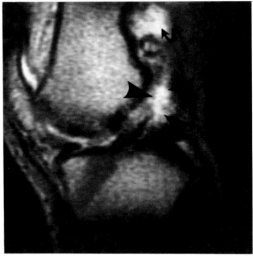

FIG. 53A.   **FIG. 53B.**

Findings: The patient has undergone prior ACL reconstruction. This is evident by the oblique band of low signal intensity traversing the proximal tibial metaphysis representing a surgically created tunnel for the reconstructed ligament. Within the tunnel the ligament is seen as a low intensity structure. The remainder of the ligament is not imaged, but rather an amorphous fluid collection is noted in the region of the expected femoral attachment (Fig. 53A, arrows). This becomes increased in signal intensity on the T2-weighted image (Fig. 53B) and presumably reflects edema and hemorrhage at the site of ligamentous tear (arrowheads). A second fluid collection is seen posterior to the proximal ACL where it exits from its femoral tunnel (curved arrow).

Diagnosis: (a) Anterior cruciate ligament reconstruction with a Gore-Tex graft; (b) retear of the proximal attachment of the reconstructed ACL.

Discussion: The subject of optimal treatment for ACL injuries remains controversial with regard to both operative versus nonoperative therapy as well as in regard to the type of procedure to be undertaken if surgery is elected. Therapy must be individualized with regard to the patient's age, level of activity, and extent of associated knee injury. Longitudinal studies evaluating the natural history of ACL-deficient knees suggest that a majority of patients with isolated ACL tears can be satisfactorily managed nonoperatively. For up to 30% of patients with chronic ACL tears, however, knee instability may require surgical ACL reconstruction.

With acute knee injuries, two situations favor a possible primary repair of the ACL tear. The most favorable situation for primary repair is the acute injury with avulsion of the tibial insertion of the ACL in patients with immature skeletons. In this setting, assessment of the osseous fragment can have particular importance for patient management. If the fragment is only minimally displaced, it may be satisfactorily reduced arthroscopically. With more extensive displacement, the fragment may escape from beneath the transverse ligament and require operative reduction. The second situation in which acute primary ACL repair may be considered is with isolated lateral bundle injuries.

With the more common intrasubstance tear, the tendon is usually insufficient in length, width, and consistency for primary repair, and tendon augmentation is required. This may be performed as either an extra- or intra-articular procedure utilizing either biological or bionic materials. One of the most common extra-articular stabilization procedures is the modified MacIntosh procedure in which a tenodesis of the iliotibial tract is performed. Compared with intra-articular reconstruction, this procedure is performed with shorter operating time and shorter interval of immobilization. The procedure does not restore normal knee stability but is particularly effective in stopping pivot-shift sprains. Intra-articular reconstructions are most commonly performed utilizing either the semimembranosis tendon, patellar tendon, or iliotibial band. Gore-Tex, as utilized in this patient, has emerged as one of the most popular of the synthetic materials and has most commonly been employed following failure of a prior reconstruction procedure. With intra-articular reconstruction, the proximal graft is either passed "over the top" of the lateral femoral condyle or through a tunnel created in the lateral metaphysis. The tibial attachment is near the anterior tubercle, and a tunnel is created in the proximal tibia for the graft to gain intra-articular access (Fig. 53C). Use of the arthroscope has facilitated isometric graft placement, which has been recently emphasized as critical to graft function and success. It is important that all damaged structures be recognized and repaired to gain joint stability.

The course of reconstructed ligaments is best appreciated on sagittal and oblique sagittal MR sequences (Figs. 53D–G). The normal graft is dark on most imaging sequences, as is the normal ACL, and runs in a similar oblique plane. When a tibial tunnel has been created, the tunnel itself with its low intensity contained ligament can be followed to the anterior aspect of the tibia. Like the normal ACL, Gore-Tex and other synthetic graft materials have a solid low signal intensity appearance across the intercondylar notch. The attachment site on the medial aspect of the lateral femoral condyle can be evaluated for isometric graft placement. Focal collections of fluid associated with a reconstructed ligament and/or discontinuity of the graft material must be regarded as overwhelmingly suggestive of a partial or incomplete graft tear as was present in the patient illustrated in this case (Fig. 53C,D). Chronic synovial effusions have recently been reported in patients with Gore-Tex grafts and may relate to an irritant effect of particles of Gore-Tex on the synovium (Fig. 53H).

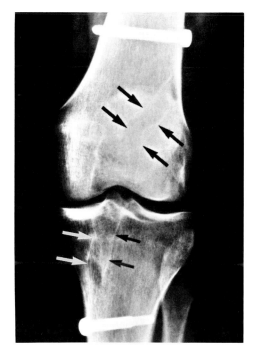

FIG. 53C. Anterior cruciate ligament reconstruction. Anteroposterior radiograph demonstrates the typical appearance of both the femoral and tibial tunnels (*arrows*) that are created for the reconstructed ligament.

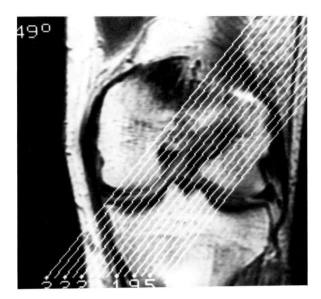

FIG. 53D. Coronal TR/TE 500/20 msec. Scout localizer. The oblique sagittal plane can optimize assessment of the reconstructed ACL.

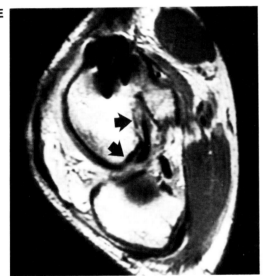

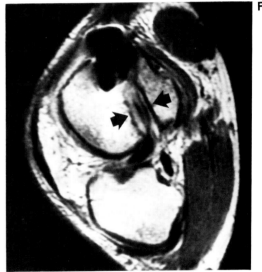

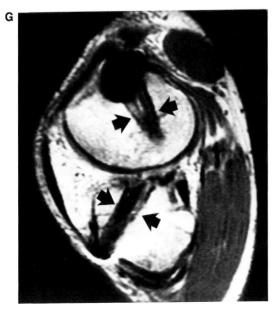

FIG. 53E–G. Oblique Sagittal TR/TE 2,000/20 msec. Three sequential sections from an oblique sagittal sequence obtained along the plane of the scout localizer illustrated in Fig. 53D. The reconstructed ligament is intact and is well demonstrated as it courses from the femur to the tibia. (*arrows*).

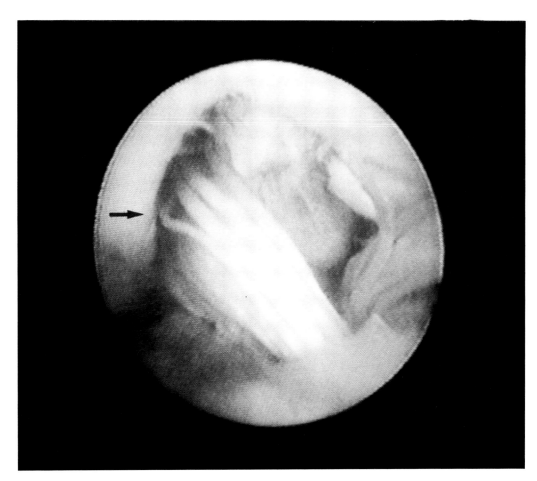

FIG. 53H. Torn prosthetic ACL. Anteroposterior view from an arthroscopic examination demonstrating separation of the individual bundles of a prosthetic (Gore-Tex) ACL at the site of exit from the lateral femoral condyle (*arrow*). This picture is from the same patient illustrated in this case. A more extensive tear was identified further proximally. (From Mink JH, Reicher MA, and Crues JV: *MRI of the knee.* Raven Press, New York, 1987.)

CASE 54

The 25-year-old man with a chronic medical problem was referred to an orthopedist because of subacute knee pain and a mass in the posterior aspect of the leg. What is the nature of this lesion outlined by arrowheads? (Fig. 54A. Sagittal MR, TR/TE 2,000/80 msec. Fig. 54B. Axial MR, TR/TE 2,000/80 msec.)

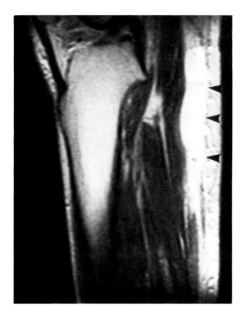

FIG. 54A.

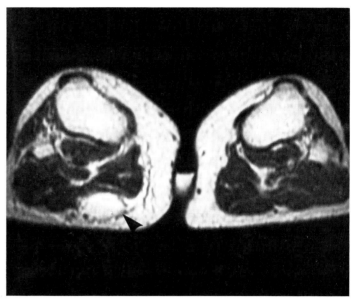

FIG. 54B.

Findings: There is an elongate fluid-filled mass extending down the leg on the dorsal aspect of the gastrocnemius muscle, extending to midcalf.

Diagnosis: Dissecting popliteal cyst, secondary to rheumatoid arthritis.

Discussion: Popliteal cysts (Baker cysts) were first described by William Baker in 1877 as synovial cysts that formed in association with a knee joint abnormality. Although a variety of cystic masses may occur around the knee, a true popliteal cyst has a constant and well-described location. The typical cyst is actually a distended gastrocnemious-semimembranosis bursa (GSB). It is located between the tendons of the medial head of the gastrocnemious and semimembranosis muscles. Communications tend to occur between the bursa and the knee joint at the thinnest part of the capsule, which is constantly 1 to 3 cm below the capsular insertion into the bone (Fig. 54C,D). It is at this point that the gastrocnemious muscle no longer adheres to and supports the capsule. It is generally believed that rupture of the capsule and subsequent communication with the bursa are most commonly due to hyperextension injury. Filling of the GSB at arthrography implies a communication with the knee joint, a condition that must be assumed to be pathologic in adults until proven otherwise. Any articular disorder, especially rheumatoid arthritis, which results in a joint effusion and increased intra-articular pressure, can result in distension of the bursa. It may extend caudally in a subcutaneous position down to the ankle and present as a mass.

On MRI, fluid within the knee joint or bursa is best depicted on a T2-weighted image. Fluid within a bursa has sharp boundaries. Occasionally, inflammation and/or sudden increases in pressures can lead to rupture of the cyst. Acute pain and tenderness are then accompanied by a decrease in the soft tissue mass. The clinical picture in such patients resembles deep venous thrombosis, a condition that has been referred to as the pseudo thrombophlebitis syndrome.

The synovial membrane of cysts is subject to the same pathologic processes as may occur within joints. Pannus secondary to rheumatoid disease, pigmented villonodular synovitis, and synovial chondromatosis have all been described as occurring in cysts. Fluid, as well as loose bodies, can migrate into cysts from the parent joint (Fig. 54E,F).

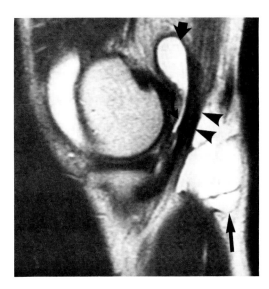

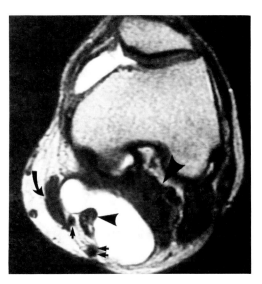

FIG. 54C. Popliteal cyst (sagittal MR, TR/TE 2,000/80 msec). A typical popliteal cyst has two components, a superior portion (arrow) that directly communicates with the joint above the semimembranosis tendon (arrowheads), and a larger component (long arrow) extending down beneath the tendon. The actual site of communication with the joint (curved arrow) can be seen.

FIG. 54D. Popliteal cyst (axial MR, TR/TE 2,000/80 msec). In the axial plane, the relationship of the cyst to the semimembranosis (arrowhead), semitendinosis (double arrow), and gastrocnemius muscles (large arrowhead) and tendons can be appreciated (arrow, gracilis tendon; curved arrow, sartorius muscle).

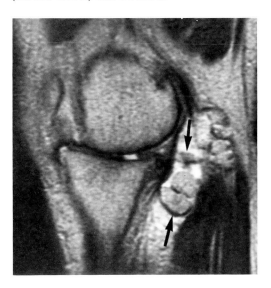

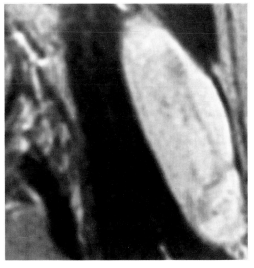

FIG. 54E. Loose bodies in a popliteal cyst (sagittal MR, TR/TE 2,000/80 msec). Multiple ossific loose bodies (arrows) are present in this popliteal cyst. The medial meniscus is not identified, and there is anterior translation of the tibia relative to the femur, indicating severe ACL insufficiency.

FIG. 54F. Pannus in a popliteal cyst (sagittal MR, TR/TE 2,000/80 msec). Poorly defined areas of decreased signal are present in this cyst. They are thought to represent pannus in this patient with rheumatoid disease.

CASE 55

This 45-year-old obese woman was being evaluated for possible internal derangement of the knee. Other MR images did demonstrate a meniscal tear, but the radiologist was overwhelmingly concerned with the pattern of signal in the medullary canal of the femur, fibula, and tibia. The lesions remained dark on long TR/TE sequence. What is the significance of the MR finding, what is the differential diagnosis, and what should be done? (Fig. 55A. Coronal MR, TR/TE 500/20 msec. Fig. 55B. Coronal MR, TR/TE 500/20 msec.)

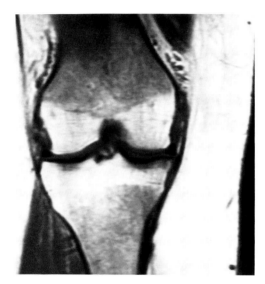 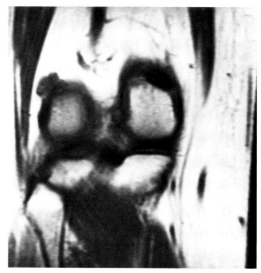

FIG. 55A.   **FIG. 55B.**

Findings: There is a diffuse decrease signal in the femoral and tibial meta-diaphysis that ends abruptly at the physeal scar. The fibula is similarly involved. There is no evidence of cortical destruction or soft tissue abnormality.

Diagnosis: Hematopoietic hyperplasia.

Discussion: In the fetus, the marrow is entirely hematopoietic, but replacement with fatty marrow begins in the distal end of the extremities soon after birth (Fig. 55C). By early adulthood, hematopoietic marrow is confined to the axial skeleton, proximal humeri, and femurs (Fig. 55D). On spin-echo imaging, normal hematopoietic marrow demonstrates low to intermediate signal on T1- and T2-weighted images in contrast to fatty marrow, which is bright on T1-weighted images. Although hematopoietic requirements are satisfied by axial red marrow, certain stresses can result in reconversion of fatty marrow to active hematopoietic marrow. Gaucher disease, myeloma, and metastatic disease are among the causes of an increased demand for hematopoietic marrow because of the replacement of the normal marrow spaces. Chronic hemolytic anemias are another common cause of marrow expansion.

In the course of routine imaging, hematopoietic hyperplasia may be occasionally encountered. A recent report detailed the characteristics of hematopoietic hyperplasia. Nearly all affected patients are women, most are obese, many are smokers, and some have a mild persistent leukocytosis, but there is no unifying explanation for marrow recruitment. In this patient, a biopsy of the femur and the posterior iliac crest at the time of arthroscopy showed hypercellular but otherwise normal marrow. The normal distal femoral marrow has <5% hematopoietic elements. In this patient, the percentage was nine times normal. The primary differential consideration must be a chronic leukemia; diffuse metastatic disease, myelofibrosis, and chronic hemolytic anemias are usually excluded by a simple blood count and a careful clinical history. In the proper clinical setting and with a MR appearance identical to the one above, invasive investigative procedures are probably not warranted.

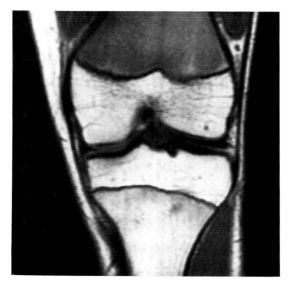

FIG. 55C. Normal marrow (coronal MR, TR/TE 500/20 msec). The femoral medullary canal is filled with hematopoetic elements in this normal 14 year old, but replacement by fatty marrow is nearly complete in the tibia.

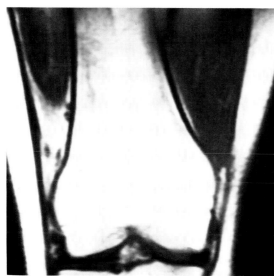

FIG. 55D. Normal marrow (coronal MR, TR/TE 500/20 msec). This pattern of hematopoetic marrow, with small strips in the peripheral and posterior aspect of the femur, is frequently found.

CASE 56

The patient experienced a sudden strenuous abduction injury with pain along the medial aspect of the joint. Physical examination revealed tenderness to direct palpation over the TCL where it crosses the medial joint line. Is the meniscus normal? What is the significance of this type of injury and what are the therapeutic implications? (Fig. 56A. Sagittal MR, TR/TE 2,000/20 msec. Fig. 56B. Sagittal MR, TR/TE 2,000/80 msec.)

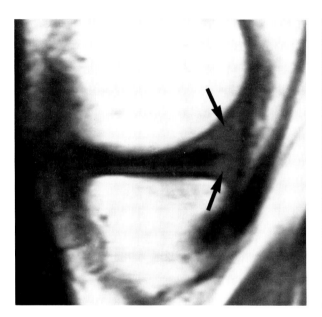

FIG. 56A.

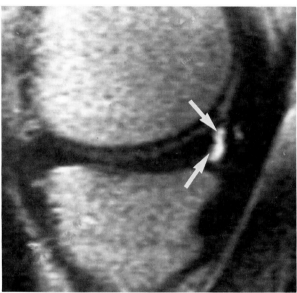

FIG. 56B.

Findings: On the proton-density image (Fig. 56A), intermediate signal intensity is seen separating the posterior (capsular) margin of the meniscus from the capsule (arrows). The tibial cartilage is minimally "uncovered." On the T2-weighted image (Fig. 56B), a continuous line of increased signal is interposed between the periphery of the meniscus and the joint capsule (arrows).

Diagnosis: Peripheral meniscal tear.

Discussion: Meniscal tears are classified into those involving the inner, middle, or outer (peripheral) thirds of the meniscus. The degenerative fringe tear is typical of inner rim tears. Longitudinal tears are most commonly located in the middle portion of the meniscus. Radial tears vary in their depth of penetration from the inner rim to the outer margin. Extreme peripheral tears in which the meniscus is in part or completely separated from its capsular attachment are referred to as menisco-capsular or meniscosynovial separations.

There are several reasons to consider peripheral tears and meniscal capsular separations as a distinct group from other meniscal tears. These tears occur through the highly vascularized periphery of the meniscus and demonstrate a unique tendency to heal by synovial ingrowth and re-epithelialization. Furthermore, there has been increasing interest among orthopedic surgeons in the repair of these lesions, and techniques have now been developed to allow this to be arthroscopically accomplished. In addition to detection it is critical to describe fully the extent of these lesions and any other associated meniscal pathology. Acute menisco-capsular separations less than 1 cm in extent in an otherwise stable knee need not be repaired or resected and can be treated by immobilization for 4–6 weeks. Acute tears greater than 1 cm in length without other associated meniscal damage represent the best candidates for meniscal repair.

Peripheral tears and menisco-capsular separations may be incomplete or complete. The posterior horn of the medial meniscus is the most commonly involved site for these lesions. The meniscus at this site is secured to the periphery of the tibia and its cartilage by the short coronary (menisco-tibial) ligaments. The relative immobility of the meniscus related to its attachments allows the meniscus to be "stretched" and predisposes it to this type of tear. On MR, the peripheral margin of the posterior horn of the medial meniscus normally aligns with the tibial articular cartilage as best depicted on the two most peripheral sagittal images. Although minor degrees of incongruence are common, greater than 5-mm displacement or grade 3 signal at the menisco-capsular junction is suggestive of the diagnosis. The extent of "displacement" is less critical than is determination of whether the meniscus is attached to the capsule. On T1-weighted images it may be difficult to differentiate these tears from fatty capsular tissue. Coronal views may be of particular value in establishing the diagnosis. The presence of fluid between the meniscus and capsule is diagnostic of a menisco-capsular separation and is best assessed on T2-weighted images (Fig. 56B). Fluid within the superior and inferior capsular recesses, however, is a normal finding and must be distinguished from the continuous band of fluid seen with menisco-capsular separation (Fig. 56C). Peripheral tears are often characterized by stellate areas of signal within the extreme periphery of the meniscus (Fig. 56D,E). These invariably can be demonstrated on both sagittal and coronal sections.

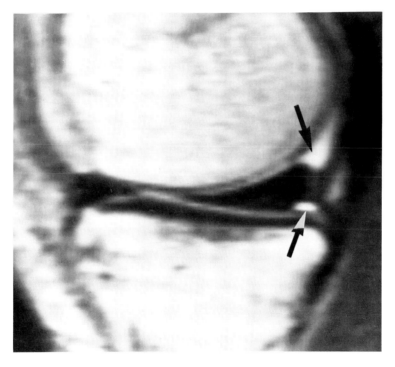

FIG. 56C. Meniscal recesses (sagittal TR/TE 2,000/80 msec). On this T2-weighted image through the posterior horn of the medial meniscus, high signal intensity fluid is seen within the superior and inferior meniscal recesses (*arrows*). No continuous line of fluid is seen between the meniscus and capsule in contrast to Fig. 56B. This meniscus is firmly applied to the joint capsule, and no potential space for fluid to accumulate between the meniscus and capsule is present. Compare with Fig. 56B.

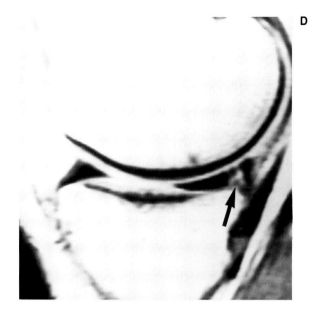

D

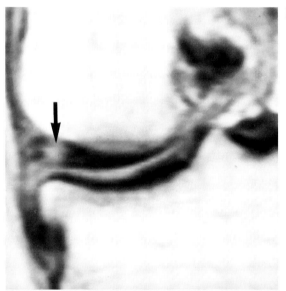

E

FIG. 56D,E. Peripheral meniscal tear (sagittal, coronal TR/TE 800/20 msec). On the sagittal sequence a triangular-shaped focus of increased signal is seen within the extreme periphery of the posterior horn of the medial meniscus (*arrow*). On the coronal sequence, a stellate focus of signal is seen with the periphery of the meniscus (*arrow*). The findings are characteristic of a small peripheral tear.

CASE 57

A 72-year-old man was referred for an MRI because of subacute, nontraumatic knee pain. What is the differential diagnosis? (Fig. 57A. Coronal MR, TR/TE 2,000/80 msec. Fig. 57B. Plain anteroposterior radiograph.)

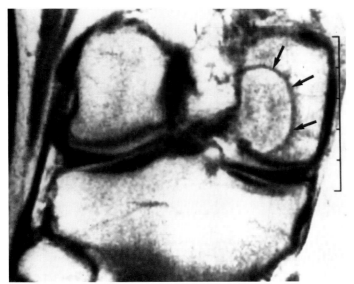

FIG. 57A.

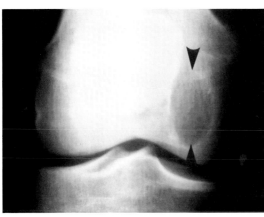

FIG. 57B.

Findings: There is a 2-cm area of bone destruction in the posteromedial femoral condyle (Fig. 57A, arrowheads). The lesion has a thin, sharply circumscribed border. It is inhomogeneous on MR but is devoid of matrix. The MRI demonstrates that the lesion has increased signal (arrows) on the long TR/TE sequence (Fig. 57B).

Diagnosis: Metastatic carcinoma of the breast.

Discussion: The patient had a mastectomy for carcinoma 7 years prior and during that period of time was felt to be tumor-free. He had never had a documented metastasis. The differential diagnosis must include a large subchondral cyst, enchondroma, lipoma, osteonecrosis, giant cell tumor, and even an intraosseous ganglion. In spite of the benign appearance, the age of the patient and the history resulted in a percutaneous biopsy. The histology was compatible with breast carcinoma identical to the original lesion. Although subchondral metastatic deposits are rare, one ought not forget that metastatic disease must be the primary differential consideration when dealing with destructive skeletal lesions in the elderly.

CASE 58

A 28-year-old man presented with a 3 to 4 week history of pain and swelling in the region of his patellar tendon. No history of trauma could be elicited. The MR appearance of abnormality was unchanged on the T2-weighted image. What are the differential considerations? How does the signal intensity of the abnormality limit the diagnostic possibilities? (Fig. 58A. Sagittal MR, TR/TE 2,000/20 msec. Fig. 58B. Axial MR, TR/TE 2,000/20 msec.)

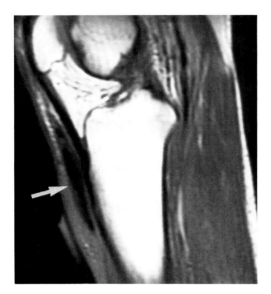
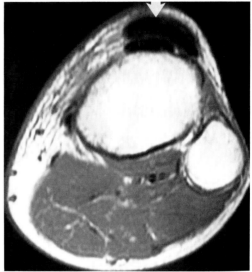

FIG. 58A.   **FIG. 58B.**

Findings: On the sagittal section (Fig. 58A), an elongate mass of diminished signal intensity is seen anterior to the distal patellar tendon and tibial tuberosity (arrow). It appears separate from the bone. On the axial section (Fig. 58B), the mass is again seen to be of similar signal intensity to the tendon and separate from bone (arrow). There is superficial soft tissue swelling.

Diagnosis: Pseudomalignant osseous tumor of soft tissue (myositis ossificans).

Discussion: The mass as depicted on the images illustrated in this case demonstrates uniformly low signal intensity similar to that of tendon and cortical bone. The signal does not change on T2-weighted sequences. These findings help to limit the differential diagnostic possibilities. Areas of calcification typically demonstrate low signal intensity on all pulse sequences. Indeed, the difficulty in depiction of small calcifications as a consequence of their low signal intensity has been a well-recognized limitation of MRI. Lesions that are predominantly fibrous in nature also characteristically demonstrate low signal intensity on all pulse sequences. Areas of hemosiderin deposition (Fig. 58C) and paramagnetic substances in high concentration may also demonstrate low signal intensity on both T1-weighted images and T2-weighted images.

The radiological and histological features of pseudomalignant osseous tumor of soft tissue are virtually indistinguishable from those of myositis ossificans traumatica. Some authorities, indeed, do not accept the distinction between the two conditions. The term pseudomalignant tumor is applied when no history of trauma can be elicited. Typically soft tissue swelling with or without pain precedes by several weeks the development of calcification and ossification. As with myositis ossificans, the lesion is characterized by a zoning phenomenon with a central connective tissue core containing proliferating fibroblasts surrounded by osteoblasts producing islands of bone. Bone maturation proceeds in a centrifugal direction. Radiographically, the lesions demonstrate a peripheral shell of developing ossification and are characteristically separate from underlying bone (Fig. 58D).

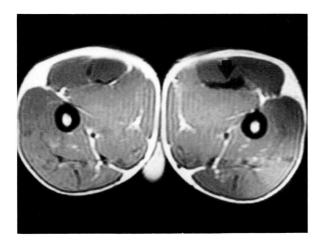

FIG. 58C. Hemosiderin deposition axial TR/TE 2,000/20 msec). On this axial section through the proximal thigh, an oblong focus of low signal intensity is seen within the rectus femoris muscle (*arrow*). This represents hemosiderin related to a prior muscle injury and hematoma. The lesion is low in signal intensity on all pulse sequences.

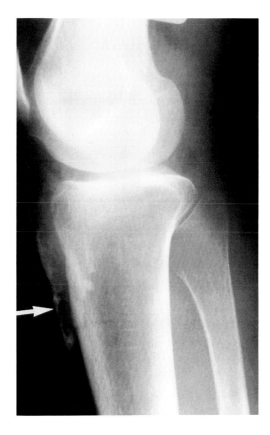

FIG. 58D. Pseudomalignant osseous tumor of soft tissue. Lateral radiograph of the knee from the patient illustrated in this case. There is a band of extensive calcification anterior to the patellar tendon and proximal tibia. Note the clear zone between the mass and underlying bone (*arrow*).

CONCLUSION

The past few years have been witness to a virtual explosion of interest in the applications of MRI in the diagnosis of disorders of the knee. The enhanced diagnostic capacity of MR has dramatically expanded the spectrum of abnormalities that can potentially be evaluated and has allowed MR virtually to replace arthrography and challenge arthroscopy as the primary diagnostic method to be utilized in patients presenting with knee complaints. It is appropriate in the conclusion to this chapter to review the data with regard to the accuracy of this method and to consider how it compares to the previously established techniques.

No critical study comparing the accuracy of knee MR to arthrography has been performed, and it is unlikely at this point that one will ever be accomplished. In the evaluation of meniscal abnormalities, arthrography has a reported accuracy of 85% to 99% for evaluation of the medial meniscus and 68% to 93% for assessment of lateral meniscal abnormalities. The technique is minimally invasive, requires contrast administration, and exposes the patient to ionizing radiation. Most critically, arthrography is highly operator-dependent, requiring a high degree of skill to obtain images of high diagnostic quality. Magnetic resonance, at a very minimum, is comparable in accuracy while suffering from none of the aforementioned limitations.

Multiple studies of varying size have compared the accuracy of MR to arthroscopy. As it is well accepted that arthroscopy is not a true "gold standard," it may be more correct to substitute "agreement of MR and arthroscopy" for the phrase "accuracy of MR." The largest and most recent study was performed at our institution and compared the preoperative MR reports of 459 virgin menisci and 252 ACLs to the findings as observed at arthroscopy. The overall agreement of MR and arthroscopy in meniscal evaluation was 93%. Thirty-seven different orthopedic surgeons participated in the review, but in a subgroup in which the surgical procedures were performed by a single knee subspecialist arthroscopist, the MR-arthroscopy agreement rate increased to 95%. Several other large studies have confirmed this extremely high agreement rate of MR and arthroscopy. Jackson and colleagues examined 87 knees that had both MR and arthroscopy; they reported an MR "accuracy" of 93% for the medial and 97% for the lateral meniscus. Mandelbaum reported an MR-arthroscopy agreement of 90%, and Polly reported a value of 94%.

One of the most important statistics for the practicing orthopedic surgeon if he or she is to rely on MR as a screening technique is the negative predictive value (NPV) of MR. The NPV is defined as the percentage of patients with a negative MR who do not have a tear arthroscopically. The four largest published series of MR of the knee have reviewed a total of 1,072 menisci that have been examined by both MR and arthroscopy. The overall NPV was 94%, or in other words, only 6% of patients with a negative MR report had a tear. Of the false-negative examinations, five of the 11 missed tears reported by Mink and two-thirds of the missed tears reported by Jackson were only minimal tears that did not require partial meniscectomy. Therefore, the likelihood of MR to fail to detect a *clinically significant* meniscal tear is very low.

False-positive MR examinations have been a greater source of error than have false-negative studies. It is worthwhile considering the underlying contributing factors as a means toward improving MR-arthroscopy agreement. Two of the potential reasons for a false-positive MR have already been discussed as possibly representing false-negative arthroscopies: (a) the operator dependence of arthroscopy, and (b) the difficulty for arthroscopists optimally to examine the posterior horn of the medial meniscus (the most common site of "false-positive MR examinations"). Additionally, true intrasubstance tears (as opposed to histologic degeneration), the closed cleavage lesion that Smillie felt was the precursor to a frank tear, may produce grade 3 MR signal but not be detected by visual inspection alone at arthroscopy. Difficulty in assessment of meniscal signal intensity by the interpreting radiologist also undoubtedly accounts for false-positive examinations. Minimal areas of signal confined to the free edge or meniscal apex may represent only fraying or fibrillation and must be distinguished from arthroscopically significant tears. Linear signal that equivocally extends to an articular surface must not be over-interpreted as representing a significant tear. Grade 3 signal that diminishes in intensity as it approaches an articular surface may represent a confined tear, and this possibility is recognized. Confirmation of the extension of intrameniscal signal to an articular surface on an orthogonal section should be sought to minimize false-positive interpretations.

Four large recent studies have evaluated the accuracy of MR in detecting tears of the ACL. In the largest, Mink and colleagues found MR to be 95% accurate; when they selected out those patients studied with T2-weighted images, the accuracy improved to 97%. Jackson, Lee, Mandelbaum, and Polly have all reported accuracy rates of greater than 97%. Arthrographic accuracy rates for ACL assessment of 85% have been reported.

If MR did nothing else but precisely assess the menisci and the cruciate ligaments, its value to the surgeon, radiologist, and patient would be assured. Magnetic resonance, however, has a unique ability to assess a wide spectrum of abnormalities affecting the knee, many of which may have previously escaped detection utilizing the established diagnostic methods of arthrography and arthroscopy. The entire extensor mechanism, with the exception of the articular surface of the patella, is extra-articular, and thus beyond the reach of the arthroscope. Disorders of the quadriceps tendon, retinaculum, patella, and patella tendon have been depicted with previously unobtainable precision with MR. Magnetic resonance has been of particular value in assessment of para-articular masses includings cysts, ganglia, bursae, blood vessels, and tumors, all of which again could not be characterized by previously available techniques. The emergence of MR as the premier imaging technique for marrow disorders has had profound impact for knee imaging. The diagnosis of osteonecrosis can be readily established well before conventional radiographic findings are present. Osteochondritis dissecans can be detected and information regarding the status of the overlying cartilage critical to management can be obtained noninvasively. Occult osseous and cartilagenous injuries are being recognized with increasing frequency. In some instances (e.g., stress fractures), these abnormalities account for the patient's symptoms and would have gone undetected had only arthrography or arthroscopy been performed. Magnetic resonance has also proven of immense value in assessment of the acutely traumatized patient in whom the limits of physical examination without resort to anesthesia have been well recognized.

Magnetic resonance, as a single test, can replace multiple (often invasive) diagnostic procedures presently utilized in the evaluation of knee pain. Magnetic resonance may assist the orthopedic surgeon in helping to select patients most optimally for surgical intervention, identify difficult-to-find lesions, and reduce unnecessary surgery, all of which contributes positively to overall patient management. Improvements in coil design and scan techniques should further enhance the already formidable diagnostic capacity of this examination.

RECOMMENDED READING

Ahlback S, Bauer GCH, Bohne WH. Spontaneous osteonecrosis of the knee. *Arthritis Rheum* 1968;11:705–733.

Arnoczky SP. Blood supply of the meniscus. In: McGinty JB, ed. *Techniques in orthopedics*, vol 5, *Arthroscopic surgery update*. Rockville, MD: Aspen Publications, 1985.

Arnoczky SP, Russel RF. Anatomy of the cruciate ligaments. In: Feagin JA, ed. *The cruciate ligaments*. New York: Churchill Livingstone, 1988.

Beltran J, Caudill JL, Herman LA, et al. Rheumatoid arthritis: MR manifestations. *Radiology* 1987;165:153–157.

Beltran J, McGhee R, Shaffer P, et al. Experimental infections of the musculoskeletal system: evaluation with MR imaging and Tc-99m MDP and Ga-67 scintigraphy. *Radiology* 1988;167:167–172.

Beltran J, Noto AM, Mosure JC, et al. The knee: surface-coil MR imaging at 1.5 T. *Radiology* 1986;159:747–751.

Berquist TH, Ehman RL, Rand JA. Musculoskeletal trauma. In: Berquist TH, ed. *Magnetic resonance of the musculoskeletal system*. New York: Raven Press, 1987;127–164.

Buckwalter KA, Braunstein EM, Wilson MR, et al. Evaluation of hyaline cartilage of the knee with MR imaging. *Radiology* 1986;161(p):139.

Burk DL, Dalinka MK, Kanal E, et al. Meniscal and ganglion cysts of the knee: MR evaluation. *AJR* 1988;150:331–336.

Burk DL, Dalinka MK, Kanal E, et al. High resolution MR imaging of the knee. In: *Magnetic resonance annual 1988*. New York: Raven Press, 1988;1.

Cailliet R: *Knee pain and disability*, 2nd ed. Philadelphia: F.A. Davis Company, 1983.

Casscells SW, ed. *Arthroscopy: diagnostic and surgical practice*. Philadelphia: Lea & Febiger, 1984.

Cohen EK, Kressel HY, Frank TS, et al. Hyaline cartilage-origin bone and soft-tissue neoplasms: MR appearance and histologic correlation. *Radiology* 1988;167:477–481.

Cohen EK, Kressel HY, Perosio T, et al. MR imaging of soft-tissue hemangiomas: correlation with pathologic findings. *AJR* 1988;150:1079–1081.

Crues JV, Mink JH, Levy TL, Lotysch M, Stoller DW. Meniscal tears of the knee: accuracy of MR imaging of the knee: first 144 cases. *Radiology* 1987;164:445–448.

Crues JV, Stoller DW. The menisci. In: Mink JH, ed. *Magnetic resonance imaging of the knee*. New York: Raven Press, 1987.

Daffner RH. Stress fractures: current concepts. *Skeletal Radiol* 1978;2:221–229.

Daffner RH, Lupetin AR, Dash N, Deeb ZL, Sefczek RJ, Shapiro RL. MRI in the detection of malignant infiltration of bone marrow. *AJR* 1987;146:353–358.

Daniel DM, Stone ML. Diagnosis of knee ligament injury: tests and measurements of joint laxity. In: Feagin JA, ed. *The cruciate ligaments*. New York: Churchill Livingstone, 1988.

Deutsch AL, Mink JH, Waxman AD. MR imaging of occult fractures of the proximal femur. *Radiology* 1989;170:113–116.

Deutsch AL, Mink JH. MRI of musculoskeletal trauma. *Radiol Clin North Am* (in press).

Deutsch AL, Mink JH. MR of articular disorders of the knee. Topics in magnetic resonance (in press).

Deutsch AL, Mink JH. MRI of the post operative meniscus. *Radiology* 1988;169(P):20.

Deutsch AL, Mink JH, Rosenfelt FP, Waxman AD. Incidental detection of hematopoietic hyperplasia on routine knee MR imaging. *AJR* 1989;152:333–336.

Deutsch AL, Resnick D, Dalinka MK, et al. 1981 Synovial plicae of the knee. *Radiology* 1981;141:627–634.

Dietz GW, Wilcox DM: Segond tibial condyle fracture: lateral capsular ligament avulsion. *Radiology* 1986;159:467.

El-Khoury GY, Bassett GS: Symptomatic bursa formation with osteochondromas. *AJR* 1979;133:985–988.

Engber WD Stress fractures of the medial tibial plateau. *J Bone Joint Surg* 1977;59-A(6):767–769.

Feagin JA: Case studies 1–15. In: Feagin JA, ed. *The cruciate ligaments*. New York: Churchill Livingstone, 1988.

Ferrer-Roca O, Vilalha C. Lesions of the meniscus. *Clin Orthop* 1980;146:289–307.

Fox JM, Sherman OH, Pevsner D: Patellofemoral problems and malalignment. In: McGinty JB, ed. *Techniques in orthopedics*, vol 5, *Arthroscopic surgery update*. Rockville, MD; Aspen Publications, 1985.

Gilley JS, Gelman MI, Edison DM, et al. Chondral fractures of the knee. *Radiology* 1981;138:51–54.

Gillies H, Seligson D. Precision on the diagnosis of meniscal lesions: a comparison of clinical evaluation, arthrography and arthroscopy. *J Bone Joint Surg* 1979;61:343.

Guerra J, Newell J, Resnick D, et al. Gastrocnemio-semimembranous bursal region of the knee. *AJR* 1981;136:593.

Hajek PC, Baker L, Sartoris D, et al. MR arthrography: anatomic-pathologic investigation. *Radiology* 1987;163:141–147.

Hayden JW: Compartment syndromes: early recognition and treatment. *Postgrad Med* 1983;74:191–202.

Herman LJ, Beltran J: Pitfalls in MR imaging of the knee. *Radiology* 1988;167:775–781.

Hohl M, Larson RL. Fractures and dislocations of the knee. In: Rockwood CA, Green DP, eds. *Fractures*. Philadelphia: JP Lippincott, 1975;1131–1286.

Hudson TM, Hamlin DJ, Enneking WF, Petterson H. Magnetic resonance imaging of bone and soft tissue tumors: early experience in 31 patients compared with computed tomography. *Skeletal Radiol* 1985;13:134–146.

Hudson TM, Springield DS, Spanier SS, et al. Benign exostoses and exostotic chondrosarcomas: evaluation of cartilage thickness by CT. *Radiology* 1984;152:595–599.

Hughston JC, Andrews JR, Cross MJ, et al. Classification of knee ligament instabilities. Part I: the medial compartment and cruciate ligaments. *J Bone Joint Surg* 1976;58:159–172.

Indelicato PA. Injury to the medial capsuloligamentous complex. In: Feagin JA, ed. *The cruciate ligaments*. New York: Churchill Livingstone, 1988.

Jackson DW, Jennings LD, Maywood RM, et al. Magnetic resonance imaging of the knee. *Am J Sports Med* 1988;16(1):29.

Jackson RW: The torn ACL: natural history of untreated lesions and rationale for selective treatment. In: Feagin JA, ed. *The cruciate ligaments*. New York: Churchill Livingstone, 1988.

Johnson LL. *Diagnostic and surgical arthroscopy*. St. Louis: Mosby, 1981.

Johnson RJ: Prevention of cruciate ligament injuries. In: Feagin JA, ed. *The cruciate ligaments*. New York: Churchill Livingstone, 1988.

Johnson RJ, Pope MH: The function of the menisci. In: Shahriaree H, ed. *O'Connor's textbook of arthroscopic surgery*. Philadelphia: JB Lippincott, 1984.

Kaplan PA, Williams SM: Mucocutaneous and peripheral soft-tissue hemangiomas: MR imaging. *Radiology* 1987:163:163–166.

Keats TE: *An atlas of normal roentgen variants that may simulate disease*. Chicago: Year Book, 1979.

Kottal RA, Volger JB III, Matamoros A, Alexander AH, Cookson JL. Pigmented villonodular synovitis: a report of MR imaging in two cases. *Radiology* 1987;163:551–553.

Kulkarni MI, Drolshagen LF, Kaye JJ, et al. Rheumatoid arthritis: MR manifestations. *Radiology* 1987;165:153–157.

Lee JK, Yao L, Phelps CT, et al. Anterior cruciate ligament tears: MR imaging compared with arthroscopy and clinical tests. *Radiology* 1988;166:861–864.

Lee KR, Cox GG, Neff JR, et al. Cystic masses of the knee: arthrographic and CT evaluation. *AJR* 1987;148:329–334.

Levy MI. Posterior meniscal capsuloligamentous complex. In: Feagin JA, ed. *The cruciate ligaments*. New York: Churchill Livingstone, 1988.

Lotke PA, Ecker ML. Osteonecrosis-like syndrome of the medial tibial plateau. *Clin Orthop Rel Res* 1983;176:148–153.

Lynch MA, Henning CE. Osteoarthritis in the ACL-deficient knee. In: Feagin JA, ed. *The cruciate ligaments*. New York: Churchill Livingstone, 1988.

Lynch TC, Crues J, Sheehan W, et al. Stress fractures of the knee; MRI evaluation. *Magn Reson Imaging* 1988;6:10 (abstract).

Matthewson MH, Dandy DJ. Osteochondral fractures of the lateral femoral condyle. *J Bone Joint Surg* 1978;60B(2):199–202.

Mesgarzadeh M, Sapega AA, Bonakdarpour A, et al. Osteochondritis dessicans: analysis of mechanical stability with radiography, scintigraphy, and MR imaging. *Radiology* 1987;165:775–780.

Milgram JW, Rogers LF, Miller JW. Osteochondral fractures: mechanisms of injury and fate of fragments. *AJR* 1978; 1130:651–658.

Mink JH: The ligaments of the knee. In: Mink JH, ed. *Magnetic resonance imaging of the knee*. New York: Raven Press, 1987.

Mink JH. Pitfalls in interpretation. In: Mink JH, ed. *Magnetic resonance imaging of the knee*. New York: Raven Press, 1987.

Mink JH, Deutsch AL. Occult osseous and cartilaginous injuries about the knee: MR assessment, detection and classification. *Radiology (in press)*.

Mink JH, Levy T, Crues JV III. Tears of the anterior cruciate ligament and menisci of the knee: MR imaging evaluation. *Radiology* 1988;167:769–775.

Mirra J. Bone Tumors.

Murbarak SJ, Hargens AR. Acute compartment syndromes. *Surg Clin North Am* 1983;63:539–565.

Norman A, Baker ND. Spontaneous osteonecrosis of the knee and medial meniscal tears. *Radiology* 1978;129:653–656.

Pollack MS, Dalinka MK, Kressel HY, Lotke PA, Spritzer CE. Magnetic resonance imaging in the evaluation of suspected osteonecrosis of the knee. *Skeletal Radiol* 1987;16:121–127.

Polly DW, Callaghan JJ, Sikes RA, McCabe JM, McMahon K, Savory CG. The accuracy of selective magnetic resonance imaging compared with findings of arthroscopy of the knee. *J Bone Joint Surg* 1988;70A;192–197.

Rao VM, Fishman M, Mitchell DG, et al. Painful sickle cell crisis bone marrow patterns observed with MR imaging. *Radiology* 1986;161:211.

Reicher MA, Hartzman S, Bassett LW, et al. Magnetic resonance imaging of the knee joint. Clinical update I: injuries to menisci, patellar tendon and cruciate ligaments. *Radiology* 1987;162:547–553.

Reicher MA, Rauschning W, Gold RH, et al. High-resolution magnetic resonance imaging of the knee joint: normal anatomy. *AJR* 1985;145:895–902.

Resnick D, Oliphant M. Hemophilia-like arthropathy of the knee associated with cutaneous and synovial hemangiomas: report of 3 cases and a review of the literature. *Radiology* 1975;114:323–326.

Rosenberg NJ. Osteochondral fractures of the lateral femoral condyle. *J Bone Joint Surg* 1964;46A(5):1013–1026.

Ross JS, Masaryk TJ, Modic MT, et al. Vertebral hemangiomas: MR imaging. *Radiology* 1987;165:165–169.

Senac MO, Deutsch D, Bernstein BH, et al. MR imaging in juvenile rheumatoid arthritis. *AJR* 1988;150:873–878.

Shahriaree H. Chondromalacia patella. In: Shahriaree H, ed. *O'Connor's textbook of arthroscopic surgery*. Philadelphia: Lippincott, 1984.

Shellock F, Mink JH, Deutsch AL, et al. Kinematic MRI for evaluating patellar tracking. *Radiology* 1988;169(P):21.

Shellock F, Mink JH, Fox JM: Patellofemoral joint: kinematic MR imaging to assess tracking abnormalities. *Radiology* 1988;168:551.

Silverman J, Mink JH, Deutsch AL: MRI of discoid menisci. *Radiology* 1988;1698:395.

Spritzer CE, Dalinka MK, Kressel HY. Magnetic resonance imaging of pigmentized villonodular synovitis: a report of two cases. *Skeletal Radiol* 1987;16:316–319.

Spritzer CE, Sussman SK, Blinder RA, et al. Deep venous thrombosis evaluation with limited-flip-angle, gradient-refocused MR imaging: preliminary experience. *Radiology* 1988;166:371–375.

Spritzer CE, Vogler JB, Martinez S, et al. MR imaging of the knee: preliminary results with a 3DFT GRASS pulse sequence. *AJR* 1988;150:597–603.

Stafford SA, Rosenthal DI, Gebhardt MC, et al. MRI in the stress fracture. *AJR* 1966;147:553–556.

Steadman RJ, Higgins RW. ACL injuries in the elite skier. In: Feagin JA, ed. *The cruciate ligaments*. New York: Churchill Livingstone, 1988.

Tang J, Gold R, Bassett L, Seeger L. Musculoskeletal infection of the extremities: evaluation with MR imaging. *Radiology* 1988:109.

Tyrell RL, Gluckert K, Parthfia M, Modic MT. Fast three-dimensional MR imaging of the knee: comparison with arthroscopy. *Radiology* 1988;166:865–872.

Unger E, Moldofsky P, Gatenby R, et al. Diagnosis of osteomyelitis by MR imaging. *AJR* 1988;150:605–610.

Vogler J, Murphy W. Bone marrow imaging. *Radiology* 1988; 168:679–693.

Weissman BNW, Sledge CB. The knee. In: Weissman BNW, Sledge CB, eds. *Orthopedic radiology*. Philadelphia: WB Saunders, 1986.

Yao L, Lee JK. Occult intraosseous fracture: detection with MR imaging. *Radiology* 1988;168:749–751.

Yuh WTC, Kathol MH, Sein MA, et al. Hemangiomas of skeletal muscle: MR findings in five patients. *AJR* 1987;149:765–768.

Yulish B, Lieberman J, Strandjord S, et al. Hemophilic arthropathy: assessment with MR imaging. *Radiology* 1987;164:759–762.

Yulish BS, Montanez J, Goodfellow DB, et al. Chondromalacia patellae: assessment with MR imaging. *Radiology* 1987;164:763–766.

CHAPTER 7

The Foot and Ankle

David J. Sartoris, Jerrold H. Mink, and Roger Kerr

Cross-sectional imaging of the foot and ankle had its inception with the development of high resolution computed tomographic (CT) equipment. During recent years, CT has been successfully applied to a wide variety of podiatric disorders, including calcaneal fractures, subtalar coalition, tarsometatarsal fracture dislocations, and primary soft tissue pathology. The advent of high resolution proton magnetic resonance imaging (MRI) has provided an alternative means for noninvasive evaluation of foot and ankle disease.

Technique: Magnetic resonance imaging has become an extremely important noninvasive diagnostic tool for the evaluation of a variety of disorders affecting the foot and ankle. This section is intended to provide practicing radiologists with the specific technical guidelines for optimal imaging of podiatric disease using this method. Although details may vary according to scanner differences and imaging time considerations, the information provided herein should, at minimum, serve as a baseline from which to develop acceptable site-specific protocols. The majority of the images depicted in this chapter were acquired on a 1.5-T superconducting system (Signa, General Electric, Milwaukee, WI).

The patient is positioned prone with the feet together and parallel to one another. Their dorsal aspects are thus in contact with and stabilized by the table top, resulting in a lesser tendency for undesired patient movement during the examination than with supine scanning. In the latter position, unconscious muscular twitches can result in slight excursion of the toes with secondary image degradation. Alternatively, a support board can be utilized with the patient supine. The feet are secured with adhesive tape, and appropriate steps are taken to maximize overall patient comfort during the study (pillow, foam pad, blanket, etc.). Careful attention to symmetrical positioning of the feet will ensure that comparable anatomical levels are depicted on individual images, greatly facilitating interpretation. Claustrophobic and pediatric patients may require sedation because motion must be avoided for as long as 20 min at a time, and if present, results in degradation of an entire series of images rather than only one image as with CT.

In general, the authors have found that satisfactory image quality is afforded by use of the same coil used routinely for MRI studies of the head. This approach eliminates the problem of signal intensity fall-off inherent in the application of surface coils and produces images of uniform homogeneity. A 20-cm field of view with a 256×256 data acquisition matrix results in a pixel size of 0.78×0.78 mm, which is equivalent to the spatial resolution of the images acquired. A single signal average with a slice thickness of 5 mm and an interslice gap of 2.5 mm creates practical imaging times and adequate coverage of the tissue volume under evaluation. Contiguous imaging may be required in certain situations involving subtle pathology but is not routinely indicated as it necessitates an objectionably longer total examination time.

Initially, a T1-weighted localization sequence should be performed in the plantar plane, using the shortest available imaging time. This generally involves a pulse repetition time (TR) on the order of 600 msec and a time to echo (TE) of approximately 25 msec (partial saturation spin-echo technique). This sequence can be completed in 5 min, and the resulting images (which may in themselves be beneficial in planning operative approaches) are then used to delineate the specific region to be evaluated subsequently in the coronal plane of the body (transverse plane of the foot). Alternatively, a gradient-echo sequence, also accomplished in a relatively short time, can be utilized for localization images. The plane perpendicular to the metatarsals is generally of greatest value for most disorders of the foot and ankle, owing to anatomical considerations and facilitation of comparison between the two sides.

The pulse sequences selected for imaging in the transverse plane will depend on the specific type of pathology suspected on the basis of prior clinical assessment and/or diagnostic imaging studies. For the detection of bone marrow disease (ischemic necrosis, osteomyelitis, malignancy), a T1-weighted imaging sequence using a TR and TE comparable to those of the localization scans usually suffices and translates to an extremely short total imaging time of approximately 10 min. All such processes generally produce decreased signal intensity in marrow-bearing areas regardless of image weighting, although recent preliminary evidence suggests that T2-weighted sequences may have merit in identifying acute areas of infarction or infection by virtue of their bright signal. Conversely, in the evaluation of soft tissue disorders affecting the foot and ankle (neurogenic and other neoplasms, abscesses, cellulitis, tendinitis, traumatic injury) by MRI, several different pulse sequences usually are required for optimal delineation and characterization of pathology. This is true because (a) the specific signal behavior of the abnormal tissue cannot be predicted before imaging, and (b) the type(s) of normal tissue (fat, muscle, tendon, fascia) with which it interfaces is similarly unknown in advance. Thus, because it is usually unclear as to which specific pulse sequence will produce optimal contrast discrimination between normal and abnormal tissue, multiple TR/TE combinations must be used. This is most conveniently achieved by initially performing a T1-weighted sequence identical to that recommended for bone marrow disease, followed by a multiple spin-echo sequence, which results in two sets of images, weighted toward proton density and T2, respectively. The latter imaging strategy generally involves a TR on the order of 2,000 msec, a first-echo TE of about 25 msec for proton-density weighted images, and a second-echo TE of approximately 70 msec for T2-weighted images. The typical multiple spin-echo sequence can be completed in about 17 min. Thus, the total examination time for adequate evaluation of a soft tissue disorder in the foot or ankle is approximately 27 min (5 + 5 + 17), not including patient positioning and scan set-up time. As a time-saving alternative or supplement to T2-weighted images, a gradient-echo sequence using short TR and TE with a low flip angle can be employed to yield a T2* effect.

The importance of multiple pulse sequences for optimal imaging of the soft tissues in the foot and ankle cannot be overemphasized. Several cases are shown in this chapter that include a TR/TE combination that renders significant pathology occult.

Images in the sagittal plane of the foot are readily obtained without moving the patient, but they contribute significant additional information only in rare instances and hence are seldom necessary. This imaging plane, however, may be optimal for disease processes oriented along the axis of one or several rays. In specific cases, obliquely oriented images can be helpful, particularly for visualizing the course of a certain tendon or muscle group, such as the peroneal tendons. Imaging in nonroutine planes always increases the total examination time and thus usually is performed using the single pulse sequence that is deemed likely to provide the greatest information based on preliminary screening of the transverse images.

On completion of the MRI examination of the foot or ankle, all images acquired should be photographed on radiographic film for permanent patient records. It is imperative, however, that the radiologist view the entire study on the scanner console, which allows for image optimization via window (contrast) and level (brightness) manipulation. The technologist cannot always be relied on to select the most diagnostic image display settings for photography. As a final technical consideration, because the method exhibits high sensitivity but only moderate specificity, it is mandatory that all MRI examinations of the foot and ankle be interpreted in conjunction with any available prior diagnostic imaging studies as well as with complete knowledge of the clinical history and physical examination findings.

Two technical problems may be encountered. Imaging in oblique orientations may not be available at this time with certain systems. This capability is particularly important for imaging joints where tendons and ligaments may not align with the x-, y-, or z-axes. In the ankle, we have encountered similar difficulties in evaluating ligament integrity after acute trauma. Even with the use only of conventional planes, however, acute ligamentous ruptures can be clearly detected in certain cases.

Dedicated surface coils for joint studies may afford images superior to those obtained with planar coils. Patients and healthy volunteers have been studied with a 3- or 5-inch circular transmit-receive surface coil. These particular coils provide very superficial signal detection, and although structures close to the coil are well depicted, deeper lesions cannot be accurately studied unless the coil is repositioned. Problems are thus encountered in the setting of marked soft tissue swelling or a cast. A rectangular receive-only surface coil designed for spinal imaging provides a more uniform signal throughout the joint.

CASE 1

This 27-year-old man has a history of chronic bilateral ankle disease and a systemic disorder. What are the findings? What is the diagnosis? (Fig. 1A. Sagittal MR, TR/TE 2,000/20 msec. Fig. 1B. Sagittal MR, TR/TE 2,000/80 msec. Reprinted with permission from Forrester DM, Kricun ME, Kerr R: Imaging of the foot and ankle. Aspen Press, Rockville, MD, 1988.)

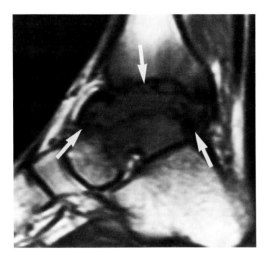 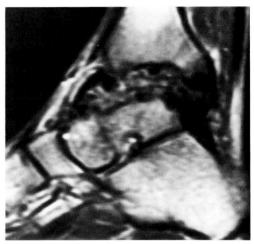

FIG. 1A.   **FIG. 1B.**

Findings: The MR images reveal material of low to intermediate signal intensity filling the tibio-talar joint space (arrows), with destructive changes and diminished signal intensity in the distal tibia and talar dome in Fig. 1A. Figure 1B demonstrates a wide range of signal intensities within the joint space compatible with fluid, inflammation, synovial hypertrophy, hemosiderin deposition, and fibrosis.

Diagnosis: Hemophilic arthropathy.

Discussion: In patients with hemophilia, multiple episodes of intra-articular bleeding often result in a progressive, destructive arthropathy. Hemarthrosis occurs most often in children and adolescents and involves joints that are most susceptible to trauma. In descending order of frequency, the most commonly involved joints are the knee, elbow, ankle, hip, and shoulder. Bleeding into muscle and bone and between fascial planes may also occur. Intraosseous or subperiosteal hemorrhage may produce a destructive, expansile lesion termed a *hemophilic pseudotumor*. This occurs most often in the ilium and femur. Soft tissue bleeding may cause nerve entrapment or a compartment syndrome.

In the foot and ankle, hemophilic arthropathy most often involves the tibio-talar and talocalcaneal joints. In addition to erosive and destructive joint changes, repeated episodes of intra-articular bleeding may also produce significant deformity of the foot and ankle. Equinus or varus deformity or fixed plantar flexion may develop because of reflex spasm or secondary to intramuscular bleeding in the posterior calf muscles. A cavus deformity can arise from intra-articular bleeding in the midfoot or from hemorrhage within the quadratus plantae muscle or the subplantar fascia. Entrapment of the common peroneal nerve has also been reported.

Most episodes of intra-articular bleeding occur spontaneously or after minor trauma. An inflammatory reaction and synovial proliferation are evoked. The hyperplastic synovium is highly vascular and predisposes the joint to recurrent bleeding episodes.

Hemosiderin deposition and fibrosis and contraction of the joint capsule follow. With chronic disease, there is destruction of articular cartilage, subchondral cyst formation, and secondary degenerative changes.

The early radiographic changes seen in hemophilic arthropathy include joint distention, soft tissue swelling, and osteoporosis. Progression of disease is characterized by osseous erosions, subchondral cysts, joint space narrowing, increased osteoporosis, and changes of secondary osteoarthritis (Fig. 1C). In the ankle, epiphyseal overgrowth and premature fusion may cause the formation of an obliquely oriented joint termed *tibiotalar slant*. This finding is not specific for hemophilic arthropathy as it may also be seen in juvenile chronic arthritis, epiphyseal dysplasia, and sickle cell anemia. The articular surface of the talus is often enlarged as osteophytes extend anteriorly over the talar neck. This bony proliferation acts to block dorsiflexion. Bony ankylosis of the talocalcaneal joints and exuberant osteophytosis at the dorsal aspect of the talonavicular joint have also been described. Widening of the tarsal sinus may be seen on an oblique view of the foot, analogous to widening of the intercondylar notch of the knee that is seen in this disease.

Hemophilic pseudotumors, within bone or soft tissue, demonstrate a variety of radiographic appearances. Bone lesions are typically well demarcated, may appear purely lytic or trabeculated, and often appear expansile. Soft tissue lesions form a fibrous capsule and may cause pressure erosion of adjacent bone.

Magnetic resonance imaging may be used to demonstrate the changes in hemophilic arthropathy. Loss of articular cartilage, bone erosions, and subchondral cysts may be identified. A subchondral cyst has a sclerotic, low signal intensity rim and its fluid contents are of low signal intensity with T1 weighting and high signal intensity with T2 weighting. Hypertrophied synovium appears as a mass of low to intermediate signal intensity, and hemosiderin deposits are of low signal intensity on both T1- and T2-weighted images. Inflamed synovium demonstrates high signal intensity on T2-weighted images. Recent hemarthrosis appears as a region of high signal intensity on both T1- and T2-weighted images. Obviously, depending on the stage and activity of joint disease, a variety of patterns of abnormal signal intensity is observed in these patients.

Areas of intraosseous or soft tissue hemorrhage may by identified with MRI before their radiographic demonstration (Fig. 1D). These lesions also show a variety of signal intensities depending on the stage of clot organization and the time elapsed since the last bleeding episode.

Magnetic resonance imaging may be used to determine the extent and severity of joint disease and for early diagnosis of pseudotumors in patients with hemophilia. Its greatest clinical utility may be in monitoring response to factor therapy and identifying patients with early articular disease who may benefit from synovectomy.

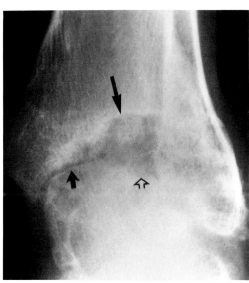

FIG. 1C. Hemophilic arthropathy. An anteroposterior (AP) radiograph of the ankle demonstrates joint space narrowing (*short arrow*) and subchrondral cysts within the distal tibia (*long arrow*) and talar dome (*open arrow*). (Reprinted with permission from Forrester DM, Kricun ME, Kerr R: Imaging of the foot and ankle. Aspen Press, Rockville, MD, 1988.)

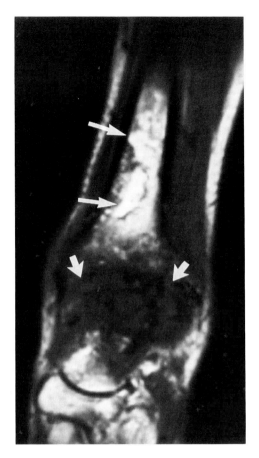

FIG. 1D. Hemophilic arthropathy (TR/TE 500/20 msec). A coronal MR image through the anterior ankle demonstrates multiple foci of high signal intensity within the distal tibia, compatible with intraosseous hemorrhage (*long arrows*). Conventional radiographs revealed a normal-appearing trabecular pattern. Material of low to intermediate signal intensity in the ankle represents synovial hypertrophy and fibrosis (*short arrows*). (Reprinted with permission from Forrester DM, Kricun ME, Kerr R: Imaging of the foot and ankle. Aspen Press, Rockville, MD, 1988.)

CASE 2

A 25-year-old man has had the gradual onset of ankle pain during the past 6 months. Physical examination demonstrated mild pain on inversion. What is the diagnosis, and what is the particular value of MR in this disorder? (Fig. 2A. Plain radiograph. Fig. 2B. Sagittal MR, TR/TE 2,000/80 msec.)

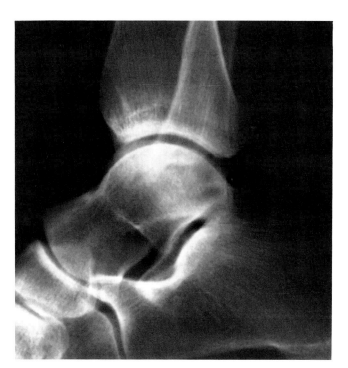

FIG. 2A.

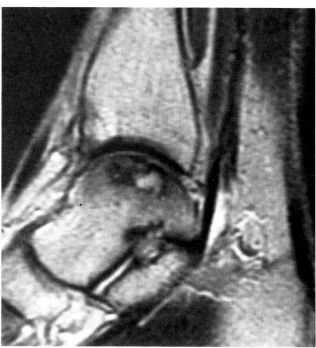

FIG. 2B.

Findings: The plain radiographs (Fig. 2A) are normal. The MR (Fig. 2B) demonstrates a large area of dark signal in the subchondral bone of the talus and a more focal area of fluid within.

Diagnosis: Osteochondritis dissecans (OCD).

Discussion: *Osteochondritis dissecans* is a term which refers to the fragmentation and often complete separation of the articular surface of a joint. Most authors accept that trauma is the etiology of OCD, although a history of a single acute event is not usually available. The term *transchondral fracture* is synonymous with OCD. The fracture occurs parallel to the joint surface and may involve cartilage alone or cartilage plus a variable amount of cortical and/or subchondral bone. Adolescents are most commonly affected, but the extremes of age are not immune from affliction.

The fate of the fracture fragment is variable. It may remain *in situ* and heal in place, or it may undergo resorption. If it becomes completely detached, it can (a) grow by means of proliferation of bone on its surface; (b) resorb secondary to osteoclastic activity; (c) implant on a synovial surface; or (d) act as a loose obstructing body.

Osteochondritis dissecans of the talar dome involves either the posterior one-third of the medial aspect or the middle third of the lateral side. As with OCD elsewhere, trauma is the primary etiology. When the foot is inverted, the lateral aspect of the talus is compressed against the medial border of the fibula; extensive and prolonged forces result in creation of an osteochondral fragment. If the foot is everted, the posteroinferior lip of the tibia compresses and fractures the talar dome medially. A medial osteochondral fracture is generally larger and deeper than a lateral lesion, which tends to have an elongate flake-like character.

Conventional radiographs may not be diagnostic, especially in cases of a small lateral lesion. There is no osseous lesion in a purely chondral lesion. Because young cartilage is quite elastic, the deforming force may be transmitted to the underlying cortical and medullary bone; the cartilage may "rebound" into a normal position. In such cases, arthrography and arthrotomography may be normal, and the subchondral abnormality may be the only manifestation of the injury.

The role of MR in assessment of osteochondral injuries of the talus has not been critically studied, but potential advantages include (a) detection of radiographically "occult" lesions (Fig. 2A,B), (b) assessment of the integrity of the overlying cartilage (Fig. 2C,D), (c) detection of loose fragments (Fig. 2E–G), and (d) precise determination of the extent and location of the lesion, all of which are critical to the choice of and timing of surgical intervention.

C

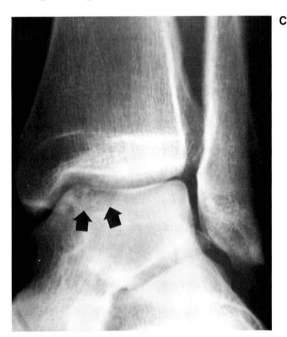

FIG. 2C,D. Talar OCD. The plain radiograph (**C**) demonstrates the lesion (*arrows*), but the surgeon requested assessment of the overlying cartilage. The sagittal MR (**D.** TR/TE 2,000/80 msec) demonstrates a large defect (*straight arrows*) eroding into the medullary bone and involving approximately 1 cm of the talar dome; a single large fragment (*curved arrow*) is seen in the bed of the defect.

D

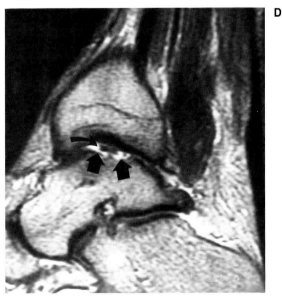

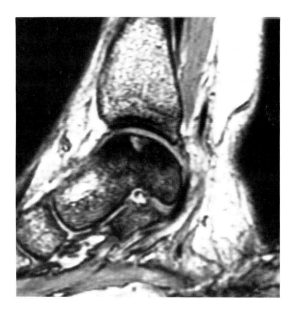

FIG. 2E. Osteochondritis dissecans (Sagittal MR, TR/TE 55/15 msec). This gradient recalled sequence demonstrates that the articular cartilage over a medial lesion is intact, although the bone is flattened. The relative abilities of gradient recalled and standard spin-echo sequences to detect cartilaginous defects await further study.

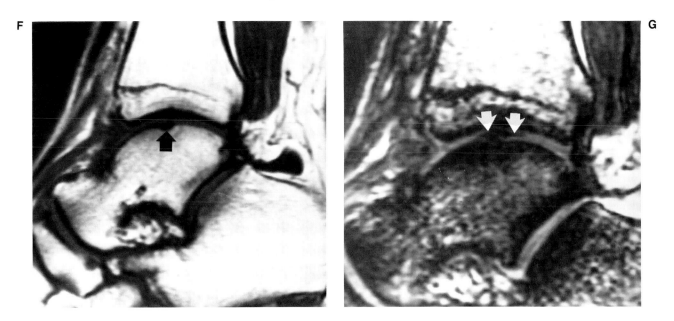

FIG. 2F,G. Osteochondritis dissecans. The sagittal spin-echo sequence (**F.** TR/TE 2,000/20 msec) demonstrates a very tiny osseous defect (*arrow*); the cartilage appears intact. The gradient recalled sequence (GRASS) (**G.** TR/TE 55/15 msec) demonstrates a small partially loose fragment (*arrows*). The bony defect appears more extensive on this GRASS sequence.

CASE 3

This 19-year-old youth had observed a swelling behind the left medial malleolus for 3 months. Lateral radiographs revealed obliteration of the left pre-Achilles fat pad by a well-defined soft tissue mass that projected toward the calcaneus but was inseparable from the musculature of the lower calf. An MRI examination was subsequently performed. What should be done next? (Fig. 3A. Sagittal MR, TR/TE 2,000/20 msec. Fig. 3B. Axial MR, TR/TE 2,000/80 msec.)

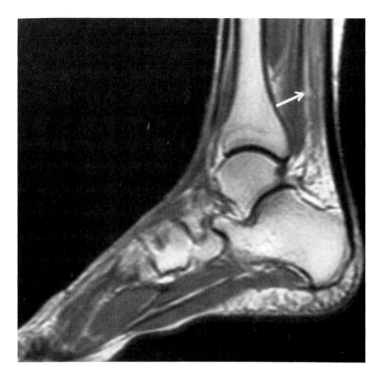

FIG. 3A.

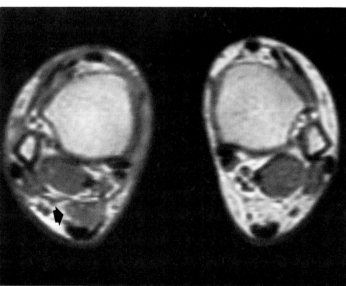

FIG. 3B.

Findings: The images reveal a tubular soft tissue mass (arrows) occupying the pre-Achilles fat pad. The lesion has a signal intensity comparable to that of muscle on both sequences.

Diagnosis: Accessory soleus muscle.

Discussion: Many anomalous muscles of the limbs have been detected in the past by meticulous dissection of cadavers. Some of these malformations are cursorily mentioned in modern anatomic textbooks but seem to be rather uncommon and have received little or no attention in clinical literature. When found *in vivo*, they are therefore not often immediately recognized and may raise various diagnostic problems. Anomalous muscles of the hand, for example, have been repeatedly mistaken for ganglions or tumors.

Anomalous muscles of the lower limbs have been diagnosed clinically in three patients reported by Dunn. In one of these cases, a hamstring muscle was aberrant, whereas in the others the soleus muscles were duplicated. The accessory muscle bundle extends into the Kager fat triangle, anterior to the Achilles tendon; it inserts onto the calcaneus, the Achilles tendon, or the flexor retinaculum.

The findings described in previous cases of supernumerary soleus muscles detected *in vivo* suggest that this malformation is recognizable on inspection and palpation of the leg. Bilateral occurrence should facilitate a correct diagnosis. However, because the malformation seems to be a rarity, its characteristic features and even its existence are not widely known. It is hardly surprising that in cases in which the malformation is unilateral, an erroneous diagnosis of tumor and surgical exploration have been made. Such errors are more understandable when one realizes that patients with an accessory soleus muscle may have pain after exercise.

Significant abnormalities suggestive of the muscular malformation can be demonstrated with xerography or conventional radiography using a technique suited for the soft tissues, but, as expected, cross-sectional imaging is much more informative. Because the major muscles and vessels of the leg can be discriminated by CT or MRI, the presence of an anomalous muscle is firmly established with either method. Moreover, these techniques reveal characteristic features sufficient to specify the observed malformation as a supernumerary soleus muscle. This is particularly true of MRI, with its multiplanar imaging capability.

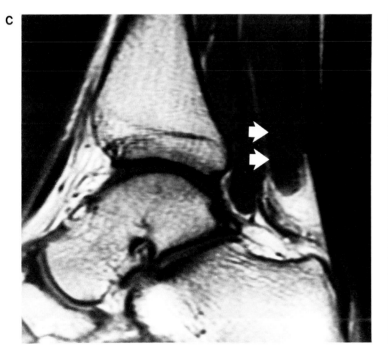 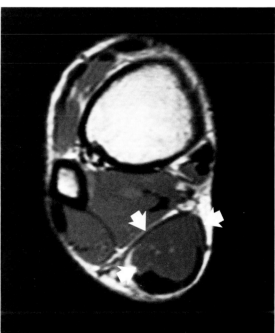

FIG. 3C,D. Accessory soleus muscle. **C.** Sagittal MR, TR/TE 600/20 msec. **D.** Axial MR, TR/TE 600/20 msec. Both images demonstrate a very large soleus muscle (*arrows*) that inserts on the Achilles tendon and the fat pad.

CASE 4

This 5-year-old boy presented with a rapidly growing soft tissue mass of the right dorsal foot, without pain. What is the diagnosis? (Fig. 4A. Axial MR, TR/TE 500/20 msec. Fig. 4B. Axial MR, TR/TE 1,500/70 msec. Courtesy of P. Kindynis, MD, Geneva, Switzerland.)

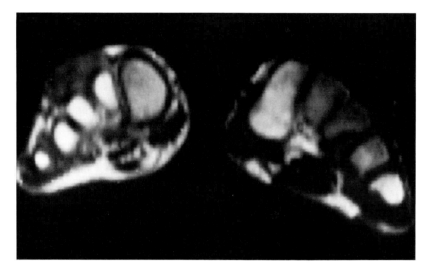

FIG. 4A.

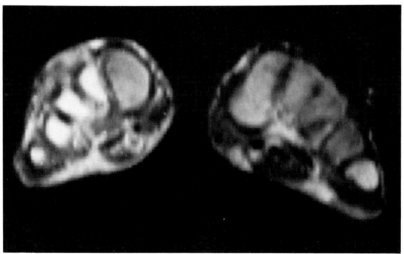

FIG. 4B.

Findings: The MR images (Fig. 4A,B) demonstrate a poorly circumscribed soft tissue mass above and between the second and third metatarsal heads. The lesion is heterogeneous with areas of diminished signal compared to the muscle in T1-weighted images, whereas it appears brighter in T2-weighted images. On CT images (Fig. 4C), this mass appears uncalcified and homogeneous with intensity identical to the muscles but demonstrates small pressure erosions of the cortex of the second metatarsal that were not visible on MR.

Diagnosis: Juvenile fibromatosis or fibrosarcoma.

Discussion: The term *fibromatosis* is applied to a group of benign fibroblastic proliferative lesions of the soft tissue that are distinguished from the self-limiting scar. These lesions can be locally aggressive but never metastasize. The term *juvenile* refers to the fact that these lesions occur predominantly during the neonatal period or childhood, with no apparent counterpart in adult life. The classification based on anatomic location and histology appearance includes fibromatosis colli, juvenile aponeurotic fibroma, recurring digital fibrous tumor of childhood, fibrous hamartoma of infancy, juvenile hyaline fibromatosis, infantile desmoid-type fibromatosis, aggressive infantile fibromatosis, and infantile myofibromatosis (originally termed *congenital generalized fibromatosis*) with three different forms: generalized, multiple, or local fibromatosis.

In this case the lesion was removed, and the diagnosis of aggressive infantile fibromatosis was made. Histologically, the level of mitotic activity typically greatly varies among the interlacing bundles of fusiform and spindle-shaped cells and reticulum and collagen fibers, making differentiation from fibrosarcoma difficult. The lesion is slightly more prominent in boys and is locally aggressive, infiltrating into muscles, vessels, nerves, fasciae, tendons, and subcutaneous fat. Unlike the other forms of infantile fibromatosis, there is no prevalence for any particular anatomic site. The radiographs may demonstrate an occasional bone deformity and scalloped defect. The tumor tends to recur following surgery, which was the case in this patient, but distant metastases are uncommon.

This case demonstrates the well-described superiority of CT over MR for assessment of the cortical extent of a tumor. Magnetic resonance, however, is clearly superior to CT in determination of intramedullary and soft tissue extent and the relationship to the neurovascular bundle.

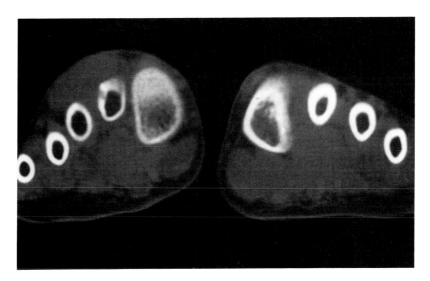

FIG. 4C. Fibromatosis. The axial CT demonstrates permeation of the cortex of the second metatarsal not appreciated on MR.

CASE 5

A 35-year-old woman presents with a history of trauma to the anterior lower leg 3 months ago. Physical examination reveals weak dorsiflexion. What is the diagnosis? (Fig. 5A. Sagittal MR, TR/TE 1,800/20 msec. Fig. 5B. Axial MR, TR/TE 1,800/80 msec.)

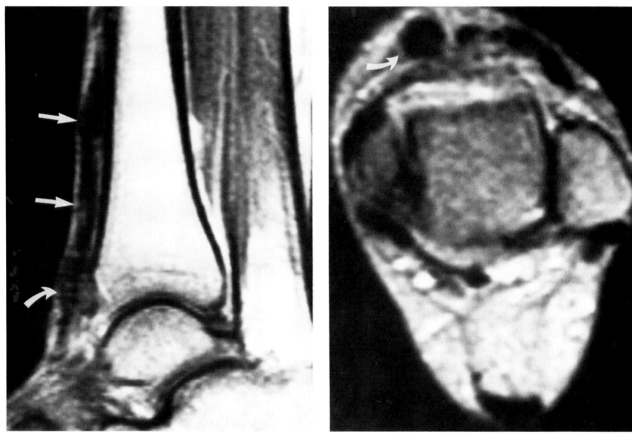

FIG. 5A.   **FIG. 5B.**

Findings: The MR images of the ankle demonstrate diffuse thickening of the anterior tibial tendon (ATT) (arrows) that is most prominent at the level of the ankle joint (curved arrows).

Diagnosis: Chronic partial rupture of the ATT.

Discussion: Magnetic resonance imaging is a useful technique for evaluating tendons. Acute and chronic tendon ruptures and tenosynovitis are reliably demonstrated. In the past, tendon injuries in the foot and ankle were imaged with tenography (an invasive procedure) or CT (utilizing ionizing radiation). Because of its superior soft tissue contrast and multiplanar imaging capability, MRI has supplanted these techniques as the method of choice for evaluating tendon pathology.

   With CT, it is often difficult to distinguish a tendon from surrounding edema or to detect a small amount of fluid within a tendon sheath or within a partially torn tendon. In contrast, with MRI a tendon is readily distinguished from tenosynovial fluid and small peritendinous or intratendinous fluid collections are easily demonstrated. This results in a more precise characterization of tendon pathology. A potential shortcoming of MRI is occasional difficulty in distinguishing tendon from adjacent cortical bone.

On MRI, a normal intact tendon appears as a smooth, well-defined hypointense structure. A recent partial tear typically demonstrates fluid within the tendon and tendon sheath, best seen on T2-weighted images. Subacute hemorrhage produces increased signal intensity on both T1- and T2-weighted images. A complete tear manifests itself as total discontinuity of tendon substance with retraction of the torn tendon (Fig. 5C). This is best demonstrated on axial cross-sectional images in which the tendon is absent from its usual position. A chronic tear or tendinitis appears as a diffusely thickened tendon, often with an associated tenosynovial fluid collection. Magnetic resonance imaging may also demonstrate peritendinous soft tissue abnormalities such as an ankle joint effusion or retrocalcaneal bursitis (Fig. 5D,E).

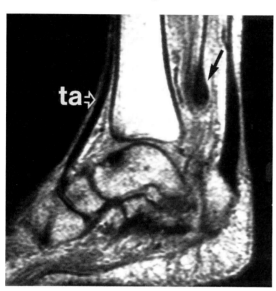

FIG. 5C. Rupture of the flexor hallucis longus tendon (sagittal MR, TR/TE 1,800/20 msec). The MR images demonstrate thickening of the proximal aspect of the flexor hallucis longus tendon (*black arrow*) caused by retraction of tendon fibers secondary to a near complete tear in a patient with a complex calcaneal fracture (not seen). Anterior tibial tendon, ta (*open arrow*).

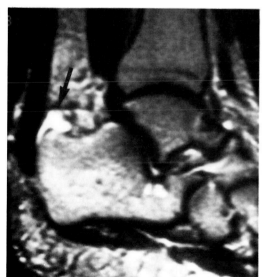
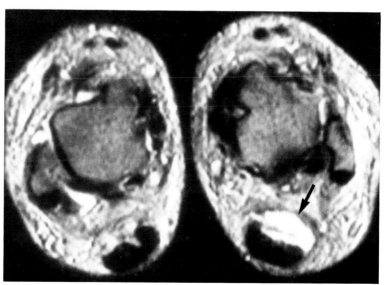

FIG. 5D,E. Retrocalcaneal bursitis (**D.** Sagittal MR, TR/TE 2,500/80 msec; **E.** Axial MR, TR/TE 2,500/80 msec). Magnetic resonance images reveal fluid within the retrocalcaneal bursa (*arrows*) caused by bursitis.

CASE 6

This 5-year-old boy complained of a medial ankle mass. What is the diagnosis? (Fig. 6A. Coronal MR, TR/TE 600/20 msec. Reprinted with permission from Forrester DM, Kricun ME, Kerr R: Imaging of the foot and ankle. Aspen Press, Rockville, MD, 1988.)

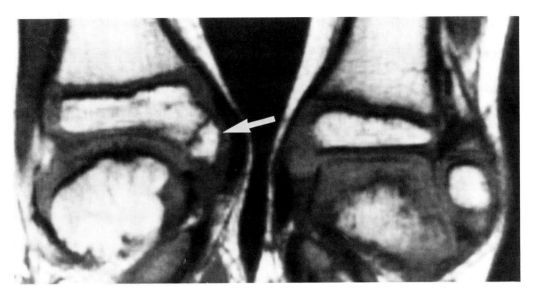

FIG. 6A.

Findings: The MR image demonstrates overgrowth of the medial distal tibial epiphysis caused by an osteocartilaginous exostosis (arrow). There is a rounded deformity of the articular surfaces of the distal tibia and the talar dome.

Diagnosis: Trevor disease (dysplasia epiphysealis hemimelica).

Discussion: Trevor disease is a congenital growth disorder characterized by cartilaginous overgrowth of a portion of one or more epiphyses. Only one limb is involved, and usually only the medial or lateral side is affected. There is predominant lower extremity involvement, and the medial side is affected twice as often as the lateral. The most common sites of involvement are the talus, distal femur, and distal and proximal tibia. This disorder usually becomes evident between the ages of 2 and 14 years and affects boys three times as often as girls. Patients may present with a painless mass, limitation of motion, valgus or varus deformity, and limb length discrepancy. Pain is not a common complaint but is seen more frequently with ankle lesions than with those arising in the knee.

Multiple lesions are identified in roughly two-thirds of patients with Trevor disease. Depending on the degree of involvement, this disorder may be classified into one of three forms. The localized form describes monostotic involvement. In the classical form, more than one area in a single extremity is involved (Fig. 6B–E). In the generalized or severe form, there is involvement of an entire lower extremity from the pelvis to the foot or ankle.

The radiographic appearance of Trevor disease is characteristic and consists of multiple irregular ossification centers arising from one side of an epiphysis. These centers coalesce as they enlarge to form a lobulated osseous mass protruding from the epiphysis or from a tarsal or carpal bone. The final radiographic appearance (and the microscopic appearance) is identical to that of an osteochondroma. The term *epiphyseal osteochondroma* has been applied to this lesion. Widening of the metaphysis and involvement of an entire epiphysis have also been reported. Deformity of the involved bone and of an adjacent bone or articular surface may be induced by the abnormal epiphyseal growth.

A limitation of conventional radiography in evaluating Trevor disease is that the unossified cartilage of the epiphysis is not visualized. Arthrography may be used to determine the configuration of the articular surface; however, the full extent of the nonossified portion of the lesion may not be demonstrated. With CT, bone is clearly shown, but unossified cartilage is of soft tissue density and may not be distinguishable from adjacent muscle. Depending on the joint involved it may be difficult to obtain images perpendicular to the articular surface with CT.

Magnetic resonance imaging is well suited to evaluating the immature skeleton. The nonossified cartilage is clearly shown, and therefore, the shape of a growing bone may be determined. The multiplanar imaging ability of MRI facilitates determination of lesion extent and assessment of the articular surface. Magnetic resonance imaging may be useful in assessing other congenital disorders, such as clubfoot, at an early stage. In the first few years of life, the shape and axis of the bones of the feet are difficult to determine radiographically as only a small portion of bone is ossified. With MRI, the nonossified portion of bone is demonstrated, and the position, shape, and axis may be determined.

The treatment of Trevor disease varies, depending on the extent of the lesion. Surgical intervention is usually reserved for lesions that produce significant deformity or limitation of motion; surgery may prevent development of deformity or degenerative arthritis. Determination of the advisability or type of surgery is probably best accomplished with MRI. Malignant transformation has not been reported in Trevor disease.

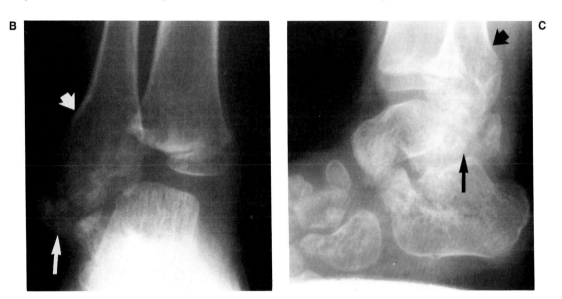

FIG. 6B,C. Trevor disease. **(B)** Anteroposterior and **(C)** lateral views of the ankle demonstrate enlargement of the entire distal fibula (megaepiphysis, *short arrow*) with multiple irregular centers of ossification (*long arrow*). The tarsal bones are also involved and appear deformed on the lateral view. (Courtesy of Donald Sauser, MD, Loma Linda, CA). **Fig. 6D,E.** *Continued.*

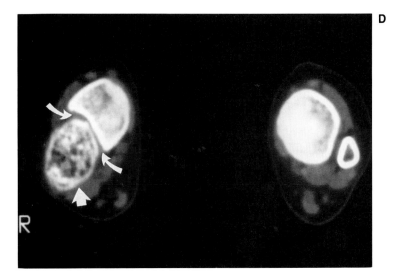

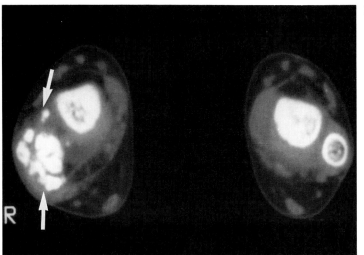

FIG. 6D,E. Trevor disease (*continued*). Axial CT images **(D)** through the metaphyseal-epiphyseal junction and **(E)** the distal epiphysis demonstrate diffuse expansion of the fibula (*short arrow*), causing a modeling deformity of the adjacent tibia (*curved arrows*). Multiple ossification centers are identified distally within the osteochondroma (*long arrows*). (Courtesy of Donald Sauser, MD, Loma Linda, CA.)

CASE 7

A 22-year-old female patient attended her physician because of a large mass located in the foot. The radiographs showed a large soft tissue mass and destruction of the second through fourth metatarsals as well as the cuneiform bones. A MRI study was performed to evaluate further the extension and nature of the lesion findings. (Fig. 7A. Sagittal MR, TR/TE 450/20 msec. Fig. 7B. Axial MR, TR/TE 2,500/100 msec. Courtesy J. Bloem, MD, Leiden, The Netherlands.)

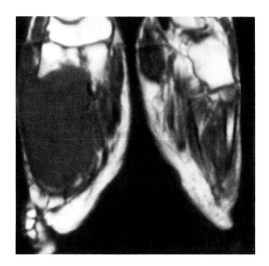

FIG. 7A.

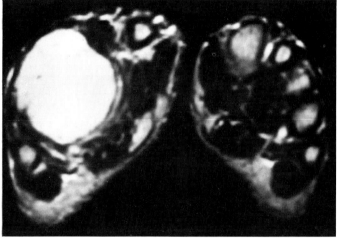

FIG. 7B.

Findings: The tumor displays a homogeneous low signal intensity on the T1-weighted images and a fairly homogeneous high signal intensity on the proton-density, T2-weighted spin-echo and GRASS images. These signal intensities were not typical for a specific histologic tumor type, but the homogeneous appearance and the well-defined margin with very little reactive changes favored the diagnosis of a benign tumor.

Diagnosis: Chondromyxoid fibroma.

Discussion: Although chondromyxoid fibroma is most frequently (98%) found in the lower extremity, the foot is not very often involved. This benign tumor is usually found in patients under the age of 30 years. Presenting symptoms include swelling, pain, and restricted motion. The radiographic findings of chondromyxoid fibroma in the foot are osteolysis, bone expansion, coarse trabeculae, and scalloped osseous erosions of surrounding bone. The dimensions of the tumor may be quite large before medical attention is sought, and the lesion may be mistaken for a sarcoma such as a synovioma. The histology of the tumor may vary considerably but contains chondroid, myxomatous, and fibrous areas.

The histology of this well-differentiated tumor is reflected by its rather nonspecific MRI characteristics. As in most neoplasms the prolonged T1 and T2 relaxation times account for the abnormality seen on the MR images. The MR images therefore do not allow a specific diagnosis to be made. It is, however, of paramount importance that the tumor is resected completely as the recurrence rate of 25% is thought to be caused by inadequate removal of tumor. Magnetic resonance imaging is the modality *par excellence* in demonstrating the extension of bone neoplasm. The main indication for MRI in these patients is therefore local staging.

CASE 8

This is a 32-year-old woman with pain between the toes of her right foot on weight bearing. What diagnosis can be suggested from this MR image? (Fig. 8A. Coronal MR, TR/TE 600/20 msec. Fig. 8B. Coronal MR, TR/TE 2,000/80 msec. Case courtesy of H. Adams, MD, San Diego, CA.)

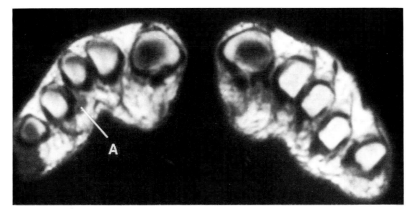

FIG. 8A.

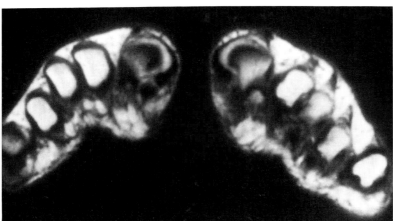

FIG. 8B.

Description: A coronal image through the forefoot at the level of the metatarsal heads, with the patient lying prone, demonstrates the flexor and extensor tendons as low signal intensity areas inferior and superior to the metatarsals surrounded by high signal intensity fat. Arising between the inferior surfaces of the third and fourth metatarsal heads is a mass of intermediate signal intensity (A, lesion). A similar lesion is not seen in the left foot. On the T2-weighted images the mass remains relatively low in signal intensity.

Diagnosis: Morton's neuroma (Morton's toe).

Discussion: Metatarsalgia was described by Morton as a specific process, occurring more commonly in women than men, manifested typically by plantar foot pain on weight bearing. The condition has been ascribed to excessive weight or tight fitting shoes in many patients but may be idiopathic in others. It is found most commonly between the third and fourth metatarsal heads. Morton's neuroma is not a true neoplasm but a reactive fibrotic process associated with nerve degeneration; support for an ischemic theory of pathogenesis is found in lesions in which arteries show occlusion.

Histologically, there is a variable loss (or degeneration) of myelinated fibers with thickening and fibrosis of the epineurium and perineurium.

The usual neuroma (which literally means *tumor of nerve*), originating in the neurolemma, contains an overabundance of Schwann cells or perineurial cells, and appears brighter than fat on T2-weighted images. The fibrosis identified histologically in Morton's neuromas is likely responsible for their relatively low signal intensity on all pulse sequences.

Amputation stump neuromas present in patients who have recurrent symptoms following initial surgical excision. They are characterized by an overgrowth of axis cylinders and Schwann cells separated by fibrocollagenous tissue. It is possible that these lesions will have MR signal characteristics similar to a Morton's neuroma. Thus identification of a mass with signal intensity similar to that of an interdigital neuroma on MR images in the area of a previously excised nerve should suggest the diagnosis of amputation stump neuroma.

CASE 9

A 25-year-old man suffered an injury to his left foot in an auto accident 2 months prior. He had polytomography and a CT scan, which had demonstrated a talo-navicular dislocation, as well as a number of other less severe fractures. Why might an orthopedist order an MR? (Fig. 9A. Sagittal MR, TR/TE 500/20 msec.)

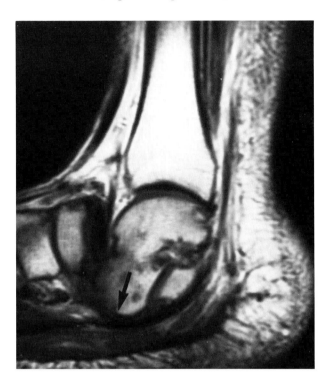

FIG. 9A.

Findings: The single image demonstrates the midfoot dislocation at the talonavicular joint. The talus is rotated slightly counterclockwise so its distal end (arrow) is directed plantar. The navicular and rest of the foot is dorsal. The signal of the talus is normal on this T1-weighted image.

Diagnosis: No evidence of osteonecrosis in this midfoot dislocation.

Discussion: The talus is a critical component of the hindfoot mechanism. It is unique in that it has multiple articular surfaces: the talotibial and talofibular above, the three talocalcaneal (subtalar) below, and the talonavicular anteriorly. No major tendon or muscle directly attaches to the talus, which is covered by articular cartilage over 60% of its surface. There is, therefore, limited access of blood vessels to the talus.

The main blood supply to the talus is via the artery of the tarsal canal, a branch of the posterior tibial artery. The peroneal and anterior tibial artery also contributes branches, as do the synovial attachments from adjacent bones. Therefore, talar dislocations or neck fractures may interrupt this tenuous blood supply and lead to osteonecrosis. The highest incidence (100%) of osteonecrosis occurs in total talar dislocations in which the subtalar, talonavicular, and tibio-talar joints are all disrupted. Type II neck fractures (a vertical fracture line extending into the subtalar joint between the middle and posterior facets, and the subtalar joint is subluxed) interrupts two and usually all three sources of blood supply to the talus. The incidence of osteonecrosis is as much as 42% in this group. Any peritalar subluxation or dislocation can, however, jeopardize a favorable outcome.

Osteonecrosis becomes evident on routine plain radiographs approximately 2 months following the injury. It is manifest as area(s) of increased density. Demineralization of the talar dome results from osteoclasis, a process that occurs acutely only in the presence of an intact blood supply (Hawkin's sign). The role of MR in predicting the viability of the talus has not been critically studied, but MR may prove as useful as it is at the hip (Fig. 9B,C).

B

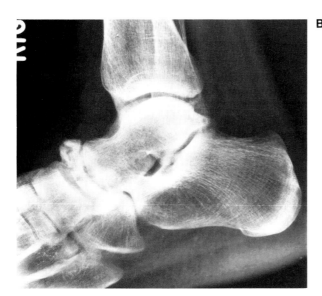

FIG. 9B,C. Navicular osteonecrosis. This patient with known systemic lupus erythematosis has taken steroids. She presented with foot pain. The plain radiograph, gallium, and bone scans (not shown) were thought to indicate the presence of an infection of the navicular. Biopsy and culture demonstrated necrotic bone without infection. The sagittal MR (TR/TE 800/20 msec) demonstrates that much of the navicular is dark, as one might expect with collapse and sclerosis. Infarcts of the tibia, talus, and calcaneus not present on the conventional radiograph are well seen on MR.

C

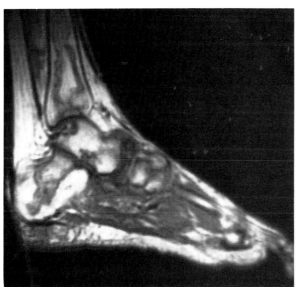

CASE 10

A 27-year-old professional football player was uncovered at the bottom of a pileup after an attempted fumble recovery. He was holding his ankle and was in pain but could not detail the exact mechanism of injury. What is the probable diagnosis? (Fig. 10A. Axial MR, TR/TE 2,000/80 msec.)

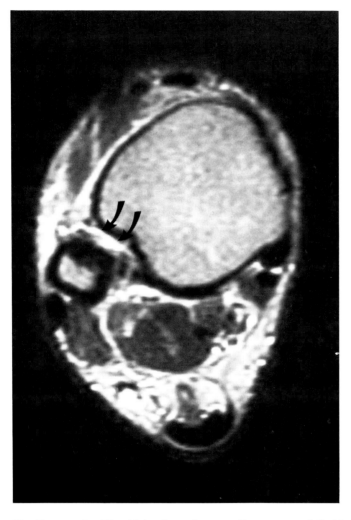

FIG. 10A.

Findings: In Fig. 10A, there is a small amount of fluid (*arrows*) situated between the tibia and fibula at the level of the distal tibiofibular joint. The fluid extends anterior to the fibula. The fibula is slightly anterior relative to the tibia, and the normal dark anterior tibiofibular ligament is absent (see Fig. 10B).

Diagnosis: Tear of the anterior inferior tibiofibular (ATIF) ligament.

Discussion: Arthrography is capable of evaluating the ligamentous integrity of the ankle with a high degree of accuracy. Specific patterns of contrast extravasation from the ankle joint permit assessment of the ATIF, the anterior and posterior talofibular, the calcaneofibular, and the deltoid ligaments.

The ATIF extends from the anterior and lateral aspects of the tibia to the adjacent anterior aspect of the fibula. Disruption of this ligament results in varying degrees of tibiofibular diastasis and is often accompanied by injuries to the other lateral ligaments. In the normal arthrogram, contrast extends into the syndesmotic recess, which rises approximately 2 cm above the ankle joint between the tibia and fibula. When the ATIF is torn, contrast extends well above the normal thin recess and tracks anterior to the fibula. An arthrogram confirmed the MR impression. There are however no controlled studies determining the accuracy of MR in assessing ligamentous injuries. The disorganization of the soft tissues between the tibia and fibula, the presence of fluid above the expected syndesmosis, and the slight subluxation of the fibula strongly suggest a tear of the ATIF (see Case 19).

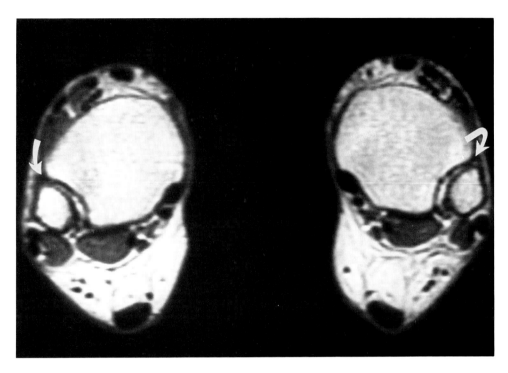

FIG. 10B. Normal ATIF (axial MR, TR/TE 2,000/80 msec). The curved arrows point to dark bands representing the ATIF, bilaterally extending from fibula to the anterior tibia. Contrast the position of the fibula in these images with those in Fig. 10A.

CASE 11

This 40-year-old woman presented with a progressive painful soft tissue swelling of several months duration in the plantar aspect of the foot. Injuries regarding any history of trauma were noncontributory. No associated skin changes or other areas of musculoskeletal involvement were noted. (Fig. 11A. Coronal MR, TR/TE 1,500/25. Fig. 11B. Coronal MR, TR/TE 2,000/80 msec. Courtesy of V. Cervilla, MD, Santiago, Chile.)

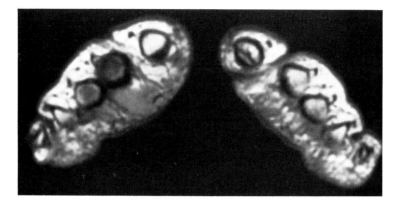

FIG. 11A.

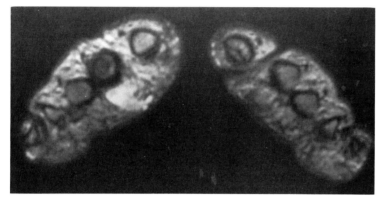

FIG. 11B.

Findings: The coronal images through the forefoot demonstrate a homogeneous, well-demarcated mass situated in the intermediate plantar compartment of the foot, adjacent to the second metatarsal head. The mass measures 2.0 × 1.5 cm, and its margins are relatively smooth and somewhat lobulated. It shows an intermediate signal intensity on the balanced image and increased signal on a T2-weighted image, most likely representing fluid. Note also the partial volume effect of bone and joint cartilage in the second and third metatarsal heads, creating the impression of an intramedullary abnormality.

Diagnosis: Nonspecific tenosynovitis of the long flexor of the second toe.

Discussion: Synovial lining of the tendon sheaths may become inflamed in various rheumatologic disorders, especially rheumatoid arthritis and less frequently seronegative spondyloarthropathies. Tenosynovitis can occasionally be seen as an isolated manifestation in these conditions. Inflammation of a flexor tendon sheath of the foot may be the result of an infective process originating either from a distant focus or most commonly following infection. In some cases no factors regarding the origin of the inflammation can be identified.

In the early stages, plain radiographs can be normal or show a nonspecific soft tissue swelling; as the inflammation progresses, osteopenia and erosions may become evident.

Typical changes of arthritis about the adjacent joints may be helpful in establishing the underlying cause of tenosynovitis. Computed tomography is able to depict soft tissue masses and image tendons; it can also detect small amounts of calcification and early cortical involvement. Soft tissue discrimination capability is somewhat limited with this modality, however.

Experience has demonstrated that MRI presents the greatest advantages in imaging normal anatomy and pathologic conditions of tendons (Fig. 11C,D). High magnetic field strength as well as dedicated surface coils are necessary to produce high resolution images. Tendons, as collagenous structures, show low signal intensity in all spin-echo pulse sequences, displaying high contrast against almost any surrounding tissue. Therefore, in most instances, MRI will be able to differentiate when a tendon is surrounded by edema, hemorrhage, or fibrosis, although characterization of certain tissues and fluids remain difficult to achieve with current techniques. Magnetic resonance is not capable of distinguishing serous from purulent fluid.

A lobulated and fluid-filled mass in continuity with the flexor tendon of the second toe was encountered at the time of excisional biopsy. Histologic sections of the specimen revealed inflammatory changes and focal proliferation of the synovium, consistent with chronic inflammation. Cultures were negative.

Magnetic resonance is capable of noninvasively evaluating a number of disorders of tendons. Tendinitis, partial tears, and complete ruptures of a major tendon, as represented by the Achilles tendon, will be discussed in Case 20. Postsurgical scarring is manifest by irregularity of the tendon surface associated with loss of the normal peritendinous fat. Tenosynovitis may be secondary to diabetes, gout, rheumatoid disease, and systemic lupus erythematosus. Tenosynovitis and rupture of the flexor hallucis longus tendon has been reported in ballet dancers and soccer players.

C

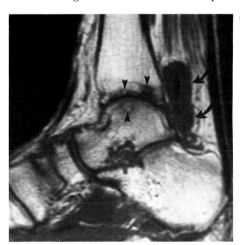

FIG. 11C,D. Tenosynovitis of the flexor hallucis longus. The sagittal image **(C)** (TR/TE 800/20 msec) demonstrates a lobulated mass (*arrows*) posterior to the talocalcaneal joint. There is narrowing of the ankle joint as well as subchondral erosions (*arrowheads*) in the adjacent bones. The differential diagnosis must include pigmented villonodular synovitis (PVNS). The axial plane **(D)** (TR/TE 2,000/80 msec) clearly demonstrates a fluid-filled tubular structure surrounding the flexor hallucis longus tendon (*arrows*). The patient is known to have chronic Achilles tendinitis and has rheumatoid disease.

D

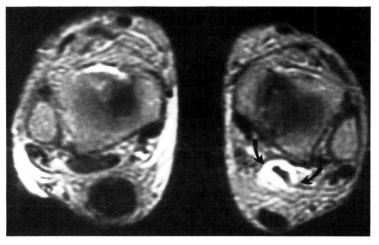

CASE 12

This 67-year-old man with a long history of drinking problems and diabetes mellitus recognized the sudden appearance of purulent discharge from a lesion on his forefoot. What is the diagnosis? What is the value of MR in this case? Fig. 12A. Axial MR, TR/TE 2,000/80 msec. Fig. 12B. Axial MR, TR/TE 2,000/20 msec. Courtesy of N. Friederich, MD, Bruderholz, Switzerland.)

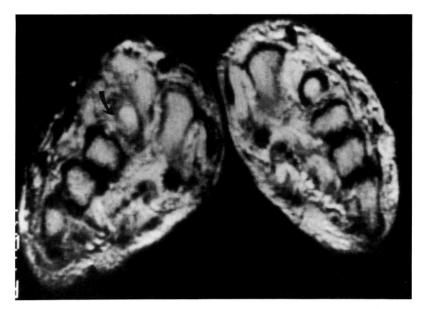

FIG. 12A.

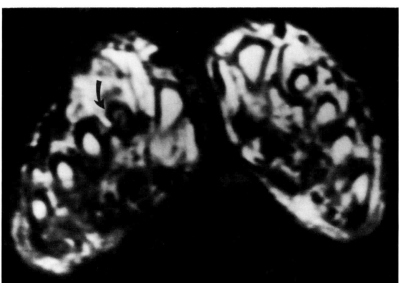

FIG. 12B.

Findings: The second right metatarsal (arrow) demonstrates diminished signal on the proton-density image, most evident when compared to the noninvolved bones. The cortical bone of the second right metatarsal (arrow) has a slightly increased signal, which, with the marrow, increases on the long TR/TE sequence.

Diagnosis: Osteomyelitis of the second right metatarsal.

Discussion: Osteomyelitis is a well-recognized disease of the elderly diabetic patient. It can often be found in conjunction with polyneuropathic disease of the extremities. The metatarsal bones are most often involved. Minor trauma, bunions, and poor fitting footwear predispose to infections that may require IV antibiotics, local debridement, and amputation for control.

In osteomyelitis the first structure involved is the spongy bone. Pathologically one finds hyperemia and increased fluid content, which decreases the fat content. Later in the course of the disease, the infection speads via haversian canals into the periosteum and leads eventually to the destruction of the cortical bone.

Normal marrow fat has a high signal on T1-weighted images. On T2-weighted images the signal is low. Therefore proton-density and spin-lattice relaxation (T1-weighted) imaging has been shown to be of most diagnostic value in the detection of osteomyelitis.

Treatment of the disease consists of appropriate antibiotic therapy, good antidiabetic therapy, and occasionally operative measures. Realignment of the metatarsals (to distribute the pressure on the forefoot more evenly) and/or amputation of the digits or the forefoot, depending on the involvement of the soft tissues, may be necessary.

Magnetic resonance imaging can be of utmost importance in decision making, because there is no other modality available to show the extent of the disease at an early stage. Radiologic changes usually are not evident for weeks after the onset of the disease, and conventional radiography does not show the extent of soft tissue involvement.

Bone scanning has been shown to be extremely sensitive to the detection of osteomyelitis, but its efficiency in determining whether an infection is in the bone or confined to the soft tissues is only 84%, as compared to 95% for MR. This difference in specificity is due to the greater spatial resolution of MR, its axial plane imaging capabilities, and its greater anatomic detail. Additionally, MR is capable of determining whether a soft tissue infection is a phlegmon or a drainable abscess (Fig. 12C); radionuclide imaging has no capability of making this important surgical distinction.

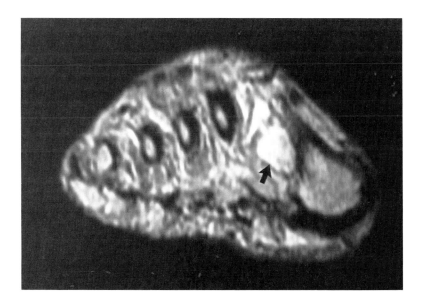

FIG. 12C. Soft tissue abscess (axial MR, TR/TE 2,000/80 msec). This diabetic patient had a fever and foot pain. The MR revealed a soft tissue abscess (*arrow*), which was drained. The marrow and cortical signal of the adjacent metatarsals was normal; there was no bone infection at surgery.

CASE 13

A 25-year-old woman IV drug abuser has noted an enlarging, mildly tender mass in her foot during the past 11 months. She is afebrile. Can you distinguish tumor from infection here? Is there bony involvement? What might be the value of MR? (Fig. 13A. Plain radiograph. Fig. 13B. Axial CT scan of foot. Fig. 13C. Sagittal MR, TR/TE 800/20 msec. Fig. 13D. Sagittal MR, TR/TE 2,000/80 msec. Fig. 13E. Axial MR, TR/TE 800/20 msec. Fig. 13F. Axial MR, TR/TE 2,000/80 msec. Case courtesy of Dean Berthoty, MD, Las Vegas, NV.)

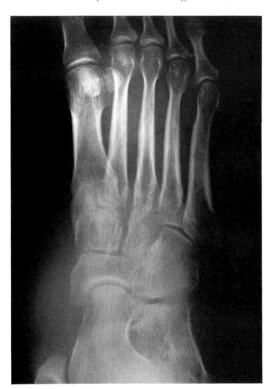

FIG. 13A.

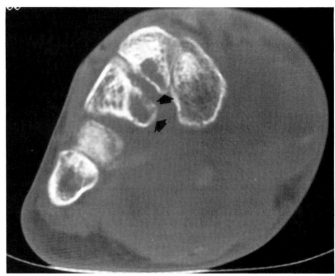

FIG. 13B.

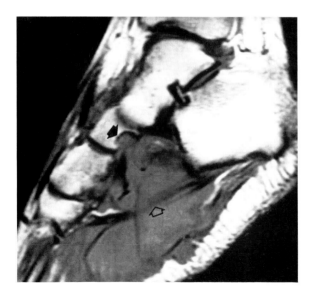

FIG. 13C.

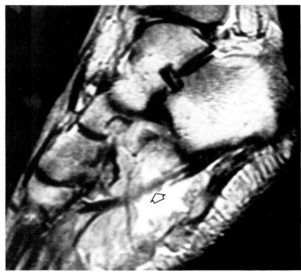

FIG. 13D.

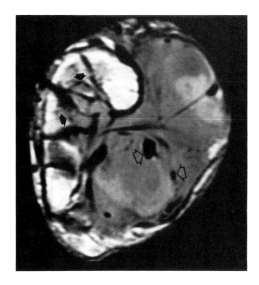 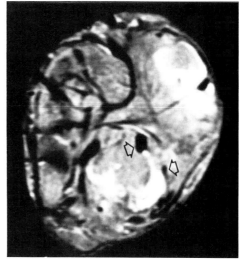

FIG. 13E.   FIG. 13F.

Findings: All studies demonstrate a large soft tissue mass in the medial and plantar aspects of the foot. The plain radiograph and CT scan were prospectively interpreted as showing local osteoporosis with no direct bone invasion by tumor in as much as the cortex appeared intact (arrows, Fig. 13B). The MR images (Fig. 13C,D), however, demonstrate the invasion of the mass into the cuneiform and navicular bones (black arrows). The T2-weighted images (TR/TE 2,000/70 msec) show the internal heterogeneity of the tumor. Both pulse sequences (Fig. 13E,F) show tumor wrapping around the flexor tendons (open arrows).

Diagnosis: Synovial sarcoma enveloping tendons and invading the cuneiform and navicular bones.

Discussion: Synovial sarcomas affect young adults without gender preference. They account for 10% of soft tissue sarcomas. The commonest site of involvement is in the lower extremity, usually around the knee. These tumors have a synovial cell origin but uncommonly are intra-articular and do not have any connection to the joint to which they are invariably adjacent. Synovial sarcomas commonly present as a large painful mass. Plain radiographs may demonstrate calcifications within the mass.

Although CT is well recognized as being the most sensitive method for detecting cortical bone destruction, T1-weighted or proton-density weighted MR images are both more sensitive and specific for *marrow* involvement. This is well illustrated in this case in which the bone invasion was missed on the high quality CT scan because of the appearance of an intact cortex and marrow rarefaction similar to that caused by osteoporosis. Cortex may appear relatively intact if bone extension is by a tumor that is so aggressive that it simply grows through the cortex without significantly destroying it. Detection of marrow disease in this instance becomes a more sensitive indicator of bone invasion. T1-weighted and proton-density weighted images are preferred as there is highest contrast differential between tumor (dark) and marrow fat (bright) with these sequences. Because of the bone invasion seen by MR, the diagnosis of a malignant tumor is overwhelming. In general, however, MR is poor at distinguishing benign from malignant soft tissue lesions as judged by signal characteristics or margination. Like almost all neoplasms of other than fatty origin, synovial sarcomas have low signal on T1-weighted images and an inhomogeneous but increased signal on T2-weighted images (Fig. 13G–J). Infection and tumor may be difficult to distinguish by MR; sarcomas usually have a well-defined mass and rather little edema.

The differential diagnosis of a lesion such as this is very limited. Without the history of trauma or signs of infection, a malignant neoplasm is by far the most likely consideration. Biopsy of soft tissue masses such as this one should be done under very controlled circumstances, such that the biopsy tract is well demarcated and can be included in the final resection *en bloc*. Disrupting the tumor bed in an uncontrolled fashion seeds the resection site and is known to increase the incidence of local recurrence. Soft tissue sarcomas commonly look encapsulated on MR images and may seem to "shell out" at surgery. Under histologic scrutiny, the apparent capsule is in fact a pseudo-capsule, with microinvasion into the surrounding soft tissues. For this reason, treatment of high grade soft tissue sarcomas consists of wide local excision with limb salvage or amputation. Adjunctive chemotherapy and/or radiation may also be used. Prognosis is best for synovial sarcomas less than 5 cm in size located distally in the limbs.

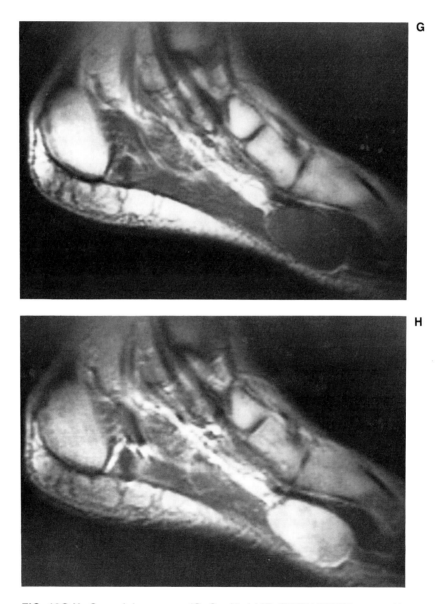

FIG. 13G,H. Synovial sarcoma (**G.** Sagittal MR, TR/TE 300/20 msec; **H.** Sagittal MR, TR/TE 1,500/80 msec). In another patient there is an oval-shaped mass in the medial forefoot that is isointense with muscle on T1-weighted images and increases in signal on the long TR/TE sequence. The mass is inhomogeneous. There is no surrounding edema. (Case courtesy of Wanda Bernreuter, MD, Birmingham, AL.)

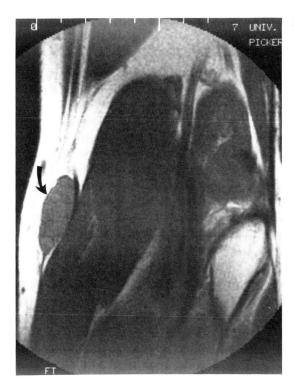

I

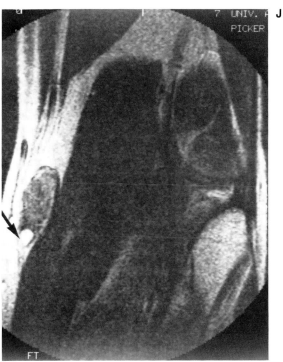

J

FIG. 13I,J. Synovial sarcoma (**I.** Coronal MR, TR/TE 450/26 msec; **J.** Coronal MR, TR/TE 1,500/80 msec). A well-circumscribed mass (*curved arrow*) is seen to lie medial to the gastrocnemius in the medial calf of this 27-year-old man. The majority of the mass is near muscle intensity on both sequences. Only the most inferior portion (*straight arrow*) gets distinctly brighter with T2 weighting. Like most synovial sarcomas, this one was high grade. At histology, the most cellular portion of the mass was at the inferior extent. (Case courtesy of Wanda Bernreuter, MD, Birmingham, AL.)

CASE 14

This 43-year-old man complains of pain and swelling in the midfoot of 1 month duration. There is no history of prior trauma. What is the diagnosis? (Fig. 14A. Plain radiograph. Fig. 14B. Coronal MR, TR/TE 2,000/100 msec.)

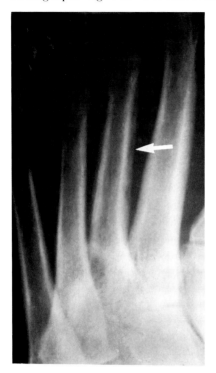

FIG. 14A.

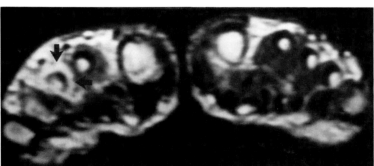

FIG. 14B.

Findings: The plain film demonstrates focal osteopenia of the third metatarsal with cortical thinning (arrow). The MR image demonstrates increased signal intensity within the marrow and cortex of the third metatarsal and within the adjacent soft tissues (arrows).

Diagnosis: Transient regional osteoporosis (subtype: partial transient osteoporosis, radial form).

Discussion: Transient regional osteoporosis is a condition of unknown etiology characterized by rapid, severe, localized osteoporosis, pain, and a self-limited clinical course. It is probably closely related to or a variant of reflex sympathetic dystrophy syndrome (RSDS). This disorder typically arises in the absence of an inciting event or following minor trauma. Two major forms of transient regional osteoporosis have been recognized: regional migratory osteoporosis and transient osteoporosis of the hip.

Regional migratory osteoporosis is characterized by the rapid onset of local pain and swelling (usually in a lower extremity) that often persists for 6 to 9 months. It usually occurs in middle-aged adults and may cause significant disability caused by severe pain on weight bearing. The patient usually complains of pain about one of the major joints of the lower extremity (i.e., the ankle, knee, or hip) or about the foot. The affected region is usually slightly tender to touch. When the ankle or foot are involved, diffuse erythema, swelling, and increased heat are often encountered. In other locations, soft tissue signs are usually minimal or absent. Classically, while the symptoms resolve in one area, successive attacks occur in other regions of the same or opposite lower extremity. This cycle of symptoms may last for several years. In this regard regional migratory osteoporosis differs from RSDS, in which multiple episodes are rare. These episodes are self-limited and respond variably to analgesics or corticosteroid therapy. A gradual pattern of recovery is typical.

On radiographs, localized osteoporosis becomes evident during the first 4 to 8 weeks. Osteoporosis may predominantly affect trabecular or cortical bone, and linear or wavy periosteal reaction has been described. A subchondral, periarticular distribution of osteoporosis about a painful joint may simulate a septic or inflammatory synovitis. Regional migratory osteoporosis, however, is distinguished by a lack of joint space narrowing or bone erosion and preservation of subchondral cortical bone. Although the clinical and radiographic findings may strongly suggest the diagnosis of regional osteoporosis, if a joint effusion is present, aspiration is advisable to exclude infection. A bone scan reveals increased activity in the area of involvement.

Biopsy of the affected bone typically reveals viable bone and thin, widely spaced trabeculae. Evidence of bone necrosis, increased bone turnover, and inflammatory changes have also been reported. The marrow and adjacent synovium may reveal hyperemia and edema.

A variant of regional migratory osteoporosis termed *partial transient osteoporosis* is characterized by a more focal pattern of osteoporosis. It has been divided into two forms. In the zonal form a portion of one bone is involved, such as one femoral condyle or one quadrant of a femoral head. In the radial form, only one or two rays of the hand or foot are involved. Both forms may eventually extend to a more generalized, periarticular pattern of osteoporosis. Migration to other sites in either lower limb may also be observed. The clinical presentation of pain and disability parallels that of regional migratory osteoporosis. The case presented here is best classified as the radial form of partial transient osteoporosis.

Transient osteoporosis of the hip was first described as a form of regional osteoporosis occurring in women during the third trimester of pregnancy. It has since been described in young and middle-aged adults of both sexes and predominantly affects men. The clinical, histologic, and radiologic findings are nearly identical to those seen in transient regional osteoporosis. Successive involvement about other joints has been described. The etiology of these disorders and of RSDS is uncertain. A neurogenic pathogenesis, with overactivity of the sympathetic nervous system and local hyperemia, has been proposed.

The MRI characteristics of transient osteoporosis affecting the hip and knee have been described. The affected region demonstrates low signal intensity on T1-weighted images and high signal intensity with T2 weighting, suggesting the presence of increased free water. It has been proposed that these findings represent bone marrow edema, which is consistent with biopsy results of prior studies. The pathophysiologic mechanism by which marrow edema is produced remains obscure. In the present case, MRI revealed abnormal signal in both the marrow and the adjacent soft tissues, consistent with edema. As noted above, when transient osteoporosis affects the foot or ankle, soft tissue swelling and edema are common.

CASE 15

A 24-year-old woman presents with a history of ankle pain and limitation of motion of several months' duration. What is the finding? What is the likely diagnosis? (Fig. 15A. Sagittal MR, TR/TE 500/30 msec. Fig. 15B. Axial MR, TR/TE 2,000/80 msec. Fig. 15C. Axial MR, TR/TE 2,000/40 msec.)

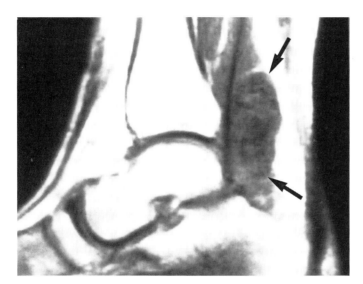

FIG. 15A.

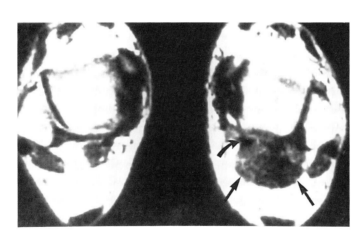

FIG. 15B.

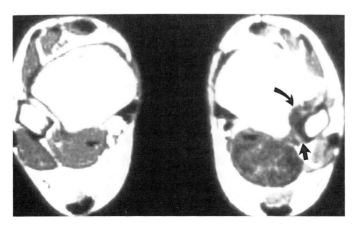

FIG. 15C.

Findings: The MR images (Fig. 15A,B) demonstrate a low to intermediate signal intensity mass (arrows) posterior to the ankle and adjacent to the flexor hallucis longus tendon (curved arrow). In Fig. 15C, the mass is seen to communicate with the joint space, extending between the distal tibia and fibula (short arrow), causing erosion of the lateral aspect of the tibia (curved arrow).

Diagnosis: Pigmented villonodular synovitis (PVNS).

Discussion: Pigmented villonodular synovitis represents a family of synovial lesions characterized by a fibrous stroma, histiocytic infiltrate, pigment or lipid deposition, and giant cells. These lesions demonstrate varied clinical manifestations and growth characteristics, depending in part on anatomical location. They arise in the synovial membrane of a joint, tendon sheath, or bursa.

Intra-articular lesions may be solitary and localized (localized nodular synovitis) or diffuse and villous (PVNS). The tendon sheath lesion is typically localized and discrete (giant cell tumor of tendon sheath or nodular tenosynovitis). The rare diffuse tendon sheath lesion probably represents extra-articular extension of a primary intra-articular process.

The etiology of PVNS remains controversial. In one study, the histologic features were interpreted as consistent with a neoplastic origin. Most investigators, however, believe that PVNS is an inflammatory hyperplastic process secondary to trauma, hemorrhage, or an unknown stimulus.

The microscopic appearance of PVNS varies from lesion to lesion depending on the proportion of giant cells, mononuclear cells, the degree of collagenization, and the amount of fat and hemosiderin deposition. Macrophages laden with lipid and hemosiderin deposits give the lesion a yellow or brownish appearance on gross examination. Hemosiderin, a breakdown product of hemoglobin, may be deposited within the stroma, macrophages, and synovial lining cells and signifies prior hemorrhage. It is usually stated that these lesions become less cellular and more fibrotic with time, although some investigators have found no relationship between the degree of fibrosis and the duration of symptoms or lesion size.

The clinical presentation of a patient with PVNS depends on the location and extent of the lesion. Affected patients are usually in the third through fifth decades of life. Localized giant cell tumor of tendon sheath occurs predominantly at the flexor and extensor tendons of the fingers of the hand (Fig. 15D). Less commonly, the foot, ankle, and knee are involved. Finger lesions usually present as a slowly enlarging, painless mass. A radiograph may reveal a soft tissue mass and, in roughly 25% of patients, erosion of the underlying bone. When this lesion occurs in the foot, it frequently produces bone erosion. The lesion does not calcify. The diffuse form of giant cell tumor of tendon sheath is rarely encountered but most often arises about the knee, foot, or ankle. The lesion may involve multiple tendon sheaths and extend into the adjacent joint or bone. Symptoms of pain, tenderness, and limitation of motion are more common with diffuse lesions.

When PVNS arises in a synovial joint, it affects the knee most often (roughly 80% of cases) (Fig. 15D–F). Other joints affected, in decreasing order of frequency, include the hip, ankle, small joints of the hands and feet, shoulder, and elbow. Patients are usually symptomatic although the duration and severity of symptoms are quite variable. Swelling or a palpable soft tissue mass is common with lesions of the knee, ankle, and elbow but unusual when the shoulder or hip is involved. Patients may also complain of limitation of motion and tenderness on palpation. In the knee, nodular lesions frequently present with symptoms of internal derangement, such as locking or giving way, suggesting the presence of a meniscal tear. Aspiration of the joint characteristically yields serosanguinous fluid with a high cholesterol content, but occasionally clear fluid is obtained.

Radiographs typically reveal a joint effusion and/or soft tissue mass. Erosive or cystic lesions of bone occur in varying frequency, depending on the joint involved. A tabulation of 146 reported cases of intra-articular PVNS revealed erosive bone lesions in 93% of hips, 75% of shoulders, 63% of elbows, 56% of ankles, and 26% of knees. The bone lesions are typically juxta-articular, with thin sclerotic rims. Joint space narrowing is uncommon with PVNS and is usually seen in long-standing lesions with superimposed degenerative joint disease.

All forms of PVNS are characterized by a propensity for local recurrence. Most localized lesions are adequately treated by local excision with a small cuff of normal tissue. The treatment of diffuse lesions, however, does not always lead to a satisfactory outcome as extensive synovectomy may not be curative and often results in disabling joint stiffness. The best approach is to remove as much of the lesion as possible without producing severe disability for the patient.

The appearance of PVNS on MRI will vary, depending on the relative proportion of lipid, hemosiderin, fluid, cellular elements, and fibrous stroma. A heterogeneous pattern of varied signal intensities is commonly seen throughout the lesion on both T1- and T2-weighted images. A low signal intensity rim may be produced by a fibrous capsule or peripheral hemosiderin deposits. The greatest value of MRI is in delineating the extent of the lesion to ensure that adequate surgical excision is accomplished. Because of its excellent soft tissue contrast, MRI is the imaging procedure of choice for preoperative evaluation of this lesion.

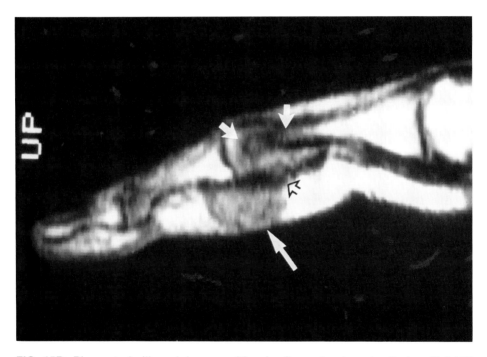

FIG. 15D. Pigmented villonodular synovitis of a flexor tendon sheath (sagittal MR, TR/TE 700/28 msec). The MR image of the finger demonstrates an intermediate signal intensity mass (*long arrow*) volar to the flexor tendon (*open arrow*) extending dorsally into the distal aspect of the proximal phalanx (*short arrows*). (Courtesy of Donald Sauser, MD, Loma Linda, CA.)

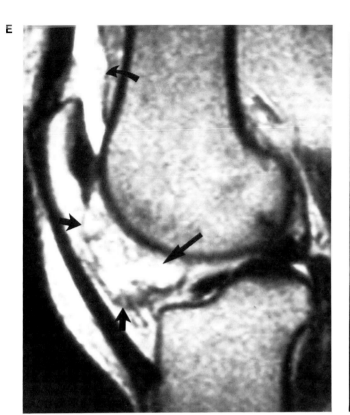 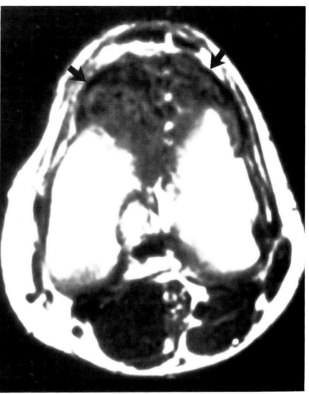

FIG. 15E,F. Pigmented villonodular synovitis of the knee (sagittal **(E)** and axial **(F)** MR, TR/TE 1,800/90 msec). Magnetic resonance images in two different patients with localized PVNS of the knee (*short arrows*) reveal varied patterns of signal intensity. **E.** Areas of high signal intensity (*long arrow*) are compatible with fluid and cyst formation. A high signal intensity joint effusion is present (*curved arrow*). **F.** Areas of very low signal intensity within the mass and a low intensity rim are consistent with hemosiderin deposition or fibrous tissue.

CASE 16

A 19-year-old marine recruit complains of pain in the foot after a marching drill. Plain radiographs appear normal. What is the diagnosis? (Fig. 16A. Sagittal MR, TR/TE 500/20 msec. Fig. 16B. Axial MR, TR/TE 2,000/100 msec. Fig. 16C. Sagittal MR, TR/TE 33/13 msec.)

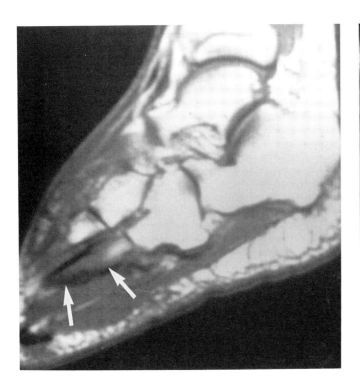

FIG. 16A.

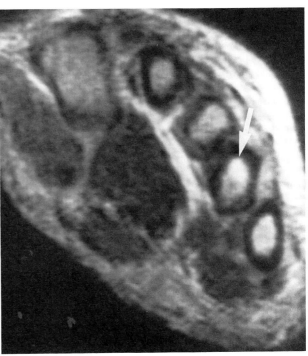

FIG. 16B.

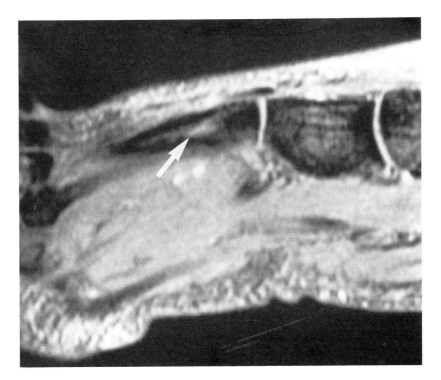

FIG. 16C.

Findings: The sagittal T1-weighted MR image (Fig. 16A) reveals diffuse diminished signal intensity within the medullary space of the fourth metatarsal (arrows). The coronal T2-weighted MR image (Fig. 16B) demonstrates an area of high signal intensity (arrow) within the fourth metatarsal. A sagittal GRASS image also demonstrates this high signal intensity region (arrow, Fig. 16C).

Diagnosis: Stress fracture of the fourth metatarsal.

Discussion: A stress fracture occurs when bone is subjected to repetitive trauma of less severity than that required to produce an acute fracture. Two types of stress fracture are recognized. A fatigue fracture occurs when normal bone is subjected to an abnormal stress to which it is unaccustomed, as occurs in military recruits and runners. An insufficiency fracture occurs when normal stress is applied to a bone that is weakened, with decreased elastic resistance, as seen in osteoporosis and Paget disease.

Most fatigue fractures are actually caused by the action of muscle on bone rather than direct bony trauma. In response to increased stress, bone first undergoes osteoclastic resorption followed by hypertrophy and buttressing. During the early resorptive phase, bone is vulnerable to the action of hypertrophied muscle. As muscle strength increases in response to activity, it may cause microfractures to develop at sites of bone resorption. If the strenuous activity is halted or diminished in intensity, adequate healing and hypertrophy of bone will occur. If the activity is continued, a gross fracture may develop. A similar mechanism explains most insufficiency fractures, as bone is weakened and loses elastic resistance.

The most common sites of stress fracture are the tibia, fibula, calcaneus, and metatarsals. Tibial and fibular stress fractures usually result from long distance running. In the foot, calcaneal stress fractures are often seen in marine recruits and result from jumping and parachuting. They are often bilateral. Metatarsal stress fractures usually occur with marching and prolonged standing and in ballet dancers. Almost 90% of metatarsal stress fractures involve the second and third metatarsals. Stress fracture of the fourth or fifth metatarsal is uncommon. Tarsal navicular stress fractures are relatively uncommon and occur in basketball players and runners. Navicular stress fractures are often subtle, occurring in the sagittal plane in the medial third of the bone.

Patients with stress fractures present with pain in the affected region. Mild swelling and tenderness may be evident on physical examination. In the acute phase, radiographs are normal and the diagnosis is usually accomplished using scintigraphic techniques. On bone scan, an early stress fracture appears as a diffuse region of low to moderate uptake of radionuclide. In the later stage of gross fracture, there is intense, discrete fusiform concentration of activity. In a patient with an appropriate history, these findings indicate stress fracture, and a negative scan excludes the diagnosis. Radiographic findings reflect reparative changes and may include callus formation, sclerosis, and endosteal or periosteal thickening. If present, a complete fracture often requires computed or conventional tomography for its detection. This is especially true of tarsal navicular fractures.

The MR appearance of stress fractures affecting various bones has been described. (Fig. 16D–I) Magnetic resonance imaging appears to have a sensitivity equal to that of radionuclide scintigraphy in identifying stress fractures and may demonstrate the lesion when radiographs appear normal. Stress fractures are associated with diminished marrow signal intensity on T1-weighted images and increased signal on T2-weighted images. A similar pattern of signal intensity, consistent with edema, may be seen in the soft tissue adjacent to the fracture and in the subperiosteal space. In a report of five stress fractures, T1-weighted images demonstrated band-like areas of very low signal intensity extending from the medullary space to the adjacent cortex. These areas were thought to represent microfracture or reparative bone sclerosis. They were surrounded by an area of mild to moderate decreased signal intensity, which converted to increased signal intensity on T2-weighted images. This most likely represented an area of hemorrhage or edema. These high intensity findings became less prominent when imaging was performed 4 weeks after the onset of symptoms.

The role of MRI in the diagnosis of stress fracture has not been established. Radionuclide scintigraphy has been used to document an abnormality in the affected bone, often within 24 hr after the onset of symptoms. Magnetic resonance imaging is probably as sensitive in detecting an abnormality and does not involve exposure to radiation. It is similar to bone scanning, however, in that the findings are nonspecific, and other etiologies such as tumor, infection, or ischemic necrosis cannot be excluded. Magnetic resonance imaging may be of particular value in evaluating elderly patients with insufficiency fractures. In these patients there is often a delay of several days before abnormal radionuclide accumulation is observed. With MRI the fracture and surrounding edema and hemorrhage should be discernable within 24 to 48 hr, permitting early detection of the fracture and appropriate conservative measures to be instituted. It is recognized, however, that the sensitivity of MRI in detecting an early stress fracture or insufficiency fracture has not been established.

D

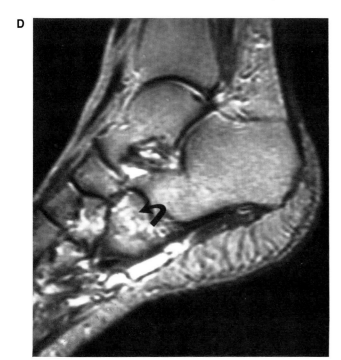

E

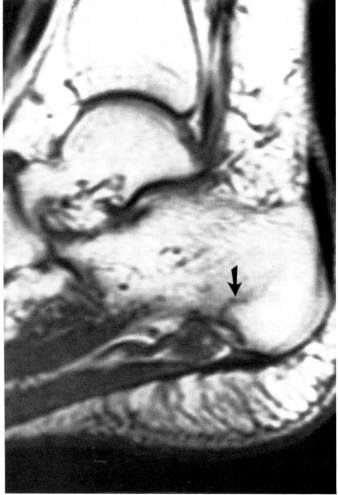

FIG. 16D,E. Stress fractures. This severely osteopenic, elderly woman presented with pain in the left foot and, after 1 year, had new pain in the right foot. The sagittal MR (TR/TE 2,000/80 msec) **(D)** demonstrates a band-like area of increased signal (*arrow*) in the cuboid, consistent with a stress fracture. The image in **(E)** (TR/TE 500/ 20 msec), made 1 year later, demonstrates a stress fracture (*arrow*) of the opposite calcaneus.

F

G

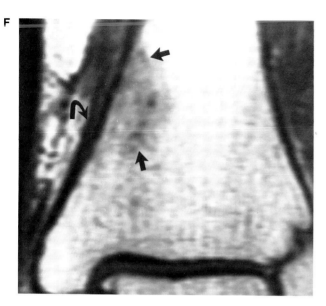
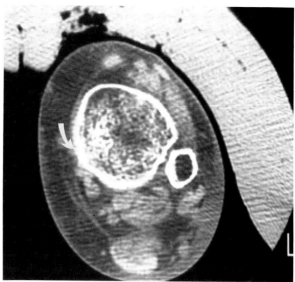

FIG. 16F,G. Stress response. A plain radiograph had shown a minimal degree of periosteal new bone along the distal medial tibia in an older man with ankle pain. The MR **(F)** (TR/TE 800/20 msec) in the coronal plane demonstrates a poorly defined area of decreased signal (*arrows*) and probable periosteal thickening (*curved arrow*). A CT **(G)** showed endosteal (*arrow*) and periosteal new bone. The clinical history and resolution were compatible with a stress response.

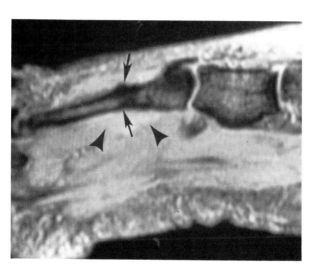

FIG. 16H. Stress fracture (sagittal MR, TR/TE 33/13 msec). There is thickening of the cortex of the fourth metatarsal caused by the presence of the callus (*arrows*). Edema (*arrowheads*) is present in the soft tissues. (Courtesy of Ann Bjorkengren, MD, Lund, Sweden.)

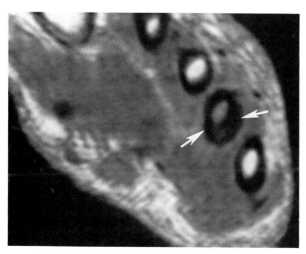

FIG. 16I. Stress fracture (axial MR, TR/TE 500/20 msec). In another patient, the typical cortical thickening, nearly circumferential in distribution, is seen (*arrows*). The decreased signal of the medullary canal is probably due to endosteal new bone and/or edema. (Courtesy of Ann Bjorkengren, MD, Lund, Sweden.)

CASE 17

A 55-year-old woman has had the insidious onset of medial foot pain. On physical examination, she is unable to stand on the ball of her affected right foot and has tenderness below the medial malleolus. What are the two primary differential considerations? (Fig. 17A. Axial MR, TR/TE 2,000/80 msec. Fig. 17B. Axial MR, TR/TE 2,000/80 msec.)

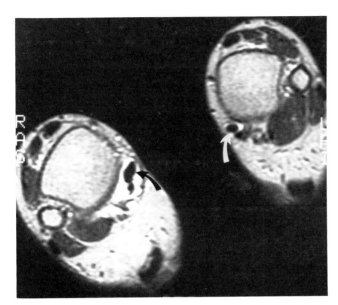

FIG. 17A.

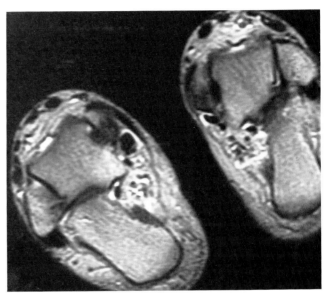

FIG. 17B.

Findings: Both images demonstrate considerable fluid in the sheath of the right posterior tibial tendon (PTT); the small amount of fluid on the left is normal. On both images, the right PTT (curved black arrow) is 2 times the size of the left (curved white arrow). Other images demonstrated a slight increase in signal within the tendon. The normal appearance of the PTT is shown in Fig. 17C–F.

Diagnosis: Posterior tibial tendon rupture (type I).

Discussion: The PTT is one of the main stabilizers of the hindfoot, preventing valgus. The muscle belly arises from the posteromedial surfaces of the tibia and fibula and courses caudally. The muscle-tendon junction is several centimeters above the medial malleolus under which the tendon turns on its course to insert on the navicular, the cuneiforms, and the bases of the second, third, and fourth metatarsals.

Chronic rupture of the PTT is a disease of women in their fifth decade. They present with pain and tenderness over the tendon between the medial malleolus and the navicular as well as a flat foot. Affected patients are unable to stand on the ball of the foot, and with weight bearing, significant flat foot deformity occurs. There is a lowering of the medial longitudinal arch and the hindfoot goes into valgus. Posterior tibial tendon rupture is a painful condition and is associated with significant disability.

The etiology of this disorder is multifactorial. Rheumatoid arthritis and the seronegative spandyloarthropathies, by virtue of the tenosynovitis with which they are associated, can lead to rupture of the tendon. In most patients, however, the cause of PTT rupture is thought to be secondary to mechanical erosion of the tendon where it passes under the medial malleolus. Patients who have flat arches before the onset of symptoms have increased stretching forces acting on the tendon and are predisposed to tenosynovitis and eventual rupture.

Posterior tibial tendon rupture is subclassified into three surgical groups. Type I is a partial tear secondary to longitudinal splits; the tendon has an increased girth. Type II (Fig. 17G,H) is a more severe but still partial tear in which the tendon is severely attenuated. Type III (Fig. 17I,J) is a complete rupture with retraction of the proximal tendon and a palpable gap. Selection of treatment is dependent on the size and extent of the tear. Orthotics may provide symptomatic relief in lesser degrees of injury by supporting the medial longitudinal arch. Direct surgical repair and augmentative tendon transfers, usually with the flexor digitorum longus, may be valuable in lesser degrees of injury; severe pain, degenerative arthritis, and marked deformity necessitate subtalar arthrodesis for stabilization.

Subtle conventional radiographic signs of PTT rupture include periostitis (Fig. 17K) of the medial distal tibia and subtalar arthritis, but CT or MR are the primary modalities permitting direct assessment of the tendon. On MR, the PTT is a homogeneously dark, ovoid to rounded structure passing posterior to the medial malleolus. At this point, the intimate relationship of the dark tendon and the dark cortical bone of the tibia makes assessment of the thickness of the tendon somewhat difficult. The PTT is normally twice as large as the flexor digitorum longus or the flexor hallucis longus tendons, both of which are in close proximity.

There are three radiographic patterns of PTT rupture identifiable by both CT and MR; they correlate well with the surgical subtypes described above. In a type I tear, the PTT enlarges in caliber and assumes a much more rounded appearance. This change in shape is most readily evident when the affected side is compared to the asymptomatic side, and therefore, it is more desirable to examine both feet side by side in cases of suspected PTT injury. The partially ruptured tendon demonstrates foci of increased signal, which are best seen on short TR/TE proton-density images; quite frequently, these foci "disappear" on long TR/TE sequences, but the fluid that is invariably present in the tendon sheath permits easy assessment of tendon size.

As opposed to the type I tear, which is described as a hypertrophic lesion, type II PTT ruptures are atrophic. The tendon is markedly attenuated just behind the medial malleolus, and its size approximates that of the flexor digitorum longus tendon. Hypertrophy of the segments just proximal and distal to the tear is common.

The effect of apparent summation of the thickness of the tendon with the adjacent cortical bone can result in overestimation of tendon size and consequently a misinterpretation of an atrophic Type II tear. This effect can be minimized by slightly plantar flexing the foot, and selecting an axial plane directly perpendicular to the PTT (Fig. 17F). Type III tears are manifest by a gap in the tendon with retraction of the proximal muscle-tendon complex. Synovial fluid is nearly universally present in cases of PTT rupture, but small amounts of fluid are also found in asymptomatic individuals.

In a recent study, MR proved to exhibit a high degree of accuracy in assessing the PTT; the sensitivity of MR was 95%, the specificity 100%, and the accuracy was 96%. MR did, however, misclassify some tears and in general underestimated the severity of the rupture.

The PTT normally widens just proximal to its insertion on the navicular, and quite frequently, there will be foci of signal within it at this point (Fig. 17L). This must not be mistaken for a type I tear. Errors will be avoided by remembering (1) that tears occur near the malleolus, not usually near the insertion and (2) that symmetry is the rule, and comparison with the opposite side is extremely helpful.

The main differential of PTT rupture is the tarsal tunnel syndrome. In this disorder, the posterior tibial nerve is chronically irritated where it passes beneath the flexor retinaculum at the malleolus. Patients present with numbness and/or burning paresthesias. The etiology of the tarsal tunnel syndrome includes retinacular fibrosis secondary to trauma, tenosynovitis, diabetes, rheumatoid disease, and hindfoot malalignment.

C

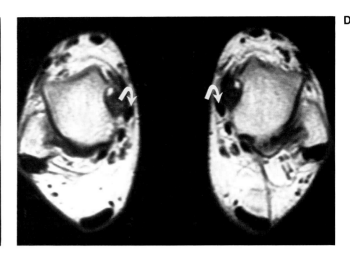

D

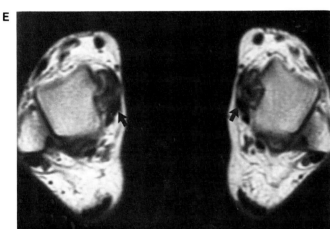

E

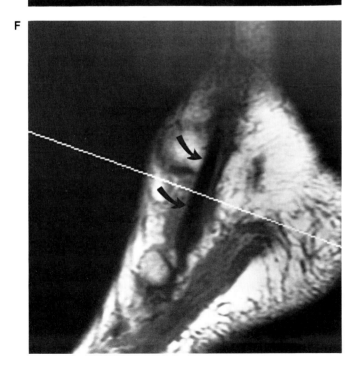

FIG. 17C–F. Normal PTT (**C–E.** Axial MR, TR/TE 2,000/20 msec; **F.** Sagittal MR, TR/TE 800/20 msec). **C.** The PTT (*curved arrows*) lies in intimate contact with the posterior aspect of the medial malleolus which it grooves. **D.** Five millimeters caudally, the tendon (*curved arrows*) moves anteriorly just under the malleolus. It is at this point that partial volume effect of the dark PTT and the dark cortical bone of the malleolus can be troublesome as in **E**, where the PTT (*arrows*) is difficult to distinguish from the malleolus. By selecting a nonorthogonal axial plane perpendicular to the tendon (**F**, *arrows*) and plantar flexing the foot, one can minimize the effects as seen by comparing Fig. 17D and 17E, made on the same patient. In all images, the PTT is twice as large as the adjacent flexor hallucis longus tendon (**C**, *open arrows*).

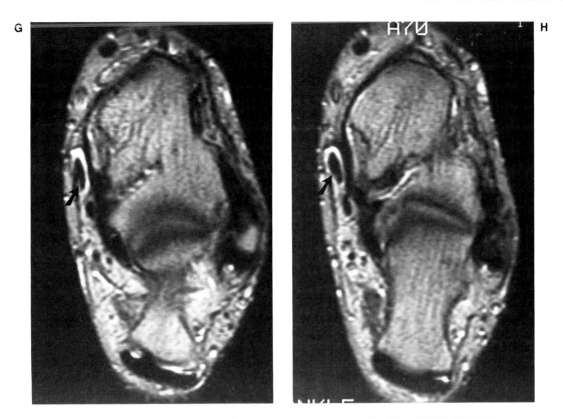

FIG. 17G,H. Type II PTT rupture (adjacent 5-mm sections, axial MR, TR/TE 2,000/80 msec). The PTT is focally thinned (*arrows*), a finding best appreciated when compared with the image just distal. There is fluid in the tendon sheath.

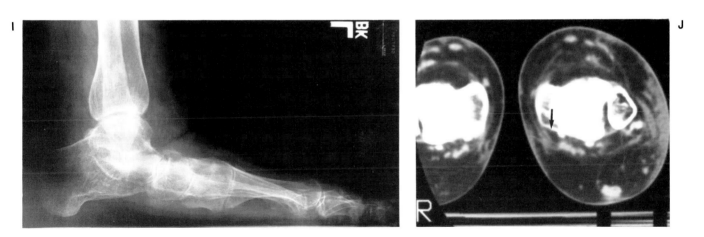

FIG. 17I,J. Type III PTT rupture. The medial longitudinal arch of the foot is flat on this conventional radiograph **(I)**. The axial CT scan **(J)** demonstrates a normal right PTT but complete absence of the left PTT (*arrow*) in this middle-aged woman with a type III rupture.

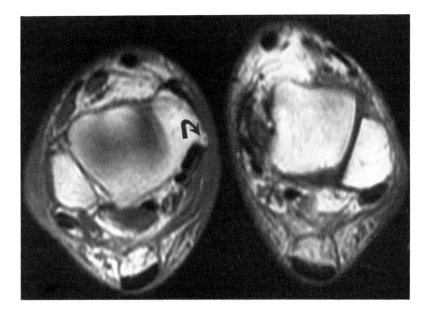

FIG. 17K. Tibial periostitis with a normal PTT (axial MR, TR/TE 2,000/20 msec). Periostitis and osteophyte formation along the distal medial tibia (*curved arrow*) should raise the question of a PTT tear. In this case, the PTT was intact; the soft tissue swelling medially and laterally was thought to be secondary to venous insufficiency, and clinically, the patient did not have a tear.

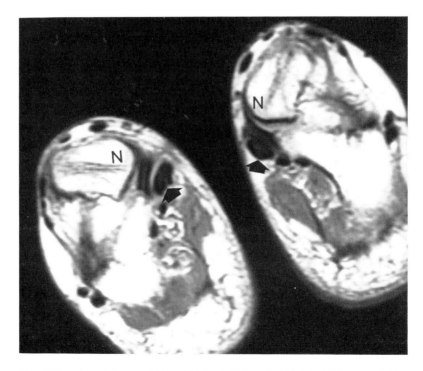

FIG. 17L. Pseudo-tear of the PTT (axial MR, TR/TE 2,000/20 msec). The PTT (*arrows*) is rounded and wider than expected and has signal within it. The pattern is symmetric, and this section is made just proximal to the PTT insertion onto the navicular. This appearance must not be mistaken for a type I tear.

CASE 18

This 42-year-old man presented with localized swelling and mild pain in the plantar surface of his left foot. What is the diagnosis? (Fig. 18A. Axial images, TR/TE 2,000/20 msec. Fig. 18B. Axial MR, TR/TE 2,000/80 msec. Courtesy of B. Schwaighoter, MD, Vienna, Austria.)

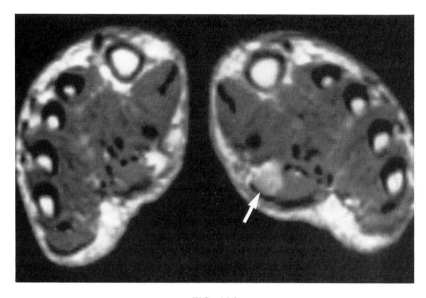

FIG. 18A.

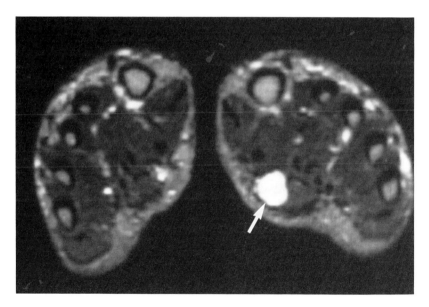

FIG. 18B.

Findings: The MR images demonstrate a well-demarcated lesion (arrows) in the plantar aspect of the left foot that is hyperintense as compared to muscle (and slightly hypointense as compared to subcutaneous fat) on the balanced sequence. On the T2-weighted sequence, the lesion increases in signal intensity. The mass exhibits signal intensity comparable to that of adjacent normal muscle on T1-weighted images (not shown). The lesion is located immediately plantar to the flexor hallucis longus muscle.

Diagnosis: Neurilemmoma (schwannoma) in the plantar aspect of the right foot.

Discussion: Considerable confusion exists regarding the terminology of neurogenic tumors involving peripheral nerves. Various terms such as neurinoma, neurilemmoma, perineural fibroblastoma, schwannoma, and neurofibroma are often used synonymously to describe tumors involving nerves. Two types of tumors are generally recognized. The first type, arising from the sheaths around nerves, is best termed a schwannoma or neurilemmoma. A neuroma is not a true neoplasm (Fig. 18C,D).

A schwannoma (Fig. 18E–H) is generally a solitary, well-encapsulated tumor that arises eccentrically from the nerve sheath. It displaces the nerve, and therefore it generally can be resected without sacrificing the nerve. The neurofibroma, however, arises from and involves neural tissue, and quite frequently nerve resection is necessary. Neurofibromata are occasionally multiple and are unencapsulated. Except in cases of neurofibromatosis, these tumors do not usually become malignant.

The detection of these lesions by conventional radiography is difficult unless they are large and calcified or affect adjacent structures such as bone, leading to osseous erosions. Computed tomography and MRI are more effective imaging modalities, especially for detecting small lesions. On such scans, tumors are generally well demarcated and closely associated with an adjacent nerve. A distinct advantage of MRI is the ability to delineate lesions without the use of contrast material, as they exhibit signal hyperintensity on T2-weighted images. Surgical excision may be required in symptomatic patients with larger lesions. Tumor recurrence is distinctly unusual.

FIG. 18C,D. Plantar neuroma (**C.** Axial MR, TR/TE 2,000/20 msec; **D.** Axial MR, TR/TE 2,000/80 msec). This man suffered a deep laceration of the foot 6 months earlier. The wound healed rapidly. Foot compression induces burning pain. The signal characteristics of this lesion suggest a neurilemmoma, but the history raises the question of an amputation stump neuroma. The latter is probably not a true neoplasm and usually does not dramatically increase in signal on long TR/TE sequences (see Case 8). There is no histology available in this case. (Courtesy of N. Friederich, MD, Bruderholz, Switzerland.)

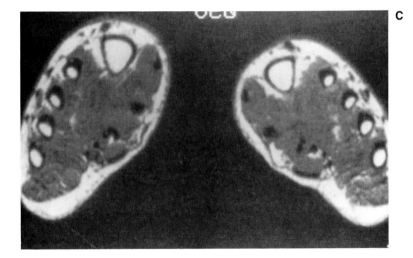

C

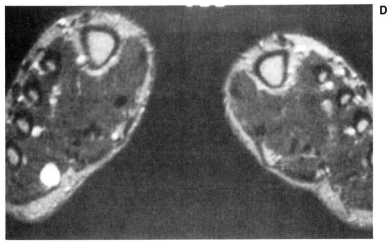

D

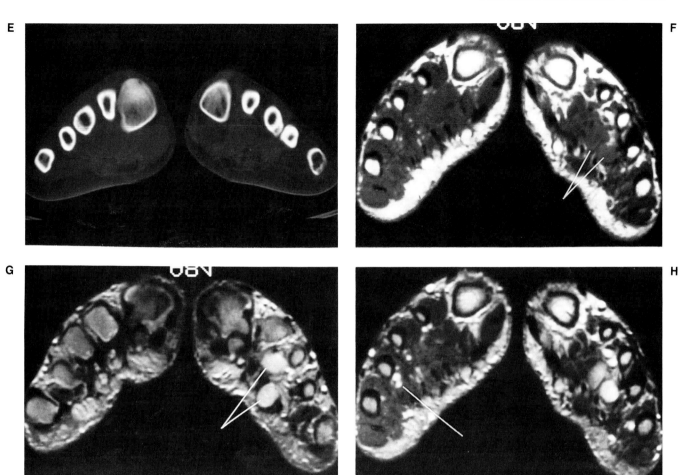

FIG. 18E–H. Multiple schwannomas. **E.** The CT scan demonstrates reactive bone formation on the left third metatarsal. **F.** The intermediate TR/TE sequence MR (TR/TE 2,000/20 msec) demonstrates two soft tissue masses in the plantar soft tissues of the left foot (*lines*). **G,H.** The long TR/TE sequences (TR/TE 2,000/80 msec) reveal another lesion in the right foot (*lines*). All lesions increase markedly in signal intensity. Multiple schwannomas should raise the question of neurofibromatosis. (Courtesy of H. Adams, MD, San Diego, CA.)

CASE 19

Several images from different patients were selected to demonstrate normal anatomic details. To what structure does each arrow point? (Fig. 19A–C. Axial MR, TR/TE 2,000/80 msec.)

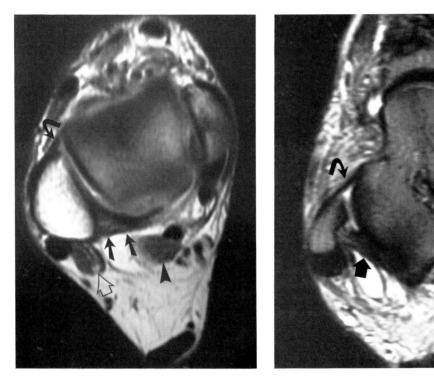

FIG. 19A. **FIG. 19B.**

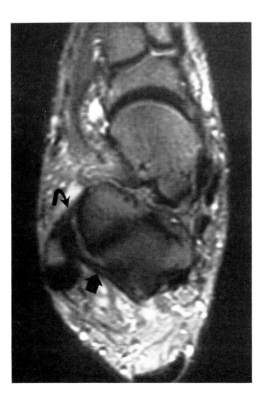

FIG. 19C.

Findings: In Fig. 19A, the curved arrow depicts the inferior lateral extent of the anterior tibiofibular ligament. The tibial attachment would best be seen 5 mm higher. The posterior tibiofibular ligament is outlined by two straight arrows. Signal is always present in this ligament. The peroneus brevis muscle (open arrow), and the flexor hallucis longus muscles (arrowhead) are also seen.

In Fig. 19B, the anterior talofibular (ATAF) ligament (curved arrow) and posterior talofibular ligaments (PTAF) (arrow) are well seen, arising from their respective surfaces on the distal fibula.

In Fig. 19C, the inferior end of the ATAF ligament (curved arrow) is still visible. The PTAF ligament seen just above is "replaced" by the calcaneofibular (CF) ligament (arrow), attaching to the calcaneus just behind and below the posterior subtalar joint.

Diagnosis: Normal lateral ligamentous anatomy of the ankle.

Discussion: Although detection of disruptions of the normal ligamentous anatomy of the ankle by MR has been reported, there are no studies determining the accuracy of MR in the assessment of such lesions. As at the knee, ligamentous tears are manifest by disorganization of tissue in the expected position of the ligament, high signal within/ around the ligament, absence of the ligament, and extravasation of joint fluid beyond its expected position. While presently assessment of ligamentous integrity is not an established indication for MR, other posttraumatic disorders are investigated, and ligamentous injuries may be found incidentally. It is worthwhile to review the normal lateral ligamentous anatomy of the ankle.

The ATAF ligament is the most commonly injured ligament at the ankle. It arises from the anterior surface of the distal aspect of the fibula and extends to the talar neck. Arthrographically, an ATAF tear is manifest by contrast extravasation from the ankle joint inferior and lateral to the distal fibula on a frontal radiograph and anterior to the fibula on a lateral exposure. On MR, the ligament can be seen on axial images as a 2 to 3 mm thick dark band extending obliquely anteromedially from fibula to tibia.

The CF ligament, the strongest of the lateral ligaments, is a thick band originating from the posterior aspect of the fibula to insert on the superior aspect of calcaneus. Its origin on the fibula is quite close to that of the PTAF ligament, and their insertions are separated by the posterior facet of the subtalar joint. When the CF tears, the common tendon sheath of the peroneals also tears, allowing the peroneal sheath to come into communication with the ankle joint. Unlike other ligament tears, a disruption of the CF can be diagnosed long after the injury, because once communication between two synovial compartments (tendon sheath and ankle joint) is established, it tends to persist. Tears of the CF commonly occur with tears of the ATAF; isolated PTAF tears are uncommon. On axial MR, the PTAF and CF are somewhat difficult to separate and are distinguished by finding the posterior subtalar facet; the CF is more inferior than the PTAF.

CASE 20

This is a 41-year-old woman with 3 months of heel pain. She had recently taken up jogging. What is your diagnosis? Which sequence shows the abnormality best? (Fig. 20A. Sagittal MR, TR/TE 2,000/20 msec. Fig. 20B. Sagittal MR, TR/TE 2,000/80 msec. Fig. 20C. Axial MR, 2,000/20 msec. Fig. 20D. Sagittal MR, TR/TE 50/13 msec. Courtesy of D. Berthoty, MD, Las Vegas, NV.)

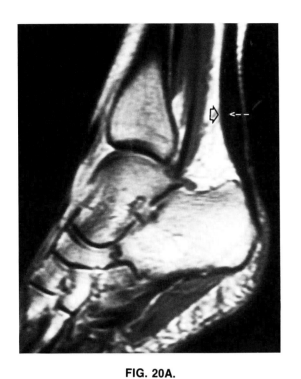

FIG. 20A.

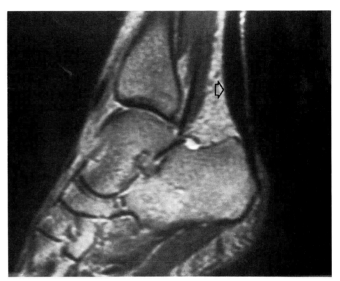

FIG. 20B.

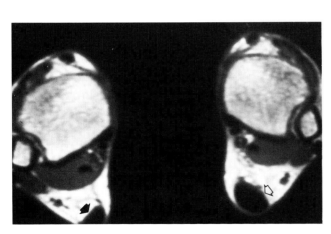

FIG. 20C.

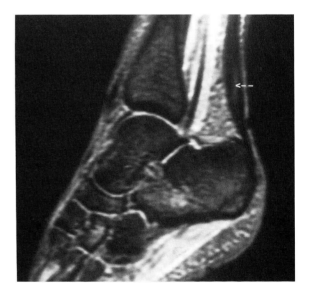

FIG. 20D.

Findings: The sagittal and axis spin-echo images show diffuse fusiform thickening of the supracalcaneal portion of the Achilles tendon (open arrows). The tendon is continuous, with only subtle increased signal within the thickened portion (dashed white arrows, Fig. 20A). With the fast scan, the intratendinous signal is much more obvious (dashed white arrow, Fig. 20D). No peritendinous signal abnormality is present. The black arrow denotes the normal right Achilles tendon.

Diagnosis: Achilles tendinitis.

Discussion: Multiple factors predispose to Achilles injuries, including a sudden increase in athletic training frequency or intensity (especially running on hills and slippery or uneven terrain) and excessive pronation of the forefoot. Gout, rheumatoid disease, systemic steroid use, and chronic renal failure predispose to tendon rupture. Local steroid injections have been implicated in causing disruption. Men are affected 80% of the time. Most Achilles tendon injuries occur 3 to 5 cm proximal to its site of attachment on the back of the calcaneus. At this point, the tendon fibers cross. The posterior fibers course medial to lateral and the anterior fibers lateral to medial. Additionally, the vascularity is less abundant at this location than in the rest of the tendon. These factors are thought to predispose to rupture.

Achilles tendon rupture is associated with excruciating, unremitting pain. Depending on the extent of the lesion, the belly of the gastrocnemius may retract up in the calf, and a palpable defect in the tendon may be present. Although this is a major injury, the exact nature may be misdiagnosed in 25% of patients.

Imaging of suspected Achilles injuries can be accomplished by low kvp and xeroradiographic techniques and ultrasound, but MR provides more direct information about the tendon and surrounding structures.

The Achilles tendon is best examined in the axial plane by MR. The tendon has a flat to concave anterior margin and rounded medial and lateral aspects, resulting in a crescentic shape (Fig. 20E,F). At all pulse sequences, the tendon is signalless, and any increase in signal must be regarded as abnormal. Symmetry is the rule. The axial plane provides the easiest way of comparing right to left, facilitating detection of differences in thickness. The sagittal plane most optimally demonstrates the longitudinal extent of a tear, and the coronal plane is rarely of value. When examining the Achilles tendon, one should be careful not to allow the patient to plantar flex the foot, a position that will produce pseudo-thickening and buckling.

Achilles injuries of varying extents can be depicted by MR. Complete tendon tears are characterized by gross areas of signal increase (on long TR/TE sequences), discontinuity of the tendon, widening of the tendon edges, and retraction of the muscle/tendon junction up into the calf. Tendinitis is manifest by small focal areas of abnormally increased signal associated with slight widening of the tendon but without discontinuity (Fig. 20G,H). Because the Achilles tendon does not have a true synovial sheath, edema is manifest in the loose connective tissue anterior to the tendon (paratenon). Partial tears demonstrate changes quite similar to tendinitis, and occasionally the differential diagnosis is made only by history. Partial tears usually do manifest some areas of discontinuity (Fig. 20I,J).

Because the diagnosis of tendinitis and partial tears may be made on the basis of subtle signal increase, imaging sequences should be optimized to detect them. In this case, the T2-weighted fast scan with low flip angle (15°) virtually eliminates T1 effects and best demonstrates the intrasubstance signal. This sequence also has the advantage of short imaging time (61 sec/slice). This compares with 8 min and 41 sec for the multislice spin-echo sequence.

Treatment of Achilles tendinitis is conservative, including heel support in shoes, gentle stretching, and oral anti-inflammatory medications until the tenderness subsides. This often requires as long as 8 weeks. Surgical excision of necrotic tissue is reserved for cases refractory to 6 months of conservative therapy. The treatment in complete ruptures remains ambiguous. The rate of rerupture of conservatively treated complete ruptures is 10%, whereas that for surgically treated cases is 4%. However, only 72% of conservatively treated patients regain normal strength. MR may help guide the length of the period of immobilization by demonstrating the degree and rate of healing and by detecting if a partial tear progresses to a complete tear.

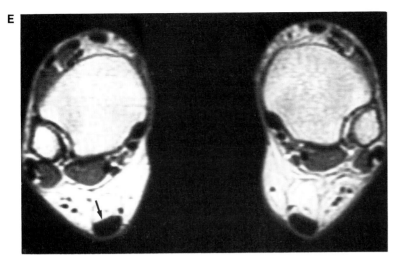

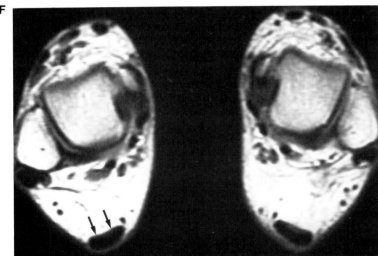

FIG. 20E,F. Normal Achilles tendon (axial MR scans, TR/TE 500/20 msec). Two representative images are shown, one **(E)** at the level of the tibial metaphysis and the other **(F)** near the calcaneal insertion. In the upper section the anterior aspect (*arrows*) of the tendon is flat, whereas in the lower section it is more concave. The tendon is wider inferiorly. Symmetry is the rule.

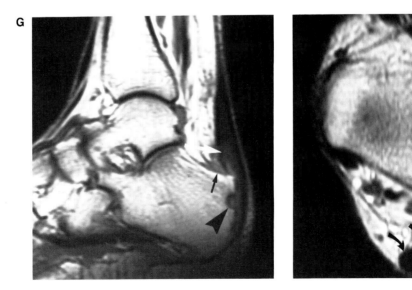

FIG. 20G,H. Achilles tendinitis in rheumatoid disease. **G.** Sagittal MR, TR/TE 2,000/20 msec. There is signal in the tendon (*white arrow*), thickening of the pre-Achilles bursa (*arrow*), and calcaneal erosions (*arrowhead*). **H.** Axial MR, TR/TE 2,000/20 msec. The tendon has a rounded anterior margin (*arrows*), and it is abnormally thickened.

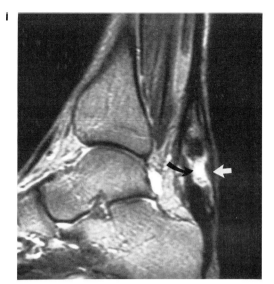

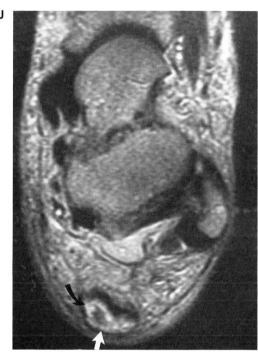

FIG. 20I,J. Achilles tear without retraction (**I.** Sagittal, 2,000/80 msec; **J.** Axial, 2,000/80 msec). Four days prior, this patient injured his Achilles tendon. Although a very large high signal focus (*curved arrow*) is present within a grossly widened tendon, a small portion of the anterior and posterior (*arrow*) fibers remain intact.

PERSPECTIVE

Tendon disorders are an important clinical problem that have received little attention in the radiologic literature. Available imaging methods include low kilovoltage plain radiography, xeroradiography, tenography, ultrasound, CT, and MRI. All of these techniques have some disadvantages. MRI has the greatest potential utility in the evaluation of tendons because of its various advantages, which include superior differentiation of tendons from surrounding inflammatory changes and hemorrhage, multiplanar imaging capability, the depiction of structures located deep in the body, and the ability to evaluate osseous as well as soft tissue abnormalities.

Although MRI of tendons using surface coils is a relatively recent innovation, early experience indicates that the improved soft tissue contrast and resolution represent significant advantages over CT and ultrasound in evaluating traumatic, inflammatory, and infectious tendon abnormalities. The depiction of an entire anatomic area, including osseous structures, is a significant advantage of MRI as compared to ultrasound. In some patients referred for evaluation of possible tendon lesions, MRI can exclude a tendon abnormality and demonstrate other unsuspected pathology, including stress fractures. With software improvements, it is expected to be feasible to visualize tendons along their long axes without having to align them with the orthogonal planes in the magnet, a technique that is time-consuming and subject to errors.

Magnetic resonance imaging, because of its inherent soft tissue contrast, is well suited for differentiation between the various components of a joint and its surrounding structures (e.g., cortical and cancellous bone, hyaline articular cartilage and fibrocartilage, fluid, abnormal synovium, ligaments, and muscle). The joints have been of particular interest to MRI researchers because their various components cannot otherwise be thoroughly evaluated by noninvasive means. Preliminary results in the foot and ankle have been encouraging.

Multiplanar MRI is particularly advantageous when evaluating the complex three-dimensional anatomy of the ankle joint. Direct coronal and sagittal imaging is especially helpful in demonstrating the tibio-talar joint and the three facets of the subtalar joint. The soft tissue structures of the ankle also are well demonstrated on MRI. All of the tendons and many of the ligaments can be identified. Effusions are depicted, even when small and in joints difficult to evaluate clinically. Cartilage integrity can be assessed clearly because of the relatively high signal intensity of the hyaline cartilage as compared to fibrocartilage or subchondral cortical bone. Inflammatory conditions such as juvenile rheumatoid arthritis have been demonstrated in the ankle. Based on preliminary experience, MRI of the ankle in acute and chronic trauma has proven to be clinically valuable. The method allows direct visualization of soft tissue lesions in all patients, with excellent demonstration of ligamentous and articular cartilage damage.

Magnetic resonance imaging usually defines the extent of soft tissue masses better than CT. This is particularly true in areas such as the foot and ankle where fat planes are poorly identified by CT. The extent of intramedullary masses is usually easier to determine by MRI than by CT. Magnetic resonance imaging often is more useful than CT for evaluating vascular and joint involvement by tumor. Computed tomography is superior to MRI for detecting small areas of demineralization and early cortical bone erosion. Visibility of small demineralized areas by MRI is dependent on the specific imaging sequence. Occasionally, MRI in nontransaxial planes helps to solve specific diagnostic problems. Simple fluid masses and benign lipomatous masses are equally well identified by CT and MRI. The latter technique is more useful in characterizing fibrous and angiomatous lesions. Transaxial gradient-echo images are useful in defining separate compartments and fascial division, important in the evaluation and staging of neoplasms affecting the foot and ankle.

In the foot and ankle, MRI has permitted successful identification of osteomyelitis (acute, subacute with Brodie abscess, chronic, and acute with septic arthritis) and of cellulitis or soft tissue abscess in the absence of osteomyelitis. Active osteomyelitis also can be excluded; both T1- and T2-weighted images are needed to identify such infective foci. Magnetic resonance images provide more accurate and detailed information regarding the extent of involvement than do radionuclide bone scans, CT scans, or standard radiographs. They also permit the differentiation of septic arthritis or cellulitis from osteomyelitis. In limited experience to date, MRI has been particularly useful for seeking foci of active infection in areas of chronic osteomyelitis complicated by surgical intervention or fracture.

In summary, this chapter has endeavored to enlighten the practicing radiologist with regard to the indications for cross-sectional imaging by MRI in the evaluation of specific clinical problems affecting the foot and ankle. In most instances, MRI will answer the diagnostic questions posed, although CT may provide complementary information and thus both may be required in certain instances. Direct multiplanar images of the foot and ankle can be obtained with both methods, although patient positioning is more easily achieved with MRI. As a general rule, MRI is preferred over CT in the setting of suspected soft tissue pathology or disorders with early involvement of bone marrow, because of its superior contrast discrimination capabilities and exquisite sensitivity. Because cortical bone contains a relative paucity of mobile protons and generates a weak MRI signal, however, CT is preferred in the assessment of osseous alterations that do not begin in the marrow space. For younger patients, MRI should be the procedure of choice because it delivers no ionizing radiation and involves no known biological hazard.

The disadvantages of MRI are not inconsequential. It is an expensive test and not yet widely available. Magnetic resonance imaging at low magnetic field strength is not capable of producing high resolution images such as those obtained with a 1.5-T magnet. Dedicated surface coils are useful to achieve high quality images, and commercial production of such coils is still limited. The absence of signal from calcified soft tissue is another disadvantage, although cortical and trabecular bone can be seen, owing to the contrast of the surrounding subcutaneous fat and fatty marrow. The characterization of different types of fluid is not possible with current techniques. For example, MRI is unable to distinguish between purulent and serous fluid collections. Finally, claustrophobia and motion are occasional problems; patients may be unable to hold their feet rigid for the entire series of images because of uncomfortable positioning.

RECOMMENDED READINGS

Aisen AM, Martel W, Braunstein EM, et al. MRI and CT evaluation of primary bone and soft tissue tumors. *AJR* 1986;146:749–756.

Alexander IJ, Johnson KA, Berquist TH. Magnetic resonance imaging in the diagnosis of disruption of the posterior tibial tendon. *Foot Ankle* 1987;8:144–147.

Aronstam A, Browne RS, Wassef M, et al. The clinical features of early bleeding into the muscles of the lower limb in severe hemophiliacs. *J Bone Joint Surg* 1983;65B:19–23.

Atlan H, Sigal R, Hadar H. Nuclear magnetic resonance proton imaging of bone pathology. *J Nucl Med* 1986;27:207–215.

Azouz EM, Slomic AM, Marton D, et al. The variable manifestations of dysplasia epiphysealis hemimelica. *Pediatr Radiol* 1985;15:44–49.

Beltran J, Noto AM, Herman LJ, et al. Tendons: high field-strength, surface coil MR imaging. *Radiology* 1987;162:735–740.

Beltran J, Noto AM, McGhee RB. Infections of the musculoskeletal system: high-field-strength MR imaging. *Radiology* 1987;164:449–454.

Beltran J, Noto AM, Mosure JC, et al. Ankle: surface coil imaging at 1.5 T, *Radiology* 1986;161;203–211.

Berndt AL, Harty M. Tranchondral fractures (osteochondritis dissecans) of the talus. *J Bone Joint Surg* 1959;41A:988.

Berquist TH, ed. *Radiology of the foot and ankle.* New York: Raven Press, 1989.

Berquist TH, Brown ML, Fitzgerald RH, et al. Magnetic resonance imaging: application in musculoskeletal infection. *Magnetic Resonance Imaging* 1985;3:219–230.

Blei LC, Nirschi RP, Grant EG. Achilles tendon: US diagnosis of pathologic conditions. *Radiology* 1986;159:765–767.

Blickenstaff LD, Morris JM: Fatigue fracture of the femoral neck. *J Bone Joint Surg* 1966;48A:1031–1047.

Bloem JL, Falke THM, Doornbos J, et al. Osteomyelitis in children: detection by magnetic resonance imaging. *Radiology* 1984;153:263–264.

Bloem J. Transient osteoporosis of the hip: MR imaging. *Radiology* 1988;167:753–755.

Bloem JL, Bluemm RG, Taminiau AHM, et al. MR imaging of primary malignant bone tumors. In: Partain CL, Price RP, Patton JA, et al., eds. *Magnetic resonance imaging.* Philadelphia: Saunders, 1988;633–654.

Burger PC, Vogel FS, eds. *Surgical pathology of the nervous system and its coverings.* New York: John Wiley & Sons, 1975.

Bydder GM, Payne JA, Collins AG, et al. Clinical use of rapid T2 weighted partial saturation sequences in MR imaging. *J Comput Assist Tomogr* 1987;11(1):17–23.

Clement PB, Taunton JE, Smart GW. Achilles tendinitis and peritendinitis: etiology and treatment. *Am J Sports Med* 1984;12:179–184.

Daffner RH. Stress fractures: current concepts. *Skeletal Radiol* 1978;2:221–229.

Daffner RH, Riemer BL, Lupetin AR, et al. Magnetic resonance imaging in acute tendon ruptures. *Skeletal Radiol* 1986;15:619–621.

Danielsson L, Theander G. Supernumerary soleus muscle. *Acta Radiol (Diag)* 1981;22:365–368.

Dorwart RH, Genant HK, Johnston WH, et al. Pigmented villonodular synovitis of synovial joints: clinical, pathologic, and radiologic features. *AJR* 1984;143:877–885.

Downey DJ, Simkin PA, Mack LA, et al. Tibialis posterior tendon rupture: a cause of rheumatoid flat foot. *Arthritis Rheum* 1988;31.

Duncan H, Frame B, Frost H, et al. Regional migratory osteoporosis. *South Med J* 1969;62:41–44.

Dunn AW. Anomalous muscles simulating soft-tissue tumors in the lower extremities. Report of three cases. *J Bone Joint Surg* 1965;47:1397–1400.

Ehman RL, Berquist TH. Magnetic resonance imaging of musculoskeletal trauma. *Radiol Clin North Am* 1986;24:291–319.

Enneking WF, Spanier SS, Goodman MA. A system for the surgical staging of musculoskeletal sarcoma. *Clin Orthop* 1980;153.

Enterlone H. Histopathology of sarcomas. *Semin Oncol* 1981;8(2):133–155.

Enzinger FM, Weiss SW. *Soft tissue tumors.* St. Louis: CV Mosby Co, 1983;502–518.

Fisher MR, Hernandez RJ, Poznanski AK, et al. Case report 262. Dysplasia epiphysealis hemimelica—Trevor disease. *Skeletal Radiol* 1984;11:147–150.

Fox JM, Blazina ME, Jobe EW, et al. Degeneration and rupture of the Achilles tendon. *Clin Orthop* 1975;107:221–224.

Goldner JL, Keats PH, Basset FH, et al. Progressive talipes equinovalgus due to trauma or degeneration of the posterior tibial tendon and medial plantar ligaments. *Orthop Clin North Am* 1974;5:39–51.

Granowitz SP, D'Antonio J, Mankin HL. The pathogenesis and long-term end results of pigmented villonodular synovitis. *Clin Orthop* 1976;114:335–351.

Greaney RB, Gerber FH, Laughlin RL. Distribution and natural history of stress fractures in U.S. marine recruits. *Radiology* 1983;146:339–346.

Griffith JC. Tendon injuries around the ankle. *J Bone Joint Surg [Br]* 1965;47:686–689.

Hadju SI, Shul MH, Fortner JG: Tendosynovial sarcoma, a clinicopathological study of 136 cases. *Cancer* 1977;39:1201–1217.

Ho AMW, Blane CE, Kling TF Jr. The role of arthrography in the management of dysplasia epiphysealis hemimelica. *Skeletal Radiol* 1986;15:224–227.

Inglis AE, Scott WN, Sculco TP, et al. Ruptures of the tendo achillis. *J Bone Joint Surg* 1976;58(A):990–993.

Jahss MH. Spontaneous rupture of the tibialis posterior tendon: clinical findings, tenographic studies, and a new technique of repair. *Foot Ankle* 1982;3:158–166.

Jelinek JS, Kransdorf MJ, Utz JA, et al. Pictorial essay. Imaging of pigmented villonodular synovitis with emphasis on MR imaging. *AJR* 1989;152:337–343.

Keats TE. Dysplasia epiphysealis hemimelica. *Radiology* 1957;68:558–563.

Kettelkamp DB, Campbell CJ, Bonfiglio M. Dysplasia epiphysealis hemimelica. A report of fifteen cases and a review of the literature. *J Bone Joint Surg* 1966;48A:746–766.

Key JA. Partial rupture of the tendon of the posterior tibial muscle. *J Bone Joint Surg [Am]* 1953;35:1006–1008.

Kneeland JB, Macrander S, Middleton WD, et al: MR imaging of the normal ankle: correlation with anatomic sections. *AJR* 1988;151:117–123.

Kottal RA, Vogler JB, Matamoros A, et al. Pigmented villonodular synovitis: a report of MR imaging in two cases. *Radiology* 1987;163:551–553.

Kulkarni MV, Drolshagen LF, Kaye JJ, et al. MR imaging of hemophiliac arthropathy. *J Comput Assist Tomogr* 1986;10:445–449.

Langloh ND, Hunder GG, Riggs BL, et al. Transient painful osteoporosis of the lower extremities. *J Bone Joint Surg* 1973;55A:1188–1196.

Lee JK, Yao L. Stress fractures: MR imaging. *Radiology* 1988;169:217–220.

Lequesne M, Kerboull M, Bensasson M, et al. Partial transient osteoporosis. *Skeletal Radiol* 1977;2:1–9.

Levy JM. Stress fractures of the first metatarsal. *AJR* 1978;130:679–681.

Mink JH, Deutsch AD. Occult osseous and cartilaginous injuries at the knee: MR detection, classification and assessment. *Radiology* 1989;170:823–829.

Naides SJ, Resnick D, Azaifler NJ. Idiopathic regional osteoporosis: a clinical spectrum. *J Rheum* 1985;12:763–768.

Nidecker AC, von Hochstetter A, Fredenhagen H: Accessory muscles of the lower calf. *Radiology* 1984;151:47–48.

Noto AM, Cheung Y, Rosenberg ZS, et al. MR imaging of the ankle: normal variants. *Radiology* 1989;170:121–124.

Powers JA. Magnetic resonance imaging in marrow diseases. *Clin Orthop* 1986;206:79–85.

Quinn SF, Murray WT, Clark RA, et al. Achilles tendon: MR imaging at 1.5 T. *Radiology* 1987;164:767–770.

Rao AS, Vigorita VJ. Pigmented villonodular synovitis (giant-cell tumor of the tendon sheath and synovial membrane). A review of eighty-one cases. *J Bone Joint Surg* 1984;66A:76–94.

Reis ND, Lanir A, Benmair, et al. Magnetic resonance imaging in orthopaedic surgery. *J Bone Joint Surg* 1985;67B(4):659–664.

Resnick D, Kyriakos M, Greenway GD. Tumors and tumor-like lesions of bone: imaging and pathology of specific lesions. In: Resnick D, Niwayama G, eds. *Diagnosis of bone and joint disorders.* Philadelphia: Saunders, 1988;3697–3701.

Richardson ML. Optimizing pulse sequences for magnetic resonance imaging of the musculoskeletal system. *Radiol Clin North Am* 1986;24(2).

Rosenberg ZS, Cheung Y, Jahss MH. Computed tomography scan and magnetic resonance imaging of ankle tendons: an overview. *Foot Ankle* 1988;8:297–307.

Rosenberg ZS, Cheung Y, Jahss MH, et al. Rupture of posterior tibial tendon: CT and MR imaging with surgical correlation. *Radiology* 1988;169:229–235.

Rosenberg ZS, Feldman F, Singson RD, et al. Computed tomography of ankle tendons. *Radiology* 1987;166:221–226.

Sartoris DJ, Resnick DR. Imaging of the musculoskeletal system: current and future status. *Am J Roentgenol* 1987;149:457–467.

Sierra A, Potechen EJ, Moore J, et al. High field magnetic resonance imaging of aseptic neurosis of the talus. *J Bone Joint Surg* 1986;68A:927–931.

Smith FW, Runge V, Permezel M, et al. Nuclear magnetic resonance (NMR) imaging in the diagnosis of spinal osteomyelitis. *Magnetic Resonance Imaging* 1984;2:53–56.

Spritzer CE, Dalinka MK, Kressel HY. Magnetic resonance of pigmented villonodular synovitis: a report of two cases. *Skeletal Radiol* 1987;16:316–319.

Stafford SA, Rosenthal DI, Gebhardt MC, et al. MRI in stress fracture. *AJR* 1986;147:553–556.

Sundaram M, McGuire MH, Fletcher J, et al. Magnetic resonance imaging of lesions of synovial origin. *Skeletal Radiol* 1986;15:110–116.

Swezey RL. Transient osteoporosis of the hip, foot and knee. *Arthritis Rheum* 1970;13:858–868.

Totty WG, Murphy WA, Lee JK. Soft-tissue tumors: MR imaging. *Radiology* 1986;160:135–141.

Weinberger G, Levinsohn EM. Computed tomography in the evaluation of sarcomatous tumors in the thigh. *Am J Roentgenol* 1978;130:115–118.

Wilson AJ, Murphy WA, Hardy DC, et al. Transient osteoporosis: transient bone marrow edema? *Radiology* 1988;167:757–760.

Wilson DA, Prince JR. MR imaging of hemophilic pseudotumors. *AJR* 1988;150:349–350.

Wright PH, Sim FH, Soule EH, et al. Synovial sarcoma. *J Bone Joint Surg* 1983;64-A:112–122.

Yulish BS, Lieberman JM, Strandjord SE, et al. Hemophilic arthropathy: assessment with MR imaging. *Radiology* 1987;164:759–762.

Zimbler S, McVerry B, Levine P. Hemophilic arthropathy of the foot and ankle. *Orthop Clin North Am* 1976;7:985–997.

CHAPTER 8

Tumors

Thomas H. Berquist and Andrew L. Deutsch

CASE 1

A 57-year-old man presented with a palpable mass in the posterior aspect of his left arm. There was no history of trauma. What is the nature of the lesion? Can you predict the histologic features? What MR criteria are established to allow one to predict whether the mass is benign or malignant? (Fig. 1A. Axial TR/TE 2,000/60 msec. Fig. 1B. Sagittal TR/TE 500/20 msec.)

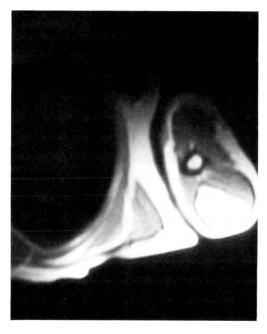

FIG. 1A.

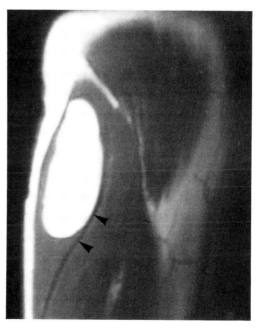

FIG. 1B.

Findings: A homogeneous well-circumscribed mass located within the region of the triceps muscle is identified. The mass is of increased signal intensity on both the T1- and T2-weighted images. The signal intensity of the mass parallels that of the subcutaneous fat. No internal septations or areas of significant inhomogeneity are present. The fascial plane is displaced (arrowheads, Fig. 1B), but there is no evidence of muscular invasion.

Diagnosis: Benign lipoma.

451

Discussion: Lipomas represent local collections of adipose tissue and are among the most common and most widely distributed soft tissue tumors of mesenchymal origin. The lesions are most commonly solitary, although multiple lipomas have been reported in 7% of patients. Deep lipomas may be subclassified as intramuscular (grow within muscles) or intermuscular (grow between muscles). Subcutaneous and intermuscular lipomas are usually well defined, with smooth or slightly lobular surfaces. In contrast, intramuscular lipomas may be ill-defined and infiltrate into adjacent muscle. Histologically, lipomas are arranged in lobules, with many fibrous septa and a surrounding capsule. Most lipomas occur in the trunk, thigh, or upper arms. Whether subcutaneous or deep, malignant change within lipomas is rare.

Lipomas can often be depicted on conventional radiographs as circumscribed radiolucent masses. Intramuscular lipomas may have an inhomogeneous appearance secondary to foci of muscle bundles of higher density traversing the mass. On CT, benign lipomas appear as homogeneous, well-defined, low attenuation masses that do not enhance with contrast. On MR, lipomas demonstrate high signal intensity on T1-weighted images and intermediate to high intensity on T2-weighted images. The appearance parallels that of the subcutaneous fat. Although well marginated as a rule, slight irregularity along a portion of the tumor margin is not uncommon and should not be interpreted as evidence of malignancy. Additionally, it is not uncommon for either MR or CT to identify sharply defined internal fibrous septa within benign lipomas. A multilobular appearance may also be seen (Fig. 1C). A lipoma could be difficult to differentiate from a subacute hematoma based on its appearance on a T1-weighted image. (Fig. 1D).

Liposarcomas can usually be differentiated from benign lipomas by their more inhomogeneous appearance (Fig. 1E). Depending on tumor cellularity, liposarcomas may demonstrate either increased or decreased (more hypercellular tumors) signal intensity on T1-weighted images. (Fig. 1F,G). Focal areas of malignancy within predominantly lipomatous lesions may demonstrate decreased signal intensity on T1-weighted images and increased intensity compared with surrounding fat on T2-weighted images.

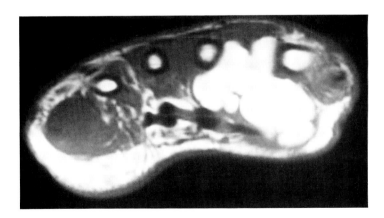

FIG. 1C. Multilobular lipoma (axial TR/TE 700/20 msec). A large multilobular lipoma is seen infiltrating within the deep intrinsic musculature of the hand.

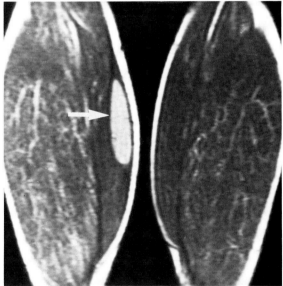

FIG. 1D. Chronic hematoma (coronal TR/TE 800/20 msec). An elliptical-shaped focus of high signal intensity is seen within the medial gastrocnemius muscle (*arrow*). Differentiation from a simple lipoma could not be accomplished based on the imaging characteristics alone.

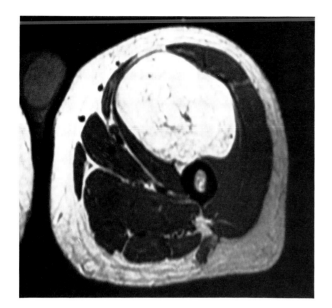

FIG. 1E. Atypical lipoma (axial TR/TE 2,000/20 msec). A large, relatively circumscribed mass is identified within the vastus musculature. The mass demonstrates considerable inhomogeneity and was considered suspicious for low grade malignancy. Excisional biopsy was accomplished, and no evidence of liposarcoma was identified.

F

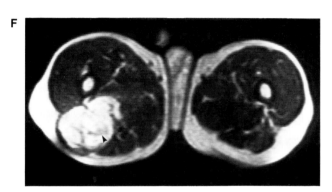

G

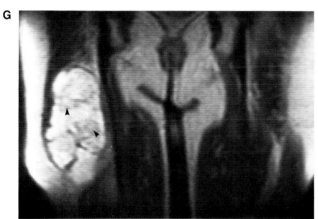

FIG. 1F,G. Low grade liposarcoma. F: Axial TR/TE 500/30 msec. On this T1-weighted image a well-circumscribed high intensity mass is identified. The intensity is similar to subcutaneous fat. Focal well-defined low signal intensity areas are identified (*arrowhead*). G: Coronal TR/TE 500/30 msec. The majority of the lesion is the same intensity as fat. Numerous areas of low signal (*arrowheads*) are present. The appearance is typical of liposarcoma.

CASE 2

A 67-year-old woman presents with a painless, palpable mass in the region of the right shoulder. The mass is firm, movable, and has been present several months. Is the mass more likely benign or malignant? What are the MR characteristics of benign soft tissue masses? (Fig. 2A. Coronal TR/TE 500/20 msec. Fig. 2B. Axial TR/TE 2,000/30 msec. Fig. 2C. Axial TR/TE 2,000/60 msec.)

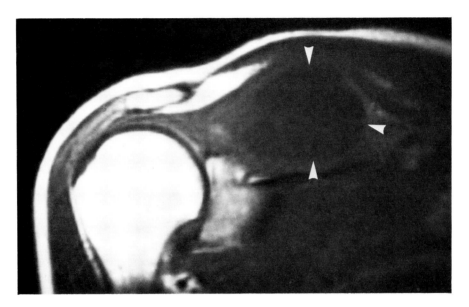

FIG. 2A.

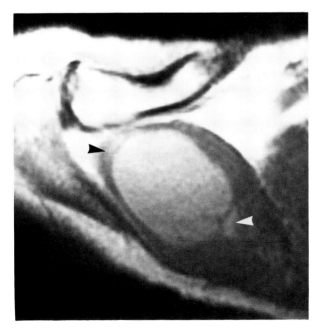

FIG. 2B.

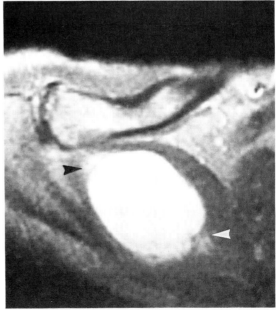

FIG. 2C.

Findings: On the coronal section (Fig. 2A), a well-circumscribed mass is identified within the supraspinatus muscle. It is of slightly lower signal intensity than muscle on this T1-weighted image (arrowheads). On the axial images (Fig. 2B,C), the mass demonstrates minimal irregularity medially (white arrowheads) and laterally (black arrowheads) where the muscle bundles converge. The mass increases uniformly in signal intensity on T2-weighted image.

Diagnosis: Intramuscular myxoma.

Discussion: Benign myxomas are uncommon soft tissue neoplasms, representing only about 0.5% to 1.0% of all soft tissue tumors. They occur primarily in patients 40 to 70 years of age. Myxomas are rare in young adults and virtually nonexistent in children. The masses characteristically enlarge slowly, are painless, and are present for many months before clinical presentation. The most common locations for the tumor are the large muscles of the thigh, shoulder, buttocks, and upper arm. Most intramuscular myxomas are solitary. Multiple intramuscular myxomas have been associated with fibrous dysplasia of bone, which is typically found in a similar anatomic distribution to the myxomas (e.g., same extremity).

Myxomas are most often unencapsulated, and although they appear grossly circumscribed, they often infiltrate the adjacent musculature or are surrounded by edematous muscle. Although the size of the lesions varies, most are 5 to 10 cm in diameter. Histologically the lesions demonstrate a fairly uniform consistency, are richly mucoid in nature, and are most commonly hypovascular. Fluid-filled spaces may occasionally be seen but are never a prominent finding, a feature in contrast to soft tissue ganglion and other superficially located myxoid lesions (Fig. 2D,E). Benign intramuscular myxomas must also be differentiated from other deeply located myxoid soft tissue tumors, most importantly from myxoid liposarcoma and myxoid malignant fibrous histiocytoma. These lesions, although richly myxoid, are both more cellular and vascular than myxomas and would be expected to demonstrate a more inhomogeneous pattern on MR.

On CT, myxomas appear as relatively well-demarcated homogeneous masses with attenuation coefficients in the range of 10 to 60 Hounsfield units. On MR, intramuscular myxomas demonstrate imaging characteristics typical of benign neoplasms. They are well circumscribed and uniform in signal intensity. The lesions are of decreased signal intensity in relation to surrounding muscle on T1-weighted images and demonstrate progressively increased signal intensity on the intermediate and T2-weighted images. The signal within the lesion is uniform. Benign intramuscular myxomas that infiltrate adjacent muscle may demonstrate slightly more irregular margins or be surrounded by edematous muscle.

The MR features most important in differentiation of benign from malignant masses principally relate to the peripheral margins of the mass as well as to the homogeneity of the internal architecture. Inhomogeneity within a solid mass must suggest the possibility of malignancy. This often correlates with the degree of cellularity of the lesion and may at times be subtle on MR (Fig. 2F,G). Inhomogeneity should be distinguished from internal septations, which may commonly be seen with a variety of benign lesions. Irregularity of the margins of a mass must also raise the spectre of malignancy, although as previously commented, some benign lesions (intramuscular myxoid tumors, desmoid tumors) may demonstrate infiltrative patterns, in which case differentiation from malignancies may not be possible. The location of a mass may also help suggest its benign or malignant nature. Involvement of more than one compartment, extension beyond fascial planes, and extension beyond a structure such as the interosseous membrane as in the leiomyosarcoma illustrated in Fig. 2F and G is more characteristic of a malignant rather than a benign mass. Direct involvement of adjacent bone also strongly suggests malignancy.

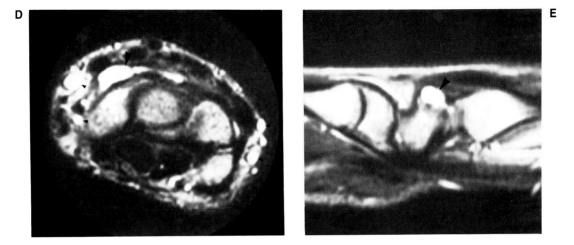

FIG. 2D,E. Soft tissue ganglion. **D.** Axial TR/TE 2,000/60 msec. Multiple well-defined areas of high signal intensity are seen around the wrist. The medial and lateral circular high intensity areas are vessels (*small arrowheads*). Dorsally, adjacent to the scaphoid is an elliptical, well-defined area of high signal intensity (*large arrowhead*). **E.** Sagittal TR/TE 2,000/60 msec. The circular to ovoid mass (*arrowhead*) is well depicted dorsal to the scaphoid. The mass was of low and uniform signal intensity on a T1-weighted image.

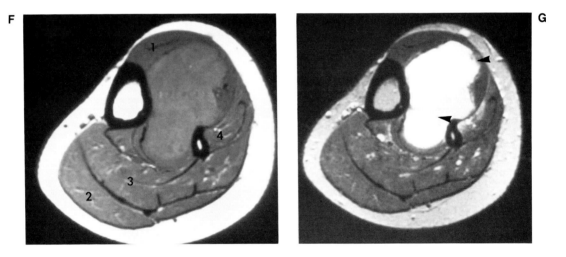

FIG. 2F,G. Leiomyosarcoma. A 50-year-old man presents with a clinically palpable soft tissue mass. **F.** Axial TR/TE 500/20 msec. A large mass involves both the anterior and deep posterior compartments of the lower leg. The lesion is nearly isointense with muscle on the T1-weighted image. The mass extends across the interosseous membrane. (1) Tibialis anterior; (2) medial gastrocnemius; (3) soleus; (4) peroneal musculature. **G.** Axial TR/TE 2,000/60 msec. The mass demonstrates significantly augmented signal intensity on the T2-weighted image. There are central areas of subtle inhomogeneity (*arrowheads*). The anterior margins of the mass are irregular. The findings are characteristic of a malignant neoplasm.

CASE 3

A 65-year-old woman with a prior history of left hip fracture presented with a soft tissue prominence anterior to the hip on her right. She was referred for MR to exclude a soft tissue neoplasm. What accounts for the clinical findings? Are the muscle groups intrinsically normal? What accounts for the image distortion in the region of the left hip on the axial section? (Fig. 3A. Axial TR/TE 500/20 msec. Fig. 3B. Coronal TR/TE 500/20 msec.)

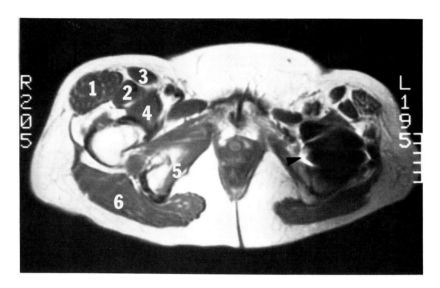

FIG. 3A.

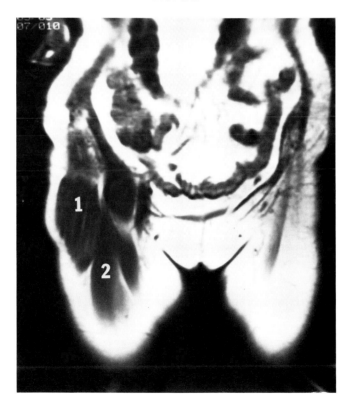

FIG. 3B.

Findings: The muscular anatomy of the hip is well demonstrated on both the axial and coronal sections. There is enlargement of the tensor fasciae latae (1), which accounts for the clinical findings. The muscle appears normal in signal intensity. No abnormal signal was identified on T2-weighted images. The other muscles [rectus femoris (2), sartorius (3), iliopsoas (4), obturator internus (5), and gluteus maximus (6)] are larger than their counterparts on the left but otherwise appear intrinsically normal. The musculature on the left is diminutive, reflecting the effects of disuse atrophy. An artifact created by an internal fixation device causes the black/white image distortion of the left hip (arrowhead).

Diagnosis: (1) Disuse atrophy of the left hip musculature; (2) hypertrophy of the tensor fasciae latae.

Discussion: The capability of MR to characterize normal and pathological states of soft tissue in addition to providing excellent anatomic detail represents a significant advantage over other imaging techniques including CT. The MR findings in this case allow confident exclusion of a significant mass. The information is obtained without recourse to invasive procedures and does not necessitate the administration of any contrast agent. T2-weighted images (not illustrated) are required for complete assessment, as pathological processes within muscle are commonly isointense with muscle on T1-weighted images.

Muscle atrophy may result from lack of use (disuse atrophy) as well as from loss of denervation (denervation atrophy). Both processes have been the subject of extensive study and can lead to significant muscle loss. As little as 1 to 2 months of disuse from any of a variety of causes (bed rest, limb immobilization, weightlessness) can lead to muscles that have atrophied to half their normal size. Disuse atrophy can be fairly rapidly reversed after short periods of immobilization. The longer the atrophy exists the longer it takes to reverse the process. After too long a period (greater than 4 months), muscle degeneration from disuse is no longer reversible. Ultimate replacement by fat cells is observed (Fig. 3C,D). Denervation atrophy has been the subject of experimental investigation with MR. Early changes in skeletal muscle following denervation are detectable utilizing MR by virtue of measured alterations in both T1 and T2. The findings appear to correlate with relative shrinking of the myoplasm and compensatory increases in the extracellular fluid spaces.

Muscle hypertrophy occurs only when a muscle contracts with a greater force than that to which it is accustomed. Hypertrophy is related to the intensity and not duration of the effort, and not until the force of contraction exceeds 70% of maximum does hypertrophy ensue. Histologically, the muscle enlargement results from an increase in the diameters of the individual fibers, with each enlarged fiber containing an increased number of myofibrils. The total number of fibers does not increase.

The localized signal void in the region of the patient's left hip prosthesis is typical of the metallic artifact seen with orthopedic hardware. The degree of image distortion is dependent on the composition, size, and orientation of the device. As can be seen, the effect is typically localized in contrast to metal-related artifact in CT and often does not preclude evaluation of adjacent structures. Magnetic resonance is extraordinarily sensitive to metallic foreign bodies. Even microscopic metallic shavings can be detected. The characteristic appearance is that of a low signal intensity focus with a partial high signal intensity border (Fig. 3E).

C

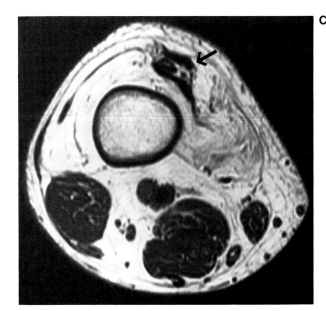

FIG. 3C,D. Muscle atrophy with fatty replacement. **C.** Muscle atrophy (axial TR/TE 800/20 msec). On this axial section through the distal femur, there is complete atrophy of the muscles of the extensor mechanism, with marked fatty replacement. The quadriceps tendon is thickened (*arrow*). The posterior musculature is normal, demonstrating the selective effect of this posttraumatic induced atrophy. **D.** Sagittal TR/TE 2,000/20 msec. The distal quadriceps tendon is mildly thickened. The gastocnemius musculature is normal without evidence of atrophy.

D

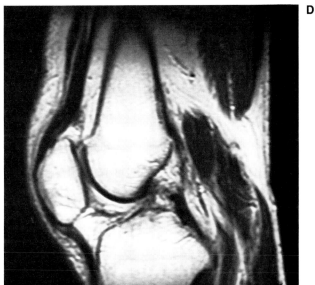

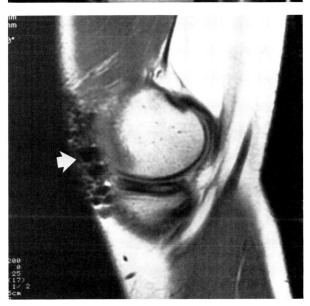

FIG. 3E. Metallic artifact (sagittal TR/TE 800/20 msec). The typical appearance of a small metal induced artifact is demonstrated (*arrow*). There is a focal area of low signal with high signal intensity border. No metal was evident on the plain radiograph. The patient has undergone arthroscopic surgery.

CASE 4

A 57-year-old woman underwent surgical resection of a soft tissue sarcoma in her left upper gluteal region 1 year ago. She recently had begun experiencing recurrent pain and swelling at the surgical site. What is the diagnosis? What MR features allow differentiation of recurrent tumor from postsurgical change? (Fig. 4A. Axial TR/TE 500/20 msec. Fig. 4B. Axial TR/TE 2,000/80 msec.)

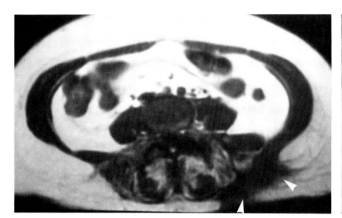 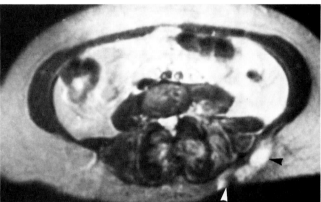

FIG. 4A.   **FIG. 4B.**

Findings: Axial T1- and T2-weighted images demonstrate an irregular mass lesion (arrowheads) in the soft tissues just superficial to the left paraspinal musculature. The lesion demonstrates low signal intensity on the T1-weighted images and significantly increased signal intensity on the T2-weighted images. The area of increased signal intensity is mildly inhomogeneous, and the margins are indistinct.

Diagnosis: Recurrent tumor.

Discussion: The evaluation of patients for possible tumor recurrence following prior surgical resection has long represented a challenging problem for diagnostic imaging. When attempting to characterize abnormalities following surgical resection, knowledge of the preoperative imaging features and extent of the lesion, time interval from surgery, and any adjuvant therapy can be critically important.

The overwhelming majority of soft tissue neoplasms are characterized by decreased signal intensity on T1-weighted images and increased signal intensity on T2-weighted images. Some tumors, however, that are predominantly collagen and/or markedly hypocellular may demonstrate low signal intensity on all pulsing sequences. These finding are independent of the histologic type of the tumor. Knowledge of the preoperative imaging features of the tumor (e.g., high or low signal on T2-weighted images) is essential in attempting to distinguish tumor recurrence from postoperative fibrosis, which is generally characterized by low signal intensity on all sequences.

In the immediate perioperative and short-term postoperative period, local inflammatory changes and hemorrhage predominate at the surgical site. Diagnostic imaging is not routinely performed in this setting except for evaluation of surgical complications such as abscess or hematoma (Fig. 4C,D). Over time, fibrosis at the surgical site develops and is characterized by progressively decreasing signal intensity. The length of time required for this to occur is dependent on multiple variables including the extent of the surgical procedure and any concomitant complications. In the majority of cases, some areas of organized scar tissue are evident by 3 months postsurgery. By 6 months postresection, the scar becomes organized and is characterized by low signal intensity on both T1- and T2-weighted sequences. It has been suggested that the accuracy of MR in making the distinction of postoperative scarring from recurrent tumor can be increased by obtaining a baseline study postoperatively and then performing reevaluation at 3 to 6 months.

The presence of low signal intensity in the surgical area on both T1-weighted images and T2-weighted images suggests postoperative scar tissue but unfortunately does not entirely exclude recurrent tumor (Fig. 4E,F). The low signal intensity allows prediction of a prominent fibrous component throughout the lesion with a marked paucity of cells. In conditions in which a reactive fibrous tissue (desmoplastic response) predominates, it may be particularly difficult to differentiate postsurgical scar from tumor intermixed with fibrosis. This problem is more typically encountered with epithelial neoplasms (colorectal carcinoma) than with soft tissue sarcomas, but the diagnostic difficulty posed remains far more than a theoretical concern. The identification of areas of increased signal intensity at the surgical site 6 months or more after resection must be considered highly suspicious for recurrent tumor (Fig. 4G,H). Enlarging areas of high signal as demonstrated in the patient illustrated in this case are invariably indicative of residual disease. The role of contrast enhancement with agents such as gadolinium in evaluation of recurrent soft tissue neoplasm has not yet been established. In primary neoplasms, contrast enhancement has been of value in identifying areas of viable tumor for biopsy.

The effect of any adjuvant therapy must also be considered in the evaluation for recurrent disease. In particular the effects of radiation therapy have been the subject of critical study with MR. Inflammatory changes secondary to radiation may manifest increased signal intensity on T2-weighted images. These inflammatory changes may be seen for years following the completion of therapy. In one recent study, the presence of only low signal abnormalities in patients who had undergone surgery and/or radiation for soft tissue sarcoma correctly predicted the presence of scar rather than recurrent tumor. The presence of high signal intensity areas, however, were seen with both recurrent tumor and radiation changes and could not accurately differentiate between the two. Inflammatory changes of postradiation therapy may be seen for years following the completion of therapy.

C

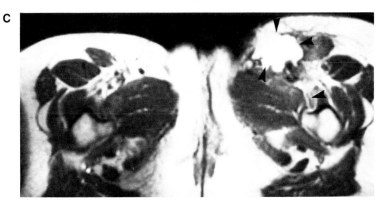

D

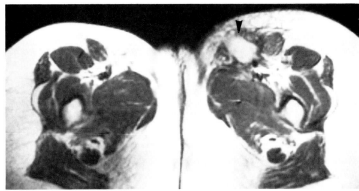

FIG. 4C,D. Postoperative hematoma. **C.** Axial TR/TE 500/30 msec. A homogeneous mass of high signal intensity is seen in the left inguinal region in this patient, status post–recent resection of a tumor (*arrowheads*). The signal characteristics of the mass allowed the clinically suspected diagnosis of hematoma to be confirmed. **D.** Axial TR/TE 500/30 msec. Follow-up study performed 3 weeks later demonstrates decreased size as well as signal intensity of the mass (*arrowhead*).

E
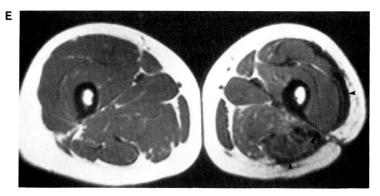

F
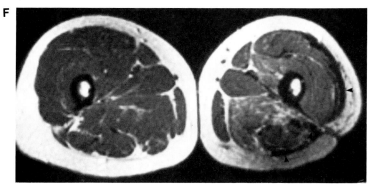

FIG. 4E,F. Postoperative fibrosis. **E.** Axial TR/TE 2,000/20 msec. There is a band of low signal intensity (*arrowheads*) seen subcutaneously along the lateral aspect of the thigh and extending deep to the femur along the incision site. **F.** Axial TR/TE 2,000/80 msec. On the T2-weighted image there is no significant enhancement (*arrowheads*). The findings are typical of postoperative scarring.

G
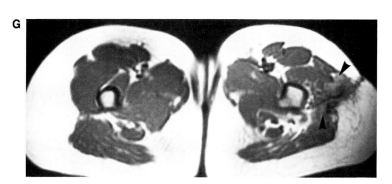

H
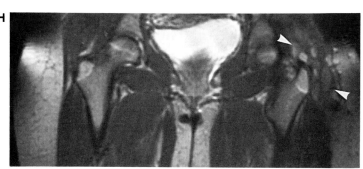

FIG. 4G,H. Recurrent tumor. **G.** Axial TR/TE 2,000/20 msec. A 20-year-old woman returned for routine follow-up 1 year after resection of a desmoid tumor in her left hip. On the axial image a focus of slightly increased signal intensity compared with the surrounding musculature is identified (*arrowheads*). **H.** Coronal TR/TE 2,000/60 msec. The focus of recurrent tumor demonstrates augmented signal intensity on this relatively T2-weighted image (*arrowheads*).

CASE 5

A 34-year-old woman presented with progressive swelling and pain in her right foot. A routine lateral radiograph (Fig. 5A) demonstrated soft tissue swelling and slight increased soft tissue density along the plantar aspect of the forefoot. Can you define the nature of this lesion with regard to malignancy? What are the principal differential considerations? (Fig. 5B. Axial TR/TE 500/20 msec. Fig. 5C. Axial TR/TE 2,000/60 msec. Fig. 5D. Sagittal TR/TE 2,000/60 msec.)

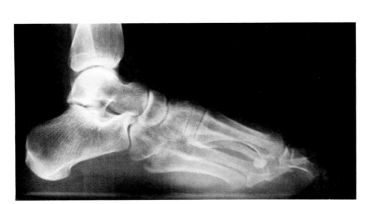

FIG. 5A.

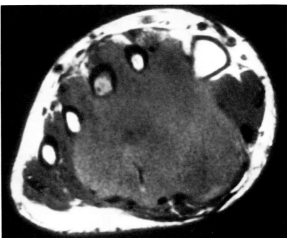

FIG. 5B.

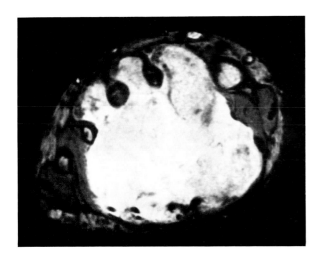

FIG. 5C.

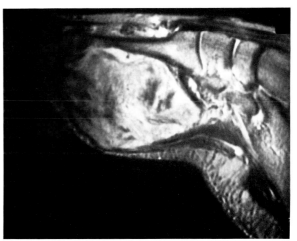

FIG. 5D.

Findings: On the axial T1-weighted images, there is a large soft tissue mass that extends from the plantar to the dorsal compartment. The mass is of slightly higher signal intensity than muscle on this sequence. No obvious bone involvement is evident. On the T2-weighted images, the lesion demonstrates inhomogeneous increased signal intensity. The sagittal section demonstrates the longitudinal extent of the lesion to advantage.

Diagnosis: Desmoid tumor (aggressive fibromatosis).

Discussion: The term *aggressive fibromatosis* encompasses a group of benign fibroblastic soft tissue neoplasms that share with other fibromatoses the propensity for local invasion and recurrence. Extra-abdominal desmoids are to be distinguished from abdominal desmoids, which originate most commonly from the aponeuroses of the rectus abdominus muscle, and from desmoplastic fibromas, which arise as primary intraosseous lesions. Patients with Gardner's syndrome, particularly postcolectomy patients, may present with mesenteric as well as musculoskeletal lesions.

Aggressive fibromatosis most commonly presents as a deep mass that rarely involves the skin. The lesions present most frequently in the third and fourth decades and predominate in women. Most cases involve the lower extremities, shoulder girdle, pelvis, or neck. Although considered benign (nonmetastasizing), the lesions are characteristically relentlessly infiltrative and recur following local excision in the majority (reported up to 77%) of instances. Surgical cure may require debilitating and often mutilating procedures.

The MR appearance of these tumors reflects their aggressive nature and represents a notable exception to the usual criteria for benign neoplasms such as smooth margins, homogeneity, and lack of neurovascular and/or osseous involvement. The MR appearance of aggressive fibromatosis is indeed more suggestive of malignancy, with irregular margination, neurovascular encasement, and osseous involvement not uncommon. The MR signal characteristics of aggressive fibromatosis reflect the degree of cellularity of the lesion. Whereas in the present case the tumor demonstrated the typical MR pattern of increased signal intensity on the T2-weighted images, many cases of aggressive fibromatosis and abdominal desmoid tumors are of low signal intensity on all pulsing sequences (Fig. 5E–I). In these instances, the lesions are characterized histologically by dense foci of paucicellular collagen, accounting for the low signal intensity.

Computed tomography has been previously utilized in evaluation of patients with musculoskeletal desmoid tumors. The infiltrative nature of the lesion and its frequent isodensity with muscle limit the precision of definition of tumor boundaries utilizing CT. Magnetic resonance is well suited to assessment of these lesions and in particular to establishment of proximity to adjacent neurovascular and osseous structures. This information is critically important to selection of radiation portals and/or preoperative planning. Wide local excision is presently the preferred therapy even in cases of recurrence.

E

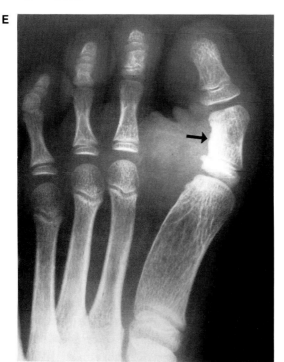

FIG. 5E–G. Desmoid tumor. **E.** Antero-posterior radiograph of the foot demonstrates a soft tissue mass between the first and second toes with pressure erosion of the fibular aspect of the first proximal phalynx (*arrow*). **FIG. 5F,G.** *Continued* on next page.

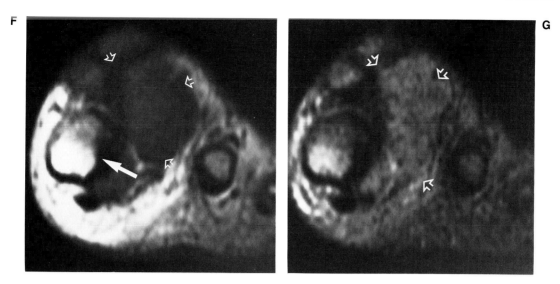

FIG. 5F,G. *Continued.* Desmoid tumor. **F.** Axial TR/TE 2,000/20 msec. The soft tissue mass is of relatively uniform decreased signal intensity (*open arrows*). Erosion of cortical bone is evident (*white arrow*). **G.** Axial TR/TE 2,000/60 msec. The mass (*open arrows*) is nearly isointense with surrounding soft tissue. This appearance is characteristic.

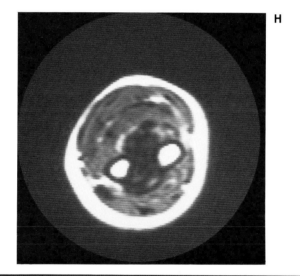

FIG. 5H,I. Desmoid tumor. **H.** Axial TR/TE 500/20 msec. An inhomogeneous, slightly irregular mass of low signal intensity is seen crossing the interosseous membrane between the radius and ulna. **I.** Coronal TR/TE 500/20 msec. The majority of the mass has low signal intensity caused by the fibrous nature of the lesion. The appearance did not change on T2-weighted images.

CASE 6

A 29-year-old man presented with pain and swelling in his left hip and gluteal region. Where is the lesion located? What differential possibilities are suggested by the foci of increased signal intensity on both the T1-weighted image and T2-weighted image? (Fig. 6A. Axial TR/TE 2,000/60 msec. Fig. 6B. Coronal TR/TE 500/20 msec.)

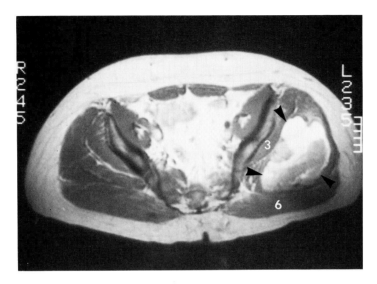

FIG. 6A.

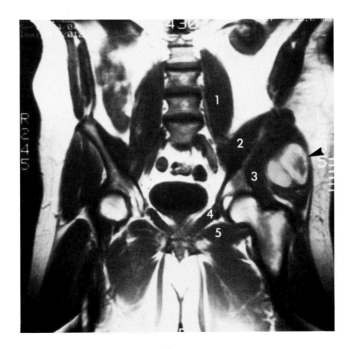

FIG. 6B.

Findings: A large, partially inhomogeneous mass is identified within the gluteus medius muscle (arrowheads) on the left. The lesion expands the muscle and displaces the gluteus maximus and minimus. There are areas of increased signal intensity within the periphery of the mass on both imaging sequences. The foci of increased signal are less intense than normal subcutaneous fat on the T1-weighted image and more intense than fat on the T2-weighted image. The findings suggest subacute hemorrhage. Muscles: (1) psoas; (2) iliacus; (3) gluteus minimus; (4) obturator internus; (5) obturator externus; (6) gluteus maximus.

Diagnosis: Malignant hemangiopericytoma.

Discussion: Hemangiopericytomas are uncommon vascular neoplasms of the soft tissues that originate from the capillary wall (pericytes of Zimmermann). The tumors most often present in adults (90% over the age of 20 years) with the peak incidence in the fifth and sixth decades. The deep soft tissue of the thigh is the most common location. The pelvis, as in this case, and the trunk are also common locations. The lesion most commonly presents as a slowly enlarging painless mass and frequently attains a large size (6–7 cm median).

The histologic features of the tumor assist in understanding its MR appearance. The mass is usually well circumscribed with a margin of large plexiform vessels. Large sinusoids filled with blood account for well-circumscribed areas of increased signal intensity on both T1- and T2-weighted sequences. The peripheral arterial branching contributes a characteristic arteriographic appearance. The peripheral vessels may be visualized on MR as small foci of signal void on spin-echo sequences and increased signal on flow sensitive sequences (gradient recalled sequences).

The MR appearance described does not allow for a specific diagnosis (Fig. 6C,D). Grossly similar areas of increased signal may be seen in synovial sarcomas, soft tissue chondrosarcomas, benign hemagiopericytomas, and other lesions in which there typically are collections of blood. The marked inhomogeneity of the lesion suggests a malignant process in this case.

C 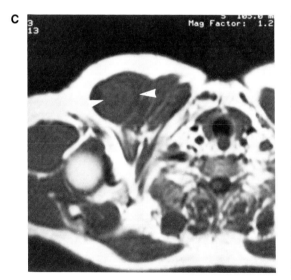 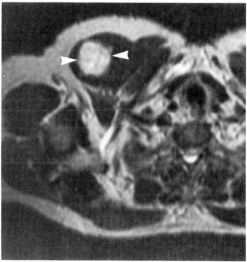 D

FIG. 6C,D. Metastatic hemangiopericytoma. **C:** Axial MR, TR/TE 300/15 msec. On this T1-weighted image a circumscribed mass is barely discernible within the musculature of the anterior chest wall (*arrowheads*). **D:** Axial MR, TR/TE 2,500/80 msec. On the T2-weighted image the mass demonstrates marked increased signal intensity. The margins are slightly lobular (*arrowheads*). The appearance is nonspecific and could be reflective of a wide number of soft tissue neoplasms.

CASE 7

A 32-year-old man presented with swelling and pain anterior to his right knee. The pain had been present for nearly a year before the swelling was noted. What is the nature of this lesion? Does the location help in differential diagnosis? Are the image features more suggestive of a benign or malignant process? (Fig. 7A. Sagittal TR/TE 500/20 msec. Fig. 7B. Sagittal TR/TE 2,000/60 msec.)

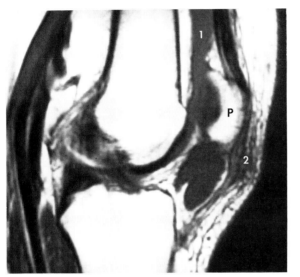

FIG. 7A.

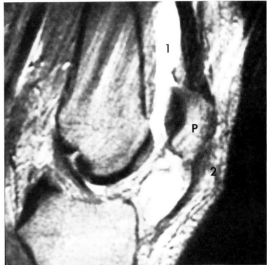

FIG. 7B.

Findings: There is a moderate-size joint effusion (1) seen within the suprapatellar bursa. A multilobular mass is seen within the infrapatellar fat pad (P) characterized by relatively uniform decreased signal on the T1-weighted image and increased intensity on the T2-weighted image. The intensity is similar to that of the synovial fluid, suggesting a "cystic" lesion. Superficial edema is seen within the prepatellar soft tissues (2).

Diagnosis: Synovial sarcoma.

Discussion: Synovial sarcomas are uncommon malignant neoplasms that account for up to 8% of all soft tissue sarcomas. Patients are usually young, male, and most often in their third or fourth decade. The tumors demonstrate a distinct predilection for the lower extremities, and the knee is the most commonly affected site. The lesions are most commonly extra-articular with less than 10% of cases arising within the joint. Conventional radiographs may demonstrate a soft tissue mass with evidence of calcification reported in approximately one-third of cases. Tumors exhibiting extensive calcification/ossification have been reported to have a slightly improved prognosis. Local recurrence occurs frequently after surgical resection. Metastasis to lung, bone, and other soft tissue locations is common. Five-year survival may be greater than 50%, but 10-year survival is only 11%.

The MR features of synovial sarcoma are similar to other soft tissue sarcomas. Malignant features include inhomogeneity, irregular margins, and encasement of neurovascular structures. Synovial sarcoma exhibits a propensity for tracking along tendon sheaths and for local invasion of bone. These characteristics were of value in prospective diagnosis of a series of lesions involving the ankle and foot that were imaged by MR. In our experience, the majority of synovial sarcomas have demonstrated a "cystic" appearance similar to that illustrated in this case (Fig. 7C). When arising within the Hoffa fat pad, the lesion could superficially resemble other conditions that have been observed in this location, including infrapatellar bursitis and pigmented villonodular synovitis (PVNS) (Fig. 7D–F). Critical observation will usually reveal irregularity along a portion of the tumor margin (upper and posterior margin, Fig. 7B). Scattered areas of signal void corresponding to calcifications can be identified but are better evaluated with radiographs and CT.

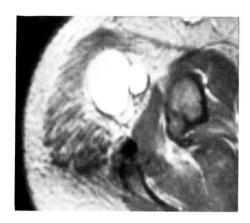

FIG. 7C. Axial TR/TE 2,000/80 msec. A "cystic"-appearing lesion is seen within the soft tissues adjacent to the lesser trochanter. The lesion is well defined and relatively homogeneous. Despite this benign appearance, the lesion was a synovial sarcoma.

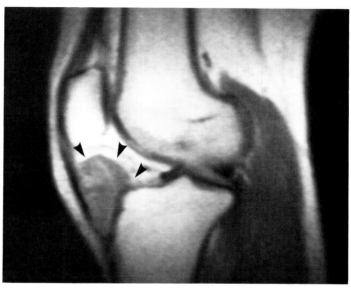

FIG. 7D. Infrapatellar bursitis (sagittal TR/TE 600/20 msec). A homogeneous low signal intensity mass (representing fluid) is seen arising from the infrapatellar bursa (*arrowheads*). Its origin from the bursa as opposed to within the fat pad assists in distinguishing between the two conditions.

E
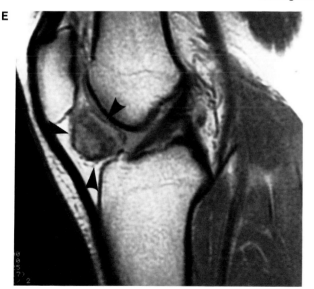
F
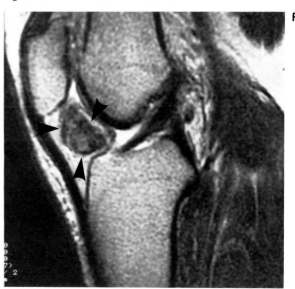

FIG. 7E,F. Pigmented villonodullar synovitis (PVNS). **E.** Sagittal TR/TE 2,000/20 msec. A circumscribed mass of low to intermediate signal intensity is seen within the infrapatellar fat pad (*arrowheads*). On this image it resembles the synovial sarcoma illustrated in Fig. 7A. **F.** Sagittal TR/TE 2,000/80 msec. On the T2-weighted image, the mass does not increase in signal intensity (*arrowheads*), a finding characteristic of PVNS and one of particular value in differential diagnosis.

CASE 8

A 43-year-old woman presented with pain and muscle weakness in her extremities. There were no other symptoms. Are the image features more suggestive of a benign or malignant process? What is the most likely diagnosis? (Fig. 8A. Axial TR/TE 2,000/ 60 msec.)

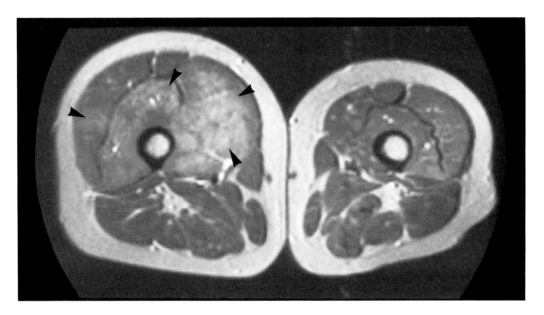

FIG. 8A.

Findings: An axial T2-weighted image of the thighs shows asymmetry with the musculature of the right leg larger than the left. There is obvious prominence of the anterior compartment on the right with scattered areas of increased signal intensity within the vasti musculature (arrowheads).

Diagnosis: Polymyositis.

Discussion: Inflammatory muscle disease may be on an infectious or noninfectious basis. Polymyositis represents an idiopathic inflammatory myopathy. The term *dermatomyositis* is applied in the presence of a characteristic skin rash (up to one-third of patients). The disease occurs most commonly in women, often black, between the ages of 40 and 70. Polymyositis may be associated with collagen vascular disease (overlap syndromes). Although not proven in any critical manner, an association between malignancy and polymyositis/dermatomyosis has been observed. The tumors found are those most common in the age group (e.g., breast, colon).

The most remarkable clinical feature of the condition is weakness, which is commonly profound and often out of proportion to the degree of muscle wasting. Symmetric weakness of the limb-girdle muscles and neck flexors, which progresses over weeks to months, is observed. The most frequent finding on muscle biopsy is necrosis of skeletal muscle fibers.

The MR features in this condition are often subtle. The superior soft tissue contrast of MR allows the affected muscles to be clearly identified. The images most commonly demonstrate either mottled or infiltrative increased signal intensity on the T2-weighted image. T1-weighted images may look entirely normal. The areas of inflammation are characteristically scattered and noncontiguous in distinction to the more confluent appearance of neoplasms. A well-defined mass lesion is not seen unless as a complication of recent biopsy with development of a hematoma. Additionally the adjacent medullary and cortical bone is within normal limits. The MR features of polymyositis are not specific but should be able to be differentiated from malignancy in most cases. Soft tissue injuries and compartment syndromes could have similar appearances but can be readily differentiated clinically (Fig. 8B). Magnetic resonance findings can allow precise localization for diagnostic biopsy. Additionally, MR can be used to follow the response to steroid administration, which is the most widely utilized therapy.

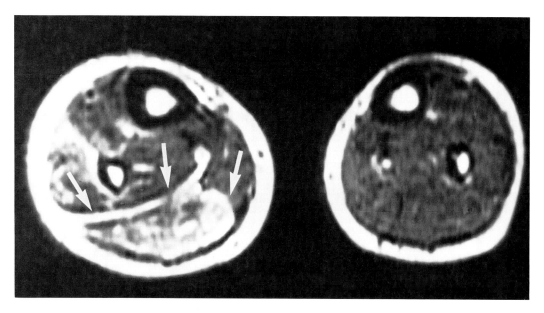

FIG. 8B. Compartment syndrome (axial TR/TE 2,000/80 msec). A professional football player received a direct blow to the back of his calf. There is diffuse increased signal intensity (edema) within the superficial posterior compartment (*arrows*).

CASE 9

A 60-year-old woman presented with pain and swelling of the proximal portion of her left thigh. The patient had undergone vascular surgery within the past month. How do the signal characteristics assist in the differential diagnosis? What is the most likely diagnostic possibility? (Fig. 9A. Axial TR/TE 500/20 msec. Fig. 9B. Axial TR/TE 2,000/60 msec.)

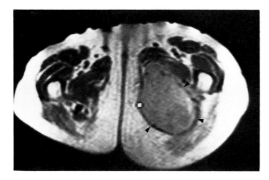
FIG. 9A.

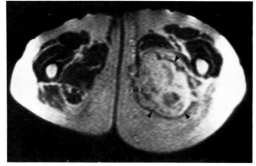
FIG. 9B.

Findings: There is marked asymmetry in size of the proximal thighs with extensive edema and infiltration of the subcutaneous fat on the left. A moderately circumscribed mass is identified (arrowheads). The mass is of slightly higher signal intensity than muscle on the T1-weighted image. On the T2-weighted image, the mass increases in signal intensity and is moderately inhomogeneous. There is a thick low signal intensity margin (arrowheads). The cortical and medullary bone of the femur is not involved.

Diagnosis: Infected hematoma.

Findings: The MR appearance in this case is entirely nonspecific, and without the clinical history, both inflammatory and neoplastic considerations could be reasonably entertained in the differential diagnosis. It is highly improbable, however, that a neoplasm could attain this size during the course of 3 to 4 weeks. The history of recent surgery suggests a postoperative complication as the most likely diagnostic possibility. The signal intensity of the lesion would be compatible with a subacute hematoma, and the peripheral rim could be secondary to hemosiderin-laden macrophages seen with chronic hematomas. Simple hematomas most often are more homogeneous and often by 3 to 4 weeks demonstrate higher signal intensity on T1-weighted images (Fig. 9C). An abscess with increased protein and areas of fibrin deposition can characteristically demonstrate an inhomogeneous appearance. The increased protein content can shorten the T1 and T2 relaxation times so that the fluid has a signal intensity slightly higher than that of muscle on the T1-weighted image and slightly less than that of fluid (approximately equal to that of fat) on the T2-weighted image. A thick surrounding capsule is commonly seen with an abscess. The extensive surrounding edema can be seen with an abscess or with a neoplasm and is not helpful in itself in differentiating between the two. In this patient, the most likely diagnosis was hematoma, with or without associated infection. The patient had a low grade fever, and aspiration yielded bloody fluid, which on culture of the specimen was positive for *Staphylococcus*.

Hematomas and hemorrhage vary in appearance on MR depending on their age and the field strength of the imaging unit. At low field strengths (less than 0.3 T), hemorrhages may have a high intensity infiltrative or interstitial appearance on T2-weighted images and a low intensity (less than muscle) appearance on T1-weighted images. No mass effect is seen as with true hematomas. At high field strength (greater than 1.0 T), acute hemorrhage generally will be seen as high intensity on T2-weighted images and slightly greater intensity than muscle on T1-weighted images. Over time, the signal intensity gradually decreases on both pulsing sequences at all field strengths.

The appearance of hematomas at different field strengths and at different stages of evolution has been studied by Gomori, Ehman, and others. Although the age of the hematoma is difficult to establish, there are certain trends that are useful in evaluating hematomas and distinguishing them from other mass lesions. As an example, on partial saturation sequences, almost all early (less than 30 days) hematomas have higher signal intensity than muscle at low and high field strengths. Neoplasms would most commonly be isointense or of diminished intensity with this pulse sequence.

As hematomas age, their water content decreases and their protein content increases, leading to decreases in T1 and T2 relaxation times. A peripheral rim of high or low intensity also is commonly noted. This high intensity ring is seen early on MR with either high or low field strength. The paramagnetic effect of methemoglobin tends to cause high intensity (short T1) margins on T1-weighted images. This effect usually is seen earlier, accounting for the bright ring at both field strengths in hematomas less than 30 days old. In older hematomas, a low intensity margin may be noted at high field strength as a result of the presence of hemosiderin-laden macrophages. A central low intensity focus is common within hematomas at high field strength (Fig. 9C). This appearance does not correlate well with the age of the lesion. This appearance on the T2-weighted image is related to the gradient created across intact erythrocyte membranes in cells with deoxygenated hemoglobin, which alters the magnetic susceptibility of the cytoplasm, and thus is thought to alter the magnetic gradient across the cell membrane. The effect is 100 times stronger at high field strength and is no longer evident following lysis of the erythrocytes.

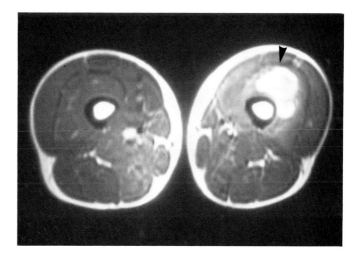

FIG. 9C. Hematoma (axial TR/TE 2,000/60 msec). A circumscribed mass of high signal intensity is seen within the vastus intermedius muscle. A central focus of diminished signal intensity is seen within the center (*arrowhead*) and relates to the paramagnetic effects of deoxygenated hemoglobin within intact erythrocytes.

CASE 10

A 23-year-old woman presented with a small movable lump in her calf. There was point tenderness and a tingling sensation when pressure was applied to the area. Can you identify the lesion? Could this represent an enlarged varix? Is the lesion more likely benign or malignant? What is the differential diagnosis? (Fig. 10A. Axial TR/TE 500/20 msec. Fig. 10B. Axial TR/TE 2,000/60 msec.)

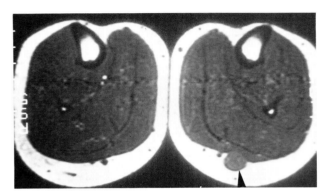
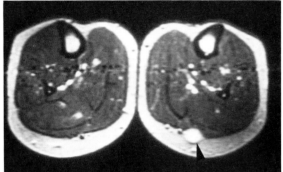

FIG. 10A.   **FIG. 10B.**

Findings: A small, well-circumscribed mass (arrowheads) is identified in a superficial location in the calf. The mass is of slightly greater signal intensity than muscle on the T1-weighted image. The mass is of uniform increased signal intensity on the T2-weighted image. No evidence of infiltration is noted.

Diagnosis: Benign neurilemoma.

Discussion: There are three principal benign peripheral nerve lesions, and the terminology utilized in their classification is often confusing. The two most important tumors are the neurilemoma and neurofibroma. Neuromas, the third common lesion, are not true neoplasms but rather result from nerve repair most commonly following repeated trauma. A classic example is that of the Morton neuroma, which is located between the third and fourth toes. This is believed to develop as a result of chronic entrapment injury to the medial or lateral plantar nerve near the transverse intertarsal ligament. Neuromas are also common at amputation sites and often lead to chronic pain.

Neurilemomas (benign schwannoma, neurinoma) are most commonly seen in patients of both sexes between the third and fifth decades. The tumor usually arises as a solitary, well-circumscribed lesion from the sheath (Schwann cell) of a peripheral nerve. The tumor grows tangential to the nerve, and it is this feature of nerve displacement rather than incorporation that distinguishes the schwannoma from a neurofibroma. Schwannomas are commonly encountered on the flexor surfaces of the upper and lower extremities, as in the present case. The lesions are typically oval to fusiform in shape and are watery or mucoid in nature, accounting for their appearance on MR (long T1, long T2). Neurilemomas are benign; tumor recurrence is unusual, and malignant transformation is rare.

Neurofibromas are more common than neuromas and neurilemomas and tend to occur in younger patients. They may be seen in all areas of the body, and within the extremities, they tend to be superficial and commonly within the subcutaneous tissue. They may occur as solitary lesions unassociated with neurofibromatosis (von Recklinghausen's disease) or as multiple lesions. Neurofibromas demonstrate low to intermediate signal intensity on T1-weighted images and are of high signal on T2-weighted images (Fig. 10C). Plexiform lesions are commonly uniformly bright in signal and characteristically extend along neural bundles in a lobulated manner (Fig. 10D,E). The nerve that harbors the plexiform neurofibroma appears as a rope braided from the individual distended fasicles. Both neuromas and neurofibromas may have more collagen and fibrous tissue, resulting in decreased signal intensity on T2-weighted images (Fig. 10F) compared with the neurilemoma illustrated in this case. Differentiation is generally not critical, as all three are benign lesions.

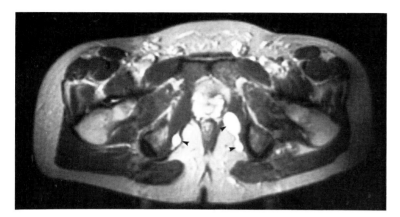

FIG. 10C. Neurofibromatosis (axial TR/TE 2,000/60 msec). On this T2-weighted image through the pelvis, multiple circumscribed high intensity neurofibromas (*arrowheads*) are identified.

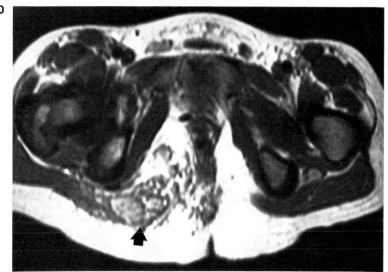

D

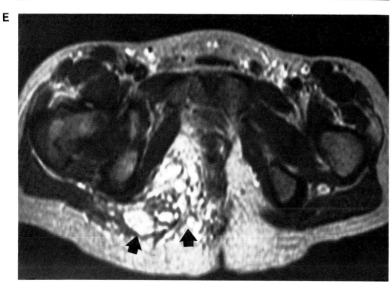

E

FIG. 10D,E. Plexiform neurofibroma. **D.** Axial TR/TE 2,000/20 msec. A classic plexiform lesion is seen in the right gluteal region of this 8-year-old boy with neurofibromatosis (*arrow*). **E.** Axial TR/TE 2,000/80 msec. The lesion demonstrates significantly increased signal intensity on the T2-weighted image (*arrows*). Interestingly, the lesion did not increase in signal intensity following administration of gadolinium.

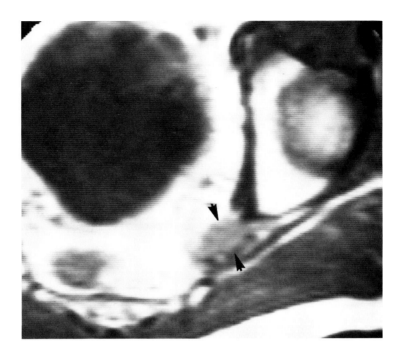

FIG. 10F. Neurofibroma (axial TR/TE 2,000/60 msec). A small circumscribed mass (*arrowheads*) is identified posterior to the hip in the distribution of the sciatic nerve. The signal intensity is only slightly greater than surrounding muscle.

CASE 11

A 73-year-old woman complained of pain within the left hip of several weeks duration. There was no history of trauma or any significant systemic disease. What is the most likely diagnosis? (Fig. 11A. Anteroposterior radiograph left hip. Fig. 11B. Anteroposterior projection from radionuclide bone scan. Fig. 11C. Coronal TR/TE 800/20 msec.)

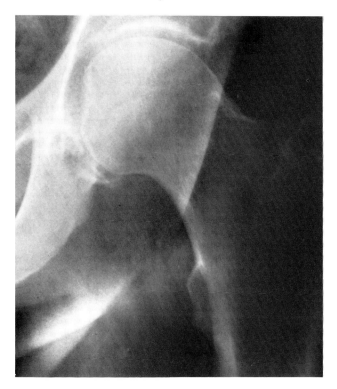

FIG. 11A.

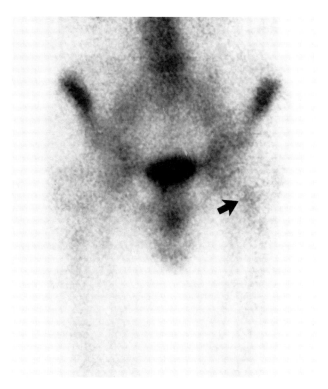

FIG. 11B.

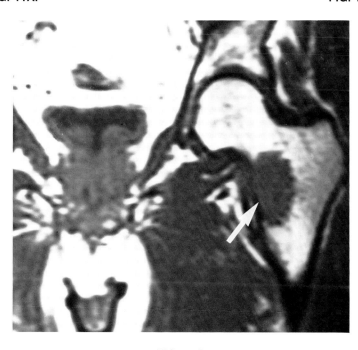

FIG. 11C.

Findings: The anteroposterior radiograph of the left hip demonstrates no significant abnormality. No trabecular interruption is noted nor is there evidence of sclerosis to suggest occult impacted fracture. The radionuclide bone scan demonstrates an equivocal area of minimally increased uptake in the region of the proximal femoral neck on the left (arrow). The coronal T1-weighted MR image reveals an elliptical-shaped focus of decreased signal intensity in the region of the lesser trochanter (arrow).

Diagnosis: Metastatic breast carcinoma.

Discussion: The skeleton is one of the most frequent sites of tumor metastasis, and metastatic lesions represent the most common type of malignant bone tumor. The vast majority of skeletal metastases occurs in adults with carcinoma of the prostate, breast, lung, and kidney, accounting for more than three-quarters of all cases. Carcinoma of the breast alone accounts for nearly 70% of skeletal metastases in women, and carcinoma of the prostate accounts for approximately 60% of cases in men. Metastatic lesions demonstrate a propensity for involvement of the axial skeleton, paralleling the distribution of hematopoietic marrow.

It is well recognized that conventional radiographs are insensitive with regard to depiction of the majority of skeletal metastases. In certain sites such as the vertebral body, extensive bone destruction is required before a metastasis will be radiographically evident. The problem is further compounded when metastatic deposits are localized to the medullary space in which few trabeculae are present. Metastatic lesions that primarily involve the cortex of bone are less common but are more readily depicted by conventional techniques. Scintigraphic techniques have been widely utilized in the evaluation of skeletal metastases. The basis for radionuclide imaging depends on changes in regional blood flow and bone turnover for focal tracer uptake. Although the potential for false-negative bone scans has long been recognized, no data is available on the incidence of this occurrence. Any tumor that is unaccompanied by ongoing new bone formation can lead to a "cold" lesion or normal scan. This situation has been most widely recognized with multiple myeloma. Additionally, the lack of specificity of scintigraphy is well known, and correlation with additional radiographic tests is frequently required for patient management.

The capability of MR to depict sensitively a broad spectrum of disorders involving bone marrow has become well established in multiple investigations. A wide number of radiographically occult abnormalities ranging from fractures to neoplasm have been detected utilizing MR. T1-weighted spin-echo sequences are presently most commonly employed to depict marrow abnormalities. These provide excellent contrast between the low signal intensity (long T1) of most pathologic processes and the normal high signal of adult marrow. T1-weighted images provide optimal contrast between fat and tumor/edema. T2-weighted images are utilized for increasing diagnostic specificity in evaluation of marrow processes and in the differentiation of tumor from surrounding muscle. Recent interest has centered on the use of a technique called short T1 inversion recovery (STIR) in assessment of marrow disorders. With this technique, an inversion time interval is chosen to suppress (null) the signal from medullary fat. In STIR imaging, T1 and T2 contrast are additive, and areas of abnormal marrow demonstrate increased signal intensity and are contrasted against a low signal intensity background of suppressed marrow fat. Recent investigations have suggested an increased sensitivity to depiction of marrow disorders utilizing this pulse sequence (Fig. 11D–F). A present drawback of STIR imaging is that the number of slices that can be simultaneously obtained is limited.

D

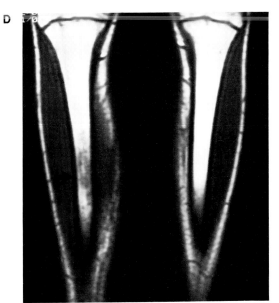

E

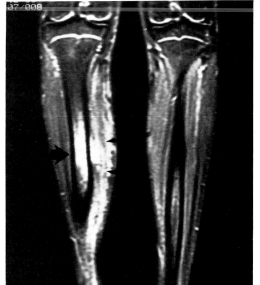

F

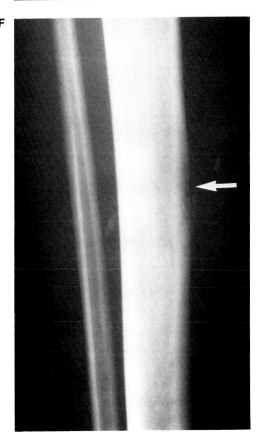

FIG. 11D–F. Ewing's sarcoma. **D.** Coronal TR/TE 800/20 msec. A 15-year-old boy underwent MR for diagnostic staging of a Ewing's sarcoma. On the coronal T1-weighted image, an area of mildly decreased signal (*arrow*) emanates from the tibial diaphysis. **E.** TR/TE 1,500/30, TI 500 msec (STIR). A STIR sequence was performed and demonstrates far more extensive involvement of the shaft than predicted with the T1-weighted image (*large arrow*). Marked increased signal within the soft tissue is also seen and represents both tumor and edema (*small arrows*). **F.** Conventional radiograph of the tibia in this patient demonstrates only a localized and subtle area of periosteal reaction (*arrow*).

CASE 12

A 23-year-old woman presented with swelling of her right thigh. There was no history of trauma, and the mass was not painful. There was no known systemic disorder. What is your diagnosis? (Fig. 12A. Axial TR/TE 2,000/80 msec.)

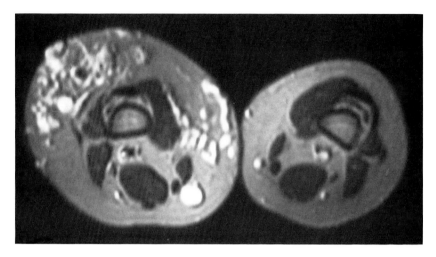

FIG. 12A.

Findings: There is marked swelling of the right thigh predominantly on the basis of extensive hypertrophy of subcutaneous fat. There is mild muscle atrophy compared with the opposite side. Numerous serpiginous and circular to ovoid areas of high signal intensity are seen within the subcutaneous fat.

Diagnosis: Hemangioma.

Discussion: Hemangiomata are benign lesions classically divided into capillary and cavernous types, depending on their predominant composition. Capillary hemangiomata are composed of capillaries with a sparse stroma, whereas cavernous hemangiomata contain large dilated sinusoidal blood-filled spaces. A large number of hemangiomata demonstrate a continuous spectrum of changes. The accumulation of adipose tissue is common with these mixed (cavernous-capillary) hemangiomata. The lesions are most commonly found in the skin and subcutaneous tissues and may be small and subtle or extensive, as illustrated in this case. Hemangiomata are also relatively common within skeletal muscle with a predilection for the limbs distal to the elbow and knee. Deep muscular hemangiomata may be more infiltrative and can be confused with angiosarcomas.

Magnetic resonance is well suited to depict the spectrum of findings that can be seen with these vascular lesions. Cohen described vascular channels separated by fibrous or fatty septa. Serpiginous vessels can be commonly identified. The lesions are often best defined on T2-weighted images (Fig. 12B,C). Yuh described the typical appearance of skeletal muscle hemangiomata. These typically demonstrated high signal intensity on both T1-weighted images and T2-weighted images, with the intensity less than that of fat on the T1-weighted images and greater than fat on T2-weighted images. The majority of patients had focal muscle atrophy. Large areas of angiomatous tissue, if solitary, could be confused with a soft tissue sarcoma. The lesions, however, are usually multiple and separated, which allows suggestion of the correct diagnosis. Small capillary hemangiomas may be subtle. In our experience, several small (less than 1 cm) angiomas have been overlooked preoperatively but could be identfed in retrospect. Although the extent of hemangiomata can be clearly identified with MR, the differentiation of veins, arteries, and lymphatics is not always possible. Angiography often is utilized if surgical resection of large lesions is contemplated.

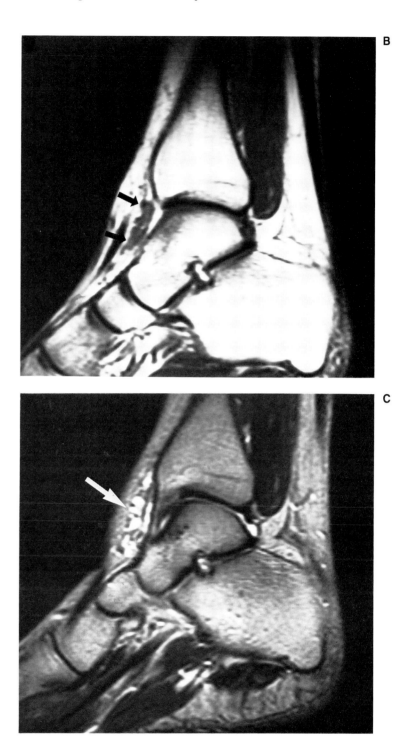

FIG. 12B,C. Superficial hemangioma. **B.** Sagittal TR/TE 800/20 msec. A superficial soft tissue mass could be palpated along the dorsum of the foot of a 20-year-old woman. A lobular mass of mixed low and high signal intensity is identified (*arrows*). **C.** Sagittal TR/TE 2,000/80 msec. There is marked increased signal intensity on the T2-weighted image. The serpiginous nature of the mass is more evident (*arrow*).

CASE 13

A 40-year-old man presented with pain and swelling in his knee. What is the most likely diagnosis? What causes the low signal intensity areas in the mass? (Fig. 13A. Sagittal TR/TE 500/20 mesec. Fig. 13B. Sagittal TR/TE 2,000/60 msec.)

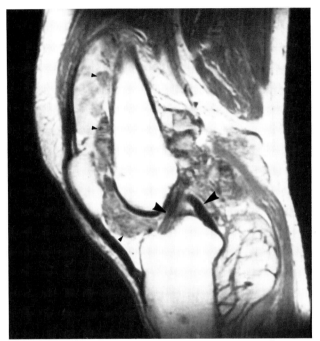 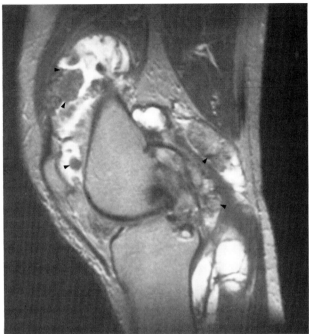

FIG. 13A.   **FIG. 13B.**

Findings: A large inhomogeneous mass involves the suprapatellar bursa and posterior capsular region of the joint. The popliteal component dissects into the calf. There are multiple low intensity areas (small arrowheads) in the lesion. The osseous structures and cruciate ligaments (large arrowheads) are normal.

Diagnosis: Pigmented villonodular synovitis.

Discussion: Pigmented villonodular synovitis is an uncommon disorder that is characterized by synovial proliferation and hemosiderin deposition into the synovial tissues of affected joints. It is encountered most often in the third through the sixth decades. Pigmented villonodular synovitis is invariably monoarticular. Patients complain of intermittent pain and swelling, often accompanied by decreased range of motion. The joint fluid may demonstrate a xanthochromic or serosanguinous appearance; classic "chocolate brown" effusions are seen in 30% of cases.

The pathogenesis of PVNS is much debated; etiologic theories include repeated trauma, neoplasm, and disorders of lipid metabolism. Two forms are classically recognized. The diffuse form involves the entire joint. A more localized form affects only a portion of a joint and is common in the hand, where it is referred to as giant cell tumor of the tendon sheath. The knee is the most frequently involved articulation, accounting for nearly 80% of cases. Other common locations include the hip, ankle, and elbow. Radiographic changes include erosions that are seen more often in confined joints such as the hip than in the more capacious joint capsule of the knee.

Pigmented villonodular synovitis is histologically characterized by synovial inflammation and macrophages containing lipids and hemosiderin. The MR appearance of PVNS can be directly related to the histologic components of the lesion, particularly hemosiderin and fat. The appearance of any particular lesion can be expected to vary, depending on the relative proportion of these components (Fig. 13C,D). The fat contained within the lesion can account for areas of high signal on T1-weighted images. Increased signal intensity on T1-weighted images may also result from subacute hemorrhagic areas in the synovium. Areas of decreased signal intensity result from the decreased T2 relaxation time created by the paramagnetic effects of hemosiderin and/or chronic synovial proliferation. The preferential T2 shortening secondary to hemosiderin deposition is particularly evident on high field strength (1.5 T) images and is well demonstrated by the large low intensity areas illustrated in the present case (arrows, Fig. 13B).

The differential diagnosis of PVNS includes synovial sarcoma, hemophilia, synovial chondromatosis, synovial hemangioma, and rheumatoid arthritis. Synovial chondromatosis is most often distinguished from PVNS on the basis of calcifications. Synovial sarcoma is often para-articular and may also contain calcifications although in a minority of cases. Synovial hemangioma, rheumatoid arthritis, and hemophilia may all be associated with hemorrhage and hemosiderin deposition but should be able to be differentiated on the basis of clinical findings.

C
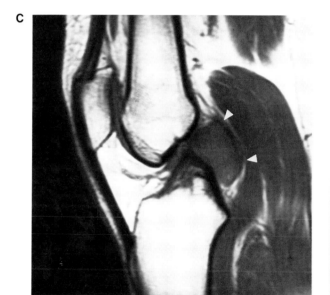

D
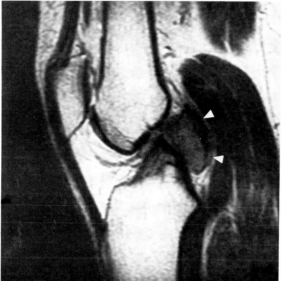

FIG. 13C,D. Pigmented villonodular synovitis. **C.** Sagittal TR/TE 2,000/20 msec. A relatively circumscribed mass of low signal intensity (*arrowheads*) is identified immediately posterior to the PCL. The findings are more localized than in the previous case. **D.** Sagittal TR/TE 2,000/80 msec. The mass (*arrowheads*) demonstrates no significant increase in signal intensity, a finding characteristic of PVNS.

CASE 14

A 14-year-old boy presented with a 3-week history of pain and swelling in the region of his mid tibia. The overlying skin was slightly warm to touch, and soft tissue swelling was present. The patient had a low grade fever. Describe the MR findings. Are they more suggestive of an inflammatory or neoplastic process? What is your diagnosis? (Fig. 14A. Coronal TR/TE 800/20 msec. Fig. 14B. Axial TR/TE 2,000/20 msec. Fig. 14C. Axial TR/TE 2,000/80 msec.)

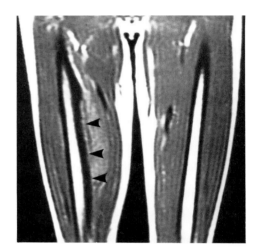

FIG. 14A.

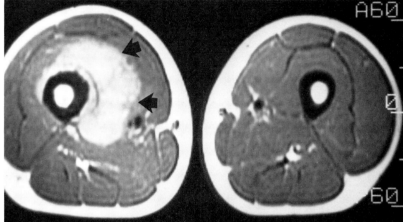

FIG. 14B.

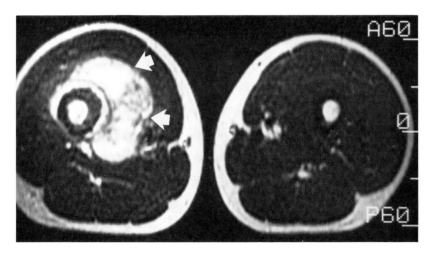

FIG. 14C.

Findings: On the coronal T1-weighted image, there is thickening of the low signal intensity band corresponding to the tibial diaphysis. The finding suggests periosteal reaction (arrowheads). On the axial intermediate and T2-weighted image, there is a relatively well-circumscribed soft tissue mass of intermediate to high signal (arrows). The mass is inhomogeneous in several areas. The cortex of the diaphysis is thickened and demonstrates inhomogeneous increased signal. No marked signal alteration of the medullary bone is evident.

Diagnosis: Ewing's sarcoma.

Discussion: Ewing's sarcoma is a relatively common malignant tumor that is usually identified during the first three decades of life, with the peak incidence in the second decade. There is a slight male predominance, and the tumor is exceedingly rare in Blacks. The patients usually experience localized pain and soft tissue swelling, and/or a discrete mass is usually evident by the time of clinical presentation. The presence of systemic symptoms including fever, leukocytosis, and increased erythrocyte sedimentation rate may simulate the clinical findings of infection.

Ewing's sarcoma most commonly involves the pelvis (sacrum, innominate bone) and bones of the lower extremity. The tumor is relatively uncommon within vertebral bodies above the sacrum as a primary lesion, although this site is frequently favored by metastatic disease (Fig. 14D,E). Within the vertebral column, Ewing's sarcoma is one of a small number of malignancies with a propensity to cross the disk space, mimicking the appearance of an infection. Although classically taught that involvement within long bones is predominantly diaphyseal, more recent experience suggests that the tumor is more commonly meta-diaphyseal in location. Epiphyseal extension may be seen in up to 10% of cases. On gross pathologic assessment, an associated extraosseous neoplastic component is present in the overwhelming majority of cases and commonly is larger than the intraosseous component of the neoplasm.

On conventional radiographs, common features include poorly marginated bone destruction (commonly permeative type), soft tissue mass, and laminated periosteal reaction. Osteosclerosis is not an uncommon feature and may predominate in some lesions, causing diagnostic confusion with osteosarcoma. The MR features of Ewing's sarcoma are not specific, and the lesion cannot be distinguished from other neoplasms based on imaging findings alone. The findings can be confused with osteomyelitis, particularly early when no large soft tissue mass is present (Fig. 14F,G). The principal use of MR is in the staging of the extent of involvement. Both T1-weighted and T2-weighted images are commonly utilized. T1-weighted images provide excellent contrast between tumor and marrow. There has been recent interest in the use of STIR sequences for assessment of marrow involvement, and these may demonstrate more extensive involvement than expected from routine spin-echo imaging (Fig. 14H,I) (also see Case 15). T2-weighted images or more recently gadolinium-enhanced T1-weighted images are utilized for assessment of soft tissue extension. It is important that the entire extent of the long bone be imaged to detect small skip lesions. Care must be taken to avoid mistaken diagnosis secondary to partial volume effects. It is often advantageous to perform oblique imaging in the plane of greatest interest (e.g., oblique coronal) to avoid this potential pitfall. This requires first obtaining a localizer image in an orthogonal plane but is often well worth the additional effort (Fig. 14J,K).

D

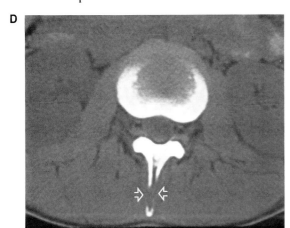

E

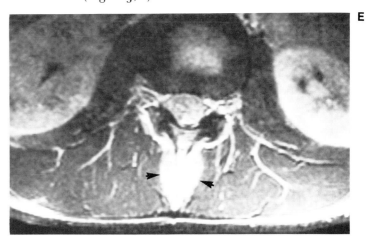

FIG. 14D,E. Metastatic Ewing's sarcoma. **D.** Axial CT section demonstrates a small area of bone destruction involving the spinous process (*arrows*). **E.** Axial TR/TE 2,000/60 msec. The soft tissue component of the metastasis is more graphically depicted (*arrowheads*). The spine is a common site of metastatic involvement in Ewing's sarcoma.

F

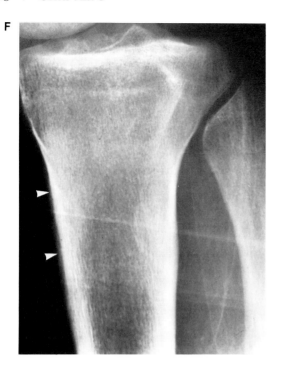

G

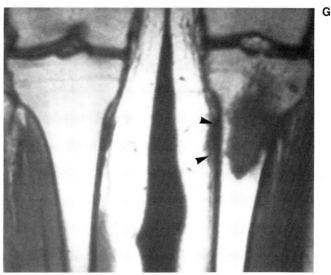

FIG. 14F,G. Osteomyelitis of the tibia. **F.** Plain radiograph of the knee of a young man presenting with knee pain of several weeks duration. There is subtle periosteal reaction (*arrowheads*). **G.** Coronal TR/TE 500/20 msec. There is a large low signal intensity mass replacing the normally high signal intensity marrow of the proximal tibia. There is associated periosteal reaction along the proximal medial tibia (*arrowheads*). The findings would be difficult to differentiate from malignancy based on the MR alone.

H

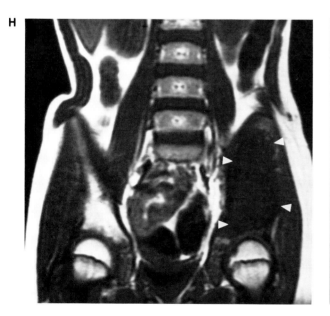

I

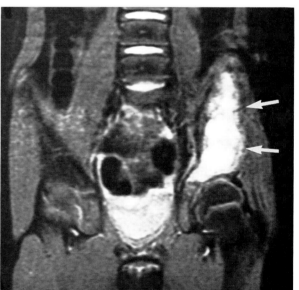

FIG. 14H,I. STIR imaging in Ewing's sarcoma. **H.** Coronal TR/TE 800/20 msec. The coronal T1-weighted image demonstrates the extensive marrow replacement involving the left ilium as marked decreased signal intensity compared to surrounding fatty marrow (*arrowheads*). **I.** Coronal TR/TE 1,500/30, TI 140 msec. On this STIR sequence the signal from marrow fat is nulled. The effects of T1 and T2 are additive. The tumor is seen as a focus of marked increased signal intensity (*arrows*). Both sequences demonstrate the tumor well in this case.

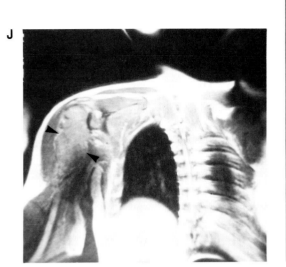

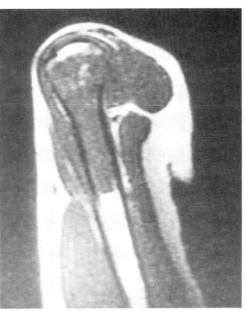

FIG. 14J,K. Ewing's sarcoma. Value of oblique imaging. **J.** Coronal TR/TE 2,000/30 msec. A large mass (*arrowheads*) is seen involving the proximal humerus. The distal extent is difficult to evaluate secondary to volume averaging. **K.** Oblique-coronal TR/TE 500/20 msec. The true extent of the lesion and its interface with normal bone is better demonstrated. The entire bone, however, must be imaged.

CASE 15

A healthy 25-year-old man was referred for an MR because of a bone scan abnormality. What are the most likely diagnosis and differential considerations? (Fig. 15A. Coronal TR/TE 800/20 msec. Fig. 15B. Axial TR/TE 2,000/80 msec.)

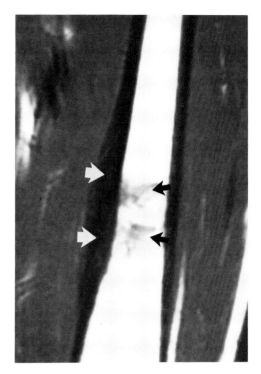

FIG. 15A.

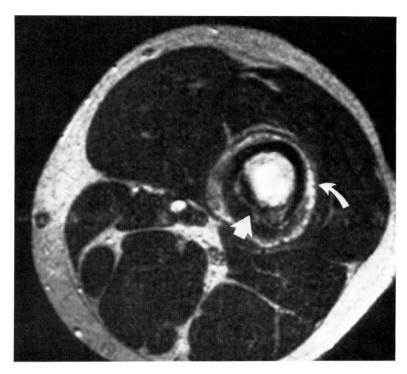

FIG. 15B.

Findings: There is a poorly defined area of decreased signal in the medullary canal of the left femur on the T1-weighted image (arrows). On the axial image there is nearly circumferential thickening of periosteal and cortical bone interleaved with linear areas of increased signal. A ring of increased signal is present in the soft tissues completely around the femur (curved arrow).

Diagnosis: Stress response.

Discussion: The patient is a police recruit in training who had pain in both legs, which led to a bone scan. Areas of increased activity were present in both proximal tibiae as well as in the left femoral lesion depicted above. Radiographs of the tibiae were normal. Although there is no histologic proof in this case, the history, clinical appraisal, bone scan, and MR were felt to be compatible with stress response. The patient has been followed for 6 months with decreasing symptomatology. A follow-up MR revealed a decrease in the soft tissue and bone marrow edema.

The origin of a stress fracture and the microscopic changes preceding it are discussed in Case 14 of Chapter 6. Most stress fractures appear as linear areas of decreased signal in the medullary bone, with a surrounding zone of edema. The lesion is perpendicular to the adjacent cortex and is meta-epiphyseal in location (Fig. 15C–E). Occasionally, only the marrow edema that precedes skeletal failure is present. The stress response in this case is somewhat unusual. The diaphyseal cortex is thickened and is "infiltrated" by increased signal on T2-weighted images. The increased signal, also present in the soft tissues, probably represents edema. The marked periosteal response, lack of a fracture line, soft tissue abnormalities, and extensive medullary component can create confusion of a stress response with infection or even malignancy. Sarcomas of bone may demonstrate bone and soft tissue changes somewhat similar to those seen in stress fractures, but tumors have soft tissue mass and rather little edema. Whereas the location of the lesion and the age of the patient might raise the question of a periosteal osteosarcoma, the lack of a mass and the clinical history of abnormal stresses overwhelmingly suggest the diagnosis of a stress response.

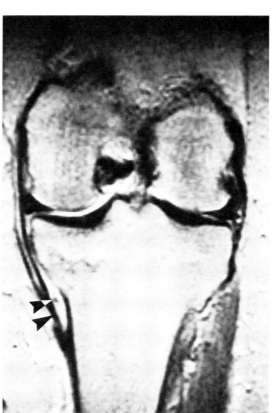

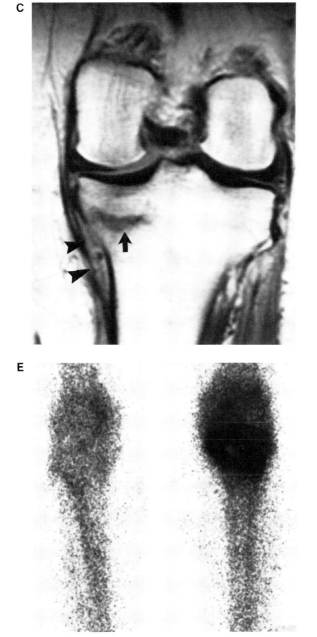

FIG. 15C–E. Stress fracture. **C.** Coronal TR/TE 2,000/30 msec. The typical appearance of a stress fracture as depicted on MR is illustrated. A low signal intensity line perpendicular to the adjacent cortex is seen crossing the proximal tibia (*arrow*). Overlying soft tissue swelling is identified (*arrowhead*). **D.** Coronal TR/TE 2,000/60 msec. On the T2-weighted image, the area surrounding the fracture line increases in signal intensity, suggesting edema. Edema in the overlying soft tissues is again noted (*arrowheads*). **E.** Radionuclide bone scan. There is diffuse increased activity within the proximal tibia. The scan is nonspecific in contrast to the MR, which provided fracture line depiction.

CASE 16

A 20-year-old man presented with soft tissue swelling above his left knee. Routine radiographs demonstrated an expansile lesion that was considered suspicious for malignancy. Magnetic resonance was performed to evaluate the extent of the lesion. Is the mass more likely benign or malignant? Can you make a diagnosis based on these images? (Fig. 16A. Axial TR/TE 2,000/60 msec.)

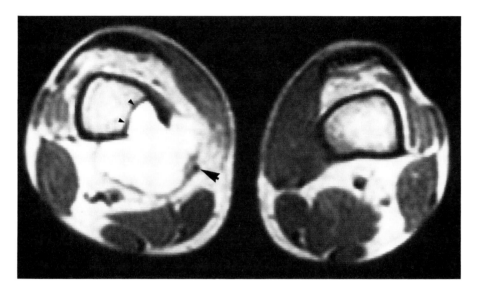

FIG. 16A.

Findings: On this T2-weighted image, there is a well-circumscribed high signal intensity mass extending through the medial femoral cortex into the adjacent soft tissues. There is a low signal intensity rim separating the lesion from the normal signal intensity marrow (small arrowheads). The margins are well defined. An adjacent vessel is displaced (large arrowhead) rather than encased, a finding suggesting a benign lesion.

Diagnosis: Aneurysmal bone cyst.

Discussion: Aneurysmal bone cysts are expansile nonneoplastic lesions characterized by thin-walled blood-filled cavities. The lesions are most commonly encountered in the first three decades with 80% of patients under 20 years of age. The spine and long tubular bones are the most common sites of occurrence. Within long tubular bones, aneurysmal bone cysts are nearly exclusively metaphyseal in location. The classic radiographic appearance is that of an eccentric, expansile, osteolytic mass centered within the metaphysis of a long bone. Vertebral involvement is most common in the posterior elements. Extension from one vertebral body to another can mimic infection.

Aneurysmal bone cysts share common histologic features with reactive processes such as giant cell reparative granuloma and hyperparathroidism. Aneurysmal bone cysts are seen in association with a variety of benign skeletal processes, and some authorities believe that they principally occur as a secondary event in association with another osseous process including trauma. Patients most commonly present with pain and swelling. The typical aneurysmal bone cyst consists of a meshwork of cysts of varying size. The cysts may contain blood and are separated by thin fibrous septae between which are located the variable solid components of the lesion that frequently contain multinucleated giant cells. The MR appearance of aneurysmal bone cyst has been well described and parallels its histologic features. A well-defined low signal intensity line characteristically surrounds the lesion at both the bone-tumor and soft tissue-tumor interface (Fig. 16A). This well-defined margin corresponds to the periosteal outer margin and has been described as an indicator of a benign process. Multiple well-defined cystic spaces may be defined, demonstrating a wide range of signal intensities. Areas of increased signal are often present on both T1- and T2-weighted images, reflecting subacute hemorrhage. Fluid/fluid levels within individual cysts may also be seen, as has been described utilizing CT (Fig. 16B,C).

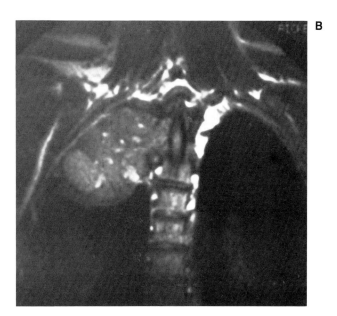

B

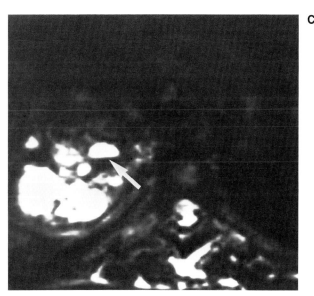

C

FIG. 16B,C. Aneurysmal bone cyst. **B.** Coronal TR/TE 2,000/80 msec. A large partially circumscribed mass projects into the chest. The mass arises from the posterior costovertebral junction. Punctate foci of high signal intensity are identified. **C.** Axial TR/TE 2,000/80 msec. Multiple areas of markedly high signal intensity are present. A fluid/fluid level is present (*arrow*). (Case courtesy of R. Lachman, MD, Los Angeles, California.)

CASE 17

A 20-year-old man presented with pain and swelling in his right knee of several weeks duration. What is the most likely diagnosis? Is there intra-articular extension? (Fig. 17A. Coronal TR/TE 2,000/60 msec. Fig. 17B. Axial TR/TE 2,000/60 msec.)

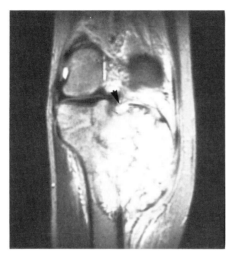

FIG. 17A.

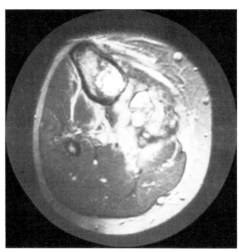

FIG. 17B.

Findings: There is an extensive high signal intensity mass within the proximal tibial metaphysis that extends to involve the epiphysis and appears to violate the joint space at the level of the tibial spine (arrowhead). The mass extends beyond the posterior lateral cortex, and a large high signal intensity soft tissue component is present.

Diagnosis: Osteosarcoma.

Discussion: Osteosarcoma is the second most common primary malignant neoplasm of bone, following only myeloma in incidence. The most common sites of involvement are the metaphyseal regions of the tubular bones of the extremities with 50% to 75% of all cases developing in the osseous structures about the knee. The peak incidence is in the second decade, and the tumor is more common in boys and men. The tumor may contain osteoid, fibroid, or chondroid elements but must contain osteoid-producing cells to be classified as an osteosarcoma.

Conventional radiographs usually suffice for lesion identification. The radiographic appearance is variable and depends on the degree of bone production. A mixed pattern consisting of both osteolysis and osteosclerosis is most common, with purely osteolytic (telangiectatic osteosarcoma) and osteosclerotic lesions less commonly encountered. Although it has long been considered that the open physis represented a barrier to epiphyseal extension, microscopic studies have suggested that transphyseal extension is not uncommon. Epiphyseal extension not evident on radiographs has repeatedly been documented on MR. Whereas most tumors exhibit the common pattern of decreased signal on T1-weighted images and increased signal on T2-weighted images, specific histological components (e.g., chondroid, fibroid) may affect the signal intensity. For example, telangiectatic osteosarcomas, as a consequence of hemorrhagic components, may demonstrate areas of increased signal on both T1- and T2-weighted images. The signal intensity also varies with the amount of osteoid production. Areas of sclerosis on radiographs will be seen as low intensity regions on both spin-echo pulse sequences (Fig. 17C).

Accurate staging is particularly important with the current approach to therapy. Excellent correlation has been reported between the extent of marrow involvement as depicted by MR and gross pathologic specimens. It is essential to image the entire extent of the affected extremity to exclude "skip" lesions, which have been reported in up to 25% of cases. Preoperative tumor response to chemotherapy can be monitored on MR (Fig. 17D,E). Before treatment, CT examinations of the chest are performed in conjunction with MR of the extremity to exclude pulmonary metastasis.

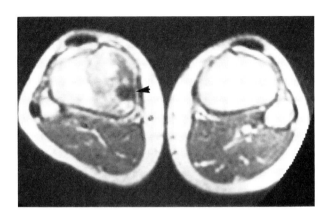

FIG. 17C. Osteogenic sarcoma (axial TR/TE 2,000/30 msec). A circular low signal intensity focus (*arrow*) corresponds to an osteoblastic area on the conventional radiograph.

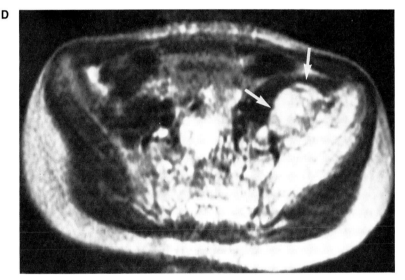

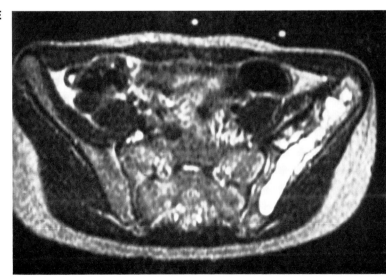

FIG. 17D,E. Ewing's sarcoma. Response to therapy. **D.** Axial TR/TE 2,000/80 msec. Motion degrades image quality. There is extensive increased signal intensity within the ilium with a large soft tissue mass (*arrows*). The patient underwent chemotherapy before planned surgical resection. **E.** Axial TR/TE 2,000/80 msec. Postchemotherapy MR demonstrates a marked decrease in the extraosseous component of the tumor. Extensive abnormal signal remains within the ilium and represented tumor at surgery.

CASE 18

A 60-year-old man presented with the recent onset of knee pain following a fall. What is your diagnosis? What are some of the complications associated with this condition? (Fig. 18A. Coronal TR/TE 500/20 msec. Fig. 18B. Coronal TR/TE 2,000/30 msec. Fig. 18C. Coronal TR/TE 2,000/60 msec.)

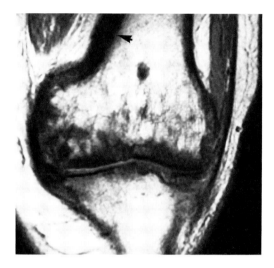

FIG. 18A.

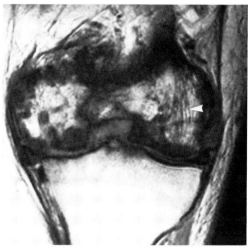

FIG. 18B.

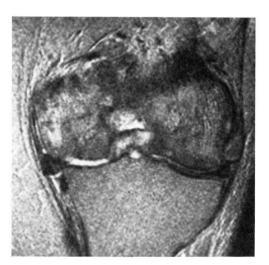

FIG. 18C.

Findings: There is marked medial compartment narrowing with virtual destruction of the medial meniscus and bone-on-bone apposition of the tibia and femur. There are mixed areas of increased and decreased signal intensity predominantly seen within the femoral condyles. There is marked thickening of the femoral cortex (arrowhead, Fig. 18A), and the trabeculae are prominent (arrowhead, Fig. 18B). There is mild increased signal within the subchondral bone of the distal medial condyle (Fig. 18C). No soft tissue mass is present.

Diagnosis: (1) Paget's disease; (2) osteoarthritis.

Discussion: Paget's disease is a common disorder of middle-aged and elderly patients, characterized by abnormal remodeling of bone. The condition most commonly affects the axial skeleton and the proximal long bones, particularly the femur. The condition evolves through various phases of activity characterized by an initial osteolytic phase, with marked resorption of normal bone followed by an osteoblastic phase that produces bone that is architecturally abnormal. Osteoclastic activity declines after a variable time, followed by eventual cessation of osteoblastic activity. The quiescent phase is characterized by thickened and sclerotic bone. The conventional radiographic features are virtually pathognomonic, and the pathologic stages described have their radiographic counterparts (Fig. 18D).

On MR, the abnormal woven bone of Paget's disease appears as areas of low signal intermixed with marrow fat. The resultant appearance can be markedly inhomogeneous and potentially confused with neoplasm if not compared with radiographs. Trabecular prominence and cortical thickening can be observed and can assist in establishing the correct diagnosis.

The role of MR in the evaluation of Paget's disease has not been clearly established but likely would involve assessment of the various musculoskeletal complications of this condition. Pathologic fractures are the most common orthopedic complication of Paget's disease. Insufficiency fractures are most often noted in the lower extremities, particularly the femur and tibia. Sarcomatous degeneration may occur in approximately 1% of patients with Paget's disease. The histology is most often osteosarcoma or fibrosarcoma, and the prognosis is poor. Degenerative joint disease, most commonly affecting the hip and knee, is a common articular complication of Paget's disease. Neurologic complications are encountered and may result from compression of the spinal cord secondary to changes at the base of the skull and within the spine. These would be assessed to advantage with MR.

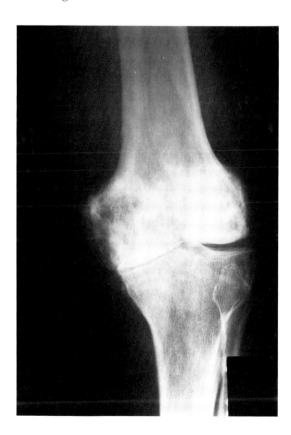

FIG. 18D. Conventional radiograph demonstrating the typical features of Paget's disease. There is marked medial compartment narrowing consistent with advanced osteoarthritis.

CASE 19

A 43-year-old man presented with progressive swelling in the region of his right hip and buttock, which had slowly increased during the past year. What is the most likely diagnosis? What are the low signal intensity areas within the lesion? (Fig. 19A. Axial TR/TE 2,000/60 msec.)

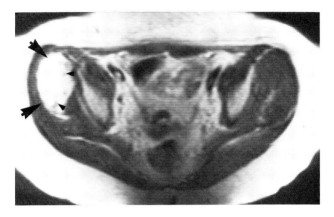

FIG. 19A.

Findings: An axial section through the pelvis at the superior acetabular level demonstrates a moderately large, relatively circumscribed high signal intensity mass within the gluteal muscles (large arrowheads). The mass is contiguous with the underlying bone. There are multiple circular to ovoid areas of low signal intensity within the mass (small arrowheads).

Diagnosis: Chondrosarcoma.

Discussion: Chondrosarcomas are approximately half as common as osteosarcomas and occur within adults and elderly patients. Pain is the most common presenting symptom, and the average duration of symptoms before presentation is 1 to 2 years. The long tubular bones (femur in particular) and the innominate bone account for nearly 75% of all cases. Chondrosarcomas are commonly categorized according to their location (central, peripheral, juxta-cortical), degree of histological differentiation (grade 1–4), and presence of unusual characteristics (e.g., clear cell, mesenchymal).

The tumors tend to be bulky and are often associated with large soft tissue components. The most extensive tumors involve the flat and irregular bones (innominate, scapula, ribs) where they can grow to large size before presentation. Histologically, chondrosarcomas contain proliferating cartilage with myxoid elements and/or cysts common in some lesions. Well-organized calcific rings are seen within cartilage and are commonly associated with lower grade tumors. High grade sarcomas often contain a greater degree of myxoid material and demonstrate large areas of noncalcified matrix.

On MR, the appearance of chondroid matrix tumors reflects their underlying histology and degree of cellularity. Most lesions demonstrate decreased signal on T1-weighted images and marked increased signal on T2-weighted images reflecting the water content of the matrix. Areas of calcification may be seen as foci of low signal intensity (Fig. 19A, arrowheads) but often are difficult to recognize on MR.

Skeletal lesions that involve flat bones such as the ilium, scapula, and ribs and lesions that contain calcification or predominant cortical destruction on conventional radiographs are often studied more precisely and their histological features predicted more accurately with CT. Figure 19B is from a CT examination on the same patient accomplished at a slightly higher level. The soft tissue calcification (arrowheads) and cortical involvement are readily detected and are useful in predicting the histological features of the lesion. The soft tissue extent of the tumor is better demonstrated on MR but otherwise the MR study lacks specificity. In this case, the CT examination was adequate, and the MR was, at most, complementary.

The prognosis for patients with chondrosarcoma correlates with the histological grade of the lesion. Treatment of chondrosarcoma differs from that of osteosarcoma because of the general slow clinical evolution and tendency to metastasize late. Local resection, often radical, is the treatment of choice and underscores the need for precision in preoperative assessment of tumor extent.

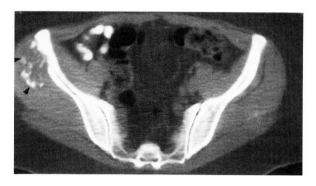

FIG. 19B.

CASE 20

A 10-year-old boy presented with dull aching pain in his left hip of several months duration. The pain increased during ambulation, and he had a detectable limp. Is the hip joint normal? What is the likely diagnosis? What other techniques might be useful for further evaluation in this setting? (Fig. 20A. Axial TR/TE 500/20 msec. Fig. 20B. Coronal TR/TE 500/20 msec.)

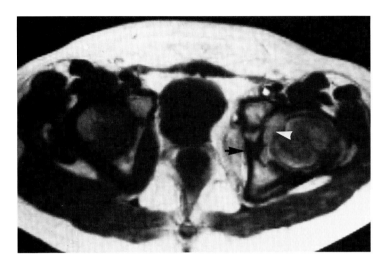

FIG. 20A.

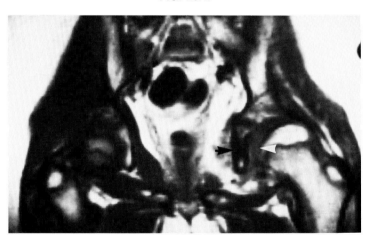

FIG. 20B.

Findings: On both the axial and coronal images, there is widening of the joint space on the left. There is a mass of intermediate signal intensity, suggesting synovial proliferation (white arrowheads). There is an associated small circumscribed low intensity area within the acetabulum medially (black arrowheads).

Diagnosis: Osteoid osteoma.

Discussion: Osteoid osteomas are benign osteoblastic tumors that are composed of a circumscribed region of new bone formation with a highly vascularized stroma of osteogenic connective tissue. The central nidus may be within the cortex, cancellous bone, or subperiosteal in location. Intracortical lesions are most common, and the central nidus is usually less than 1 cm in diameter and is surrounded by extensive sclerotic new bone. The lesions are most commonly encountered in patients less than 25 years of age and are more frequent in boys and men. Pain, often dull and aching in nature, is the clinical hallmark of this condition. This pain is often worse at night and frequently responds to small doses of salicylates. The most common sites of involvement are the proximal femur, tibia, and posterior elements of the spine.

Intra-articular osteoid osteomas, as illustrated in this case, are uncommon and may present with clinical manifestations, suggesting a primary articular disorder. In addition to pain, soft tissue swelling and a joint effusion may contribute to restriction of joint motion. A lymphofollicular synovial response may develop and can lead to irreversible joint destruction. Diagnosis is often delayed.

The classic radiographic features are well known and consist of a centrally located radiolucent area (nidus) measuring less than 1 cm surrounded by a zone of intense sclerosis. The lesions, however, are commonly subtle, and radionuclide bone scans and tomography have been successfully utilized to establish a diagnosis. On MR the area of sclerosis demonstrates decreased signal intensity on T1-weighted images. Associated cortical thickening may be observed. The nidus, as in this case, is often not detected but when demonstrated reveals an intermediate signal intensity on T1-weighted images that increases in signal intensity on T2-weighted images. With intra-articular lesions, joint effusions and areas of synovial proliferation may be recognized (arrows, Fig. 20A,B).

The MR characteristics of osteoid osteoma are less specific than features demonstrated with CT (Fig. 20D) and conventional radiographs. The reaction created by these tumors may cause large areas of marrow abnormality, which can be confused with infection and malignancy (Fig. 20C). For these reasons we do not presently prefer MR as the primary technique for localizing or confirming the diagnosis. It is obviously important, however, to be familiar with the appearance of osteoid osteoma on MR.

C
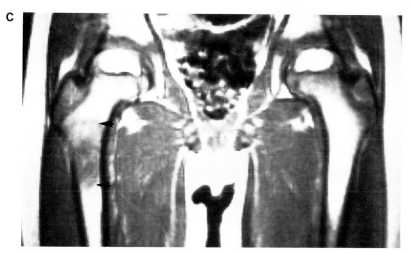

D
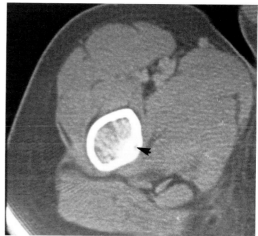

FIG. 20C,D. Osteoid osteoma. **C.** Coronal TR/TE 500/20 msec. A large region of decreased signal intensity (*arrowheads*) is seen within the femoral neck. The findings are entirely non-specific and could represent osteomyelitis, reactive change from a stress fracture, or malignancy. **D.** The axial CT demonstrates the characteristic features of osteoid osteoma and allows precise localization of the tumor nidus (*arrowhead*).

CASE 21

A 50-year-old man presented with peripheral neuropathy. Routine radiographs of the spine and pelvis were negative. A monoclonal protein peak was noted on laboratory studies. (Fig. 21A. Coronal TR/TE 500/20 msec. Fig. 21B. Sagittal TR/TE 500/20 msec.)

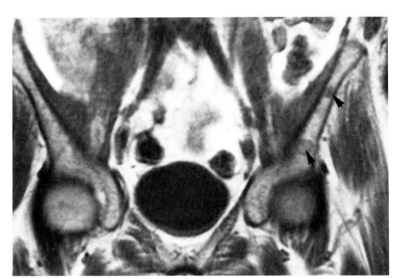

FIG. 21A.

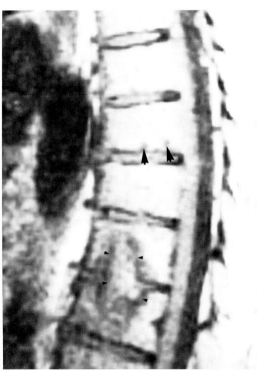

FIG. 21B.

Findings: On the sagittal section of the mid dorsal spine (Fig. 21B), two discrete low signal intensity foci are seen within the inferior aspect of the T9 vertebral body (large arrowheads). These did not change in signal intensity on the T2-weighted image (not illustrated). Postoperative changes are seen at the T11–12 disk space (small arrowheads) (not related). On the coronal section through the pelvis Fig. 21A, additional small foci of decreased signal intensity are identified (arrowheads).

Diagnosis: Plasma cell myeloma (osteosclerotic-type).

Discussion: Multiple myeloma was the most commonly encountered neoplasm in Dahlin and Unni's series of 8,542 cases; representing 43% of all primary malignant bone tumors seen at the Mayo Clinic. Patients with myeloma are usually older (50–70 years), and the disease is slightly more common in men. Most patients demonstrate multiple or diffuse areas of skeletal involvement, although solitary lesions (plasmacytoma) occur in approximately 5% of cases. Interestingly, there is a striking male predominance in plasmacytoma, and myeloma overall is particularly common among Blacks. Bone pain represents the most persistent and incapacitating symptom. Systemic manifestations include fever, bleeding, weight loss, and neuropathy. Coexistent amyloidosis is seen in 15% of patients with plasma cell myeloma and is considered responsible for many of the articular manifestations of the disease.

The radiographic features of myeloma have been well described. The predominant pattern is that of osteolysis, with multiple small radiolucent "punched-out" lesions seen principally within the axial skeleton. The lesions are characteristically similar in size, a feature useful in differential diagnosis from metastatic disease. Diffuse skeletal osteopenia without discrete areas of bone destruction can also be observed. Osteosclerotic myeloma is being detected with increasing frequency. Seventy-five percent of patients with this form of myeloma present with peripheral neuropathy and sclerotic skeletal lesions. The radiographic findings are often subtle and can be overlooked, particularly when clinical history is not available (Fig. 21C). When sclerotic lesions are more diffuse, diagnostic confusion with osteoblastic metastasis, myelofibrosis, and renal osteodystrophy can occur. Patients with osteosclerotic myeloma and neuropathy commonly have a more indolent course.

The limitations of radionuclide examinations in detection of myeloma have been generally recognized (Fig. 21D). Magnetic resonance, in contrast, is particularly well suited to depiction of marrow abnormalites and is an excellent method for evaluation of patients with myeloma. The lesions of classic osteolytic myeloma demonstrate the typical pattern of decreased signal intensity on T1-weighted images and increased signal intensity on T2-weighted images. Marrow involvement is also displayed to advantage, utilizing STIR sequences in which the lesions will demonstrate increased signal intensity contrasted against the low signal of suppressed marrow fat. Plasma cell myeloma may also present as marked inhomogeneity of the marrow on spin-echo sequences without discrete or confluent lesions. In this setting, differentiation from focal fatty deposition within the marrow can be difficult, and special techniques including chemical shift imaging can be employed. Osteosclerotic myeloma most often demonstrates decreased signal on all pulsing sequences; an imaging feature also demonstrated by the majority of osteoblastic metastasis. Computed tomography is particularly useful in demonstrating these small areas of involvement (Fig. 21E).

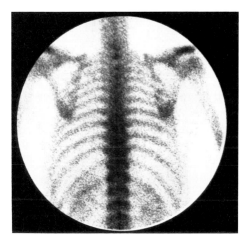

FIG. 21D. Radionuclide bone scan. Posterior view of the thoracic spine obtained in the patient illustrated in the case. The study is normal.

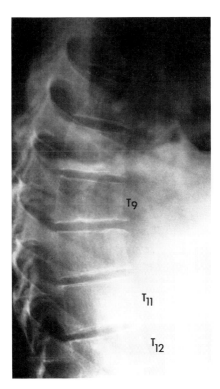

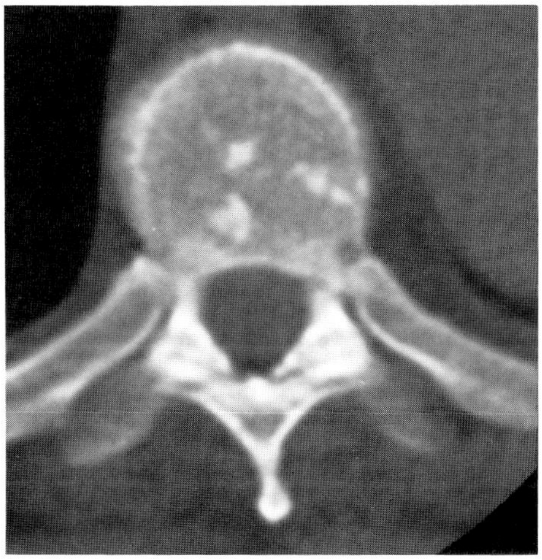

FIG. 21E. Osteosclerotic myeloma. Axial CT scan through the T9 vertebral body of the patient illustrated in this case. The small sclerotic lesions are well depicted.

FIG. 21C. Osteosclerotic myeloma. Plain radiograph of the thoracic spine of the patient illustrated in this case. The lesions are radiographically occult.

CASE 22

A 13-year-old boy presented with bilateral knee pain. His age and the relationship to sporting activities suggested the diagnosis of Osgood-Schlatter disease clinically. What is the diagnosis? Is the signal intensity within the distal femurs within normal limits for the patient's age? (Fig. 22A. Anteroposterior radiograph. Fig. 22B. Coronal TR/TE 500/30 msec.)

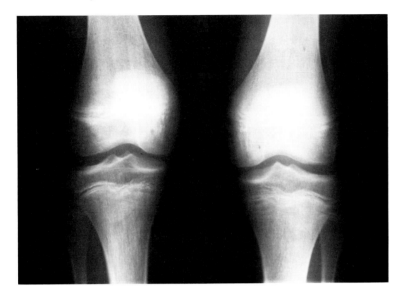

FIG. 22A.

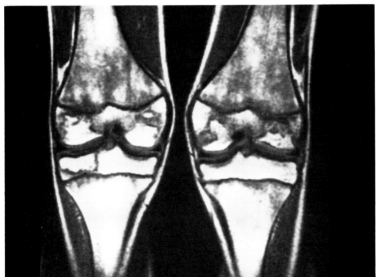

FIG. 22B.

Findings: On the coronal T1-weighted image of both knees, there are marked foci of decreased signal intensity with the femoral epiphysis bilaterally. No subchondral collapse is noted. There is striking inhomogeneous decreased signal within the distal metaphysis of both femurs. Only minimum foci of decreased signal are seen within the proximal tibiae.

Diagnosis: (1) Osteonecrosis; (2) leukemia.

Discussion: Magnetic resonance has become rapidly established as the premier imaging method for depiction of abnormalities of bone marrow. Knowledge of the normal appearance and distribution of red (hematopoietic) and yellow (fatty) marrow is fundamental to accurate diagnosis. In the fetus, bone marrow is entirely hematopoietic. After birth, conversion to fatty marrow begins in the distal extremities, followed by the epiphyses and midshaft of the long bones. By early adulthood, hematopoietic marrow is confined to the axial skeleton, proximal humeri, femoral neck, and intertrochanteric regions. On spin-echo imaging, hematopoietic marrow demonstrates uniform low signal intensity on T1-weighted images and low to intermediate signal on T2-weighted images. Yellow marrow, in contrast, demonstrates increased signal on T1-weighted images, reflecting its predominant aliphatic composition. On T2-weighted images, yellow marrow demonstrates intermediate to high signal. On gradient-echo sequences, there is poor discrimination between red and yellow marrow. STIR sequences are designed to suppress the signal from yellow marrow, which consequently appears low in intensity.

Under conditions of stress, yellow marrow can undergo reconversion to hematopoietic marrow. The yellow marrow elements do not themselves actually convert, but rather the conditions permit for hyperplasia of the existing hematopoietic elements. This phenomenon is most commonly observed in chronic anemias (particularly hemolytic types) and infiltrative marrow disorders both benign and malignant, such as Gaucher's disease, myelofibrosis, and myeloma (Fig. 22C). Recently, the presence of focal deposits of hyperplastic hematopoietic marrow in the distal femur has been reported as an incidental finding in patients undergoing knee MR examination for suspected internal derangement. Fatty marrow "reconversion" begins proximally and extends distally, the reverse of physiological marrow conversion.

The use of MR in assessment of leukemia has been described. The MR appearance of marrow abnormalities will depend on whether the marrow in the affected body part is red or yellow. On routine T1-weighted images, infiltration of fatty marrow appears as decreased signal intensity compared to the high signal of normal fatty marrow. This signal often increases in signal intensity on T2-weighted images. When lesions involve sites of normal hematopoietic marrow they are often isointense and thus less conspicuous. In children with leukemia, the vertebral marrow demonstrates decreased signal on T1-weighted images that increases in signal on T2-weighted images. Because the normal marrow of children has decreased signal on T1-weighted images, the findings alone are not diagnostic. Measurements of T1 relaxation times can overcome this limitation and separate normal from pathologically involved marrow. T1 relaxation times have also been of value in assessment of patients in remission.

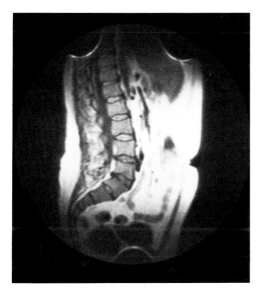

FIG. 22C. Gaucher's disease (sagittal TR/TE 600/20 msec). There is mild uniformly decreased signal intensity from all the vertebrae in this patient with known Gaucher's disease. The findings represent extensive infiltration, but their uniformity may allow them to go undetected.

CASE 23

A 12-year-old boy presented with leg pain after trauma. A routine radiograph showed a lucent lesion in the mid fibula. Is this malignant? What is the most likely diagnosis? (Fig. 23A. Sagittal TR/TE 500/20 msec. Fig. 23B. Sagittal TR/TE 2,000/30 msec.)

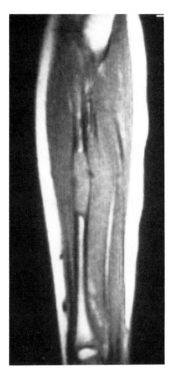
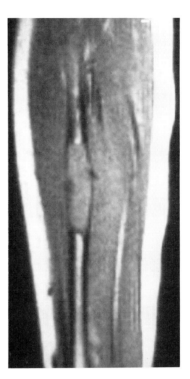

FIG. 23A.   **FIG. 23B.**

Findings: Sagittal spin-echo images show a homogeneous lesion in the mid fibula with signal intensity similar to that of muscle. The lesion causes expansion and thinning of the cortex. There is no soft tissue mass.

Diagnosis: Fibrous dysplasia.

Discussion: Lesions of fibrous dysplasia may be solitary or multiple (polyostotic). Of the 471 cases reported by Dahlin and Unni, 418 (89%) were solitary lesions, as in this patient. The polyostotic form may be accompanied by skin pigmentation and endocrine abnormalities—Albright's syndrome.

The radiographic features of fibrous dysplasia usually are sufficiently characteristic that further studies are not required (Fig. 23E). However, in atypical cases or when symptomatic lytic lesions are noted, further evaluation may be necessary. In bones with little marrow and thin cortex, CT may be preferable to MR. However, in this patient, MR was requested, and the findings were characteristic of fibrous dysplasia. The MR appearance of fibrous dysplasia is dependent on the predominance of the various tissues that compose a given lesion. When hypocellular fibrous tissue, some of which is poorly mineralized, is the dominant component, MR will demonstrate decreased signal intensity on both T1- and T2-weighted images. Cartilaginous tissue has also been found in fibrous dysplastic lesions, a reflection of their hamartomatous nature. Cartilage has a rather long T2 value secondary to its high water content. The MR signal is therefore quite intense on T2-weighted image. In this case, MR shows an expanding lesion with intermediate signal intensity on both short and intermediate TR sequences. There is no soft tissue involvement. These findings are typical of fibrous dysplasia in the ribs or fibula.

C

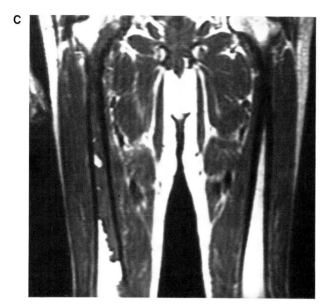

D

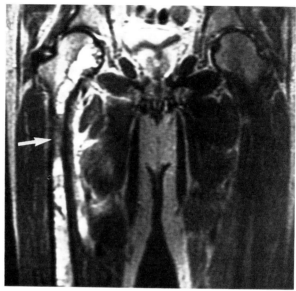

E

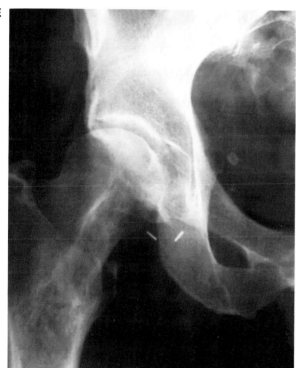

FIG. 23C–E. Fibrous dysplasia. **C.** Coronal TR/TE 2,000/20 msec. There is an elongate lesion within the right femur that is of low signal intensity on this sequence. No soft tissue component is identified. **D.** Coronal TR/TE 2,000/80 msec. Most of the lesion demonstrates marked increased signal intensity on the T2-weighted image, suggesting that there is a large amount of cartilage in the lesion. A single focus of presumed fibrous tissue remains of low signal (*arrow*). **E.** Anterposterior radiograph of the right hip in this patient. The proximal extent of the lesion is seen and has a characteristic appearance of fibrous dysplasia.

CASE 24

A 20-year-old man presented with the recent onset of pain in a previously painless hard lump behind his knee. What is the diagnosis? (Fig. 24A. Axial TR/TE 1,000/40 msec. Fig. 24B. Sagittal TR/TE 1,000/60 msec.)

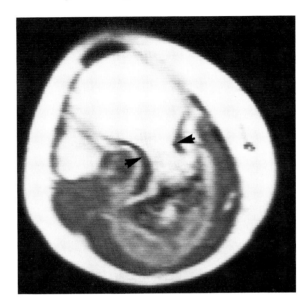

FIG. 24A.

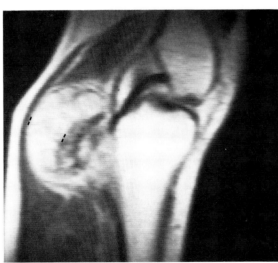

FIG. 24B.

Findings: Axial and sagittal images show a large lesion in the upper tibia extending from the marrow (arrowheads). The proximal portion of the mass is of low intensity and the peripheral cap is of higher intensity than normal cartilage. The cap measured 3 cm on the sagittal image (dark lines).

Diagnosis: Chondrosarcoma, grade 1.

Discussion: Osteochondromas are common benign lesions of the skeleton that represent 36% of all benign bone tumors. Men are affected more than women. The majority of lesions are discovered before the age of 30.

Symptoms are related to the effect that osteochondromata have on adjacent structures. Pressure effect on nerves or on arteries (resulting in development of a pseudoaneurysm) may necessitate resection. Malignant degeneration occurs in less than 1% of solitary osteochondromata, and transformation must be considered in patients with pain or growth of the lesion following closure of the nearest physis. Fracture or infarction, however, may also cause pain.

Osteochondromas have a characteristic radiographic appearance. The mass is continuous with the adjacent cortex and medullary bone, with variable degrees of calcification or ossification in the pedunculated portion of the lesion. These features are also evident on MR and CT.

The cartilaginous cap of an osteochondroma usually is only several millimeters thick (Fig. 24C). However, thicknesses of up to 3 cm have been reported during the years of active growth. There is controversy regarding the value of measuring the cartilage cap as a method of predicting malignant degeneration. Generally, when the cap is larger than 2 cm, the patient should be followed, and resection should be considered in symptomatic patients. Magnetic resonance is very effective in evaluating the thickness of cartilage; even thin caps, measuring 3 mm, can be detected.

The importance of increased signal intensity (more than normal cartilage) is unclear. However, in our experience, when the signal intensity is increased and the cap is larger than 2 cm, a low grade chondrosarcoma should be considered, especially if there is a history of recent onset of pain.

In this setting, the MR features and clinical symptoms led to resection, and the diagnosis of low grade chondrosarcoma was established. Resection should be contemplated in patients who have pain, enlarging lesions, or radiographic features suggesting malignancy.

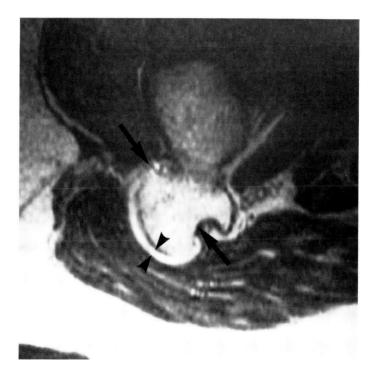

FIG. 24C.

CASE 25

A 52-year-old man presented with localized pain and swelling in his lower leg. He had no constitutional symptoms. What are the differential considerations? (Fig. 25A. Conventional radiograph of the femur; Fig. 25B. Axial TR/TE 2,000/20 msec. Fig. 25C. Axial TR/TE 2,000/80 msec.)

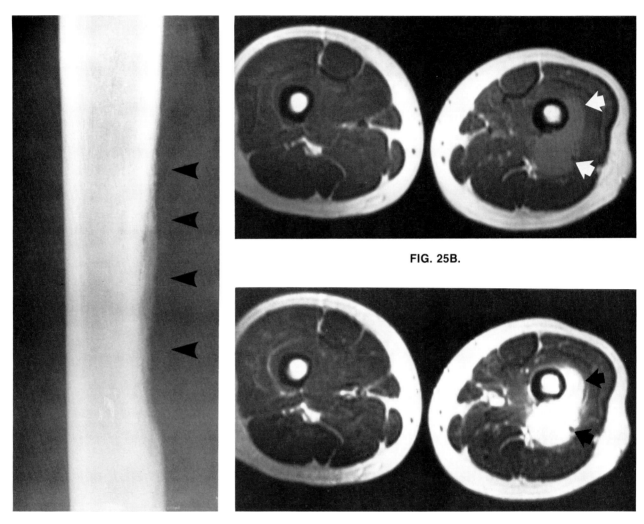

FIG. 25A.

FIG. 25B.

FIG. 25C.

Findings: On the plain radiograph, there is a soft tissue mass and evidence of saucerization of the cortex of the posterior femoral diaphysis (arrowheads). On the axial proton-density image, there is a relatively homogeneous and circumscribed elliptical-shaped soft tissue mass immediately contiguous with the cortex of the femur (arrows). The marrow signal is grossly normal. On the T2-weighted image, there is marked increased signal from the soft tissue mass (arrows).

Diagnosis: Non-Hodgkin lymphoma (primary), with no other organ systems involved.

Discussion: In contrast to leukemias, which represent myeloproliferative neoplasms of the reticuloendothelial system, lymphomas constitute lymphoreticular neoplasms and arise from lymphocytic, reticulum, or primitive precursor cells. Lymphomas are commonly classified as non-Hodgkin or Hodgkin lymphomas, although more currently adopted systems emphasize cell type (e.g., lymphocytic) and degree of differentiation.

Lymphoma may involve bone secondarily as a result of hematogenous dissemination or may arise as a primary process. On rare occasion the tumor may arise in soft tissue and involve bone by direct extension. The most common lymphoreticular neoplasm to involve bone primarily is diffuse histiocytic lymphoma. Skeletal involvement with histiocytic lymphoma is nearly twice as common (21% versus 12%) as with lymphocytic lymphoma. In general, bone involvement, either hematogenously or by direct spread, is more commonly observed the more immature the cell line comprising the tumor is. Although skeletal involvement is common in all of the lymphomas, conventional radiographs underestimate the extent of disease compared with autopsy studies. As prognosis is directly related to extent of disease, accurate staging is essential. Primary osseous involvement with non-Hodgkin lymphoma is considered stage 1, whereas skeletal involvement associated with disease in other sites is considered stage 4 disease. In addition to detection and staging, MR may be of value in assessing the response to therapy (Fig. 25D).

Involvement of bone in non-Hodgkin lymphoma is much more commonly a reflection of dissemination than primary involvement. With disseminated disease, involvement of the axial skeleton predominates and most often is the result of hematogenous spread. With primary skeletal non-Hodgkin lymphoma, the appendicular skeleton is the most common site of occurrence with a predeliction for the lower extremities. With this form, systemic symptoms are characteristically absent, and patients typically present with pain and swelling. The most common appearance on radiographs is that of an osteolytic metaphyseal lesion with poorly defined margins.

The MR appearance of lymphoma shares many imaging characteristics with other marrow infiltrative lesions. The marrow is of decreased signal intensity on T1-weighted images and demonstrates moderate increased signal on T2-weighted images (Fig. 25E,F). Involvement of the skeleton with diffuse lymphoma, however, is often more nodular or patchy than with other infiltrative processes, and this pattern may allow the diagnosis to be suggested. With primary lymphoma of bone, the presence of a soft tissue mass is common, and the lesion is difficult to differentiate from other primary osseous sarcomas based on imaging characteristics alone.

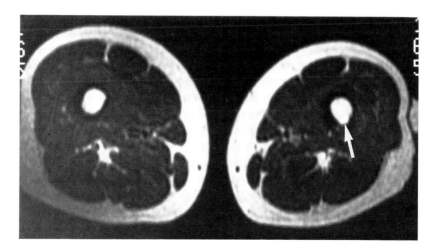

FIG. 25D. Six-month follow-up exam of patient illustrated in this case (axial TR/TE 2,000/80 msec). There has been marked interval resolution of the soft tissue component. A small focus of increased signal intensity remains within the posterior cortex (*arrow*).

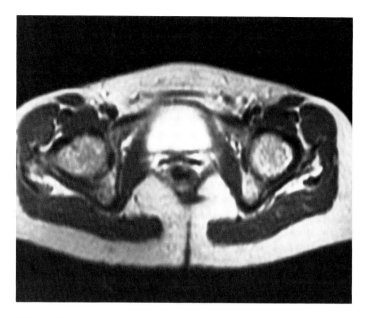

FIG. 25E. A 35-year-old man with non-Hodgkin lymphoma (axial TR/TE 2,000/20 msec). There is diffuse low signal intensity with the femoral head and ischium.

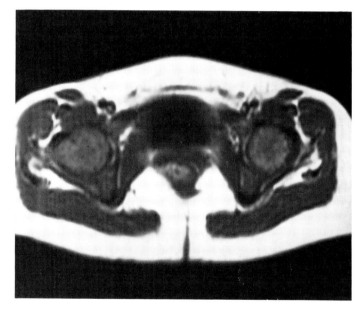

FIG. 25F. Axial TR/TE 2,000/80 msec. There is patchy mild increased signal intensity within the involved bones.

CASE 26

A 63-year-old woman presents with a history of localized back pain for 3 weeks. She has a prior history of breast cancer but has no known evidence of metastatic disease. Is this more likely a benign (e.g., osteoporotic) or malignant compression fracture? What imaging features are of help in making this distinction? What else would you do in this case? (Fig. 26A. Sagittal TR/TE 2,000/20 msec. Fig. 26B. Sagittal TR/TE 2,000/ 80 msec. Fig. 26C. Sagittal (STIR) TR/TE 1,500/30, TI 120 msec.)

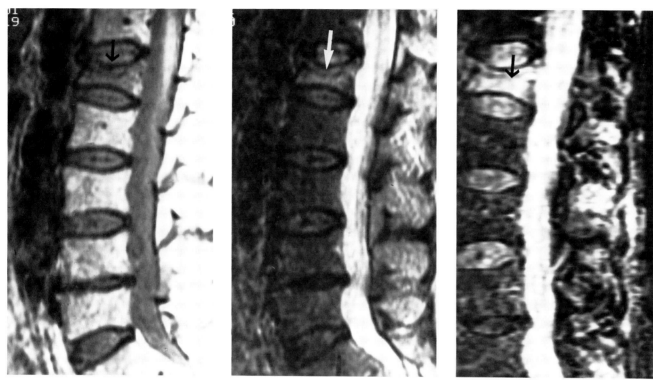

FIG. 26A.   FIG. 26B.   FIG. 26C.

Findings: On the sagittal intermediate weighted image, there is evidence of a compression fracture involving L1. The superior end plate demonstrates a relatively smooth concave border (arrow). The signal intensity is uniformly slightly diminished. On the T1-weighted image (not illustrated), the signal intensity was significantly decreased. On the T2-weighted image, there is slight inhomogeneous increase in signal intensity within the verterbral body (arrow). On the STIR sequence, the vertebral body demonstrates marked increased signal intensity (arrow).

Diagnosis: Benign vertebral body compression fracture.

Discussion: The differentiation of benign from malignant vertebral body collapse constitutes a common and often difficult diagnostic problem. In patients with a history of malignancy and vertebral collapse, tumor represents the basis for the collapse only one-third of the time. The diagnostic dilemma is particularly enhanced when, as in the present case, the lesion presents as an isolated abnormality. Given the critical importance of establishing the correct diagnosis, biopsy is ultimately required in many patients.

The spine represents the most common site for skeletal metastases, and vertebral collapse may be the initial manifestation of involvement in many patients. Carcinomas of the breast, lung, and prostate are the most common causes of collapse. Metastases predominate in the thoracic and lumbar segments, with cervical involvement observed less commonly. Location is of some value in differential diagnosis as involvement of the upper thoracic spine is more suggestive of malignancy than of a benign etiology (e.g., osteoporosis/osteomalacia). The appearance of the vertebral end plates may also be of value in differential diagnosis. Angular or irregular deformity of the end plates is suggestive of tumor (Fig. 26D). More uniform end-plate depression is characteristic of the "fish" vertebrae seen with osteomalacia and osteoporosis. Diagnostic difficulty occurs, however, when the two conditions (metastasis and osteoporosis) coexist in the same population, or when irregular end-plate involvement results from benign Schmorl's node formation. The presence of a soft tissue mass (Fig. 26E) and/or of pedicle destruction are valuable indicators of metastasis but may not be evident in early disease. Multiplicity of involvement is highly suggestive of metastatic disease.

Metastatic tumors typically demonstrate prolonged T1 and T2 relaxation times. On T1-weighted images, the low signal intensity produced by most metastatic tumors is contrasted against the higher signal intensity of adult vertebral bodies, and the conspicuity of lesions is therefore high. On T2-weighted images, most metastatic lesions demonstrate some increase in signal intensity (Fig. 26F,G). Osteoblastic lesions are an exception, demonstrating low intensity on most sequences. On STIR sequences, T1 and T2 are additive, and recent investigations have suggested an increased sensitivity for this pulse sequence in depicting marrow abnormalities. Recent vertebral collapse secondary to benign etiologies such as osteoporosis may also demonstrate low signal on T1-weighted images and increased signal on T2-weighted images and STIR sequences. The findings relate to marrow edema rather than tumor. In the absence of overt bone destruction or an extradural mass, it may be extremely difficult to differentiate benign from malignant collapse in this setting. The increased sensitivity of MR to marrow infiltration may, however, allow depiction of additional sites of previously occult involvement that can assist in establishing the correct diagnosis (Fig. 26H,I). More chronic vertebral collapse secondary to osteoporosis may demonstrate near normal signal intensity, a finding that would be unusual for malignancy. In a similar manner, although the presence of increased signal on a STIR sequence can be seen with both benign and malignant collapse, the absence of increased signal would be uncommon with tumor in our experience.

The patient in this case underwent biopsy of the involved vertebral body to allow a definitive diagnosis to be established. The biopsy was percutaneously accomplished under local anesthesia on an outpatient basis. Given the implications of the diagnosis of metastatic disease for patient prognosis and management, it is our current approach that a tissue diagnosis be obtained when feasible for every first presumed metastatic lesion.

D
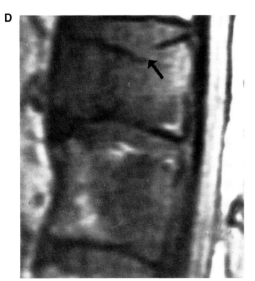
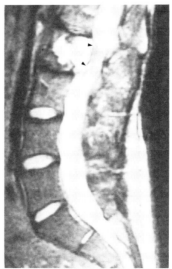
E

FIG. 26D,E. Non-Hodgkin lymphoma. **D.** Sagittal TR/TE 500/20 msec. A compression fracture of the L1 vertebral body demonstrates an angular contour to the superior vertebral end plate (*arrow*). The signal intensity is homogeneous and markedly decreased. There is involvement of the L2 vertebrae; T12 (not illustrated) was normal. **E.** Metastatic disease with extraosseous mass (sagittal TR/TE 2,000/60 msec). A metastatic lesion is seen within the L2 vertebral body and demonstrates high signal intensity. An extraosseous mass encroaches on the canal (*arrowheads*).

F

G

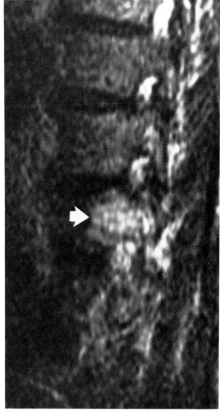

FIG. 26F,G. Pathological compression fracture. **F.** Sagittal TR/TE 2,000/20 msec. On this midline section, extensive collapse of the L2 vertebral body is noted. The inferior end plate is slightly angular. The signal is diminished. **G.** Sagittal TR/TE 2,000/80 msec. On this parasagittal section, a focus of increased signal intensity is observed in the region of the pedicle (*arrow*). A biopsy of this site demonstrated non-Hodgkin lymphoma.

H

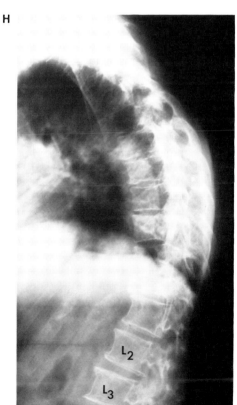

I

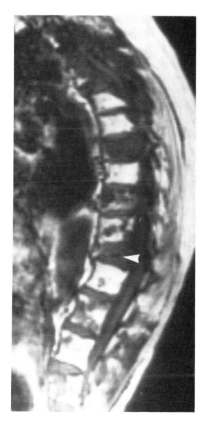

FIG. 26H,I. Metastatic disease (radiographically occult). **H.** Lateral thoracic spine film demonstrates a compression fraction of T12, a common location for an osteoporotic fracture. No other abnormalities are noted. **I.** Sagittal TR/TE 500/20 msec. Multiple metastatic lesions are identified as areas of low signal intensity (*arrowhead*) on this T1-weighted image. The detection of multiple lesions that were not suggested radiographically allows for confident diagnosis of the nature of the vertebral collapse in this patient.

CASE 27

A 65-year-old male farmer presented with a firm area distally in his thigh that was becoming larger. What is the most likely diagnosis? (Fig. 27A. Coronal TR/TE 500/20 msec. Fig. 27B. Axial TR/TE 2,000/60 msec.)

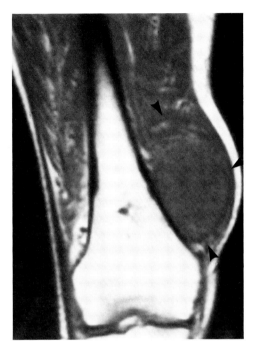

FIG. 27A.

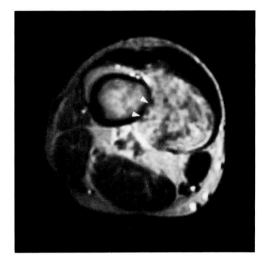

FIG. 27B.

Findings: On the coronal and axial images, a large inhomogeneous mass in the vastus medialis is identified. The margins are relatively well defined on the T2-weighted image. There is involvement of the medial femoral cortex (arrowheads). The mass is nearly isointense with muscle on the coronal image.

Diangosis: Malignant fibrous histiocytoma (MFH).

Discussion: The MFH is the most common soft tissue sarcoma of later adult life. The term encompasses a group of soft tissue tumors characterized by a storiform or cartwheel-like growth pattern histologically. The majority of cases are observed between the ages of 50 and 70. The lesions are rare in childhood, and the diagnosis should be made with caution in patients under the age of 20. The lesions are more common in men. The lower extremity is the most common site of involvement (67–75% of cases), followed by the upper extremity and retroperitoneum.

When the tumor involves an extremity, it most commonly presents as a painless enlarging mass often palpable by the patient for up to 1 year before diagnosis. Patients with retroperitoneal tumors, in contrast, present with constitutional symptoms including anorexia, malaise, and weight loss. A subgroup of MFH (inflammatory-type) may present with fever and leukocytosis dominating the clinical picture.

Although the tumor has a grossly circumscribed appearance, it often spreads for considerable distances along fascial planes or between muscle bundles. This accounts for the high incidence of local recurrence following resection. The majority of tumors are confined to skeletal muscle but, if adjacent to bone, may directly invade the cortex, as in the present case. Distant metastasis, most commonly to the lung, develops in up to 50% of patients. Five-year survival rates of 30% to 55% have been reported.

The MR features of MFH are not specific enough to suggest histologic diagnosis. Frequently there are myxoid, angiomatous, and fibrous areas within the lesion, so an inhomogeneous signal intensity, a useful indicator of malignancy, is almost always evident. Hemorrhage and necrosis are also common features of MFH. In our experience, the lesions are usually fairly well marginated. Low to intermediate signal with mixed inhomogeneity is seen on T1-weighted images, which increases in signal intensity on T2-weighted images. Areas of hemorrhage may be detected as increased signal on both T1- and T2-weighted images. The inhomogeneity and bone involvement when present are indicative of malignancy. When a lesion with malignant features (inhomogeneous signal intensity, neurovascular encasement, irregular margins, or bone involvement) is identified in the lower extremity in a patient over 45 years of age, MFH is the most likely diagnosis.

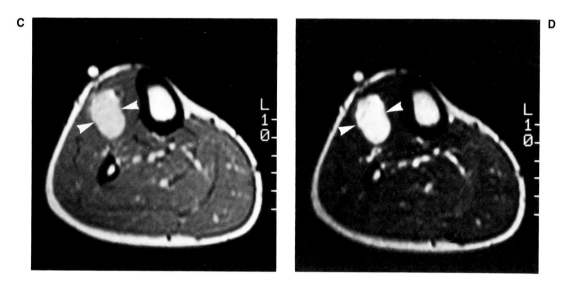

FIG. 27 C,D. Malignant fibrous histiocytoma. **C:** Axial MR, TR/TE 2,000/20 msec. On this intermediate weighted image, a well-circumscribed mass is seen centered within the anterior crural musculature (*arrowheads*). **D:** Axial MR, TR/TE 2,000/80 msec. The mass increases markedly in signal intensity on the T2-weighted image (*arrowheads*). The mass is well defined, exhibits no significant inhomogeneity, and yet represents a malignant lesion.

CASE 28

A 16-year-old boy complained of pain in his left tibia for 3 months' duration. Is the lesion most likely benign or malignant? What are the principal lesions classically associated with this location in children? (Fig. 28A. Anteroposterior radiograph. Fig. 28B. Axial CT scan. Fig. 28C. Coronal TR/TE 500/20 msec. Fig. 28D. Coronal STIR TR/TE 1,500/30, TI 120.)

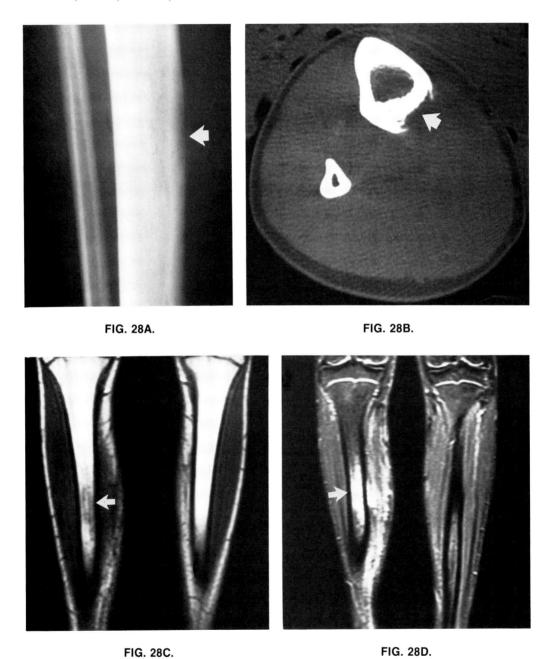

FIG. 28A.

FIG. 28B.

FIG. 28C.

FIG. 28D.

Findings: On the anteroposterior radiograph a subtle area of bone destruction is seen involving the cortex of the diaphysis of the tibia. A single line of periosteal reaction is seen (arrow). On the CT examination, a scalloped lesion of the tibial cortex is identified without definite medullary involvement (arrow). The coronal T1 weighted MR demonstrates a relatively small zone of slightly diminished signal intensity within the tibial diaphysis (arrow). The lesion is far more conspicuous on the coronal STIR sequence (arrow, Fig. 28D).

Diagnosis: Juvenile adamantinoma associated with fibro-osseous dysplasia.

Discussion: The patient was initially evaluated at a community hospital where the conventional radiographs and CT were obtained. The patient subsequently underwent a biopsy which was originally interpreted as representing Ewing sarcoma. The patient was referred for definitive therapy, and the MR was obtained as part of a prechemotherapy staging protocol. The findings on the T1-weighted images were considered unusually subtle for a Ewing sarcoma and the patient was returned for a STIR sequence. Lesion conspicuity and extent on this latter sequence were greatly increased. The soft tissue changes were considered to be on the basis of recent surgery, and no discrete soft tissue mass associated with the bone lesion was present. The initial biopsy material upon review was considered inadequate and a repeat biopsy performed revealing evidence of adamantinoma in association with fibro-osseous dysplasia. No evidence of Ewing sarcoma was demonstrated.

Adamantinoma is an uncommon neoplasm with a well-reported and rather striking predilection for the diaphysis of the tibia. While the age range of affected individuals is extremely variable, the tumor has until recently been considered exceedingly uncommon in childhood. Recent speculation has suggested that the presence of the tumor may be commonly masked by the coexistent presence of fibrous dysplasia-like lesions, which may dominate the radiographic and histologic picture.

An association between adamantinoma and fibrous dysplasia-like lesions has long been recognized but has recently been suggested to be more common than has previously been considered. These fibrous dysplasia-like lesions have variously been referred to as osteofibrous dysplasia, ossifying fibroma, and intra-cortical fibrous dysplasia. Histologically, they are characterized by the presence of woven bone trabeculae contained within an abundant fibrous stroma surrounded by rows of active osteoblasts. The predominance of the fibrous dysplasia-like pattern can mask the presence of the epithelioid neoplasm resulting in a mistaken diagnosis of isolated osteofibrous dysplasia. Failure to recognize coexistent adamantinoma has been suggested to explain the aggressive behavior of some cases of supposed osteofibrous dysplasia that have recurred in spite of repeated excisions. The coexistence of the two lesions has also led to the suggestion that the cells of origin in adamantinoma are capable of undergoing both epithelial and mesenchymal differentiation, with the latter reflected in the fibrous dysplasia-like regions. An association between adamantinoma and Ewing sarcoma-like tumor cells has been reported and may have contributed to the initial diagnostic difficulty in this case.

Adamantinomas are locally aggressive and can metastasize. Recurrence is common following inadequate resection, and frequently the behavior of the recurrent tumor is more aggressive. Radiographically, the lesions typically appear as slightly eccentric—multilocular and osteolytic—often with reactive sclerosis. The differential diagnosis usually includes fibrous dysplasia, nonossifying/ossifying fibroma, aneurysmal bone cyst, hemangioendothelioma, chondromyxoid fibroma, and eosinophilic granuloma.

RECOMMENDED READING

Agartz I, Sääf J, Wahlund L, and Wetterberg L. Magnetic resonance imaging at 0.02T in clinical practice and research. *Magn Reson Imaging* 1987;5:179–187.

Aisen AM, Martel W, Braunstein EM, et al. MRI and CT evaluation of primary bone and soft-tissue tumors. *AJR* 1986;146:749.

Bellon EM, Haacke EM, Coleman PE, et al. MR artifacts: a review. *AJR* 1986;147:1271–1281.

Beltran J, Simon DC, Katz W. Increased MR signal intensity in skeletal muscle adjacent to malignant tumors: pathologic correlation and clinical relevance. *Radiology* 1987;162:251–255.

Beltran J, Simon DC, Katz W, and Weis LD. Increased MR signal intensity in skeletal muscle adjacent to malignant tumors: pathologic correlation and clinical relevance. *Radiology* 1987;162:251–255.

Beltran J, Simon DC, Levy M, et al. Aneurysmal bone cysts: MR imaging at 1.5 T. *Radiology* 1986;158:689–690.

Berquist TH. Magnetic resonance imaging: preliminary experience in orthopedic radiology. *Magn Reson Imaging* 1984;2:41–52.

Berquist TH, Ehman RL, and Richardson ML. *Magnetic resonance of the musculoskeletal system.* New York: Raven Press, 1987.

Bland, KI, McCoy DM, Kinard RE, and Copeland EM, III. Application of magnetic resonance imaging and computerized tomography as an adjunct to the surgical management of soft tissue sarcomas. *Ann Surg* 1987;205:473–481.

Boyko OB, Cory DA, Cohen MD, et al. MR imaging of osteogenic and Ewing sarcoma. *AJR* 1987;148:317–433.

Brown SS, Van Sonnenberg E, Gerber KH, et al. Magnetic resonance relaxation times of percutaneously obtained normal and abnormal body fluids. *Radiology* 1985;154:727–731.

Burger PC, Vogel FS. *Surgical pathology of the nervous system and its coverings.* New York: John Wiley and Sons, 1976.

Chan L. The current status of magnetic resonance spectroscopy—basic and clinical aspects. *West J Med* 1985;143:773–781.

Chang AE, Matory YL, Dwyer AJ, et al. Magnetic resonance imaging versus computed tomography in evaluation of soft tissue tumors of the extremities. *Ann Surg* 1987;205:340–348.

Coerkamp E, Kroon HM. Cortical bone metastases. *Radiology* 1988;169:525–528.

Cohen EK, Kressel HY, Frank TS, et al. Hyaline cartilage-origin bone and soft-tissue neoplasms: MR appearance and histologic correlation. *Radiology* 1988;167:477–481.

Cohen EK, Kressel HY, Perosio T, et al. MR imaging of soft-tissue hemangiomas: correlation with pathologic findings. *AJR* 1988;150:1079–1081.

Cohen JM, Weinreb JC, and Maravilla KR. Fluid collections in the intraperitoneal and extraperitoneal spaces: comparison of MR and CT. *Radiology* 1985;155:705–708.

Cohen JM, Weinreb JC, Redman HC. Arteriovenous malformations of the extremities: MR imaging. *Radiology* 1986;158:475–479.

Cohen MD, Klatte EC, Baehner R, et al. Magnetic resonance imaging of bone marrow disease in children. *Radiology* 1984;151:715–718.

Cohen MD, Weetman RM, Provisor AJ, et al. Efficacy of magnetic resonance imaging in 139 children with tumors. *Arch Surg* 1986;121:522–529.

Cohen MD, McGuire W, Cory DA, and Smith JA. MR appearance of blood and blood products: an *in vitro* study. *AJR* 1986;146:1293–1297.

Daffner RH, Lupetin AR, Dash N, et al. MRI in the detection of malignant infiltration of bone marrow. *AJR* 1986;146:353–358.

Dahlin DC and Unni KK. *Bone tumors.* 4th ed. Springfield, Illinois: Charles C. Thomas, 1986.

Demas BE, Heelan RT, Lane J, et al. Soft-tissue sarcomas of the extremities: comparison of MR and CT in determining the extent of disease. *AJR* 1988;150:615–620.

Deutsch AL, Resnick D. Eccentric cortical metastases to the skeleton from bronchogenic carcinoma. *Radiology* 1980;137:49–52.

Dooms GC, et al. Lipomatous tumors and tumors with fatty component: MR imaging potential and comparison of the MR and CT results. *Radiology* 1985;157:479–483.

Dooms GC, Fisher MR, Hricak H, et al. Bone marrow imaging: magnetic resonance studies related to age and sex. *Radiology* 1985;155:429–432.

Ehman RL and Berquist TH. Magnetic resonance imaging of musculoskeletal trauma. *Radiol Clin North Am* 1986;24:291–319.

Ehman RL, Berquist TH, and McLeod RA. MR imaging of musculoskeletal system: a 5-year appraisal. *Radiology* 1988;166:313–320.

Enzinger FM, Weiss SH. *Soft tissue tumors.* St. Louis: CV Mosby Co., 1983.

Fitzgerald RH and Berquist TH. Magnetic resonance imaging (editorial). *J Bone Joint Surg [Am]* 1986;68:799–800.

Fobben ES, Dalinka MK, Schiebler DL, et al. The magnetic resonance imaging appearance of cartilaginous tumors involving the epiphysis of 1.5 Tesla. *Skeletal Radiol* 1987;16:647–651.

Fruehwald FXJ, Tscholakoff D, Schwaighofer B, et al. Magnetic resonance imaging of the lower vertebral column in patients with multiple myeloma. *Invest Radiol* 1988;23:193–199.

Gillespy TG, Manfrini M, Ruggieri P, et al. Staging of intraosseous extent of osteosarcoma: correlation of preoperative CT and MR imaging with pathologic macroslides. *Radiology* 1988;167:765–767.

Dwyer AJ, Frank JA, Sauk VJ, et al. Short TI inversion-recovery pulse sequence: analysis and initial experience in cancer imaging. *Radiology* 1988;168:827–836.

Glazer HS, Lee JKT, Levitt RG, et al. Radiation fibrosis: differentiation from recurrent tumor by MR imaging. *Radiology* 1985;156:721–726.

Gormori JM, Grossman RI, Goldberg HI, et al. Intracranial hematomas: imaging by high-field MRI. *Radiology* 1985;157:87–93.

Heiken JP, Lee JKT, Smathers RL, Totty WG, and Murphy WA. CT of benign soft-tissue masses of the extremities. *AJR* 1984;142:575–580.

Hudson TM, Hamlin DJ, Enneking WF, and Pettersson H. Magnetic resonance imaging of bone and soft tissue tumors: early experience in 31 patients compared with computed tomography. *Skeletal Radiol* 1985;13:134–146.

Hudson TM. *Radiologic-pathologic correlation of musculoskeletal lesions.* Baltimore: Williams & Wilkins, 1987.

Kaplan PA, Asleson RJ, Klassen LW, and Duggan MJ. Bone marrow patterns in aplastic anemia: observations with 1.5 T MR imaging. *Radiology* 1987;164:441–444.

Kissane JM. *Anderson's Pathology.* 8th ed. St. Louis: C. V. Mosby, 1985.

Lee JK, Yao L, and Wirth CR. MR imaging of solitary osteochondromas: report of eight cases. *AJR* 1987;149:557–560.

Mankin HJ, and Gebhardt MC. Advances in management of bone tumors. *Clin Orthop* 1985;200:73–86.

McKinstry CS, Steiner RE, Young AT, et al. Bone marrow in leukemia and aplastic anemia: MR imaging before, during, and after treatment. *Radiology* 1987;a162:701–707.

Moore SG, Gooding SA, Brasch RC, et al. Bone marrow in children with acute lymphocytic leukemia. *Radiology* 1986;160:237.

Nidecker AC, Müller S, Aue WP, et al. Extremity bone tumors: evaluation by P-31 MR spectroscopy. *Radiology* 1985;157:167–174.

Olson DO, Shields AF, Scheurich CJ, et al. Magnetic resonance imaging of the bone marrow in patients with leukemia, aplastic anemia, and lymphoma. *Invest Radiol* 1986;21:540.

Petasnick JP, Turner DA, Charters, JR, et al. Soft-tissue masses of the locomotor system: comparison of MR imaging with CT. *Radiology* 1986;160:125–133.

Pettersson H, Gillespy T, Hamlin, DJ, et al. Primary musculoskeletal tumors: examination with MR imaging compared with conventional modalities. Radiology 1987;164:237–241.

Polak JF, Jolesz FA, Adams DF. Magnetic resonance imaging of skeletal muscle: prolongation of T1 and T2 subsequent to denervation. *Invest Radiol* 1988;23:365–369.

Porter BA, Hastrup W, Richardson ML, et al. Classification and investigation of artifacts in magnetic resonance imaging. *Radiographics* 1987;7:271.

CHAPTER 9

The Temporomandibular Joint

Barry D. Pressman and Frank G. Shellock

The first applications of magnetic resonance imaging (MRI) for examination of the temporomandibular joint (TMJ) were described in the mid-1980s. Since then, this diagnostic modality has rapidly evolved to the point of now being regarded as the procedure of choice for imaging disorders that affect the TMJ. Compared to other radiographic techniques, MRI is unsurpassed for depicting important components of the TMJ including the disk, mandibular condyle, glenoid fossa, articular eminence, and associated soft tissues structures.

Initially, only static views of the TMJ were attainable by MRI, and the inability to obtain a functional assessment of jaw biomechanics and meniscocondylar coordination was considered to be a significant limitation. Consequently, MRI strategies were developed that provided a motion, or *kinematic* (a term used to describe the motion of a body without reference to force or mass), study of the TMJ so that movement-related abnormalities could be assessed.

In this chapter, static and kinematic MRI techniques used to evaluate the TMJ are described and examples of pathology that affect this joint are discussed.

TECHNICAL ASPECTS

Pulse Sequences: Because abnormalities that affect the TMJ are primarily associated with the position of the disk relative to the mandibular condyle, T1-weighted MR images that clearly delineate the anatomy are usually sufficient for identifying joint pathology. T2-weighted images may be useful in selected cases in which there is suspicion of an inflammatory process, joint effusion, arthritic condition, or tumor. The use of partial flip angle or "fast scan" pulse sequences has been proposed for dynamic (kinematic) "motion" studies. However, because of the intrinsically poor spatial resolution and susceptibility to artifacts associated with the use of partial flip angle techniques, T1-weighted imaging provides a more advantageous means of obtaining the kinematic examination of the TMJ. Therefore, some groups prefer this imaging strategy.

Dual Surface Coil MRI: A high incidence of bilateral abnormalities exists in patients with internal derangements of the TMJ. Therefore, it is advisable to image both joints to examine patients with TMJ arthralgia thoroughly. The use of dual surface coils for MRI of bilateral TMJs reduces the overall set-up and data acquisition time by approximately 50% compared to using a single surface coil (Fig. A). Furthermore, the combination of dual surface coils and kinematic MRI provides views obtained simultaneously from the right and left TMJs so that a direct comparison can be made between the two sides at the same degree of mouth opening. This is particularly important for diagnosing asymmetrical motion abnormalities.

521

Optimal Imaging Planes: The condyle of the TMJ is a relatively small structure situated in obliqued angles in both sagittal and coronal planes. Studies have indicated that the optimal orientation for images of the TMJ should be with reference to the long axis of the condylar head because a significant number of both normal subjects and patients have condyles that have abnormal configurations with respect to their size and orientation.

The recent capability of obtaining user-determined slice locations in any imaging plane (referred to by some manufacturers as "graphic prescription") enables the selection of images to be obtained in planes that are in perpendicular (obliqued sagittals) and parallel (obliqued coronals) positions relative to the long axis of the condylar head (as viewed on an axial localizer scan) (Figs. B–E). The graphic prescription of slice locations provides a standardized and consistent means of assessing the TMJ by MRI. In addition, the examination time is usually decreased because only the slice locations from the anatomy of interest are acquired.

Recent studies have suggested that coronal plane imaging of the TMJ combined with the usual sagittal views improves the accuracy of MRI for determining displacement of the disk. Medial or lateral (also referred to as "sideways" displacement) dislocation of the disk, as well as rotational displacement (i.e., a combination of anterior and medial or lateral displacement), is easily identified on coronal plane images. The above derangements of the TMJ appear to be more common than previously suspected. Coronal plane views are also useful for evaluating bony anatomy and the muscles of mastication.

Kinematic MRI: The kinematic study of the TMJ is accomplished using a nonferromagnetic positioning device (Fig. F) to open the patient's mouth at several predetermined increments, and MRI is performed at each position. Several thin slices (i.e., 3 mm or less) are typically acquired through the TMJ in an obliqued sagittal plane at the following mouth positions: (a) closed mouth (i.e., maximum intercuspation of teeth), (b) "closed" mouth with the positioning device in place (the distance the patient's mouth is opened with the positioning device in place is dependent on the length of the

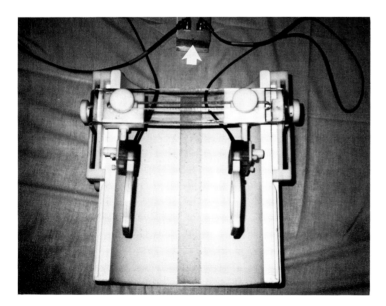

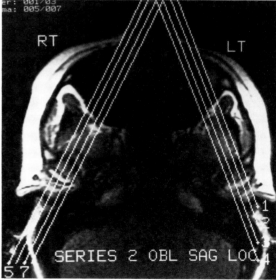

FIG. A. Dual 6.5-cm surface coils with combiner box (*arrow*) used for simultaneous bilateral imaging of TMJs.

FIG. B. T1-weighted, 3-mm axial slice demonstrating localizer scan used for determining obliqued sagittal slice locations selected in an orientation that is perpendicular to the long axis of the right and left condylar heads using graphic prescription software.

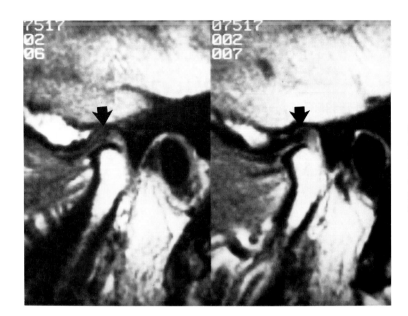

FIG. C. Consecutive T1-weighted, 3-mm slices obtained in an obliqued sagittal plane perpendicular to the long axis of the condylar head. The posterior band of the disk (*arrows*) is in a normal "12 o'clock" position relative to the mandibular condyle.

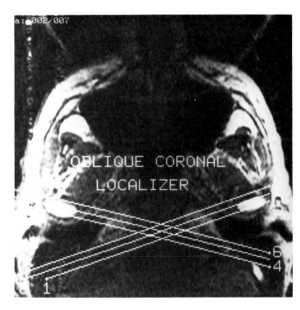

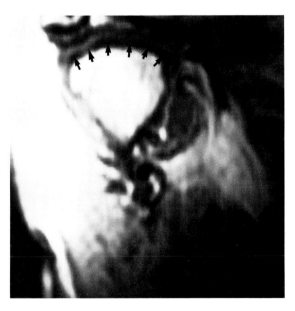

FIG. D. T1-weighted 3-mm axial slice demonstrating localizer scan used for determining obliqued coronal plane slice locations selected in an orientation that is parallel to the long axis of the right and left condylar heads using graphic prescription software.

FIG. E. T1-weighted 3-mm-thick slice obtained in an obliqued coronal plane parallel to the long axis of the condylar head. The disk (*arrows*) is in a normal "capped" position relative to the condyle.

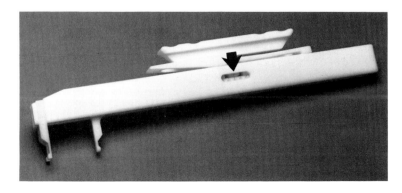

FIG. F. Positioning device used to open the patient's mouth at predetermined increments for kinematic MRI of the TMJ. A gauge (*arrow*) is used to determine the extent of mouth opening.

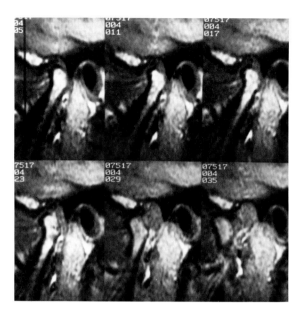

FIG. G. T1-weighted 3-mm-thick slices obtained in an obliqued sagittal plane perpendicular to the long axis of the condylar head at progressive 7-mm increments of mouth opening for the kinematic study of the TMJ. The disk (*arrowhead*) is in a normal position in the closed-mouth view. During the mouth-opening phases, the condyle moves to the intermediate zone (*curved arrow*) of the disk, and this meniscocondylar relationship is maintained as the condyle anteriorly translates to a position below the articular eminence.

incisors), and (c) repeatedly at successive increments of opening (usually 15% to 20% of the patient's mandibular range of motion) to a fully opened mouth position. The kinematic study of the TMJ may be viewed as either multiple static images (Fig. G) or, preferably, in a cine-loop format that facilitates the visualization of the various images and provides a better appreciation of asymmetrical motion abnormalities.

Kinematic MRI of the TMJ is useful for several reasons. In the evaluation of a patient with an anteriorly displaced disk, viewing the TMJ at several increments between closed and fully opened positions of the mouth is necessary to ascertain if there is reduction of the disk and the relative position of the mandible at this point, so that the proper application of splint therapy or other appropriate treatment can be implemented. In addition, because the mandible has a bilateral articulation with the cranium, both TMJs must function in synchrony, and any disordered movement noted on the kinematic study is diagnostic of an abnormality. Finally, because of the similarity between normal variants and internal derangements observed on closed mouth views of the TMJ, studying the motion aspects of the joint exhibited by kinematic MRI may help to differentiate between a normal and abnormal TMJ and, therefore, improves the diagnostic accuracy of the examination.

CASE 1

This 26-year-old man has mild pain and an opening click in his TMJs. Is the condyle in normal position in the closed-mouth study? What is the cause of the opening click?

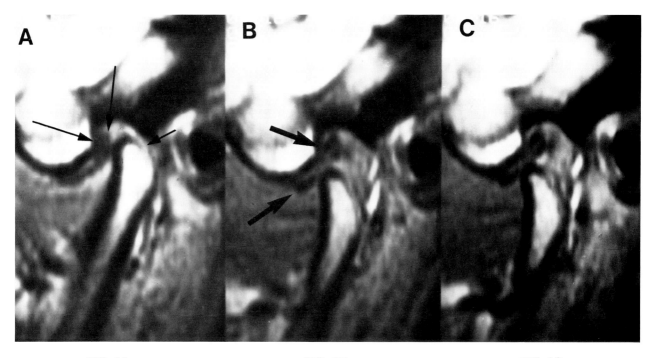

FIG. 1A.   **FIG. 1B.**   **FIG. 1C.**

Findings: The first three images of a sagittal kinematic series (Fig. 1A–C). In the first image (closed-mouth) the condyle is more posterior than normal with a narrow posterior joint space (short arrow), and the disk is anteriorly subluxated (Fig. 1A). The posterior band (long arrows) is anterior to the condyle rather than above it at the normal 12 o'clock position. In the second image (Fig. 1B), after only 6 mm of jaw opening, the disk (arrows) is in normal position relative to the condyle, with the posterior band above the condyle and the condyle "riding" on the intermediate zone, i.e., there has been reduction of the disk early in the jaw opening series.

Diagnosis: Anterior subluxation of the disk, with early reduction.

Discussion: As laxity develops in the diskal attachments the disk may dislocate, most commonly, anteriorly. Normally the posterior band is directly above the condyle when the jaw is closed. If the posterior band is anterior to this position, the disk is anteriorly dislocated. Unless the diskal attachments are very stretched (loose), the condyle will move under the posterior band as the jaw opens (recapturing the meniscus) and ride on the intermediate zone of the disk as it continues to translate anteriorly. The more normal the attachments, the earlier the point of recapture (reduction) as the jaw opens. There may be a click (opening click) as the condyle moves under the posterior band, and a "reciprocal click" often occurs as the jaw closes and the condyle slides posterior to the posterior band. In the closed-mouth position, anterior dislocation may be associated with a posteriorly positioned condyle, as in this case. In fact, this may be the only radiographic/tomographic indication of disk displacement and has been so used for many years by dental radiographers.

CASE 2

This patient has had increasing bilateral jaw pain, right greater than left, for several years. There is an opening click on the left, which MR images (not shown) demonstrate to be secondary to anterior subluxation of the disk with early reduction. How does the appearance of the right side explain the greater pain and absence of a click? To which side would you expect the jaw to rotate as it opens?

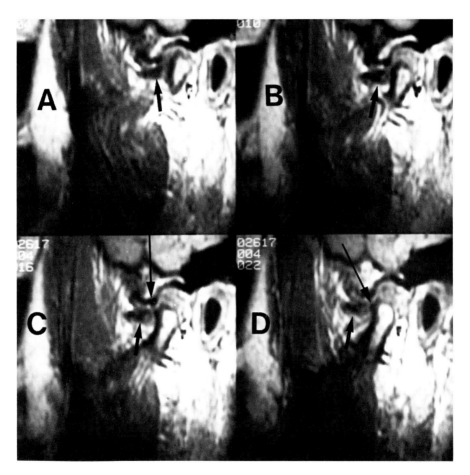

FIG. 2A–D.

Findings: A sagittal kinematic series with 18-mm maximum jaw opening (Fig. 2A–D). A small spur (long arrow, Fig. 2D) is present at the anterior aspect of the condyle, and the glenoid shows some sclerosis posterosuperiorly. The disk (short arrows) is anteriorly dislocated and does not reduce as the jaw opens and the condyle translates anteriorly. Condylar translation is limited with the condyle not reaching the inferior aspect of the articular eminence (long arrow, Fig. 2C), whereas normally it would reach this far anteriorly or even slightly anterior to the inferior aspect of the articular eminence.

Diagnosis: Anterior dislocation of the disk, without reduction.

Discussion: An opening click usually results from reduction of a dislocated meniscus as the condyle passes below and anterior to the dislocated posterior band. A reciprocal click may occur as the jaw closes and the condyle passes below and posterior to the posterior band. If there is no reduction of the meniscus, as on the right side in this patient, there is no click.

Anterior dislocation without reduction is usually preceded by a history of a click, indicating previous subluxation with reduction. If the disk does not reduce as the jaw opens, it may act as an obstruction (closed-lock) to condylar translation with resultant limitation of jaw opening. Rotation of the jaw to the side of the block often occurs.

The degenerative changes of the right condyle, articular eminence, and glenoid suggest that there is an unseen perforation of the posterior discal attachments or of the disk itself, frequently associated with anterior disk dislocation.

CASE 3

After suffering jaw trauma 6 months before this study, this 50-year-old woman noted progressively increasing noise in her TMJs as she chewed. For the past 3 months she has also had bilateral TMJ pain, left greater than the right, while chewing. What is the cause of the joint pain during chewing?

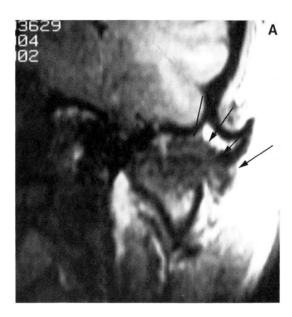

FIG. 3A.

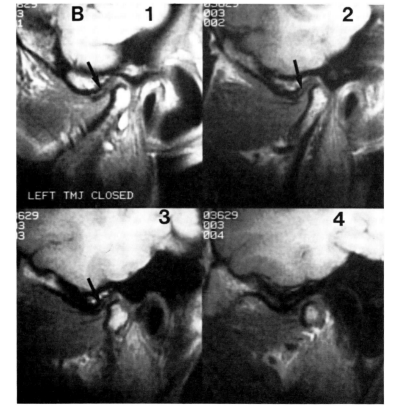

FIG. 3B.

Findings: Lateral subluxation of the disk (arrows) is obvious on the coronal closed-mouth image (Fig. 3A). This may also be recognized on the sagittal closed-mouth series (Fig. 3B[1–4]) by noting that the disk is well seen in the more lateral positions (Fig. 3B[1], [2]), poorly on an image from a more medial position (Fig. 3B[3]), and not at all on the most medial position (Fig. 3B[4]).

Diagnosis: Lateral subluxation of the disk.

Discussion: Lateral and medial subluxation are more common than previously recognized. One study showed a higher incidence of disk irreducibility and perforation with sideways dislocation compared to anterior dislocation. Anterior dislocation may or may not be associated with sideways dislocation. Although careful evaluation of the sagittal images may disclose sideways dislocation, the coronal plane makes this diagnosis easier and more definitive. The exact relationship of the pain to the disk dislocation is not known but may relate to the traction on the medial and lateral capsular attachments, which are innervated.

CASE 4

A 72-year-old woman has had limited jaw opening (25 mm) and bilateral TMJ pain for many years. On occasion she notes an opening click on one or both sides. What are the MR findings of the right TMJ that help to explain the patient's symptoms?

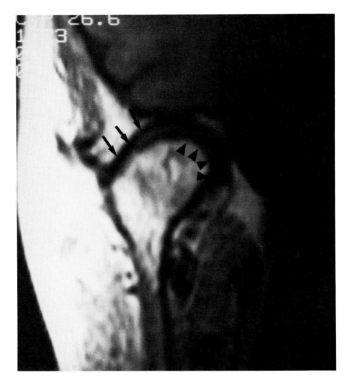

FIG. 4.

Findings: A coronal image (Fig. 4) demonstrates a very narrow superior and lateral joint space (arrows). The medial subluxation of the disk is apparent (arrowheads). Anterior dislocation was noted on the sagittal images (not shown).

Diagnosis: Medial dislocation of the disk.

Discussion: Although anterior dislocation is more common, medial or lateral dislocation is not uncommon and may be of similar clinical significance. The coronal images are particularly useful for such a diagnosis, but as seen in this case, the sagittal study may also be diagnostic. Narrowing of the lateral joint space may occur with medial disk dislocation, similar to posterior joint space narrowing seen with anterior disk dislocation. Associated anterior dislocation of the disk is not uncommon but is not necessarily present. The opening click is a clinical indication that the dislocation reduces during jaw opening.

CASE 5

This 37-year-old woman has had bilateral TMJ pain and grinding sounds for several years following jaw trauma. What are the likely causes of the pain as indicated by the findings on the static MR images?

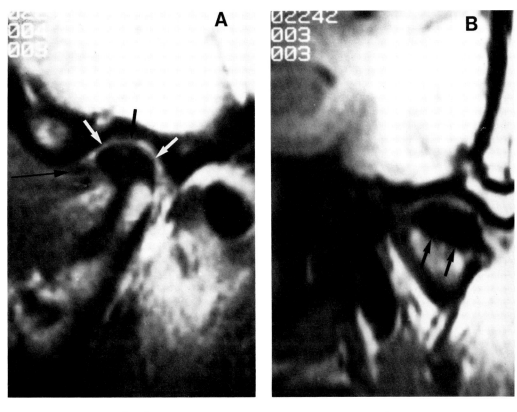

FIG. 5A.   **FIG. 5B.**

Findings: Sagittal (Fig. 5A) and coronal (Fig. 5B) closed-mouth images of the left TMJ. The signal of the condyle (white arrows on sagittal, black arrows on coronal) is very reduced, consistent with sclerosis. The joint space (short black arrow) is narrowed. The deformed disk (long black arrow) on the sagittal image is seen anterior to the condyle and joint, i.e., anteriorly dislocated on this closed-mouth image.

Diagnosis: Severe degenerative changes associated with anterior dislocation of the meniscus.

Discussion: Osseous degenerative changes of the TMJ are usually preceded by dislocation of the disk unless they are secondary to a fracture or primary inflammatory arthritis. Perforation or tears of the posterior attachments (bilaminar zone) or of the disk itself may then occur, and degenerative changes frequently will rapidly follow. These may consist of flattening or narrowing of the condyle and sclerosis or spurring of the condyle, articular eminence, and glenoid. Crepitus or grinding sounds associated with jaw opening may develop secondary to the perforation and degenerative changes.

The disk itself is not innervated, but its posterior attachment (the bilaminar zone) is innervated and very vascular. With anterior displacement there is loading on the bilaminar zone, which may cause pain, as in this patient. Further, bony degenerative changes and bone impingement on bone may also elicit pain.

CASE 6

A car accident initiated a long history of left TMJ pain in this 32-year-old woman. Severe left facial pain necessitated two surgeries resulting in removal of the condyle and the articular eminence. Symptoms were greatly relieved by the surgical procedures. Does the MR help to explain why the patient's symptoms were reduced by surgery?

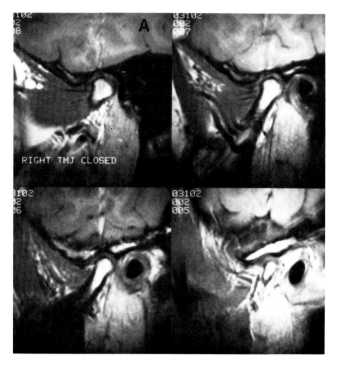

FIG. 6A.

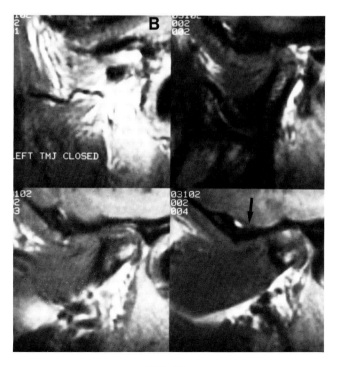

FIG. 6B.

Findings: Sagittal closed-mouth images. The right side appears normal (Fig. 6A). The left articular eminence (arrow, Fig. 6B) has been partially surgically resected and is flattened, and the condyle (curved arrow) has also been partly resected and does not enter the glenoid fossa (Fig. 6B).

Diagnosis: Postoperative changes with truncated condyle not entering the glenoid or impinging on the articular eminence.

Discussion: This case illustrates the importance of clinical history. The examination would be very difficult to interpret without the history of surgery. The MR findings suggest the presence of severe clinical symptoms, whereas the patient is only mildly symptomatic because the surgery has resulted in absence of bone grating against bone. This case is but one example of the necessity of being carefully descriptive but not judgmental in interpreting MRI of the TMJs. It is not uncommon for the symptoms and the MRI findings to be widely divergent.

CASE 7

This patient, a 61-year-old woman, reported no previous history of TMJ pain or disorder. She developed left-sided pain after eating an apple. Do the MRI findings correlate with the clinical history? On which side would you expect the pain?

Findings: Static sagittal images of left (Fig. 7A) and right (Fig. 7B) TMJ. Degenerative sclerosis of the condyle, the glenoid, and the articular eminence (Fig. 7A). Anterior displacement of the disk is demonstrated (arrow). Narrowing and prominent sclerosis of the condyle with less marked sclerosis of the glenoid (Fig. 7B). The disk is anteriorly displaced (arrow).

Diagnosis: Bilateral anterior dislocation of the disk and bilateral degenerative changes of the TMJs, right greater than left.

Discussion: In patients with symptoms of TMJ dysfunction, it is not uncommon that the clinical history and symptoms do not correlate well with the imaging studies. The clinical aphorism "treat the patient, not the x-rays" must always be remembered. However, the imaging changes cannot be disregarded and may be crucial in planning proper therapy. In this patient the MR strongly indicates a prolonged course of TMJ

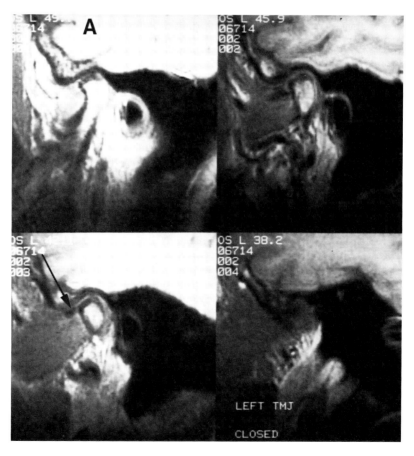

FIG. 7A. Degenerative sclerosis of the condyle, the glenoid, and the articular eminence. Anterior dislocation of the disk is demonstrated (*arrow*).

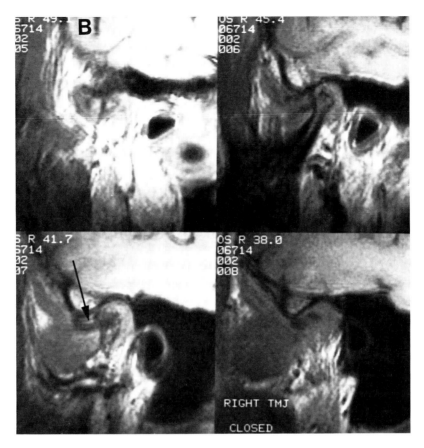

FIG. 7B. Narrowing and prominent sclerosis of the condyle with less marked sclerosis of the glenoid. The disk is anteriorly dislocated (*arrow*).

dysfunction accounting for the degenerative changes that are especially prominent on the right. Why the patient did not notice any joint pain or dysfunction over this presumed long history is an enigma, but not an uncommon one. Had there been any legal aspects to this case, the radiologist would be able to state quite definitively that there was TMJ disease preexisting the acute episode and onset of symptoms. Anterior displacement of the disk can occur acutely, but the bony degenerative changes certainly are not acute.

CASE 8

A 32-year-old man had a 10-year history of pain in the left jaw while chewing. He had been treated for a "bite" problem as a child. Taken individually, are there any abnormalities in the kinematic studies of each TMJ? Compare the two sides carefully, image by image—now are any abnormalities apparent?

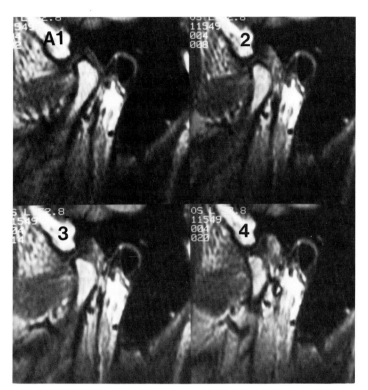

FIG. 8A.

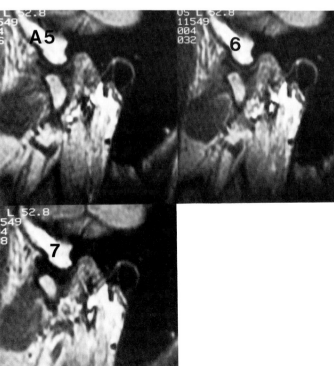

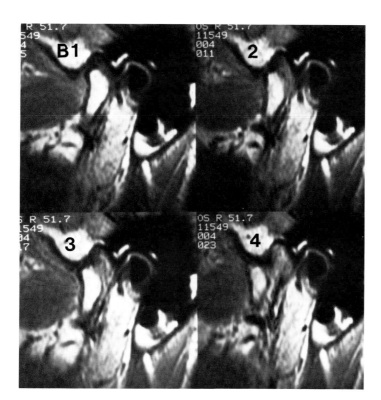

FIG. 8B.

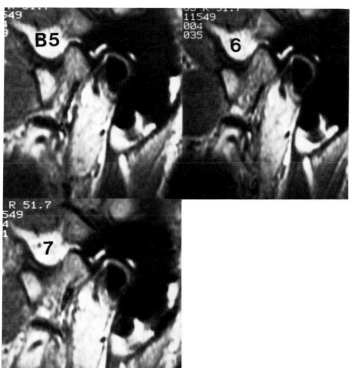

Findings: Sagittal kinematic series of the left (Fig. 8A[1–4]) and right (Fig. 8B[5–7]) TMJs. Each series appears normal if viewed individually. However, at positions 4 and 5 on the left, the condyle has not translated as much as on the simultaneously obtained right-sided positions 4 and 5. The two sides are again symmetrical at the fully open (Fig. 8A[7], B[7]) positions (range of motion, 30 mm). This asymmetry of motion could not be perceived if the two joints had not been studied simultaneously and at multiple positions of jaw opening, not just at the fully closed and the fully opened positions.

Diagnosis: Mild motion abnormality with the left side lagging behind the right during the midphase of jaw opening.

Discussion: The disk and osseous structures bilaterally have a normal MR appearance in the closed-mouth position. The fully opened position also is bilaterally symmetrical and normal. It is only because the two sides have been simultaneously imaged during the kinematic study that it is possible to determine that there is a mild asymmetry of condylar translation, with the left lagging only during mid–jaw opening. The exact explanation of this finding is not known, but it does demonstrate objective evidence of TMJ dysfunction that may account for, or be secondary to, the pain experienced during mastication.

When the MR examination using T1-weighted images (Fig. 8C) does not answer the clinical questions, consideration should be given to a T2-weighted series. T2-weighted images allow for easier diagnosis of joint effusion (Fig. 8D) and abnormal signal of the disk. They are particularly helpful if acute inflammatory arthritis is suspected.

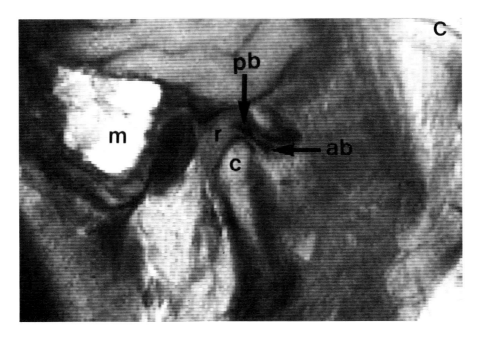

FIG. 8C. T1-weighted sagittal closed-mouth images in a different patient. The retrodiskal space is widened but the disk appears normal. [pb, posterior band; ab, anterior band; r, retrodiskal space; c, condyle; m, mastoid (surgically filled with fat).]

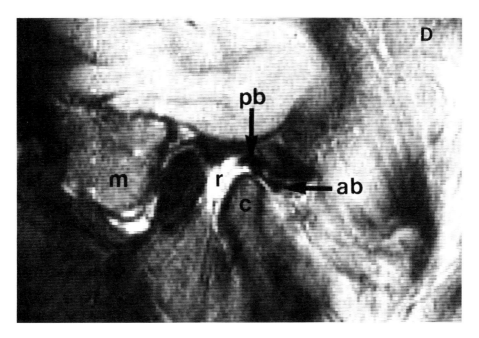

FIG. 8D. Same patient as in Fig. 8C. T2-weighted sagittal image demonstrates high signal in the joint posteriorly, indicating a joint effusion (pb, posterior band; ab, anterior band; r, retrodiskal space; c, condyle; m, mastoid.)

CASE 9

This 68-year-old woman had right-sided jaw, ear, temple, and neck pain for 1 year. There is no prior history of jaw pain, trauma, or treatment for TMJ dysfunction. Is the right disk normal? If not, how does the kinematic series clarify the diagnosis?

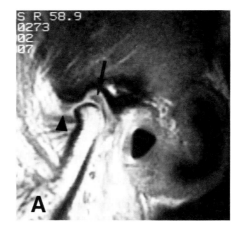

FIG. 9A.

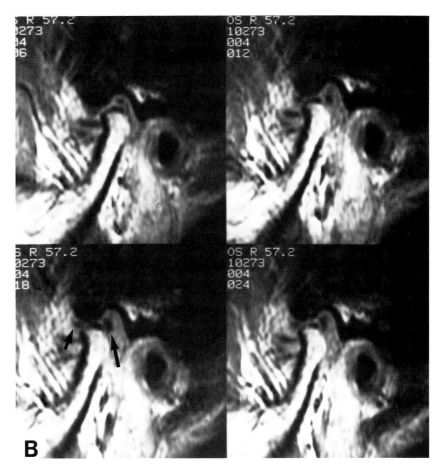

FIG. 9B.

Findings: Sagittal MR of right TMJ (Fig. 9A, B). The distance between the anterior (arrowhead) and posterior (arrows) bands is greater than normal, and no intermediate zone is visible (Fig. 9A).

The sagittal kinematic series from the right side (Fig. 9B) shows the anterior band (shorter arrow) to translate anteriorly with the condyle, whereas the posterior band (longer arrow) remains within the joint space posterior to the superior margin of the condyle.

The widened space between the anterior and posterior bands and the nonconjoint motion are indications of a torn disk.

Diagnosis: Torn intermediate zone of the right disk.

Discussion: The diagnosis of disk perforation/tear by MRI may be very difficult and, especially in the case of perforation, often impossible. Arthrography is still the procedure most dependable for the demonstration of perforations and tears. We believe that MRI has obviated the necessity of arthrography in many patients because it offers so much information relative to the joint, disk, bone, and function in a noninvasive fashion. However, if the MRI examination is inconclusive regarding a clinical question of disk perforation or tear, arthrography may be very valuable. Before the advent of MRI, arthrography was the only method available to study disk motion and still offers a more thorough evaluation thereof.

CASE 10

A 33-year-old male radiology resident has had intermittent jaw pain and clicking for many years. Not infrequently he experiences two separate clicks on opening his mouth, one early and one late. Pain is sometimes associated. His sister also has had TMJ dysfunction symptoms for many years, and both of them grind their teeth during their sleep. What is the explanation for the two clicks? Is there any congenital anatomical predisposition to the condition displayed in this patients?

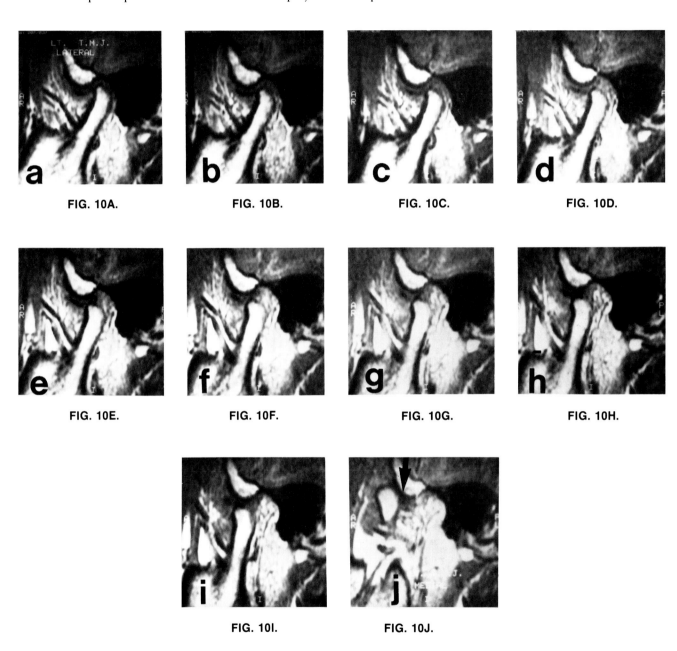

FIG. 10A.   FIG. 10B.   FIG. 10C.   FIG. 10D.

FIG. 10E.   FIG. 10F.   FIG. 10G.   FIG. 10H.

FIG. 10I.   FIG. 10J.

Findings: Kinematic series of the left TMJ (Fig. 10A–J). In the first image (Fig. 10A), the disk (curved arrow) is anteriorly dislocated but is recaptured early in the jaw opening series. At full jaw opening (Fig. 10J), the condyle dislocates anterior to the anterior band (arrow), resulting in the second click. The anterior translation is greater between the last two images (Fig. 10I, J) than between any other two images, although jaw opening increments are the same throughout the study.

Diagnosis: Anterior dislocation with reduction. Hypermobile TMJ with anterior dislocation of the condyle relative to the disk, in the widely jaw-opened position.

Discussion: The initial click is explained by the reduction of the disk as the jaw opens. The intermittent second click is apparently secondary to the hypermobility of the jaw indicated by the marked anterior translation of the condyles. The patient's jaw opening was approximately 60 mm. The kinematic series shows that the translation is much greater at the last step, although the jaw opening increments were constant at 8 mm. The condyle actually dislocates anterior to the disk itself, indicating very lax ligaments possibly related to the patient's history of grinding his teeth during sleep or to a familial tendency. Normally the condyle translates to a position close to the inferior aspect of the articular eminence, or slightly anterior to this, but not as far anterior as in this patient. The anterior dislocation of the condyle relative to the disk is not always associated with hypermobility. There is some suggestion in the literature that hyper-mobility may be more common in patients with a relatively flat articular eminence, as in this patient.

Because the patient voluntarily restricted jaw opening to prevent pain, he did not experience the terminal click as often as the initial click.

CASE 11

This patient has a long history of TMJ pain and dysfunction with limited opening. How do the two sides differ in appearance, position, and motion?

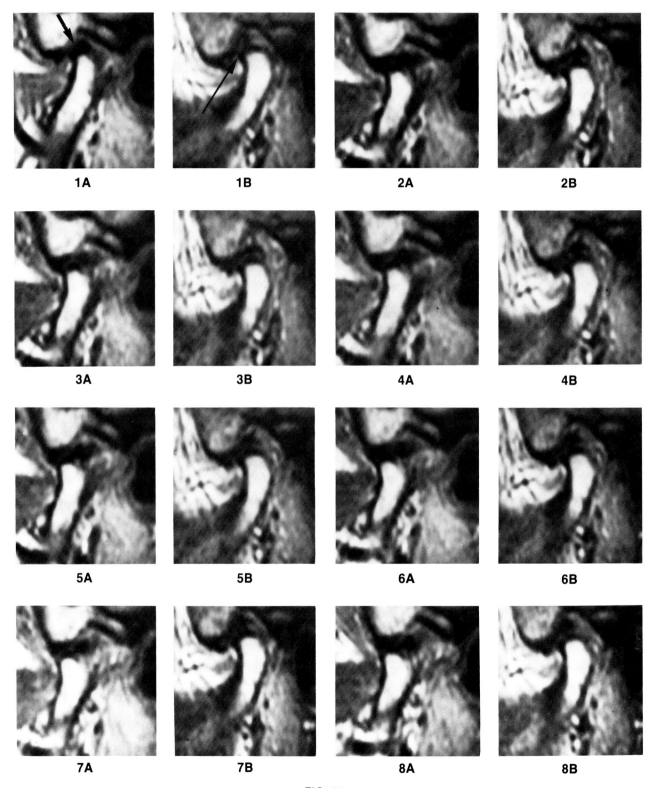

1A 1B 2A 2B

3A 3B 4A 4B

5A 5B 6A 6B

7A 7B 8A 8B

FIG. 11.

Findings: Kinematic series of the right (Fig. 11[1A–8A]) and left (Fig. 11[1B–8B]) TMJs. In the closed-mouth position (Fig. 11[1A, B]) arrows indicate sclerosis of the right condyle and a spur on the left. There is mild anterior subluxation of the left disk (the posterior band in Fig. 11[1B] is not directly above the condyle), which reduces after minimal jaw opening (Fig. 11[2B]). The left articular eminence is steeper than the right. The jaw opening series shows progressive anterior translation of the right condyle (Fig. 11[1A,1B]). However, after initial anterior translation on the left (Fig. 11[1B–3B]) there is retrograde motion (Fig. 11[4B–8B]), the condyle almost returning to a closed-mouth position within the glenoid (Fig. 11[7B, 8B]). The disks translate with the condyles.

Diagnosis: Disordered condylar translation as the jaw opens.

Discussion: Although bony degenerative changes are usually preceded by disk dislocation, this is not always the case, as in this patient with no dislocation on the right and only mild displacement with very early reduction on the left.

On the kinematic motion studies, condylar motion is very disordered. The explanation is not apparent, but the most likely possibility is muscular spasm, which is induced by pain as the jaw opens. Shortened muscles, fibrous adhesions, or fibrotic contractures are other considerations. Reduced motion of a condyle may also be secondary to obstruction related to bony degenerative changes or to soft tissue block (anterior disk displacement without reduction), neither of which is present in this patient.

There is asymmetry of the articular eminences, with the angle on the left being steeper than on the right. Several authors have suggested that an increased steepness of the articular eminence predisposes to TMJ dysfunction.

This case illustrates the necessity of simultaneous bilateral motion studies, without which it would not have been possible to assess adequately the degree of asymmetrical motion.

CASE 12

This 24-year-old male wrestler suffered a jaw dislocation during a match several years previously. Since then he has had intermittent jaw dislocation while chewing, frequently associated with a grinding sensation. Does the kinematic series indicate lateral jaw motion?

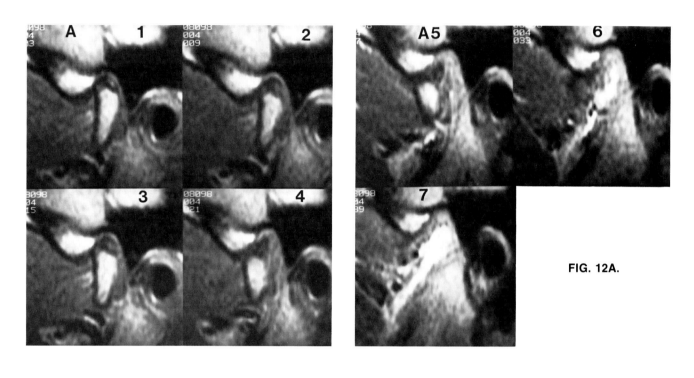

FIG. 12A.

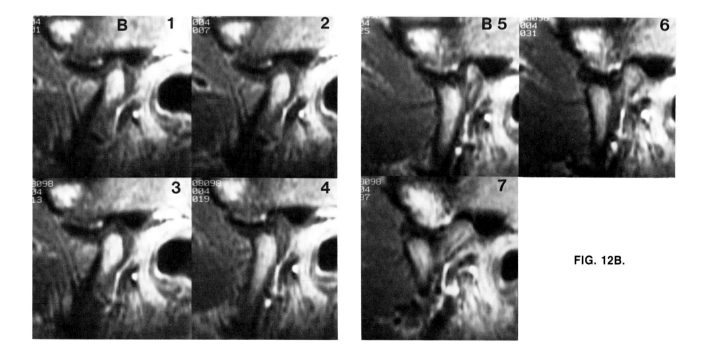

FIG. 12B.

Findings: The static images (not shown) are normal. The left kinematic series (Fig. 12A[1–7]—most lateral positions; Fig. 12B[1–7]—most medial positions) also initially appears normal. However, careful observation indicates that the left condyle is moving to the left (laterally) as the jaw opens. In the early phase of jaw opening, the condyle is well seen both laterally and medially. As the jaw opens widely (positions 6 and 7), the condyle shape is minimally changed laterally and the condyle is out of plane medially, indicating lateral (to the left) motion.

Diagnosis: Jaw instability with left lateral subluxation during jaw opening.

Discussion: Lateral jaw motion is usually clinically evident. Evaluation would be easier with coronal images, but anteroposterior head position often changes with jaw opening, resulting in inadequate coronal motion studies. It is therefore necessary to observe the sagittal kinematic images carefully to determine if the condyles are moving sideways. The kinematic study on the right (not shown) also demonstrates condylar motion to the left, resulting in the condyle being not visible on the most lateral images at the opened-mouth position.

CASE 13

A 70-year-old man had a crown cemented 2 days before developing a "locked jaw." His dentist requested MR to determine if the jaw locking was secondary to a mechanical block, in particular, disk displacement. Is the meniscus normal in appearance and position? Why is the jaw locked?

Findings: Closed-mouth sagittal images were obtained from the left side (Fig. 13A), the side of the dental procedure. The posterior and anterior bands of the disk (short arrows) are in normal positions, but there is elongation of the intermediate zone (long arrow). On the kinematic series (Fig. 13B) made at progressive 2-mm intervals of jaw opening, there is no anterior translation of the condyle or disk. Jaw opening was limited to 10 mm.

Diagnosis: Delayed trismus secondary to dental procedure with mandibular block.

Discussion: Acute onset of locked jaw may be secondary to a mechanical obstruction such as disk displacement or to muscular spasm (trismus). Delayed trismus after a dental procedure with a mandibular block is a known phenomenon. The MR excludes mechanical obstruction in this patient and suggests trismus by demonstrating the stretched disk, presumably secondary to spasm of the superior belly of the external pterygoid muscle, which is attached to the anterior joint capsule and disk.

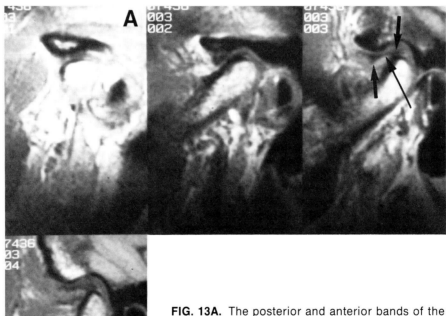

FIG. 13A. The posterior and anterior bands of the disk (*short arrows*) are in normal position on the sagittal closed-mouth series. There is elongation of the intermediate zone (*long arrow*), suggesting stretching or ventral pulling by a spastic external pterygoid muscle.

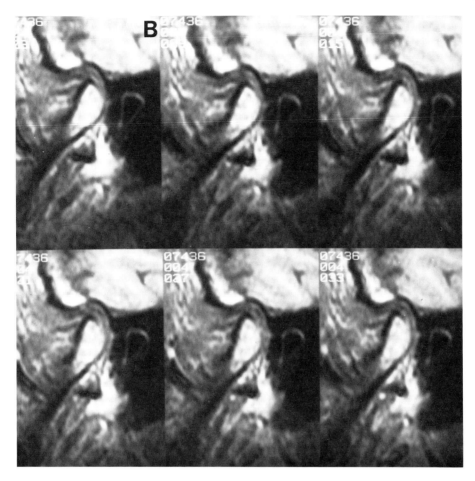

FIG. 13B. There is no anterior translation of the condyle or disk on the kinematic series made at progressive 2-mm intervals of jaw opening (maximum jaw opening of 10 mm).

CONCLUSION

Clinical evaluation and treatment of TMJ disorders may be very difficult. Many patients with TMJ dysfunction and pain can be successfully treated with conservative methods such as counseling, analgesics, and muscle relaxants, whereas others will require long-term and, possibly, surgical therapy.

A multitude of imaging techniques have been used to evaluate the TMJ. Radiography, especially tomography, delineates osseous changes, the position of the condyle, the shape of the glenoid, the angle of the articular eminence, and the joint space width (Fig. H). Soft tissue changes are not primarily visualized and may only be assumed based on the radiographic findings.

Arthrography permits direct visualization of the disk and joint spaces. Tears and perforations of the disk and bilaminar zone may be delineated. Arthrofluoroscopy has been successfully utilized to study disk and condylar motion during jaw opening and closing. Unfortunately, arthrography is invasive, may be painful, and is not without risk.

Computed tomography was the first noninvasive imaging technique that visualized the disk while also defining bony pathology. However, interpretation is very difficult, requiring the use of the blink mode and extensive radiologist involvement (Fig. I,J). It entails radiation exposure and is limited in its information content regarding the disk (mainly delineating disk dislocation) and does not allow for dynamic studies.

Magnetic resonance imaging is a major advance in TMJ imaging. Although the bone detail is not equivalent to that of CT or tomography, most clinically significant osseous abnormalities may be adequately defined. The shape of the glenoid, articular eminence, and the condyle may be delineated, and sclerosis and spurs of these structures are shown with sufficient clarity to obviate the necessity of tomography or CT in most patients studied by MR. The disk may be directly and exquisitely visualized, and its motion and that of the condyle may be evaluated. Although arthrography allows a more detailed evaluation of the joint spaces and disk motion and remains the gold standard for the evaluation of tears and perforations, it is an invasive, often painful procedure with inherent risks. Further, the information content of state-of-the-art MRI in many ways surpasses that of arthrography and will continue to improve as techniques and equipment evolve.

The technical aspects of MRI of the TMJ require careful attention to detail if optimal images are to be obtained. Bilateral simultaneous imaging, particularly for the motion studies, is a prerequisite to a quality examination, as we have shown in several case examples.

However, it is extremely important to remain cognizant of the fact that MRI is only one facet of TMJ evaluation. An excellent history and physical examination by a qualified clinician must precede and determine the necessity for imaging studies and will be required to determine the significance of the findings. Symptoms and, to a lesser extent, signs are not always clarified by imaging studies, and in fact, the images may on occasion seem incompatible with clinical findings. An astute and able clinician is necessary to weigh the importance of all the information obtained. It is not for the radiologist to try to determine the significance of specific imaging findings but rather to offer them as one piece of the puzzle.

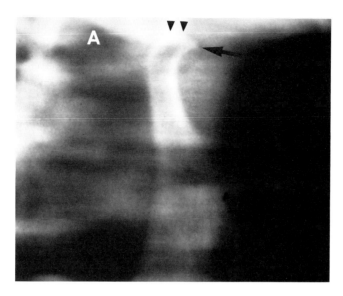

FIG. H. Sagittal linear tomogram of the mandibular condyle. Anterior spurring (*long arrow*) and superior sclerosis (*arrowheads*) are apparent.

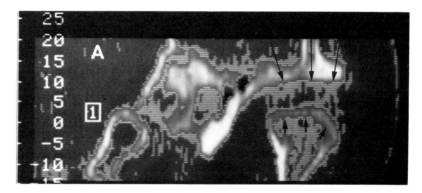

FIG. I. Coronal CT, blink mode, of a normal TMJ. The condyle (*short arrows*) is normal in shape and the disk (*long arrows*) is directly above the condyle.

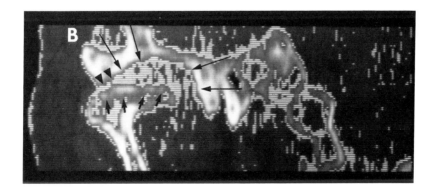

FIG. J. Coronal CT, blink mode, of medial disk displacement. The condyle (*short arrows*) is deformed, the lateral joint space (*arrowheads*) is narrowed, and the disk material (*long arrows*) extends medial to the condyle.

RECOMMENDED READING

Avrahami E, Schreiber R, Benmair J, Paltiel Z, Machtey J, Horowitz I. Magnetic resonance imaging of the temporo-mandibular joint and meniscus dislocation. *Br J Radiol* 1986;59:1153–1158.

Bell KA, Walters PJ. Videofluoroscopy during arthrography of the temporomandibular joint. *Radiology* 1983;147:879.

Burnett KR, Davis CL, Read J. Dynamic display of the temporo-mandibular joint meniscus by using "fast-scan" MR imaging. *AJR* 1987;149:959–962.

Cirbus MT, Smilack MS, Beltran J, Simon DC. Magnetic resonance imaging in confirming internal derangement of the temporo-mandibular joint. *J Prosthet Dent* 1987;57:488–494.

Donlon WC, Moon KL. Comparison of magnetic resonance imaging, arthrotomography and clinical and surgical findings in temporomandibular joint internal derangements. *Oral Surg Oral Med Oral Pathol* 1987;64:2–5.

Hardy CJ, Katzberg RW, Frey RL, Szumowski J, Totterman S, Mueller OM. Switched surface coil system for bilateral MR imaging. *Radiology* 1988;167:835–838.

Harms SE, Wilk RM. Bilateral surface coils for temporomandibular joint imaging. *Society for Magnetic Resonance in Medicine, Book of Abstracts*, vol 1. 1987;139.

Harms SE, Wilk RM. Magnetic resonance imaging of the temporomandibular joint. *Radiographics* 1987;7:521–542.

Harms SE, Wilk RM, Wolford LM, Chiles DG, Milam SB. The temporomandibular joint: magnetic resonance imaging using surface coils. *Radiology* 1985;157:133–136.

Helms CA, Gillespy T, Sims RE, Richardson ML. Magnetic resonance imaging of internal derangement of the temporomandibular joint. *Radiol Clin North Am* 1986;24:189–192.

Helms CA, Katzberg RW, Dolwick MF. *Internal derangements of the temporomandibular joint.* San Francisco: Radiology Research and Education Foundation, University of California Printing Department, 1983.

Helms CA, Richardson ML, Moon KL, Ware WH. Nuclear magnetic resonance imaging of the temporomandibular joint: preliminary observations. *J Craniomand Pract* 1984;2:219–224.

Hyde JS, Foncisz W, Jesmanowicz A, Kneeland JB, Grist TM. Parallel image acquisition with non-interacting local coils. *Soc Magn Reson Med,* Book of Abstracts. 1986;343–344.

Kaplan PA, Tu HK, Williams SM, Lydiatt DD. The normal temporomandibular joint: MR and arthrographic correlation. *Radiology* 1987;165:177–178.

Katzberg RW. Temporomandibular joint imaging. *Radiology* 1989;170:297–307.

Katzberg RW, Bessette RW, Tallents RH, Plewes DB, Manzione JV, Schenck JF, Foster TH, Hart HR. Normal and abnormal temporomandibular joint: MR imaging with surface coil. *Radiology* 1986;158:183–189.

Katzberg RW, Schenck J, Roberst D, Tallents RH, Manzione JV, Hart HR, Foster TH, Wayne WS, Bessette RW. Magnetic resonance imaging of the temporomandibular joint meniscus. *Oral Surg Oral Med Oral Pathol* 1985;59:332–335.

Katzberg RW, Westesson P-L, Tallents RH, Anderson R, Kurita K, Manzione JV, Totterman S. Temporomandibular joint: MR assessment of rotational and sideways disk displacements. *Radiology* 1988;169:741–748.

Khoury MB, Colan E. Sideways dislocation of the temporomandibular joint meniscus: the edge sign. *AJNR* 1986;7:869–872.

Laurell KA, Tootle R, Cunningham R, Beltran J, Simon D. Magnetic resonance imaging of the temporomandibular joint. Part I: literature review. *J Prosthet Dent* 1987;58:83–89.

Laurell KA, Tootle R, Cunningham R, Beltran J, Simon D. Magnetic resonance imaging of the temporomandibular joint. Part II: comparison with laminographic, autopsy, and histologic findings. *J Prosthet Dent* 1987;58:211–218.

Laurell KA, Tootle R, Cunningham R, Beltran J, Simon D. Magnetic resonance imaging of the temporomandibular joint. Part III: use of a cephalostat for clinical imaging. *J Prosthet Dent* 1987;58:355–359.

Manzione JV, Tallents RH, Katzberg RW, Oster C, Miller TL. Arthrographically guided therapy for recapturing the temporomandibular joint meniscus. *Oral Surg Oral Med Oral Pathol* 1984;57:235–240.

Pressman BD, Shellock FG. Static and kinematic MR imaging. *California Dent Assoc J* 1988;August:32–37.

Pressman BD, Shellock FG, Tourje JE. Dual coil MR imaging of temporomandibular joints: value in assessment of asymmetrical motion abnormalities. *AJNR* (abstract) 1988.

Roberts D, Schenck J, Joseph P, Foster T, Hart H, Pettigrew J, Kundel HL, Edelstein W, Haber B. Temporomandibular joint: magnetic resonance imaging. *Radiology* 1985;154:829–830.

Sanchez-Woodworth RE, Tallents RH, Katzberg RW, Guay JA. Bilateral internal derangements of temporomandibular joint: evaluation by magnetic resonance imaging. *Oral Surg Oral Med Oral Pathol* 1988;65:281–285.

Schellhas KP, Wilkes CH, Fritts HM, Omlie MR, Heithoff KB, Jahn JA. Temporomandibular joint: MR imaging of internal derangements and postoperative changes. *AJR* 1988;150:381–389.

Schellhas KP, Wilkes CH, Omlie MR, Block JC, Larsen JW, Idelkope BI. Temporomandibular joint imaging. Practical application of available technology. *Arch Otolaryngol Head Neck Surg* 1987;113:744–748.

Schellhas KP, Wilkes CH, Omlie MR, Peterson CM, Johnson SD, Keck RJ, Block JC, Fitts HM, Heithoff KB. The diagnosis of temporomandibular joint disease: two compartment arthrography and MR. *AJNR* 1988;9:579–588.

Shellock FG, Pressman BD. Dual-surface-coil imaging of bilateral temporomandibular joints: improvements in the imaging protocol. *AJNR* 1989;10:595–598.

Tallents RH, Katzberg RW, Miller TL, Manzione JV, Oster C. Arthrographically assisted splint therapy. *J Prosthet Dent* 1985;53:235–238.

Westesson P-L, Bronstein SL, Liedberg J. Temporomandibular joint: correlation between single-contrast videoarthrography and postmortem morphology. *Radiology* 1986;160:767–771.

Westesson P-L, Katzberg RW, Tallents RH, Sanchez-Woodworth RE, Svensson SA. CT and MR of the temporomandibular joint: comparison with autopsy specimens. *AJR* 1987;148:1165–1171.

Westesson P-L, Katzberg RW, Tallents RH, Sanchez-Woodworth RE, Svensson SA, Espeland MA. Temporomandibular joint: comparison of MR images with cryosectional anatomy. *Radiology* 1987;164:59–64.

Wilk RM, Harms SE, Wolford LM. Magnetic resonance imaging of the temporomandibular joint using a surface coil. *J Oral Maxillofac Surg* 1986;44:935–943.

CHAPTER 10

Kinematic MRI of the Joints

Frank G. Shellock and Bert R. Mandelbaum

In recent years, the advent of innovative diagnostic techniques, development of low morbid therapeutic interventions, and utilization of creative rehabilitation protocols have contributed to the optimal care of those afflicted with musculoskeletal disorders. In the area of diagnosis, the use of static MRI has been demonstrated to be extremely effective throughout the musculoskeletal system, and many types of studies are now applied as the primary examination to assess a variety of abnormalities. In the evaluation of patients with joint pain, the majority of patients present with arthralgia related to specific movements or activities during movement and/or forceful loading of the joint. Consequently, it is often imperative to not only define anatomic abnormalities but to evaluate the complex functional changes that may occur during the range of motion of the joint. The recent development and implementation of kinematic MRI to these specific disorders provides the clinician with an augmented data base of anatomy and motion-related information. This latest improvement in MRI technology has resulted in refined conceptional interpretations of a multitude of musculoskeletal conditions. Kinematic MRI protocols have been most effectively utilized to examine idiopathic pain syndromes that affect the patellofemoral joint, wrist, and ankle. In this chapter, the technique of kinematic MRI will be described and the applications for evaluating the above-mentioned joints will be discussed.

THE PATELLOFEMORAL JOINT

Technique for Kinematic MRI of the Patellofemoral Joint: A patient-activated non-ferromagnetic positioning device designed to permit simultaneous axial imaging of both patellofemoral joints with the knees flexed at 5, 10, 15, 20, 25, and 30° is used for the kinematic MRI study (Fig. A). A T1-weighted spin-echo pulse sequence is typically used, as follows: TR/TE 500/20 msec; number of excitations, 0.5; matrix, 256 × 128; field of view, 38 cm; slice thickness, 5 mm; interslice gap, 0.5 mm; imaging plane, axial; acquisition time, 34 sec. These parameters provide good anatomic detail so that the relationship of the patella to the femoral trochlear groove can be assessed. In addition, the data acquisition time is kept to a minimum. After completion of the scan, the patient is instructed to adjust the handle on the positioning device to move to the next increment of knee flexion. In this manner, six images are acquired at several different slice locations through the patellofemoral joint. The total examination time using the above-mentioned parameters is approximately 15 min. The acquired images are inspected, and slice locations that best depict the patella and its orientation to the femoral trochlear groove are selected and filmed for the kinematic study. Computer software is used to produce a "cine-loop" format that expedites the viewing of the multiple images. The cine-loop format also appears to enhance the ability to identify subtle patellar tracking abnormalities.

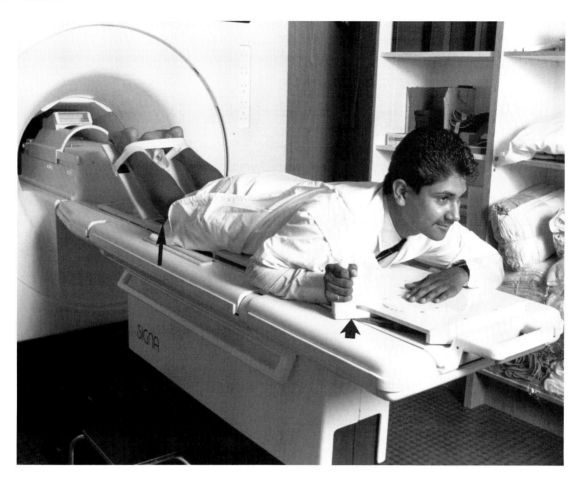

FIG. A. Positioning device made from nonferromagnetic materials used for kinematic MRI of the patellofemoral joint. This device flexes the knees up to 30° of flexion at 5° increments using a patient-activated mechanism (*short arrow*). A gauge indicates the degree of flexion. A cut-out area permits uninhibited movement of the patellofemoral joint (*long arrow*).

CASE 1

A 16-year-old girl presented with an 8-month history of anterior pain that was greater on the right compared to the left knee. Physical examination showed no remarkable ligamentous laxity or signs suggestive of meniscal derangement. Routine MRI high resolution views of both knees did not demonstrate any abnormalities. Because abnormal patellar tracking is a common cause of knee pain, particularly in young women, the patient was referred for a kinematic MRI of bilateral patellofemoral joints. What are the findings? (Fig. 1, TR/TE 500/20 msec; 5 = mm axial sections; 0.5 = mm gap; obtained at the same slice location through the patellofemoral joints with the knees flexed from 5 to 30°. Unless otherwise indicated, all kinematic studies were obtained with the above imaging parameters.)

Findings: The multiple sequential images demonstrate normal patellar alignment for both the right and left patellofemoral joints (Fig. 1). Note that because the patellae move distally during knee flexion, their size and/or shape appears to change during each progressive increment of flexion (this is more evident on the images of the left patellofemoral joint). The scattered areas of decreased signal intensity seen on the left side represent the open growth plate of this 16-year-old patient.

Diagnosis: Normal bilateral patellar tracking.

Discussion: Normal patellar tracking is displayed with the apex of the patella positioned in the femoral trochlear groove as it travels in a vertical plane during flexion of the knee, without transverse displacement of the medial or lateral facets of the patella. This orientation causes the patella to appear "centered" in the femoral trochlear groove. The kinematic MRI evaluation of the right and left patellofemoral joints ruled out abnormal patellar tracking as the potential cause of bilateral knee pain in this patient. Other possible causes of anterior knee pain (e.g., synovitis, reflex sympathetic dystrophy, inflamed plica, arthritis, or referred pain) should be considered, and further diagnostic evaluation is warranted.

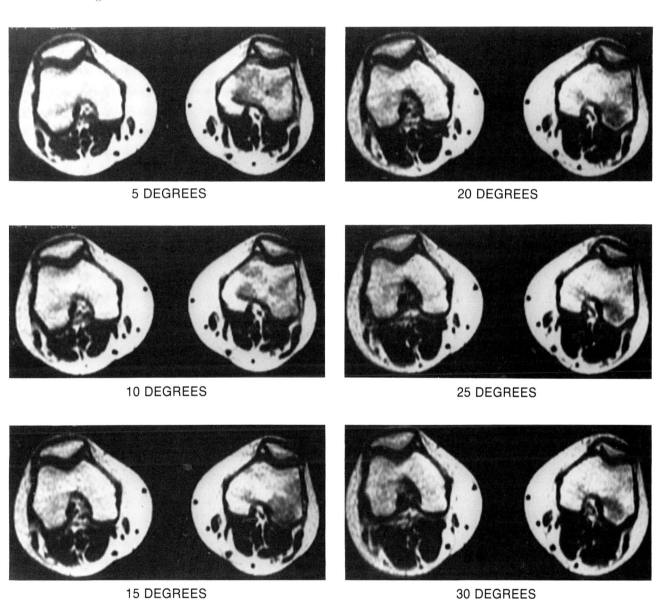

5 DEGREES

20 DEGREES

10 DEGREES

25 DEGREES

15 DEGREES

30 DEGREES

FIG. 1.

CASE 2

A 13-year-old girl with chronic lateral joint-line pain in the left knee and a normal Q angle underwent a kinematic MRI of the patellofemoral joint, which identified a slight lateral subluxation and tilt of the patella. What are the overall findings (Fig. 2A)? Does the kinematic MRI of the patellofemoral joint provide any information concerning the reason for this abnormal patellar alignment?

FIG. 2A.

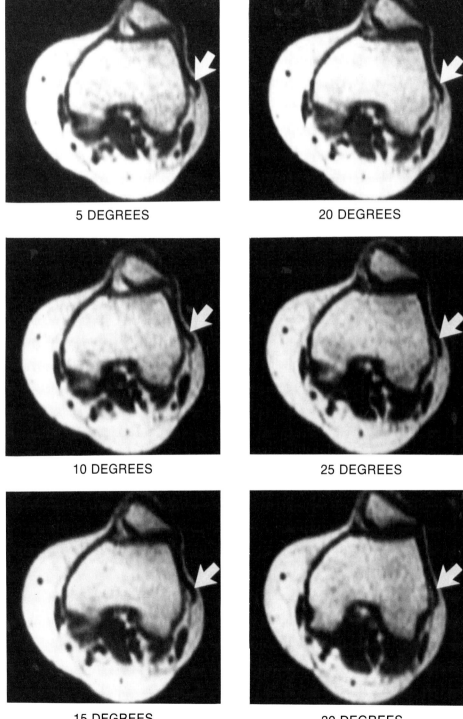

5 DEGREES   20 DEGREES

10 DEGREES   25 DEGREES

15 DEGREES   30 DEGREES

Findings: The lateral retinaculum appears redundant up to 25° of flexion (arrows), suggesting that the lateral displacement and tilting of the patella is not caused by excessive force produced by exorbitant stretching of the lateral retinaculum. In addition, the patella appears misshapened and the femoral trochlear groove is dysplastic (note that it may be necessary to inspect several slice locations through the patellofemoral joint to appreciate thoroughly the shapes of both the patella and femoral trochlea).

Diagnosis: Slight lateral subluxation and tilting of the left patella with abnormal configurations of the patella and femoral trochlear groove.

Discussion: The normal motion of the patella is maintained by the interplay between static stabilizers (i.e., ligaments such as the patellar tendon, medial patellofemoral ligament, medial meniscal patellar ligament, lateral patellofemoral ligament, lateral meniscal patellar ligament, medial retinaculum, lateral retinaculum, and facia lata), dynamic stabilizers (i.e., muscles such as rectus femoris, vastus medialis, vastus lateralis, vastus intermedius, and rectus femoris), bony structures (in particular, the shapes of the patella and the femoral trochlear groove), and the alignment of the femur and tibia (the Q angle). The disruption of one or more of the above may result in abnormal patellar tracking. Considering that the interplay between the factors that are responsible for the alignment of the patella may be simple or complicated depending on the relative involvement of each of these components, determination of the mechanism(s) causing abnormal patellar tracking is not always possible. However, clues may be obtained from the kinematic MRI examination.

Because the patella articulates with the femoral trochlear groove during knee flexion, congruous formations of these structures are crucial because they produce mechanical restraints to aberrant motion. Therefore, careful inspection and classification of the shapes of the patella and femoral trochlea may provide additional evidence concerning the presence of patellofemoral disorders. It should be noted, however, that pathologic morphology does not always preclude normal patellar tracking.

A variety of patellar forms are consistent with normal patellofemoral joint function, whereas others are associated with patellar malalignment or dislocation. The most commonly found patellar configurations considered to be normal are the Wiberg Types I (lateral and medial facets are symmetrical, Fig. 2B) and II (the lateral facet is longer than the medial facet, Fig. 2C) shapes. The Wiberg Type III patellar shape (the lateral facet is dominant with a significantly smaller medial facet, Fig. 2D) and other dysplastic forms (Fig. 2E,F) are typically observed in conjunction with patellofemoral arthrosis, chondromalacia, and recurrent dislocation.

The normal femoral trochlear shape has well-defined medial and lateral facets that are either equal in size or with a slightly larger lateral facet. Most importantly, the shape of the groove must conform to the shape of the patella for proper articulation. Abnormal shapes of the femoral trochlear groove are highly variable insofar as the medial and/or lateral aspects may be hypoplastic or dysplastic (Fig. 2F–H). With some abnormal formations of the femoral trochlear groove, not only is the width affected but also the height (Fig. 2I).

The patient in this case example has abnormal patellar and femoral trochlear groove configurations, and these are likely to be partially responsible for her abnormal patellar tracking. In addition, the presence of a redundant lateral retinaculum found in conjunction with the slightly lateral position of the patella is an important finding in this patient insofar as it suggests that doing a lateral retinacular release, which is frequently performed in an attempt to realign a laterally displaced patella, would not be beneficial. Abnormal lateral forces and/or insufficient medial forces imposed by one or more of the other static stabilizers, combined with the abnormal bony anatomy, is probably causing patellar malalignment in this patient.

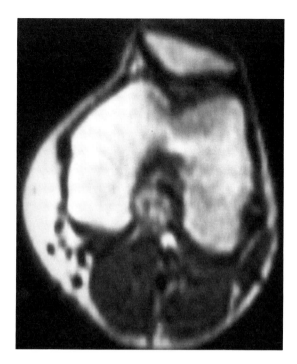

FIG. 2B. Example of Wiberg Type I patellar shape. The lateral and medial patellar facets are symmetrical.

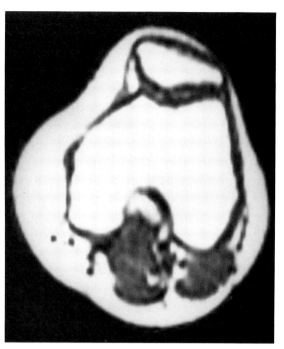

FIG. 2C. Example of Wiberg Type II patellar shape. The lateral patellar facet is longer than the medial patellar facet.

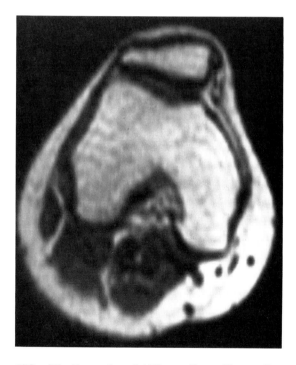

FIG. 2D. Example of Wiberg Type III patellar shape. The lateral patellar facet is dominant, with a significantly smaller medial patellar facet.

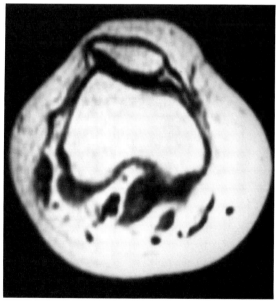

FIG. 2E. Example of a dysplastic patella. Also note that the medial aspect of the femoral trochlea is hypoplastic.

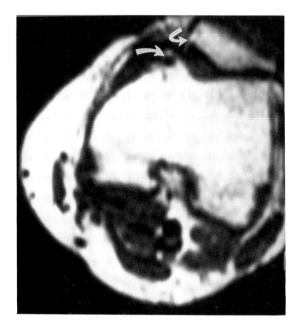

FIG. 2F. Dysplastic femoral condyle and dysplastic patella. Note the osteophytes on the medial aspects of the condyle and patella (arrows).

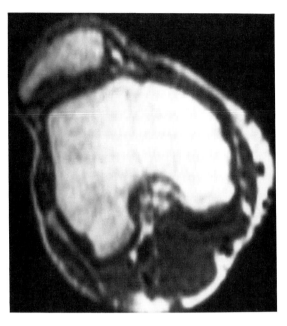

FIG. 2G. Dysplastic femoral condyle with Wiberg Type II patella.

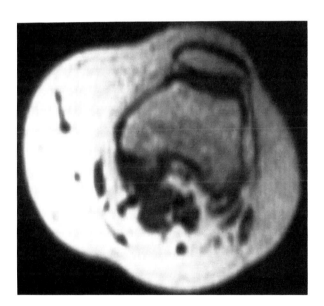

FIG. 2H. Dysplastic femoral condyle and dysplastic patella.

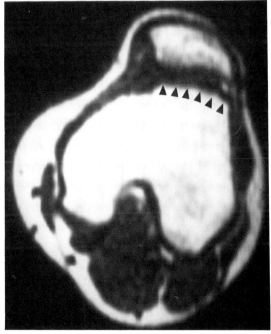

FIG. 2I. Dysplastic femoral condyle and Wiberg Type III patellar shape. Note the convex lateral femoral trochlea (*arrowheads*).

CASE 3

A 24-year-old man presents with a history of bilateral knee pain and repeated dislocations of the right and left patellae since childhood. Physical examination showed decreased joint mobility that was worse on the right compared to the left knee, lateral joint-line tenderness on the right knee, and bilateral patellar clicks. A plain film "skyline" view with the knees flexed at 45° and routine MRI studies of both knees were normal. Are any abnormalities seen on the images obtained with the knees extended (Fig. 3A) and/or the kinematic MRI study (Fig. 3B)?

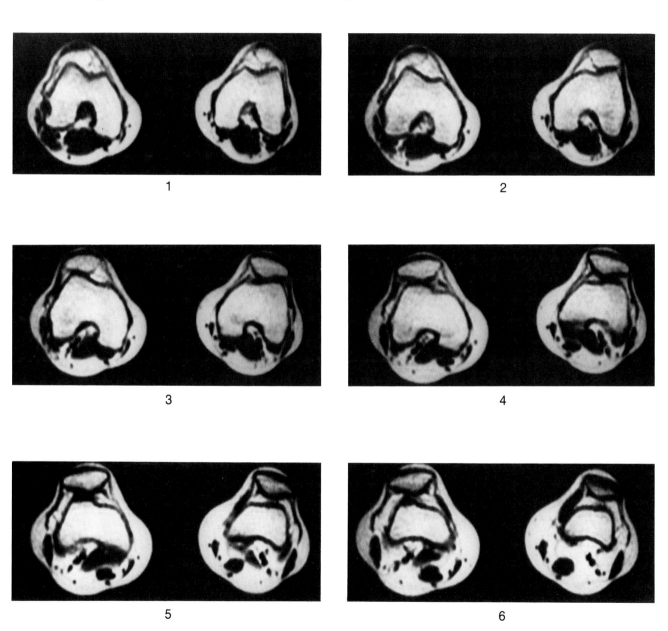

1

2

3

4

5

6

FIG. 3A.

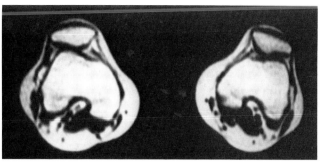

5 DEGREES

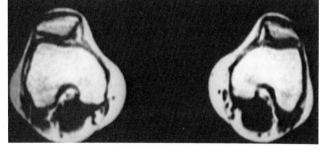

20 DEGREES

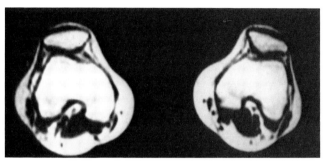

10 DEGREES

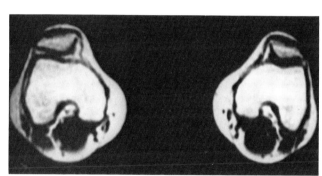

25 DEGREES

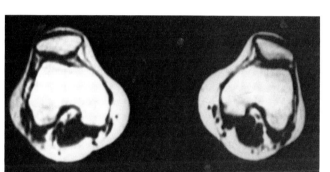

15 DEGREES

30 DEGREES

FIG. 3B.

Findings: The axial sections from the middle portions of the patellae (images 4 and 5) are positioned above the femoral trochlear grooves (images 1 and 2), demonstrating "functional" patella alta, with the right patella slightly higher than the left. The right and left patellae have Wiberg Type I shapes, and the medial aspects of the femoral trochleas appear to be hypoplastic.

The left patella shows a slight lateral subluxation at 5 and 10° of flexion but centers during the remaining degrees of flexion. The right patella is laterally subluxed from 5 to 20° and centers by 25° of flexion.

Diagnosis: Patella alta and lateral subluxation of the right and left patellae.

Discussion: Determining the reason(s) for knee pain using physical examination criteria alone may present a diagnostic dilemma because the symptoms of patellar malalignment and other types of derangements of the knee are similar. In addition, there may be combined lesions. Therefore, it is frequently necessary to obtain both routine MRI views and a kinematic MRI study of the patellofemoral joint to ascertain properly the etiology of knee pain. Of note is that plain films obtained with the knees flexed above 30° may or may not be helpful in the assessment of knee pain.

The confusing clinical presentation often derived from physical examination of the knee was the primary impetus for development of a variety of radiographic imaging procedures to show the relationship of the patella to the femoral trochlear groove. Most of these methods, however, have associated disadvantages including (a) the exposure to ionizing radiation, (b) the technical difficulty of positioning the patellofemoral joint tangentially to the x-ray beam and film, (c) the tendency for distortion of the image caused by superimposition of bony anatomy, (d) the inability to image the patellofemoral joint with the knee flexed less than 30°, and (e) the fact that usually only one image at a single joint angle is obtained and analyzed. These last two disadvantages can be especially detrimental to an accurate evaluation of patellar malalignment because subtle abnormalities may not be recognized if the initial degrees of knee flexion are not imaged. Obtaining true axial plane images of the patellofemoral joint during the early degrees of flexion using either MRI or CT will alleviate the above problems.

In this particular patient, plain film x-rays were obtained with both knees flexed at 45° and demonstrated normal alignment of the patellae relative to the femoral trochlear grooves. As the knee flexes, the patella is drawn deeper into the femoral trochlear groove by the extensor mechanism. This can cause a subluxed patella to move into a normal alignment and obscure patellar tracking abnormalities that occur during the early increments of flexion. For example, this patient's skyline view of the patellofemoral joints did not show the laterally subluxed patellae because it was obtained at 45° of flexion and both patellae were normally aligned by 25° of knee flexion. Besides the "centering" movement that may be observed during knee flexion, a patient with a laterally subluxed patella may either exhibit a persistence (Fig. 3C) or worsening (Fig. 3D) of this aberrant patellar position.

The additional finding of bilateral patella alta in this patient is significant insofar as patella alta is frequently associated with recurrent lateral subluxations and dislocations. Lateral subluxation occurs in conjunction with patella alta because the patella is not buttressed or restrained by the lateral femoral condyle when it is in this superior position.

Various radiographic criteria for diagnosing patella alta using plain films have been described. None of these techniques, however, is as useful as serial axial images for evaluating the relationship of the patella to the femoral trochlear groove and assessing the configurations of these structures. Patella alta may also be responsible for abnormal growth patterns of the patella and/or the femoral trochlear groove as seen in Fig. 3E.

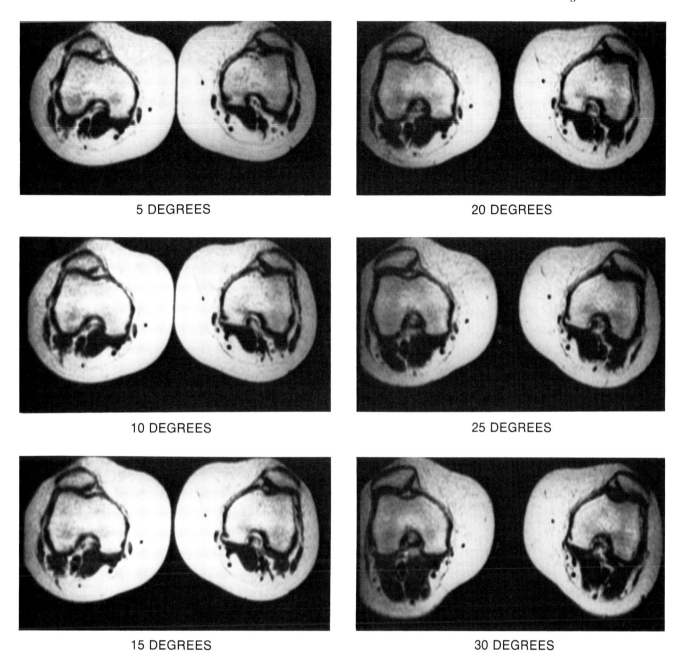

5 DEGREES

20 DEGREES

10 DEGREES

25 DEGREES

15 DEGREES

30 DEGREES

FIG. 3C. Both patellae are laterally subluxed throughout the range of motion evaluated and the relative displacement is constant. Note the gradual changes in the appearance of the patellae associated with distal movement during knee flexion.

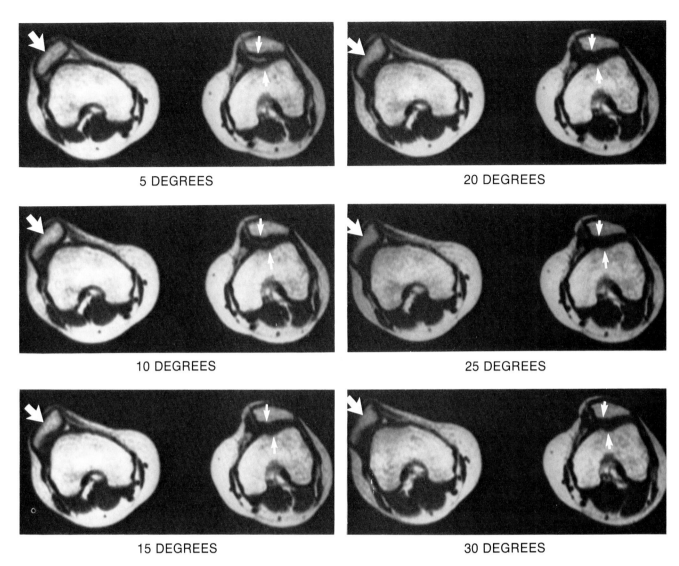

5 DEGREES

20 DEGREES

10 DEGREES

25 DEGREES

15 DEGREES

30 DEGREES

FIG. 3D. There is severe lateral subluxation of the right patella that appears to worsen with increasing increments of flexion (*large arrows*). The right trochlea is markedly dysplastic. The left patella is tilted and medially displaced (*small arrows*).

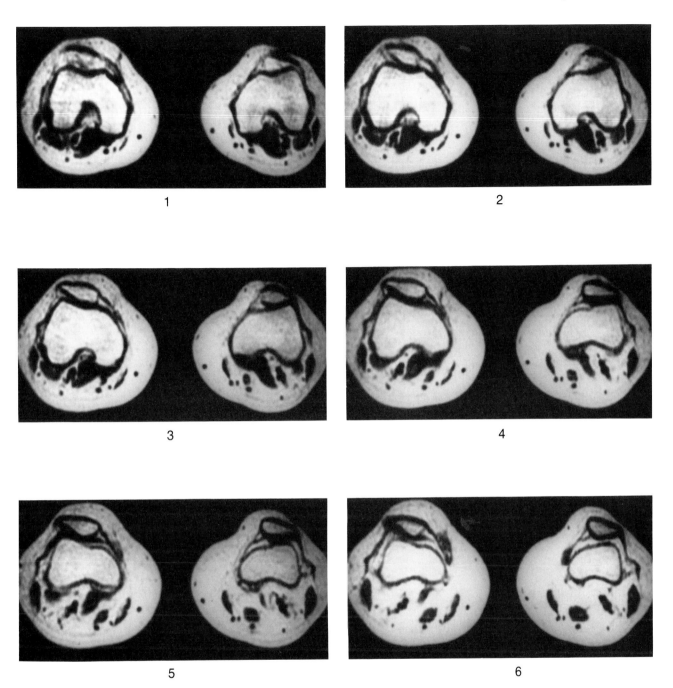

FIG. 3E. 5-mm, skip 0.5 mm, serial axial sections (TR/TE 500/20 msec) obtained with the knees extended. This patient has severe bilateral patella alta, with the patellae positioned well above the femoral trochlear grooves. The patellae are dysplastic, and the femoral trochleas do not have well-formed grooves.

CASE 4

A 25-year-old athletic man had complaints of pain in the lateral aspect of his left knee that persisted and progressively worsened for approximately 8 months. The patient stated that the pain was aggravated during activities such as running and playing racketball. Physical examination indicated that the patient had a normal range of motion and no evidence of gross joint effusion. Anterior drawer, Lachman, varus, and valgus maneuvers showed no evidence of ligamentous laxity. There was lateral joint-line tenderness, 2+ patellofemoral crepitus, and a positive patellar grind. Plain film radiographs demonstrated no bony abnormalities, and a routine MRI of the left knee displayed normal menisci and ligaments. What is the probable cause of this patient's pain (Fig. 4A)?

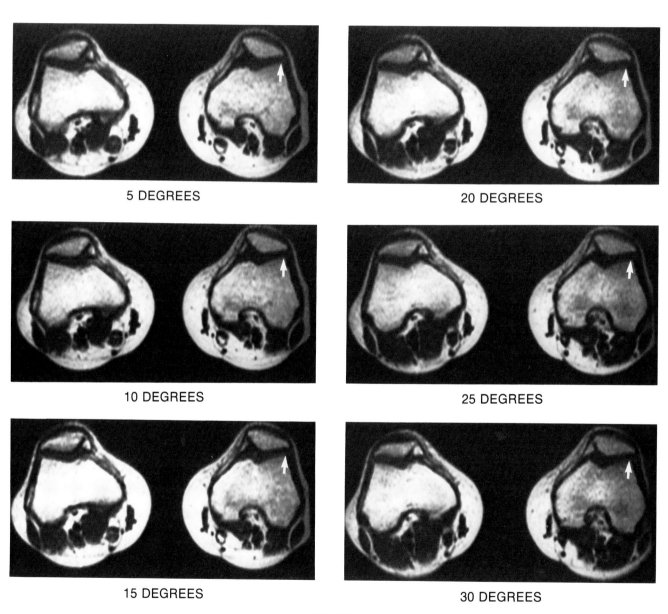

5 DEGREES

20 DEGREES

10 DEGREES

25 DEGREES

15 DEGREES

30 DEGREES

FIG. 4A.

Findings: The right patella is in a normal position during each increment of knee flexion studied. The left patella is tilted but not subluxed at 5° of flexion. During flexion from 10 to 30°, the patella demonstrates a slight lateral subluxation (i.e., the apex of the patella is laterally displaced relative to the femoral trochlear groove and the lateral facet of the patella is laterally displaced relative to the lateral aspect of the femoral trochlea, arrows) and is tilted. The increased lateral subluxation and tilting associated with the progressive increments of knee flexion indicate an inordinate amount of tension from soft tissue structures acting on the patella.

Diagnosis: Normal kinematic study of the right patellofemoral joint. Excessive lateral pressure syndrome with mild lateral subluxation of the patella on the left patellofemoral joint.

Discussion: Excessive lateral pressure syndrome (ELPS) was first described by Ficat and Hungerford and evolved from the routine use of obtaining x-rays at 30, 60, and 90° of knee flexion. This abnormality is considered to be a clinical-radiological entity, characterized clinically by patellofemoral pain and radiologically by tilting of the patella, with functional patellar lateralization usually onto a dominant lateral facet. A slight lateral subluxation of the patella may occur with the higher degrees of knee flexion, as increasing tension from one or more overly taut soft tissue structure(s) (i.e., lateral patellofemoral ligament, lateral meniscal patellar ligament, fascia lata) progressively displace the patella.

This case example demonstrates the usefulness of kinematic MRI of the patellofemoral joint in diagnosing ELPS because of the tendency for patellar tilt and/or lateral subluxation to worsen during flexion of the knee. In certain instances, ELPS may be evident by observing a lateral patellar tilt with the patella well centered in the femoral trochlear groove or with a slight medial displacement of the apex of the patella (Fig. 3D). Joint-line narrowing and bony changes are commonly found with this syndrome. The diagnostician should realize that large peripatellar effusions may also cause tilting and/or displacement of the patella, but this is not considered to be related to ELPS (Fig. 4B).

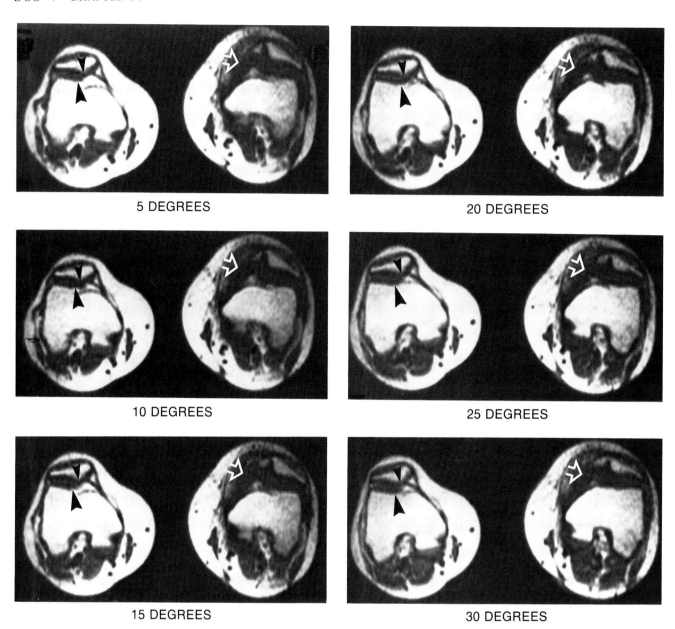

5 DEGREES

20 DEGREES

10 DEGREES

25 DEGREES

15 DEGREES

30 DEGREES

FIG. 4B. The left patella is tilted with a slight lateral displacement, possibly caused by the large peripatellar effusion (*open arrows*). There is medial subluxation of the right patella that worsens during incremental flexion (*arrowheads*).

CASE 5

A 26-year-old woman with a history of left knee pain for several years underwent an arthroscopic lateral retinacular release for correction of subluxation of the patella. Six months after the operation, the patient continued to have knee pain. Physical examination of lower extremity alignment indicated that the patient had moderate-to-severe foot pronation that was worse on the left compared to the right. There was femoral anteversion and compensatory external rotation of the left tibia. Anterior drawer, Lachman, varus, and valgus maneuvers demonstrated no evidence of ligamentous laxity. There was 1+ patellofemoral crepitus with a positive patellar grind. A plain film x-ray skyline view of the patellofemoral joint with the knee flexed at 45° showed normal alignment of the patella relative to the femoral trochlear groove, and no bony abnormalities were apparent (Fig. 5A). The patient was referred for kinematic MRI of the patellofemoral joint. What complication of lateral retinacular release is suspected to be responsible for knee pain in this patient (Fig. 5B)?

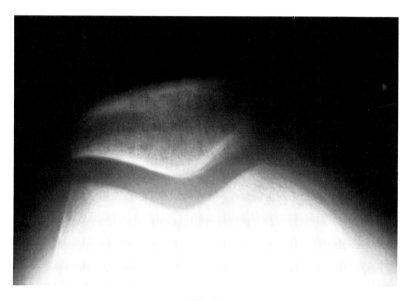

FIG. 5A.

FIG. 5B.

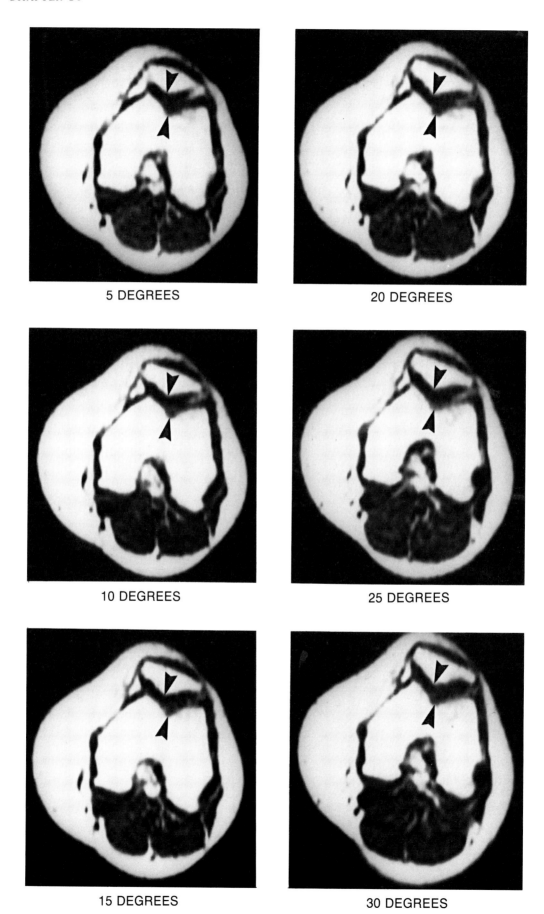

5 DEGREES

20 DEGREES

10 DEGREES

25 DEGREES

15 DEGREES

30 DEGREES

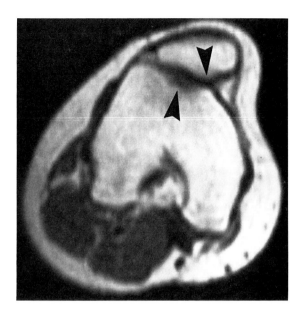

FIG. 5C.

Findings: The kinematic MRI of the left patellofemoral joint demonstrates that the medial facet of the patella is medially displaced relative to the medial aspect of the femoral trochlea, and the apex of the patella is medially displaced relative to the femoral trochlear groove from 5 to 30° of flexion (arrowheads).

Diagnosis: Medial subluxation of the patella, possibly secondary to lateral retinacular release.

Discussion: Medial subluxation of the patella has been observed in a high percentage of patients with persistent knee pain following lateral retinacular release. Diagnosis of medial subluxation of the patella by physical examination alone is obviously not a trivial task, however, because this aberrant patellar position has only recently been recognized and described. The plain film skyline view of the patellofemoral joint was not helpful in evaluating this patient. As indicated earlier in this chapter, there is often disparity between plain film radiographs of the patellofemoral joint with the knee flexed at 30° or greater and the patient's history and physical findings. Kinematic MRI of the patellofemoral joint is capable of clearly showing medial subluxation of the patella. Furthermore, this imaging technique is especially efficacious for evaluating the patient that presents with persistent knee pain following a surgical attempt to correct an abnormal patellar position.

Of note is that medial subluxation of the patella may also be found in patients without prior surgical realignment procedures (Figs. 4B and 5C). It is particularly important to identify medial subluxation of the patella and for the orthopedic surgeon to be aware of this clinical entity because classic surgical stabilization techniques such as medial transposition of the extensor mechanism and/or lateral release can further increase the medial displacement and not only lead to failure but can also exacerbate the patients' symptoms. Like other forms of aberrant patellar alignment, a medially displaced patella can cause significant localized pressure and eventually produce destruction of the cartilage (Fig. 5D). The development of chondromalacia caused by contact stress from a subluxing patella is probably the most common but frequently unappreciated etiologic factor responsible for knee pain.

In most instances, medial subluxation of the patella may be successfully treated with rehabilitation techniques. However, certain cases may require surgical intervention. For example, a young female patient with severe medial displacement of the patella underwent an arthroscopic medial retinacular release after rehabilitation for 9 months did not alleviate her symptoms. Although there was only a slight-to-moderate improvement in the medial position of the patella (Fig. 5E) compared to the position before surgery (Fig. 5F), surgical realignment did produce cessation of her knee pain.

To complicate the phenomenon of medial subluxation of the patella further, certain patients may exhibit a lateral-to-medial patellar movement. On the kinematic MRI examination, the patella starts in a slightly lateral position with the knee extended, moves into a central position during the early degrees of flexion, and is displaced medially at the higher degrees of flexion (Fig. 5G). The mechanism responsible for lateral-to-medial subluxation of the patella is complicated and not well understood. However, kinematic MRI is particularly suited to identify this unusual type of patellar tracking.

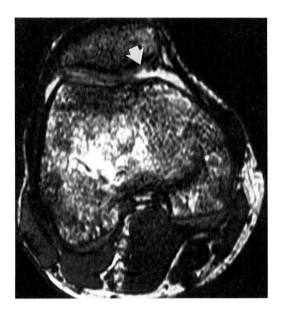

FIG. 5D. 1.5-mm axial section (TR/TE 60/15 msec, flip angle, 30°) obtained using 3DFT gradient-echo technique to optimize visualization of articular cartilage. Patient had medial subluxation of the patella. (This image was obtained with the patient's knee in an extended position; therefore, medial subluxation is not apparent on this view.) There is a defect in the cartilage localized to the medial facet of the patella that corresponds to the contact point observed on the kinematic study (*arrow*).

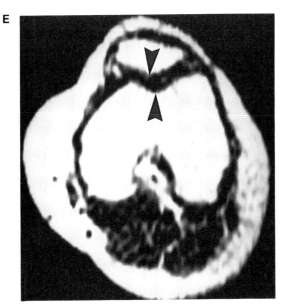

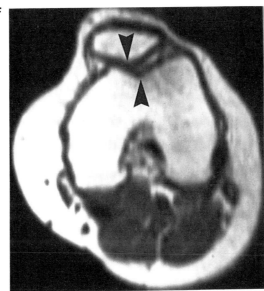

FIG. 5E, F. 5-mm axial sections (TR/TE 500/20 msec) obtained with the knee at 25° of flexion before **(F)** and after **(E)** medial retinacular release. There is less medial subluxation of the patella evident on the postsurgical image (*arrowheads*).

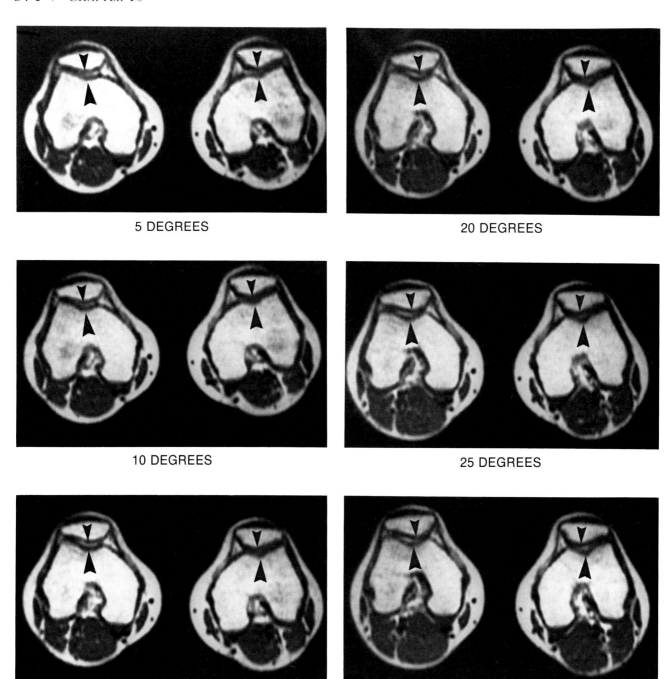

5 DEGREES

20 DEGREES

10 DEGREES

25 DEGREES

15 DEGREES

30 DEGREES

FIG. 5G. There is a slight lateral subluxation of the right patella at 5° of flexion, central, and finally, medial movement (note alignment of *arrowheads*) with increasing degrees of flexion. This kinematic pattern is indicative of lateral-to-medial subluxation of the patella. There is a slight medial subluxation of the left patella that is most apparent during the higher increments of flexion.

THE WRIST

Technique for Kinematic MRI of the Wrist: A Plexiglas device is used to place the wrist from extreme ulnar to radial deviation positions (Fig. B). The small size of this joint and anatomy of interest requires that high resolution images be obtained for the kinematic study. Therefore, either a 3″ or 5″ surface coil is suspended over the wrist, and multiple T1-weighted spin-echo images are acquired in the coronal plane for the kinematic MRI evaluation of the wrist. The patient is placed in a prone position with the elbow extended and foam padding positioned at various sites under the axilla, arm, elbow, and hand for support and comfort. Magnetic resonance imaging is performed with the joint placed in an extreme ulnar position and repeated as the wrist is incrementally placed into neutral and extreme radial deviation positions. Like other types of kinematic studies, either multiple static images or a cine-loop format is used to assess this examination.

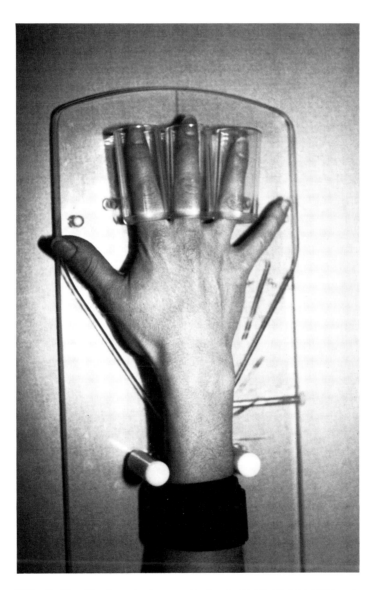

FIG. B. Positioning device made from nonferromagnetic materials used for kinematic MRI of the wrist. This device permits incremental placement of the wrist from extreme ulnar to radial deviation positions.

CASE 6

This 23-year-old male gymnast presents for evaluation of wrist pain. The patient has been competing in gymnastics for 16 years and, during the past 2 years, has developed a significant amount of pain associated with gymnastic activities localized to the dorsal ulnar aspect of the wrist. Pain is not present at rest but occurs with gymnastic activities including dorsiflexion, rotation, and compression during weight-bearing activities of the pommel horse and floor exercise. In this scenario, what is the optimal method of evaluation of the patient?

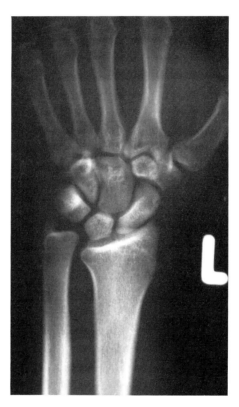

FIG. 6A.

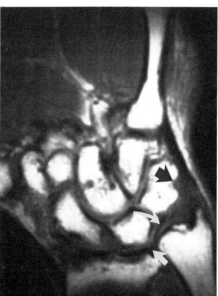

FIG. 6B.

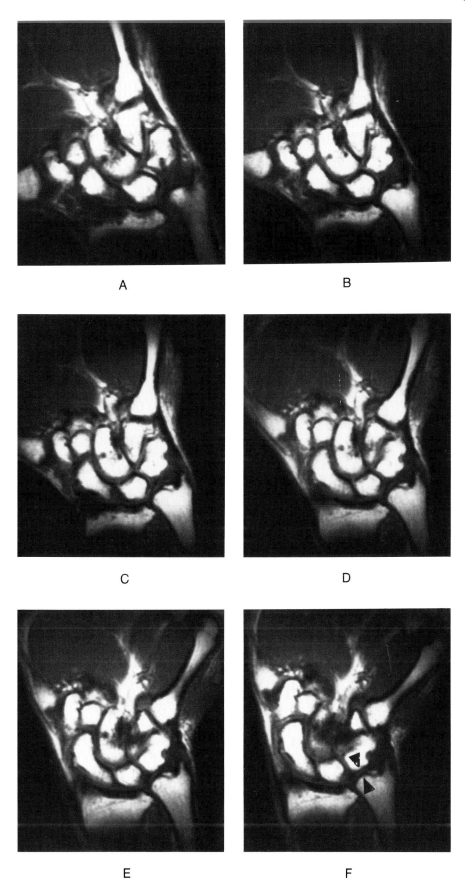

FIG. 6C.

Findings: Plain films indicate osseous changes including a 6-mm positive ulnar variance, a small triquetrial avulsion, and no evidence of alignment abnormalities (Fig. 6A). A tear of the triangular fibrocartilage is seen on arthrography. Static MRI views reveal cystic degeneration of the triquetrum (Fig. 6B, black arrow) as well as changes of the triangular fibrocartilage complex consistent with a tear (Fig. 6B—tear, white arrow; normal segment, curved arrow). In addition, kinematic MRI of the wrist indicates an impingement with ulnar deviation and loading of the distal ulnar on the triquetrum (arrowheads, Fig. 6C).

Diagnosis: Wrist pain secondary to abnormally severe positive ulnar variance with associated tear of the triangular fibrocartilage complex, erosions of the distal ulna, lunate, and triquetrum and associated triquetral degenerative cysts. Pathokinematic changes include motion-related impingement resulting in the pathoanatomy detected. This appears to result in painful motion and eventual degenerative changes along with erosion of the ulnar triquetral joint.

Discussion: Wrist pain in this athlete represents a difficult diagnostic and therapeutic problem. The upper extremity is not typically utilized for weight bearing, but during gymnastics, wrestling, and football, acute and chronic loads seem to have deleterious effects on the wrist joint. High levels of acute and chronic compression, tension, and shear forces result in a structurally maladaptive response. Resultant injury to the musculoskeletal system includes fractures; tendon, ligament, or fibrocartilage discontinuities; alignment changes; or loss of articulation. Clinically, these patients present with complaints of painful loss of motion, functional instability, weakness, and disability. The disorder can further be defined with regard to character, location, intensity, and associated symptoms that worsen it. Further elucidation of pain related to a particular position or motion is essential phenomena because, statically, there are no symptoms. Therefore, it is the integration of forces with specific motions that results in pain and associated loss of function. The development of appropriate and selective therapies is dependent on the definitive correlation between pathoanatomy, pathokinematics, and pain. It is this correlation, in view of the lack of diagnostic tools, that there is a significant challenge. Currently utilized diagnostic techniques include plain films, static and cine arthrography, and bone scintigraphy. However, these techniques do not define the complex transitions between the ulna, the triangular fibrocartilage, and the dorsal and volar ulnocarpal ligaments and triquetrum.

In this case, wrist pain is secondary to positive ulnar variance such that the ulna is 6 mm longer than the radius. The etiology and pathogenesis of this abnormal ulnar variance is chronic, repetitive upper extremity weight bearing resulting in injury to the distal radial physis. Positive ulnar variance has been associated with tears of the triangular fibrocartilage complex and articular erosions of the lunate and triquetrum. Alternatively, negative ulnar variance has been associated with progression of Kienböck's disease or avascular necrosis of the lunate.

Recent studies indicate that MRI is excellent for imaging the wrist. Pathoanatomically, static MRI provides information regarding ligament or tendon discontinuities, spatial orientations, and alignments. To date, MRI has been applied to patients with Kienböck's disease, triangular fibrocartilage (TFC) tears, dislocations, and scaphoid fractures. Impingements, instability patterns, and idiopathic pain syndromes related to movement cannot be characterized by static MRI techniques. There are several advantages of combining kinematic with static MRI of the wrist, which include:

1. The kinematic display enhances the visualization of the muscle, tendon, ligaments, and hyaline and fibrocartilage structures during controlled motion.
2. It is possible to calculate *in vivo* carpal translations from the kinematic study (Fig. 6D,E). Carpal instability patterns resulting from wrist ligament injuries can be difficult to diagnose because pain results from abnormal carpal motion. The

spectrum of abnormal kinematics has not yet been defined; however, an *in vivo* technique of kinematically assessing carpal translation allows accurate characterization of anatomic and kinematic clinical correlates.

3. Impingement entities related to motion may be better appreciated and diagnosed.
4. Alignment and orientation changes in wrist ligament injuries may be quantified.
5. Correlation of clinical data (i.e., pain, tenderness, clicks) associated with movement may be made from the kinematic study.

FIG. 6D. Example of scaphoid translation measurements during various degrees of radial/ulnar deviation.

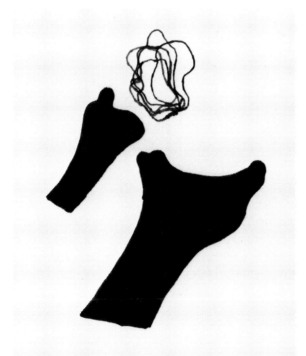

FIG. 6E. Example of triquetral translation measurements during various degrees of radial/ulnar deviation.

CASE 7

A 24-year-old male gymnast presents for further evaluation of wrist pain. Radiologic evaluation in the anteroposterior and lateral projections is not remarkable (Fig. 7A,B). Arthrography indicates a tear (arrows) of the triangular fibrocartilage (Fig. 7C). A kinematic MRI of the wrist was also performed using a T1-weighted spin-echo pulse sequence. What are the findings (Fig. 7D)?

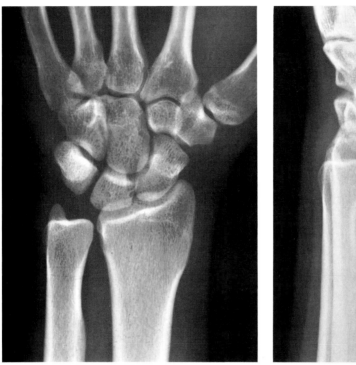

FIG. 7A.

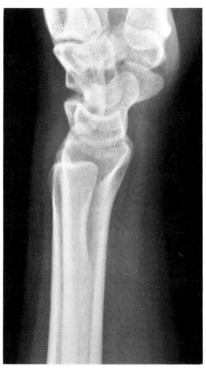

FIG. 7B.

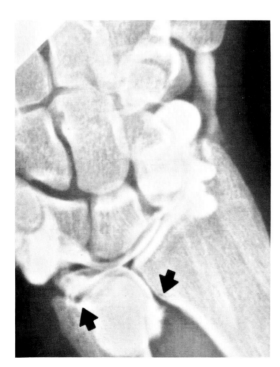

FIG. 7C.

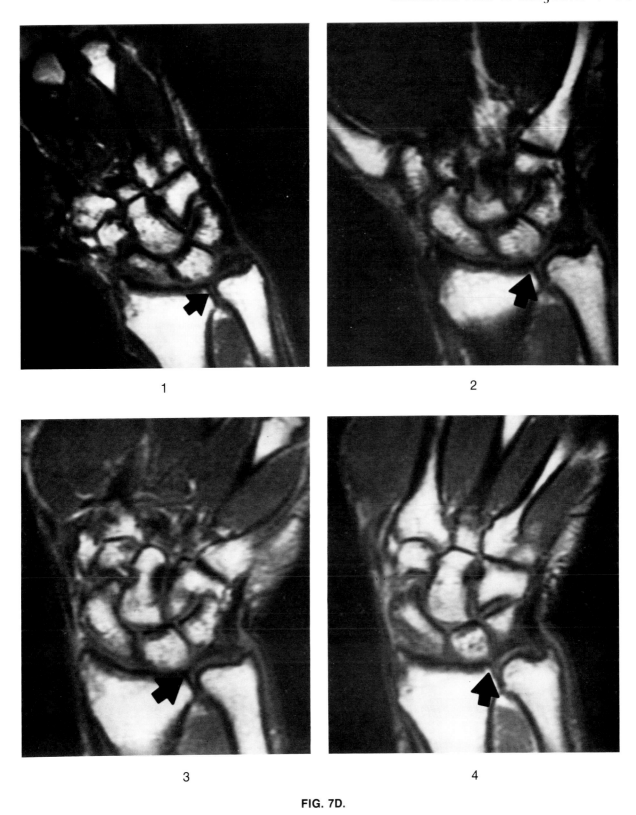

1   2

3   4

FIG. 7D.

Findings: Kinematic MRI of the wrist in the coronal plain reveals a tear of the triangular fibrocartilage with no evidence of any instability pattern or impingements (arrows, Fig. 7D).

Diagnosis: Tear of the triangular fibrocartilage with no degenerative changes or impingement.

Discussion: The major functions of the triangular fibrocartilage complex of the wrist are cushioning, stabilizing, and acting as a spacer in the dorsal ulna aspect of the wrist. A cadaveric study demonstrated that the TFC remains intact in normal status until the third or fourth decade. Of additional significance is the fact that there is an inverse relationship between the size of the triangular fibrocartilage and the length of the ulna with respect to the radius. Because the wrist has not evolved for upper extremity weight bearing, the maladaptive response typically seen in athletes that chronically loads the wrist often results in tears of the TFC and eventual pain. Pathologically, most tears occur at the radial attachment of the TFC into the sigmoid notch of the radius. The diagnostic challenge arises because these tears may not reveal arthrographic evidence of extravasation of contrast into the radioulnar joint. Furthermore, this area of the TFC has extremely complex transitions between fibrocartilage, ligament, and hyaline cartilage (Fig. 7E). In view of these transitions, the signal observed on MRI will vary depending on the slice thickness, plane of imaging, and pulse sequence used. Therefore, imaging the TFC accurately with the goal of differentiating pathological tears must be accomplished with these details in mind. Kinematic MRI of the wrist can provide additional information, negatively or positively, regarding associated impingements or instability problems.

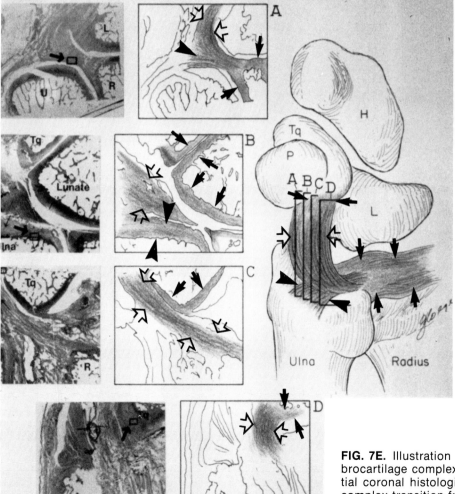

FIG. 7E. Illustration depicting triangular fibrocartilage complex as shown by sequential coronal histological sections. Note the complex transition from ligament (*open arrows*), to fibrocartilage (*arrowheads*), and to hyaline cartilage (*closed arrows*).

THE ANKLE

Technique for Kinematic MRI of the Ankle: A Plexiglas device is used to position the ankle in the sagittal plane from plantar to dorsiflexion (Fig. C). T1-weighted spin-echo images are acquired in the sagittal plane using the body coil for the kinematic MRI study of the ankle. The patient is placed in a supine position on the ankle device, and MRI is repeatedly performed as the ankle is incrementally positioned via a special handle apparatus through a range of motion from extreme plantar flexion to extreme dorsiflexion. In certain instances, active muscular force may be applied during imaging by asking the patient to apply a distally directed force while images are acquired. This maneuver appears to simulate realistically the loads on the ankle joint and is sometimes useful for determining the cause of pain related to a "dynamic" condition. Again, either static views or the cine-loop format may be used to evaluate the multiple images that are obtained for this study.

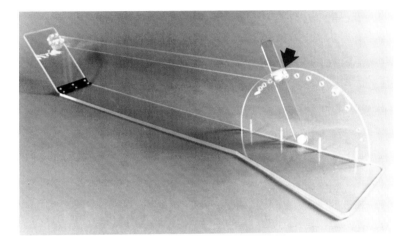

FIG. C. Positioning device made from nonferromagnetic materials used for kinematic MRI of the ankle. This device positions the ankle from plantar to dorsiflexion using a patient-activated apparatus (*arrow*).

CASE 8

A 41-year-old man presents for evaluation of ankle pain after an inversion ankle injury. The patient was treated with extensive immobilization and nonweight bearing but, despite these measures, continued to have pain. Plain films showed evidence of fracture of the talus (arrows, Fig. 8A,B). What are the findings on the kinematic MRI (Fig. 8C)?

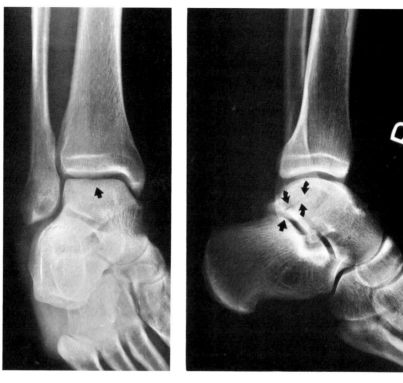

FIG. 8A.   FIG. 8B.

FIG. 8C.

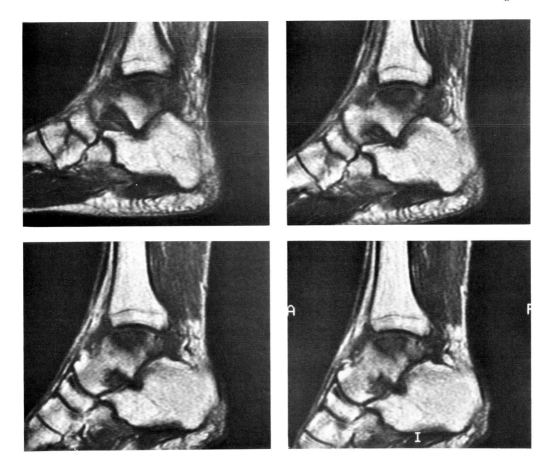

Findings: Sagittal static MRI views demonstrate significant abnormal signal in the dome of the talus. In addition, the integrity of the cortical bone and articular surface is questionable. Kinematic MRI of the ankle shows the avascular segment to be mobile in the area under the tibia and shows that this is directly under the weight-bearing surface (Fig. 8C).

Diagnosis: Avascular necrosis of the talus.

Discussion: Avascular necrosis of the talus is a difficult problem that can have disastrous consequences resulting in joint destruction. The important principles include preservation of the articular cartilage and subchondral bone. The pathophysiologic sequence of events after trauma results in obliteration of vascularity and loss of perfusion of trabecular bone and the eventual progression to joint destruction and arthrosis. The important principles include identification and detection of the process resulting in appropriate staging and therapeutic intervention. The major goal is to arrest the otherwise eventual progression of this process. Plain films indicate areas of osteolysis and perhaps areas of osteosclerosis. There is absolutely no information about the state of vascularity with this technique. Static MRI offers an opportunity to assess articular cartilage, status of subchondral bone, and the pathologic state of osteonecrosis of the subchondral bone. Furthermore, kinematic MRI shows the localized area of osteonecrosis and then, through planar motions, demonstrates the exact location in the articulation of the ankle. In addition, studying the ankle through a range of motion permits an evaluation of whether there is displacement of the segment. It is from this information that the clinician may then decide that the optimal management may include open reduction internal fixation of a free fragment, possibility of bone grafting, or intra-articular drilling to restore increased capillary ingrowth.

CASE 9

A 27-year-old male who played high school soccer presents with chronic right ankle pain during forced dorsiflexion and loading as he approaches the high jump activity. He has a history of an inversion injury. Plain films are normal. How are the static and kinematic MRI examinations of the ankle useful, and what is the diagnosis? (Fig. 9A,B. Axial plane, TR/TE 500/20 msec. Fig. 9C,D. Sagittal plane images from kinematic study, TR/TE 500/20 msec.)

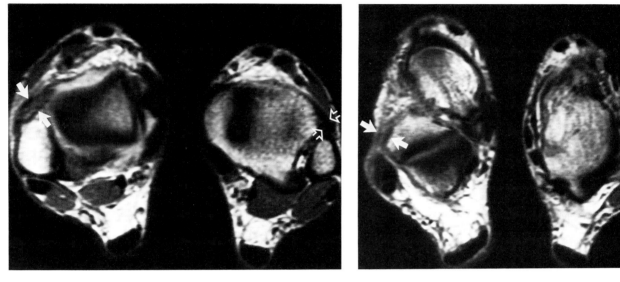

FIG. 9A.   FIG. 9B.

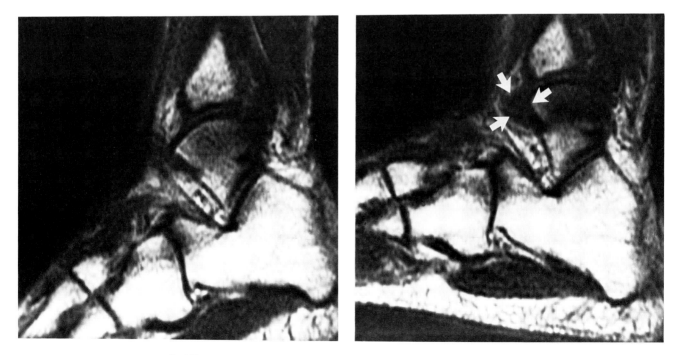

FIG. 9C.   FIG. 9D.

Findings: The static MR image (Fig. 9A) at the level of the distal tibia demonstrates a normal, dark *left* tibiofibular ligament (open arrows). On the *right* ankle, no definite ligament is seen, but rather, there is a vague soft tissue density anterior to the talus and fibula (white arrows). The axial image at a lower level (Fig. 9B) reveals a normal *left* talofibular ligament (open arrows) and no identifiable right ligament (white arrows).

Figure 9C and D are the extreme plantar (C) and dorsiflexion (D) images. The dorsiflexion image confirms the thickening of the soft tissues (Fig. 9D, arrows) anterior to and between the talus and the fibula. At surgery, insufficient lateral ligaments and lateral impingement were found.

Diagnosis: Ankle impingement syndrome secondary to capsular hypertrophy.

Discussion: Chronic ankle pain remains a difficult problem of assessment in that potential etiologies include impingements of soft tissue cartilage and osteophytes, osteochondritis dissecans, avascular necrosis, or loose bodies. Kinematic MRI offers an opportunity to assess the status of articular cartilage, subchondral bone as well as trabecular bone during active, loaded motion. It is the utilization of these techniques that allows one to create a clinical analog of loaded motion.

The most common athletic injury is an ankle sprain, and the anterior talofibular ligament is the most frequently injured ligament of the ankle. This important structure arises from the anterior surface of the distal aspect of the fibula and extends obliquely upward to the talar neck. Tears of the calcaneofibular ligament often occur in association with anterior talofibular ligament tears.

Forty percent of moderate-to-severe sprains can have residual problems including pain, functional instability, and weakness. The anterior talofibular ligament fibrocartilage is an intracapsular structure. Following a tear, healing is accompanied by scarring and capsular hypertrophy in the anterolateral space, which can result in anterior capsular impingement. Affected patients present with dorsiflexion pain and functional instability. Kinematic MRI of the ankle may demonstrate the "mass" of soft tissue that, as in this case, is caught between the talus, tibia, and fibula. The vascular tissue is richly innervated and covered by inflamed synovium, all of which make it particularly vulnerable to entrapment and a source of pain.

Hypertrophy of the anterior talofibular ligament complex is one of several causes of anterior ankle impingement. Chronic traction of the capsule on the tibia and/or talus can result in osteophytes, which can become sources of impingement. Loose bodies and osteochondritis dissecans can also prevent normal, painless ankle motion.

CASE 10

A 20-year-old female gymnast ruptured her Achilles tendon with a poorly executed dismount from the vault. This was repaired primarily as an "early motion" protocol without casting immobilization. The postoperative course was unremarkable. An investigation was undertaken to evaluate the status of the tendon to determine if early motion permitted normal ankle kinematics. Plain films revealed no abnormalities. A kinematic MRI study of the ankle was performed with the patient applying active muscular force (Fig. 10A).

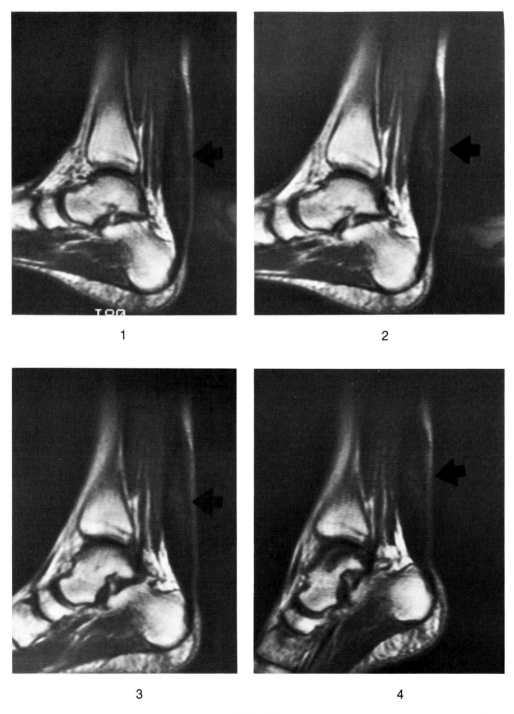

1

2

3

4

FIG. 10A.

Findings: The kinematic MRI views of the Achilles tendon defined the healing process, showing excellent continuity of the muscle-tendon and tendon-tendon junctions. Furthermore, the images demonstrate normal mobility of the tendon and normal ankle joint kinematics (arrows, Fig. 10A).

Diagnosis: Repaired Achilles tendon tear.

Discussion: As a result of significant injury to the muscle-tendon complex, the primary goal for the orthopedic surgeon is restoration of normal anatomy and function. Kinematic MRI of the ankle allows optimal characterization and determination of the surgical result. New surgical protocols including stable repair followed by early motion create new challenges of evaluating the parameters of the surgical result.

Whereas static MRI defines the continuity of the muscle-tendon junction, the kinematic MRI of the ankle provides additional information and enables the assessment of repair over a range of motion of the ankle (with and without muscle contraction) and calculation of moment arms (Fig. 10B). These data are valuable for establishing normative data of the kinematics of the ankle joint. Eventually this procedure will provide pathokinematic comparisons.

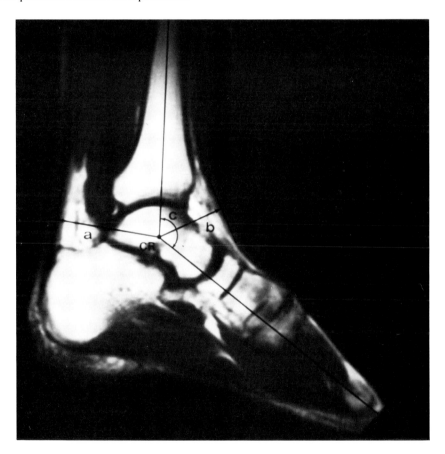

FIG. 10B. Protocol for measuring moment arms for Achilles and anterior tibial tendons.

RECOMMENDED READING

Baker LL, Hajek PC, Bjorkengren A, Galbraith R, Sartoris DJ, Gelberman RH, Resnick D. High-resolution magnetic resonance imaging of the wrist: normal anatomy. *Skeletal Radiol* 1987; 16:128–132.

Beltran J, Noto AM, Mosure JC, Shamam OM, Weiss KL, Zuelzer WA. Ankle: surface coil MR imaging at 1.5 T. *Radiology* 1986; 161:203–209.

Brattstrom H. Shape of the intercondylar groove normally and in recurrent dislocation of the patella. *Acta Orthop Scand* 1964;68:1–138.

Carson WG, James SL, Larson RL, Singer KM, Winternitz WW. Patellofemoral disorders: physical and radiographic evaluation. Part I: physical examination. *Clin Orthop* 1984;185:165–177.

Carson WG, James SL, Larson RL, Singer KM, Winternitz WW. Patellofemoral disorders: physical and radiographic evaluation. Part II: radiographic examination. *Clin Orthop* 1984;185:178–186.

Culver JE. Instabilities of the wrist. *Clin Sports Med* 1986;5:725–740.

Dehaven KE, Solan WA, Mayer PJ. Chondromalacia patellae in athletes. Clinical presentation and conservative management. *J Sports Med* 1979;7:5–11.

Delgado-Martins H. A study of the position of the patella using computerised tomography. *J Bone Joint Surg* 1979;61-B:443–444.

Ficat RP, Hungerford DS. *Disorders of the patello-femoral joint.* Baltimore: Williams and Wilkins, 1977.

Fulkerson JP. The etiology of patellofemoral pain in young, active patients: a prospective study. *Clin Orthop* 1983;179:129–131.

Hajek PC, Baker LL, Bjorkengren A, Sartoris DJ, Neumann CH, Resnick D. High-resolution magnetic resonance imaging of the ankle: normal anatomy. *Skeletal Radiol* 1986;15:536–540.

Hughston JC, Deese M. Medial subluxation of the patella as a complication of lateral release. *Am J Sports Med* 1988;16:383–388.

Hungerford DS, Barry M. Biomechanics of the patellofemoral joint. *Clin Orthop* 1979;144:9–15.

Insall J. Chondromalacia patellae: patellar malalignment syndrome. *Orthop Clin North Am* 1979;10:117–127.

Insall J, Falvo KA, Wise DW. Patellar pain and incongruence. II: clinical application. *Clin Orthop* 1983;176:225–232.

Kapandji I. The knee. In: *The physiology of the joints. Annotated diagrams of the mechanics of the human joints,* vol 2. *Lower limb.* London: Churchill Livingstone, 1970;102.

Kauer JMG. Functional anatomy of the wrist. *Clin Orthop* 1980; 149:9–20.

Kneeland JB, Macrandar S, Middleton WD, Cates JD, Jesmanowicz A, Hyde JS. MR imaging of the normal ankle: correlation with anatomic sections. *AJR* 1988;151:117–123.

Koenig H, Lucas D, Meissner R. The wrist: a preliminary report of high-resolution MR imaging. *Radiology* 1986;160:463–467.

Kummel BM. The diagnosis of patellofemoral derangements. *Primary Care* 1980;7:199–216.

Larson RL. Subluxation-dislocation of the patella. In: Kennedy JC, ed. *The injured adolescent knee.* Baltimore: Williams and Wilkins, 1979;161–204.

Laurin CA, Dussault R, Levesque HP. The tangential x-ray investigation of the patellofemoral joint: x-ray technique, diagnostic criteria and their interpretation. *Clin Orthop* 1979;144:16–26.

Lichtman DM, Noble WH, Alexander CE. Dynamic triquetrolunate instability. *J Hand Surg* 1984;9A:185–187.

Linscheid RL, Dobyns JH, Beabout JW, Bryan RS. Traumatic instability of the wrist. *J Bone Joint Surg* 1972;54A:1612–1632.

MacNab L. Recurrent dislocation of the patella. *J Bone Joint Surg (AM)* 1952;34:957–976.

Martin DF, Baker CL, Curl WW, Andrews JR, Robie DB, Haas AF. Operative ankle arthroscopy: long-term followup. *Am J Sports Med* 1989;17:16–23.

Martinez S, Korobkin M, Fondren FB, Hedlund LW, Goldner JL. Diagnosis of patellofemoral malalignment by computed tomography. *J Comput Assist Tomogr* 1983;7:1050–1053.

Merchant AC, Mercer RL, Jacobsen RH, Cool CR. Roentgenographic analysis of patellofemoral congruence. *J Bone Joint Surg* 1974;56-A:1391–1396.

Mikic Z. Age changes in the triangular fibrocartilage of the wrist joint. *J Anat* 1984;126:367–384.

Moller BN, Krebs B, Jurik AG. Patellofemoral incongruence in chondromalacia and instability of the patella. *Acta Orthop Scand* 1986;57:232–234.

Quinn SF, Murray WT, Clark RA, Cochran CF. Achilles tendon: MR imaging at 1.5 T. *Radiology* 1987;164:767–770.

Schutzer SF, Ramsby GR, Fulkerson JP. Computed tomographic classification of patellofemoral joint pain patients. *Orthop Clin North Am* 1986;17:235–248.

Shahriaree H. *O'Connor's textbook of arthroscopic surgery.* Philadelphia: JB Lippincott Co., 1984;237–262.

Shellock FG, Mink JH, Deutsch A, Fox JM. Evaluation of patellar tracking abnormalities using kinematic MR imaging: clinical experience in 130 patients. *Radiology* 1989;172:799–804.

Shellock FG, Mink JH, Deutsch A, Fox JM. Kinematic magnetic resonance imaging for evaluation of patellar tracking. *Physician Sports Med* 1989;17:99–108.

Shellock FG, Mink JH, Fox JM. Patellofemoral joint: kinematic MR imaging to assess tracking abnormalities. *Radiology* 1988;168:551–553.

Sierra A, Potchen EJ, Moore J, Smith HG. High-field magnetic-resonance imaging of aseptic necrosis of the talus. *J Bone Joint Surg* 1986;68A:927–928.

Spritzer CE, Vogler JB, Martinez S, Garrett WE, Johnson GA, McNamara MJ, Lohnes J, Herfkens RJ. MR imaging of the knee: preliminary results with a 3DFT GRASS pulse sequence. *AJR* 1988;150:597–603.

Weiss KL, Beltran J, Shamam OM, Stilla RF, Levey M. High-field MR surface-coil imaging of the hand and wrist. Part I. Normal anatomy. *Radiology* 1986;160:143–146.

Weiss KL, Beltran J, Shamam OM, Stilla RF, Levey M. High-field MR surface-coil imaging of the hand and wrist. Part II. Pathological correlations and clinical relevance. *Radiology* 1986; 160:147–152.

Wiberg G. Roentgenographic and anatomic studies on the femoropatellar joint, with special reference to chondromalacia patellae. *Acta Orthop Scand* 1941;12:319–410.

Subject Index